Seed Genomics

Seed Genomics

Edited by

Philip W. Becraft
Genetics, Development and Cell Biology Department
Agronomy Department, Iowa State University
Ames, Iowa, USA

WILEY-BLACKWELL
A John Wiley & Sons, Inc., Publication

Wiley-Blackwell is an imprint of John Wiley & Sons, formed by the merger of Wiley's global Scientific, Technical and Medical business with Blackwell Publishing.

Editorial offices: 2121 State Avenue, Ames, Iowa 50014-8300, USA
The Atrium, Southern Gate, Chichester, West Sussex, PO19 8SQ, UK
9600 Garsington Road, Oxford, OX4 2DQ, UK

For details of our global editorial offices, for customer services and for information about how to apply for permission to reuse the copyright material in this book please see our website at www.wiley.com/wiley-blackwell.

Library of Congress Cataloging-in-Publication Data is available upon request.

A catalogue record for this book is available from the British Library.

Wiley also publishes its books in a variety of electronic formats. Some content that appears in print may not be available in electronic books.

Cover design by Matt Kuhns

Set in 10.5/12 pt Times by Aptara® Inc., New Delhi, India
Printed and bound in Malaysia by Vivar Printing Sdn Bhd

1 2013

Contents

Contributors

Jose M. Barrero	CSIRO Plant Industry, Canberra, ACT, Australia
Philip W. Becraft	Genetics, Development and Cell Biology Department, Agronomy Department, Iowa State University, Ames, Iowa, USA
Judith Burstin	INRA, UMR1347 Agroécologie, Dijon, France
Colin Cavanagh	CSIRO Plant Industry, Canberra, ACT, Australia
Z. Jeffrey Chen	Section of Molecular Cell and Developmental Biology and Center for Computational Biology and Bioinformatics, The University of Texas at Austin, Austin, Texas, USA
Robert L. Fischer	Department of Plant and Microbial Biology, University of California, Berkeley, California, USA
Jennifer M. Frost	Department of Plant and Microbial Biology, University of California, Berkeley, California, USA
Karine Gallardo	INRA, UMR1347 Agroécologie, Dijon, France
Xueying Guan	Section of Molecular Cell and Developmental Biology and Center for Computational Biology and Bioinformatics, The University of Texas at Austin, Austin, Texas, USA
Frank Gubler	CSIRO Plant Industry, Canberra, ACT, Australia
Jeffery L. Gustin	Horticultural Sciences Department, University of Florida, Gainesville, Florida, USA
Tracie A. Hennen-Bierwagen	Roy J. Carver Department of Biochemistry, Biophysics, and Molecular Biology, Iowa State University, Ames, Iowa, USA
David Holding	Department of Agronomy and Horticulture, Center for Plant Science Innovation, University of Nebraska, and Beadle Center for Biotechnology, Lincoln, Nebraska, USA
Tzung-Fu Hsieh	Department of Plant and Microbial Biology, University of California, Berkeley, California, USA
Christian A. Ibarra	Department of Plant and Microbial Biology, University of California, Berkeley, California, USA

Anna M. G. Koltunow Commonwealth Scientific and Industrial Research Organization (CSIRO) Plant Industry, Urrbrae, South Australia, Australia

D. L. C. Kumari Fonseka The Integrative Cell Biology Laboratory, School of Biological and Biomedical Sciences, Durham University, Durham, UK

Christine Le Signor INRA, UMR1347 Agroécologie, Dijon, France

Keith Lindsey The Integrative Cell Biology Laboratory, School of Biological and Biomedical Sciences, Durham University, Durham, UK

Christine J. Lucas Department of Crop Sciences, University of Illinois, Urbana-Champaign, Urbana, Illinois, USA

David W. Meinke Department of Botany, Oklahoma State University, Stillwater, Oklahoma, USA

Joachim Messing Waksman Institute of Microbiology, Rutgers University, Piscataway, New Jersey, USA

Stephen P. Moose Department of Crop Sciences, University of Illinois, Urbana-Champaign, Urbana, Illinois, USA

Anna Mudge The Integrative Cell Biology Laboratory, School of Biological and Biomedical Sciences, Durham University, Durham, UK

Alan M. Myers Roy J. Carver Department of Biochemistry, Biophysics, and Molecular Biology, Iowa State University, Ames, Iowa, USA

Mélanie Noguero INRA, UMR1347 Agroécologie, Dijon, France

Odd-Arne Olsen CIGENE/Department of Plant and Environmental Sciences, Norwegian University of Life Sciences, Aas; and Faculty of Education and Natural Sciences, Hedmark University College, Hamar, Norway

Peggy Ozias-Akins Department of Horticulture, College of Agricultural and Environmental Sciences, University of Georgia Tifton Campus, Tifton, Georgia, USA

Martha Schneerman Department of Crop Sciences, University of Illinois, Urbana-Champaign, Urbana, Illinois, USA

A. Mark Settles Horticultural Sciences Department, University of Florida, Gainesville, Florida, USA

Peter R. Shewry Rothamsted Research, Harpenden, Hertfordshire, UK

Juhyun Shin Department of Plant and Microbial Biology, University of California, Berkeley, California, USA

Imran Siddiqi Centre for Cellular and Molecular Biology (CCMB), Hyderabad, India

Richard Thompson INRA, UMR1347 Agroécologie, Dijon, France

Jennifer F. Topping The Integrative Cell Biology Laboratory, School of Biological and Biomedical Sciences, Durham University, Durham, UK

Jérôme Verdier The Samuel Roberts Noble Foundation, Ardmore, Oklahoma, USA

Xiyan Yang The Integrative Cell Biology Laboratory, School of Biological and Biomedical Sciences, Durham University, Durham, UK

Han Zhao Department of Crop Sciences, University of Illinois, Urbana-Champaign, Urbana, Illinois, USA

Introduction

Philip W. Becraft

Agrarian civilization arose independently several times around the world. One of the earliest events occurred in the Fertile Crescent encompassing the Tigris and Euphrates river valleys of what is presently southeastern Turkey and northern Syria (Lev-Yadun *et al.*, 2000). This is believed to have occurred as early as 11,000 BP and to have involved cultivation of seven founder crops: einkorn wheat, emmer wheat, barley, lentil, pea, bitter vetch, and chickpea. At roughly the same time, early agriculture was occurring in the Yangtze valley of China centered on rice cultivation (Zhao, 2010) and in Mesoamerica involving primarily maize, beans, and squash (Zizumbo-Villarreal and Colunga-GarcíaMarín, 2010). It is notable how prevalent seeds and grains are among these early crops, and this is no accident but due to their high nutritional content and amenability to long-term storage without spoiling. With the advent of agriculture and the resultant stable food supplies came the ability to form permanent settlements, which led ultimately to the rise of modern civilization.

Amazingly, we remain as dependent as ever on seed crops. According to the Food and Agriculture Organization of the United Nations (FAO Statistical Yearbook 2012; http://www.fao.org/), an estimated 50% of global human dietary calories come directly from cereal grains. This figure represents a decline in recent decades, which is largely attributed to increased consumption of calories from vegetable oils, primarily derived from oilseed crops. Livestock products, including dairy, account for only about 13% of human calories, and much of that is indirectly derived from seed-based feeds. Thus, most human caloric intake derives from seed crops.

Seed science has never faced more important challenges or more exciting opportunities than at the present time. As human populations continue to grow, fuel costs soar, and climate change progresses, agriculture will face ever-increasing pressure to produce more food and biofuel, with lower inputs and under increasingly adverse environmental conditions. It is paramount that research investments be made to keep ahead of these growing challenges. New genomic technologies allow biological systems to be studied on scales and at depths not possible just a few years ago. These technologies are providing new insights into the fundamental biology of seed development and metabolism and leading to new strategies for improving seed traits through biotechnical approaches and breeding.

Seed biology is fascinating and complex. Seeds must survive a highly desiccated state and remain quiescent for an indeterminate period of time, then on sensing favorable environmental conditions, reactivate metabolic processes and initiate germination. Seed development involves the coordinated activities of three genetically distinct entities: the embryo, the endosperm, and the maternal plant. The embryo represents the next plant generation; the endosperm is a support tissue that nourishes

Seed Genomics, First Edition. Edited by Philip W. Becraft.

the embryo and, in some species, the germinating seedling; and the maternal tissues contribute the protective and dispersal functions of the seed coat and pericarp. During the morphogenetic phase, the basic body plan of the embryo is established. During filling, storage products accumulate, and finally during seed maturation, tissues acquire the highly specialized ability to survive seed drying and often develop dormancy to ensure against premature germination. The accumulated storage products include starches, oils, proteins, and minerals. They are required to nourish the germinating seedling until it can become established, produce its own photosynthate, and acquire its own mineral nutrients. These storage products are also what make seeds valuable as crops.

This book contains contributions from internationally renowned scientists who describe the application of genomic analyses to various aspects of seed research and improvement. The primary focus of the book is biological rather than technical, although a wide spectrum of technical approaches and considerations are described throughout. In Chapter 1, David Meinke, one of the pioneers of large-scale seed mutant analysis in *Arabidopsis*, provides a historical perspective on the field and his group's contributions. He discusses the SeedGenes database, which compiles a vast reservoir of community information and data on existing seed mutants and the corresponding genes. This chapter illustrates one of the most important ongoing challenges in the genomics era: storing and managing huge amounts of data and presenting it in a format that is accessible and useful to the research community.

Chapters 2 and 3 provide detailed accounts of the processes of embryogenesis and endosperm development, emphasizing their genetic regulation. The embryo produces the next generation of sporophyte plant. Embryogenesis begins with a single-celled zygote and through processes of pattern formation and morphogenesis produces an embryo containing the basic body plan that is perpetuated throughout the life of the plant. The endosperm derives from a second fertilization event and serves as a support tissue to nourish the embryo during early embryogenesis. In species with persistent endosperm, such as cereals, the endosperm also nourishes the germinating seedling until it can become established. In addition to their biological significance, both structures serve as reservoirs for seed storage compounds, which are of value to humans. Both chapters highlight the complexities of these systems, illustrating the power of single-gene mutant analyses and their inherent limitations and the need for systems biology approaches that fully integrate data to understand the interacting networks that simultaneously occur at different levels (e.g., transcriptomic, proteomic, metabolomic).

Endosperm also exhibits gene imprinting, whereby maternally inherited versus paternally inherited alleles show differential expression because of epigenetic regulation. The adaptive functions and molecular mechanisms of this phenomenon are presented in Chapter 4. It appears to be involved in regulating nutrient allocation to developing seeds with implications for seed yield as well as maintaining genome integrity by suppressing transposon activity during reproduction. One exciting aspect of imprinting is that some of the molecular machinery appears to be involved in repressing seed development until triggered by fertilization, which could relate to apomixis. Apomixis is the fertilization-independent formation of seeds that retain the identical genetic constitution of the mother plant. As discussed in Chapter 5, apomixis has enormous economic potential because of the possibility of fixing hybrid vigor, and more recent progress suggests it might soon be possible to engineer apomixis into sexual crop species.

Seeds occupy a critical phase in the plant life cycle, and seed dormancy controls the timing of germination to maximize the likelihood that seedlings will be met with favorable conditions to establish, grow, and complete their reproductive cycle. The many mechanisms of dormancy allow different species to exist in their respective ecological niches by synchronizing germination to the various limiting conditions present in different environments (e.g., temperature or moisture).

Dormancy is also a critical agronomic trait; inadequate dormancy can result in crop yield losses owing to preharvest sprouting, whereas overly dormant seeds might fail to germinate when planted resulting in poor stand establishment. Chapter 6 discusses ongoing approaches to dissect the complex regulation of seed dormancy.

As mentioned, the major value of seeds to humans comes from the storage compounds they accumulate, primarily proteins, oils, and starch as well as minerals and secondary metabolites. In addition to nourishing the germinating seedling, these compounds contribute to the nutritional value of seeds for human or livestock consumption, providing energy and protein as well as other dietary benefits such as antioxidants and fiber. These compounds have found increasing use more recently in industrial applications, including biofuels, plastics, and more. Not only is the yield of these various compounds important but also the quality. The biochemical differences in seed composition impact the end use of seeds by affecting things such as baking characteristics of flour, flavor or heat tolerance of oils, or the digestibility of starch. Chapters 7–10 discuss starches, proteins, and oils, including their metabolism and factors that affect their accumulation and quality for various end uses. A common theme for all these compounds is the surprising complexity in their metabolism and how subtle structural variation can influence their physical properties. For example, starch with nothing but polymers of glucose subunits connected by α1-4 or α1-6 glycoside bonds shows dramatic differences in things such as gelling properties and digestibility, depending on the particular arrangement of the bonds and molecular packing into granules. There is a large repertoire of enzymes, not fully understood, that confer these molecular properties to the starch molecule. Proteins and oils are similarly diverse and complex. Genetic and genomic studies, including comparative genomics of different species, are lending insights to how variation in such properties are controlled and how these storage systems evolved.

In addition to the storage compounds that accumulate in seeds, another valuable seed product is cotton fiber, which is important in the textile industry and for other uses. Chapter 11 describes genomic studies in cotton where the most important seed trait is fiber. Ongoing studies seek to understand the genomic underpinnings controlling fiber quality and yield. This also serves as a model for studying processes of plant cell growth and cell wall deposition. Studies on the establishment of fiber cell fate specification provide an excellent example of translational research where basic research in *Arabidopsis* trichome development directly contributes to the understanding of an economically important trait. Cotton is also a model polyploid system for studying the negotiations and accommodations that occur between independent genomes when they are combined.

One of the most exciting areas of crop genomic science is at the interface with crop breeding. After all, the ultimate goal of plant genomics research is for crop improvement. The Illinois Long-Term Selection Project is a unique resource where a single starting maize population has been subjected to >110 cycles of continuous selection for seed traits including protein and oil content. These selection schemes have been reversed for several subpopulations, lines have been crossed to create mapping populations, germplasm has contributed to breeding programs, and, more recently, genomic analyses have been applied to these populations. As described in Chapter 12, this has provided new insights into genome-level responses to long-term selection, which will have bearing on one of the great questions pondered by plant breeders (or probably more often by nonbreeders): "When will the genetic variation run out?"

Finally, phenotypic analysis is often cited as the bottleneck to high-throughput studies. In closing, Chapter 13 discusses various spectral imaging technologies that are being combined with computer algorithms to develop high-throughput, automated systems for analyzing seed traits. As described, these approaches afford the opportunity to gather much more information in a single measurement than is possible with manual techniques and to do it more quickly and more accurately. Some of these imaging techniques provide three-dimensional spatial information as well as compositional information. Furthermore, the data are preserved and can often be mined for additional information as new computer algorithms are developed. This area holds tremendous promise for future advancement as new imaging technologies are developed and applied to the analysis of seed traits. When combined with genomic studies, basic research on seed biology and breeding for improved seed traits can be greatly accelerated, and genetic potentials can be realized.

I thank the authors for their outstanding contributions. Their efforts make readily accessible an enormous amount of information, some of which was previously unpublished. I greatly enjoyed working on this project and found each of the chapters exciting and educational. I hope you find it valuable, too.

References

Lev-Yadun S, Gopher A, Abbo S (2000). The cradle of agriculture. Science 288: 1602–1603.

Zhao Z (2010). New data and new issues for the study of origin of rice agriculture in China. Archaeological and Anthropological Sciences 2: 99–105.

Zizumbo-Villarreal D, Colunga-GarcíaMarín P (2010). Origin of agriculture and plant domestication in West Mesoamerica. Genetic Resources and Crop Evolution 57: 813–825.

1 Large-Scale Mutant Analysis of Seed Development in *Arabidopsis*

David W. Meinke

Introduction

With advances in DNA sequencing and reports of sequenced genomes appearing at an accelerating rate, one can easily forget an important principle that first guided research in molecular plant biology 25 years ago – that genomics and proteomics are most powerful when focused on model genetic organisms. It is therefore fitting that a book devoted to seed genomics should include several chapters on the use of genetic analysis to address fundamental questions in seed biology. My objective in this chapter is not to detail all of the seed mutants analyzed to date or to describe all of the biological questions that have been addressed with these mutants. Instead, I have chosen to focus on my own professional journey, spanning the past 35 years, to isolate and characterize large numbers of embryo-defective (*emb*) mutants in the model plant, *Arabidopsis thaliana*. This choice is justified by a quick look at the numbers involved. More embryo mutants have been isolated and characterized in *Arabidopsis*, and their genes identified, than in all other angiosperms combined. Any discussion of the strategies, procedures, and conclusions drawn from the analysis of large numbers of mutants defective in seed development must therefore focus on what has been accomplished in *Arabidopsis*. This work has been performed over several decades by dozens of individuals in my laboratory, along with scores of investigators throughout the *Arabidopsis* community. The results summarized in this chapter are a testament to their combined efforts and insights. Readers unfamiliar with basic features of seed development are referred to Chapters 2 and 3 of this book.

Historical Perspective

Mutants defective in seed development have long played an important role in genetic analysis (Meinke, 1986, 1995) – from Mendel's wrinkled seed phenotype in pea, which results from transposon inactivation of a starch-branching enzyme (Bhattacharyya *et al.*, 1990), to studies by early plant geneticists on germless (embryo-specific) and defective kernel (*dek*) mutants of maize (Demerec, 1923; Mangelsdorf, 1923; Emerson, 1932) and the nature of embryo-endosperm interactions during seed development (Brink and Cooper, 1947). Large-scale mutant analysis of seed development in maize began in the late 1970s with the isolation and characterization of several hundred *dek* mutants generated following ethyl methanesulfonate (EMS) pollen mutagenesis (Neuffer and

---

Seed Genomics, First Edition. Edited by Philip W. Becraft.

Sheridan, 1980; Sheridan and Neuffer, 1980). Another 64 *dek* mutants, along with 51 embryo-specific mutants, were later described in genetic stocks known to contain the transposable element, Robertson's Mutator (Clark and Sheridan, 1991; Sheridan and Clark, 1993; Scanlon *et al.*, 1994). Although many additional mutants of this type have likely been encountered in screens of other transposon insertion lines, a global analysis of all disrupted genes associated with kernel phenotypes in maize has not been published. Attention has focused instead on a detailed characterization of selected mutants of particular interest (José-Estanyol *et al.*, 2009), including the *dek1* mutant defective in aleurone cell identity (Becraft *et al.*, 2002, Tian *et al.*, 2007; Yi *et al.*, 2011) and a number of viviparous mutants that exhibit premature germination (Suzuki *et al.*, 2003, 2006, 2008). By contrast, seed mutants in other grasses such as rice (Hong *et al.*, 1996; Kamiya *et al.*, 2003; Kurata *et al.*, 2005) have been examined in much less detail, with most genetic studies focused on other phenotypes of interest.

The isolation and characterization of embryo-lethal mutants of *Arabidopsis* was first described by Andreas Müller in Gatersleben, Germany. Müller (1963) characterized 60 mutants with different embryo phenotypes, including defects in embryo pigmentation, demonstrated that mutant and wild-type seeds could be distinguished in heterozygous siliques, and established the "Müller embryo test" to assess the mutagenic effects of ionizing radiation and chemical treatments in *Arabidopsis*. Although his attention was later directed to other systems, Müller remained particularly interested in fusca mutants, which accumulate anthocyanin during embryo maturation (Miséra *et al.*, 1994). Original stocks of the other mutants identified by Müller (1963) were not maintained.

I started to work on *Arabidopsis* as a graduate student in the laboratory of Ian Sussex at Yale University. My Ph.D. dissertation described the isolation and characterization of six embryo-lethal mutants of *Arabidopsis* and the value of such mutants in the study of plant embryo development (Meinke and Sussex, 1979a, 1979b). This work began at a time when *Arabidopsis* was known more for research in biochemical genetics than in developmental or molecular genetics. After completing a postdoctoral project on soybean seed storage proteins with Roger Beachy at Washington University in St. Louis (Meinke *et al.*, 1981), I moved to Oklahoma State University, where I focused my attention again on embryo mutants of *Arabidopsis*. My initial strategy was to analyze additional mutants isolated following EMS seed mutagenesis (Meinke, 1985; Baus *et al.*, 1986; Heath *et al.*, 1986; Franzmann *et al.*, 1989). Because some mutant seeds were capable of germinating and producing defective seedlings in culture, I adopted the term "embryo defective" (*emb*) rather than "embryo lethal" to describe the expanding collection. This nomenclature has been used ever since, although some *EMB* locus numbers were later replaced with more informative symbols (*sus*, *twn*, *lec*, *bio*, *ttn*) to indicate phenotypes of special interest.

A different approach to genetic analysis of plant embryo development was first described 20 years ago in a publication from Gerd Jürgens' laboratory in Germany (Mayer *et al.*, 1991). Rather than attempt to analyze every mutant defective in embryo development, the Jürgens group focused attention on a small number of mutants with defective seedlings that appeared to result from alterations in embryo pattern formation. As described elsewhere in this book, several of these mutants uncovered important cellular pathways associated with plant embryo development, although in many cases, the gene products were unexpected and did not appear to support the original hypothesis, based on work with *Drosophila*, that embryo patterning mutants should identify transcription factors that regulate developmental decisions. Whereas my approach was to "cast a wide net" and explore interesting stories based on the analysis of many different types of mutants, the Jürgens group focused on a limited set of phenotypes defined by a handful of genes with multiple alleles and identified gene networks associated with those phenotypes. In retrospect, both of these approaches were required to develop a comprehensive picture of the genetic control of plant embryo development.

Table 1.1 Experimental Features That Make *Arabidopsis thaliana* an Attractive System for Large-Scale Mutant Analysis of Seed Development

Arabidopsis Feature	Relevance of Feature to Genetic Analysis of Seed Development
Self-pollinated flowers	Crosses not required to maintain *emb* mutants and most genetic stocks
Indeterminate inflorescences	Mature plants contain large numbers of siliques at different stages of development, arranged in a predictable progression along each stem; facilitates identification of embryos at desired stage of development
Transparent seed coat	Wild-type seeds at the cotyledon stage are green and can be readily distinguished from unfertilized ovules and aborted seeds
Spontaneous seed abortion rare	Facilitates identification of mutant seeds in heterozygous siliques
Small seed size at maturity	Embryos within immature seeds are readily observed with Nomarski (DIC) light microscopy; optical sectioning through immature seeds possible
Siliques contain 50–60 seeds	Segregation of normal and mutant seeds readily observed in 1 silique
Short pollen-tube growth path	Facilitates recovery of mutants defective in both embryo and gametophyte development

Arabidopsis Embryo Mutant System

The advantages of *Arabidopsis* as a model system for research in plant biology are well known (Redéi, 1975; Meyerowitz and Somerville, 1994; Meinke *et al.*, 1998; Koornneef and Meinke, 2010). Important features that make *Arabidopsis* suitable for large-scale mutant analysis of seed development have also been described (Meinke, 1994). Several of these features are highlighted in Table 1.1. Recessive embryo-defective mutants are maintained as heterozygotes, which typically produce 25% mutant seeds after self-pollination. Because each silique contains 50–60 total seeds and multiple siliques are arranged in a developmental progression along the length of each stem, mutant seeds at many different stages of development can be found on a single plant at maturity. Mutant and normal seeds can be readily distinguished, based on size, color, and embryo morphology, by screening immature siliques under a dissecting microscope. Mutant embryos that have reached an advanced globular stage can be removed with fine-tipped forceps and examined further; embryos arrested at earlier stages of development are best observed under a compound microscope equipped with Nomarski (differential interference contrast [DIC]) optics. After seed mutagenesis, siliques of chimeric M_1 plants can be screened to identify flowers that arose from the mutant sector (Meinke and Sussex, 1979a, 1979b). Mature siliques derived from this sector are harvested to collect dry seeds. After germination, heterozygous and wild-type plants often segregate in a 2:1 ratio. If insertion lines are involved and the disrupted *EMB* gene is associated with a selectable marker, the appropriate selection agent can be used to identify heterozygous plants at the seedling stage, provided that there are no additional inserts located elsewhere in the genome. With EMS mutants, heterozygous plants cannot be distinguished from wild-type plants until selfed siliques have matured and are screened for defective seeds. When plants segregating for an *emb* mutation are crossed for allelism tests, parental heterozygotes must be identified before the cross can be performed, which limits the time available for crosses to be completed. Allelic mutants that fail to complement result in siliques with 25% mutant seeds; mutants disrupted in different genes typically produce siliques with all normal seeds.

Large-Scale Forward Genetic Screens for Seed Mutants

In contrast to screens for most visible phenotypes in *Arabidopsis*, which involve the identification of homozygotes in a second (M_2) generation following seed mutagenesis, forward genetic screens for embryo-defective mutations can be performed directly on M_1 plants. This approach was used to isolate most of the original *emb* mutants analyzed in my laboratory (Meinke, 1985). When Ken Feldmann developed a method for *Agrobacterium*-mediated seed transformation in *Arabidopsis* and began to grow large populations of T-DNA insertion lines at DuPont in the late 1980s (Feldmann, 1991), two different groups were involved in screening the populations for embryo-defective mutations. One group, comprising investigators associated with Robert Goldberg's laboratory at UCLA (Yadegari *et al.*, 1994), screened half of the plants; members of my laboratory screened the other half (Errampalli *et al.*, 1991; Castle *et al.*, 1993). The same strategy was used with a second population of plants that Feldmann made available several years later at the University of Arizona. My approach to the analysis of these populations was first to determine which mutants were tagged with T-DNA and which were not tagged. About two thirds of the lines that segregated for an embryo-defective mutation were not amenable to rapid gene identification because they fell into the second category. The method used to resolve tagging status involved transplanting kanamycin-resistant seedlings derived from selfed heterozygotes to soil and determining whether all of those plants produced siliques with 25% mutant seeds, as expected if a single T-DNA insert was present and disrupted an *EMB* gene. When additional inserts were involved, we identified subfamilies in future generations that contained a single insert and then proceeded with the analysis described above. For mutants examined in my laboratory, the original *emb1* to *emb69* alleles were identified after EMS (or in some cases x-ray) seed mutagenesis, *emb71* to *emb180* mutants involved the DuPont collection, and the *emb200* series was reserved for the Arizona collection. Most of these *EMB* loci are listed in Meinke (1994) and in the "Archival Data on Meinke Lab Mutants" link at the SeedGenes website devoted to genes with essential functions during seed development in *Arabidopsis* (www.seedgenes.org). In some cases, the gene responsible for the mutant phenotype has since been identified. In many cases, however, the association between mutant phenotype and gene function remains to be determined.

A major breakthrough in forward genetic analysis of seed development occurred in the late 1990s, when David Patton and Eric Ward at Ciba-Geigy, which later became Syngenta (Research Triangle Park, NC), embarked on a large-scale, forward genetic screen for essential genes of *Arabidopsis*. The rationale was that some essential gene products identified through such efforts might represent promising targets for novel herbicides. Over the next 15 years, in close collaboration with my laboratory, >120,000 T-DNA insertion lines were screened for seed phenotypes, including embryo and seed pigment defects, >1600 promising mutants were isolated and characterized, ~440 tagged mutants were identified, and ~200 gene identities were revealed (McElver *et al.*, 2001). Of equal importance, Syngenta ultimately agreed to make most of these gene identities public, after they had been evaluated in house (Tzafrir *et al.*, 2004). This provided the foundation for a large-scale NSF 2010 project in my laboratory that established, in collaboration with Allan Dickerman at the Virginia Bioinformatics Institute, a comprehensive database of all known essential genes required for seed development in *Arabidopsis* (Tzafrir *et al.*, 2003). Results of the "SeedGenes" project are described later in this chapter.

Approaches to Mutant Analysis

The belief that lethal mutants are not useful or informative because they cannot be analyzed in detail is misguided. Sometimes the terminal phenotype alone is sufficient to offer valuable insights. Abnormal suspensor (*sus*) and twin (*twn*) mutants provided early support for the idea that the embryo proper normally restricts the developmental potential of the suspensor (Marsden and Meinke, 1985; Schwartz *et al.*, 1994; Vernon and Meinke, 1994). The leafy cotyledon (*lec*) mutant phenotype (Meinke, 1992; Meinke *et al.*, 1994) indicated that the default state for cotyledons is a leaflike structure, consistent with the evolution of cotyledons as modified leaves. The initial steps in mutant analysis often involve determination of segregation ratios of mutant seeds in heterozygous siliques and the use of dissecting microscopes and light microscopy to characterize the terminal embryo phenotype. A reduced frequency or unusual distribution of mutant seeds in heterozygous siliques frequently indicates an additional role of the disrupted gene in male gametophyte development (Meinke, 1982; Muralla *et al.*, 2011). A mixture of aborted seeds and unfertilized ovules often indicates a role in female gametophyte development (Berg *et al.*, 2005). Examination of mutant seeds with Nomarski optics or sectioned material with light or electron microscopy can reveal unexpected features, such as incomplete cell walls in the *cyt1* mutant (Nickle and Meinke, 1998) and giant endosperm nuclei and enlarged embryo cells in titan (*ttn*) mutants (Liu and Meinke, 1998; Liu *et al.*, 2002; Tzafrir *et al.*, 2002). Light microscopy, in combination with gel electrophoresis of seed proteins, can also be used to determine whether mutant embryos unable to complete morphogenesis continue to differentiate at the cellular level (Heath *et al.*, 1986; Patton and Meinke, 1990; Yadegari *et al.*, 1994).

The original idea behind examining the response of mutant embryos in culture was to determine whether mutant seedlings or callus tissue could be produced for further analysis and to search for auxotrophic mutants that survived on enriched media containing the required nutrient (Baus *et al.*, 1986). This approach resulted in the successful identification of the first auxotrophic mutant (*bio1*) known to be associated with embryo lethality in *Arabidopsis*, and it helped to explain the scarcity of plant auxotrophs identified at the seedling stage (Schneider *et al.*, 1989). Another mutant (*bio2*) was later found to be blocked at a different step in the same pathway (Patton *et al.*, 1998). The chromosomal deletion associated with this mutant allele also contributed, by chance, to the identification of a closely linked gene (*FPA*) involved in floral induction (Schomburg *et al.*, 2001). Reverse genetic analysis later revealed that *BIO1* is part of a complex (*BIO3–BIO1*) locus that encodes a fusion protein responsible for two sequential steps in biotin synthesis (Muralla *et al.*, 2008). Embryo culture experiments also enabled further characterization of mutants with late defects in embryo development (Vernon and Meinke, 1995) and resulted in the identification of two mutants with especially striking phenotypes: *lec1* (Meinke, 1992) and *twn1* (Vernon and Meinke, 1994).

Another approach to mutant analysis that can occur independently of gene isolation involves mapping the chromosomal locations of *EMB* genes relative to morphological or molecular markers. We used this approach to enhance the classical genetic map of *Arabidopsis* and to facilitate the identification of potential mutant alleles suitable for genetic complementation tests (Patton *et al.*, 1991; Franzmann *et al.*, 1995). More recently, we performed allelism tests between mapped (but not cloned) mutants and cloned (but not mapped) mutants to identify new alleles of cloned *EMB* genes and reveal the identities of mapped *EMB* genes (Meinke *et al.*, 2009b). Classical genetic mapping with *emb* mutants is enhanced by the fact that heterozygous plants can be identified by screening selfed siliques for the presence of defective seeds. After completion of the genome sequence, the classical genetic map was found to have many regions inconsistent with the known order of genes along the chromosome. This finding led to the establishment of a sequence-based map of genes with mutant phenotypes to document the confirmed locations of genes previously found on the classical genetic map (Meinke *et al.*, 2003). An expanded version of this dataset, which includes 2400 *Arabidopsis* genes with a loss-of-function mutant phenotype, was recently published (Lloyd and Meinke, 2012).

Because of their small size, *Arabidopsis* seeds were once viewed as inaccessible to biochemical or molecular analysis. With continued technological advances and the development of alternative methods for detecting trace substances of interest, some of these barriers have been removed. One example from my laboratory involved the use of sensitive microbiological assays to demonstrate that arrested embryos from the *bio1* mutant of *Arabidopsis* contain reduced levels of biotin (Shellhammer and Meinke, 1990). A more recent example is the demonstration that arrested embryos from the *sus1* mutant contain altered profiles of miRNAs, consistent with the known function of the *SUS1/DCL1* gene (Schauer *et al.*, 2002) in promoting the formation of miRNAs (Nodine and Bartel, 2010). Although seed size continues to be an issue for some *Arabidopsis* experiments, all plant embryos begin as single cells, which means that analyzing trace materials during early embryo development will continue to present unique challenges, regardless of the final size of the embryo at maturity.

Ultimately, the most powerful approach to the large-scale analysis of mutants defective in seed development involves identifying the disrupted genes. Although some *EMB* genes have been identified through map-based cloning, most were identified by amplifying genomic sequences flanking insertion sites in T-DNA tagged mutants. Overall, 80% of the *emb* mutants found in the SeedGenes database were generated with T-DNA insertional mutagenesis compared with 9% with transposable elements and 9% with EMS. Advances in polymerase chain reaction (PCR)–based strategies for insertion site recovery played a critical role in identifying large numbers of genes required for seed development in *Arabidopsis*. For *EMB* genes analyzed in my laboratory, *EMB* numbers 1000 through 2750 denote genes uncovered through forward genetic screens of Syngenta insertion lines; *EMB* numbers 2761 through 2820 indicate genes uncovered through reverse genetic screens, often involving Salk insertion lines; *EMB* numbers 3002 to 3013 correspond to French insertion lines (Devic, 2008); and *EMB* numbers 3101 to 3147 correspond to lines first tested in the laboratory of Kazuo Shinozaki at the Riken Plant Science Center in Japan (Bryant *et al.*, 2011).

Strategies for Approaching Saturation

Forward genetics eventually becomes an inefficient strategy for identifying *EMB* genes because many of the new mutants examined represent alleles of known *EMB* genes. This trend has already been observed in *Arabidopsis*, with duplicate mutant alleles frequently encountered in mapped populations (Franzmann *et al.*, 1995; Meinke *et al.*, 2009b) and sequenced insertion lines (McElver *et al.*, 2001; Tzafrir *et al.*, 2004). A substantial number of mutants analyzed in detail in other laboratories have also turned out to be allelic to mutants first identified in my laboratory. About 8 years ago, we began to explore reverse genetic strategies for approaching saturation by focusing on *EMB* gene candidates not found through forward genetics. Promising candidates included *Arabidopsis* orthologs of known essential genes in other model organisms (Tzafrir *et al.*, 2004); genes encoding proteins that function in a shared biosynthetic pathway (Muralla *et al.*, 2007, 2008), cellular process (Berg *et al.*, 2005), or intracellular compartment (Bryant *et al.*, 2011) as a known EMB protein; and genes encoding a protein interactor of a known *EMB* gene product. We also analyzed hundreds of insertion lines that appeared from other studies (O'Malley and Ecker, 2010) to lack insertion homozygotes (Meinke *et al.*, 2008), which we reasoned might indicate embryo or gametophyte lethality. Although dealing with insertion lines on a large scale can be problematic, dozens of additional *EMB* genes were identified through a combination of these approaches. Reverse genetics was also used to find second alleles of genes first identified through forward genetics. When accompanied by genetic complementation tests, these additional alleles confirmed that the gene responsible for the mutant phenotype had been identified. The most difficult problem with Salk insertion lines (Alonso *et al.*, 2003) was reduced expression of the kanamycin-resistance marker, which meant that efficient methods developed for demonstrating close linkage between the disrupted gene and mutant phenotype based on selection, transplantation, and screening protocols (McElver *et al.*, 2001) were replaced by PCR genotyping, which is more expensive and subject to errors. In a substantial number of cases, the predicted insert could not be found or did not cosegregate with the phenotype. Similar problems with large-scale screens of Salk insertion lines have been described elsewhere (Ajjawi *et al.*, 2010). Populations of insertion lines with a more consistent selectable marker, including the GABI (Rosso *et al.*, 2003) and Riken (Kuromori *et al.*, 2004) collections, were more efficiently analyzed, but unexplained results were still encountered, and decisions had to be made about whether to resolve the ambiguities or move ahead with additional candidates. Some *EMB* candidates confirmed with reverse genetics also turned out to be the subject of ongoing studies in other laboratories, which meant that unwanted duplication of effort was involved. Because of these added complications, we eventually abandoned reverse genetic analysis on a large scale and began to focus instead on further analysis of the existing collection of *EMB* genes.

SeedGenes Database of Essential Genes in *Arabidopsis*

One goal of my NSF 2010 project was to establish a public database that summarized information on genes required for seed development in *Arabidopsis*. The resulting database (www.seedgenes.org) was first released in 2002 and has since been updated multiple times. Allan Dickerman at the Virginia Bioinformatics Institute assisted with construction of the database and oversees its maintenance. The most recent (eighth) database release (December 2010) includes information on 481 genes and 888 mutants. Over 60% of the mutants have been analyzed in my laboratory. Information about the remaining mutants was extracted from the literature. Three classes of mutants are included in the database: embryo defectives, mutants with a pigment-defective embryo (albino, pale green, fusca) of normal morphology, and mutants that produce 50% rather than 25% defective seeds after self-pollination. On entering the database, users encounter the "Access Page," which provides links to lists of genes and mutants found in the database, supplemental and archival datasets, additional information on mutant collections, a tutorial on analyzing embryo-defective mutants, and details on project objectives and participants. The linked "Query Page" is divided into two different parts: gene information and mutant information. Users can browse a list of all genes or mutants, determine which genes of interest are included in the database, and generate lists of genes or mutants that match desired criteria. Database terms are linked to a glossary that provides further details. Each gene is associated with a "Profile" page, which summarizes relevant gene information on the left side of the page and mutant information on the right side. Figure 1.1 shows an example of a Profile page. From this page, users can link to further details on insertion sites for Syngenta mutants, phenotype details and Nomarski images for mutants analyzed in my laboratory, relevant sequence information, and top BLAST hits. Dividing the database into distinct but connected sections for gene and mutant information was critical for data management and represents a key design feature that could be used to develop similar databases for other model plants.

Deciding how to present phenotype information in the database was a major challenge, in part because the project evolved over a period of years and involved multiple student assistants with varying degrees of expertise. For Syngenta lines and mutants that my laboratory analyzed in some detail, we established a standardized set of terminal phenotypes (Figure 1.2) based on embryo morphology as visualized under a dissecting microscope. We also captured Nomarski images of mutant embryos inside the developing seed (Figure 1.3). Details of these methods are given at the tutorial section of the SeedGenes website. Although this approach provided insights into the stage of developmental arrest and the diversity of embryo phenotypes observed, subtle differences in cell division patterns and defects that first distinguished mutant from wild-type embryos were generally not recorded. The SeedGenes database should therefore be viewed as a broad community resource

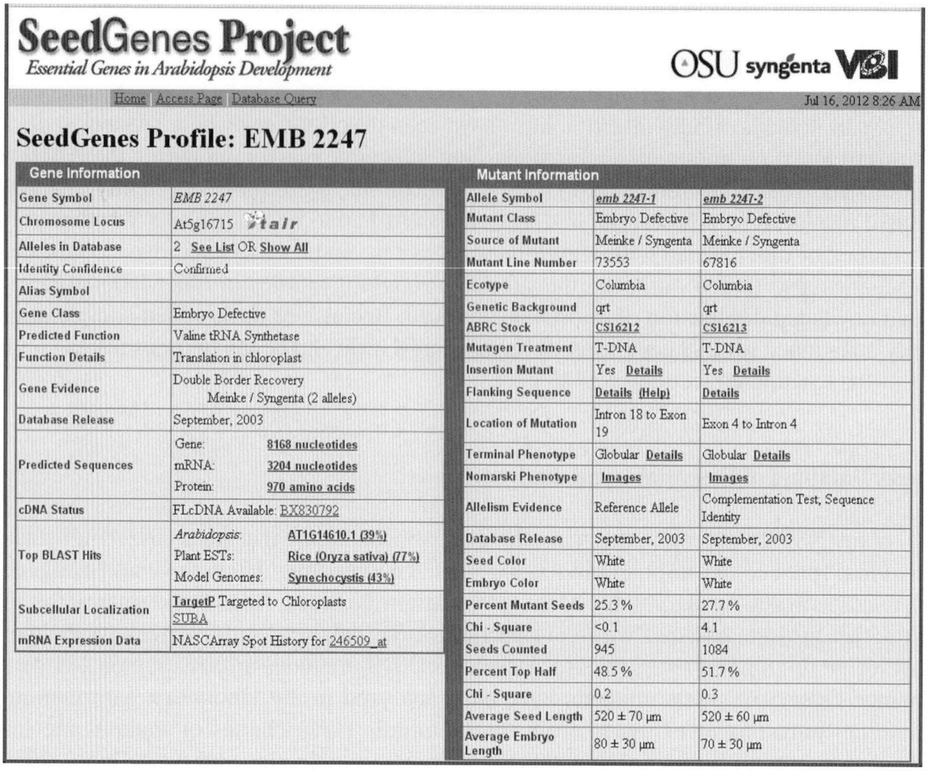

SeedGenes Project
Essential Genes in Arabidopsis Development

OSU syngenta VBI

Home | Access Page | Database Query — Jul 16, 2012 8:26 AM

SeedGenes Profile: EMB 2247

Gene Information	
Gene Symbol	*EMB 2247*
Chromosome Locus	At5g16715 tair
Alleles in Database	2 See List OR Show All
Identity Confidence	Confirmed
Alias Symbol	
Gene Class	Embryo Defective
Predicted Function	Valine tRNA Synthetase
Function Details	Translation in chloroplast
Gene Evidence	Double Border Recovery Meinke / Syngenta (2 alleles)
Database Release	September, 2003
Predicted Sequences	Gene: 8168 nucleotides mRNA: 3204 nucleotides Protein: 970 amino acids
cDNA Status	FLcDNA Available: BX830792
Top BLAST Hits	*Arabidopsis*: AT1G14610.1 (39%) Plant ESTs: Rice (Oryza sativa) (77%) Model Genomes: Synechocystis (43%)
Subcellular Localization	TargetP Targeted to Chloroplasts SUBA
mRNA Expression Data	NASCArray Spot History for 246509_at

Mutant Information		
Allele Symbol	*emb 2247-1*	*emb 2247-2*
Mutant Class	Embryo Defective	Embryo Defective
Source of Mutant	Meinke / Syngenta	Meinke / Syngenta
Mutant Line Number	73553	67816
Ecotype	Columbia	Columbia
Genetic Background	qrt	qrt
ABRC Stock	CS16212	CS16213
Mutagen Treatment	T-DNA	T-DNA
Insertion Mutant	Yes Details	Yes Details
Flanking Sequence	Details (Help)	Details
Location of Mutation	Intron 18 to Exon 19	Exon 4 to Intron 4
Terminal Phenotype	Globular Details	Globular Details
Nomarski Phenotype	Images	Images
Allelism Evidence	Reference Allele	Complementation Test, Sequence Identity
Database Release	September, 2003	September, 2003
Seed Color	White	White
Embryo Color	White	White
Percent Mutant Seeds	25.3 %	27.7 %
Chi - Square	<0.1	4.1
Seeds Counted	945	1084
Percent Top Half	48.5 %	51.7 %
Chi - Square	0.2	0.3
Average Seed Length	520 ± 70 μm	520 ± 60 μm
Average Embryo Length	80 ± 30 μm	70 ± 30 μm

Figure 1.1 Screen capture of the Profile page for a representative gene (*EMB2247*) included in the SeedGenes database of essential genes in *Arabidopsis* (www.seedgenes.org). In this case, both mutant alleles were identified through a forward genetic screen of T-DNA insertion lines generated at Syngenta. Underlined and colored terms shown here link to other pages, primarily within the SeedGenes database.

Figure 1.2 Classification system for terminal phenotypes of mutant embryos removed from seeds before desiccation. For most of the Syngenta mutants analyzed in the early stages of our NSF 2010 project, we dissected 100 mutant seeds from siliques that contained normal seeds at a mature green stage of development. We then attempted to place each mutant embryo into one of the phenotype classes shown here. For stage "X," no mutant embryo was found on dissection; this usually meant that the mutant embryo was arrested before a late globular stage. Results of these phenotype screens are found by clicking on the "Details" link in the "Terminal Phenotype" section of the SeedGenes Profile page.

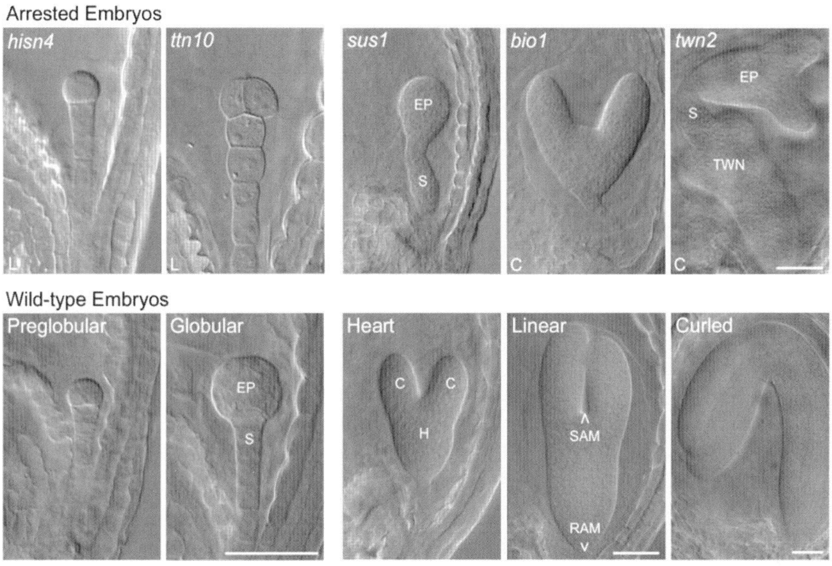

Figure 1.3 Representative collection of embryo-defective phenotypes found in the SeedGenes database. Regions of wild-type embryos include the embryo proper (EP), suspensor (S), cotyledons (C), hypocotyl (H), shoot apical meristem (SAM), and root apical meristem (RAM). Examples of aberrant development include irregular patterns of cell division, altered embryo morphology, giant suspensors, and twin embryos. The second (TWN) embryo in *twn2* arises from the suspensor (S) of the first embryo (EP). Seeds were removed from immature siliques and visualized with Nomarski (DIC) light microscopy. Arrested (mutant) embryos were obtained from heterozygous siliques at the linear (L) or curled cotyledon (C) stage of seed development. The four images on the left side are more highly magnified than the images on the right. Scale bars, 50 μm. (Modified from Meinke *et al.* [2008].)

and a starting point for additional studies rather than a definitive source of detailed phenotype information on individual mutants of interest.

Embryo Mutants with Gametophyte Defects

The more we began to characterize mutants defective in embryo development, the more important it became to distinguish between embryo and gametophyte mutants. Some gametophyte mutants of *Arabidopsis* are leaky, resulting in embryo lethality whenever fertilization takes place. In addition, some embryo mutants exhibit reduced transmission of the mutant allele and noticeable defects in gametophyte development. This raises a fundamental question: How can mutant (*emb*) gametophytes survive if an essential function required throughout the life cycle is disrupted? In other words, why do some essential gene disruptions in *Arabidopsis* result in gametophyte lethality, whereas others lead to embryo lethality?

To address this question, we first had to establish a comprehensive dataset of gametophyte essential genes in *Arabidopsis* that could be compared with the embryo dataset presented at SeedGenes. My laboratory recently created such a dataset, further edited the SeedGenes collection of *EMB* genes, and established several different categories of embryo and gametophyte mutants to facilitate comparative studies (Muralla *et al.*, 2011). The edited *EMB* dataset, which excluded six problematic SeedGenes loci, provides detailed information, including terminal phenotype classes, for 396 *EMB* genes in *Arabidopsis*. This dataset includes 352 “true *EMB*” genes required for seed development but without a known gametophyte defect and 44 genes assigned to the EMG (Embryo-Gametophyte) subclass of embryo and gametophyte loci, which produce at least 10% defective seeds following self-pollination of heterozygotes and have a reduced frequency of mutant seeds overall, too few mutant seeds at the base of the silique, or an excessive number of aborted ovules, all of which indicate a secondary role in male or female gametophyte function. Genes assigned to the GEM (Gametophyte-Embryo) subclass of gametophyte mutants have a more significant defect in gametophyte function, with heterozygotes known or predicted to produce 2%–10% mutant seeds. The GAM (Gametophyte) subclass of mutants is characterized by even more severe defects in gametophyte function, with <2% mutant seeds expected from selfed heterozygotes. Other gametophyte mutants have more variable or less well-defined defects or give rise to viable homozygotes.

To examine the functional differences among these mutant classes, we compared 70 GAM genes with reduced transmission efficiency, 352 true *EMB* loci, and 72 EMG and GEM genes with defects in both embryo and gametophyte development (Muralla *et al.*, 2011). The difference between embryo and gametophyte mutants could not be explained based on protein function alone, although distinctive features of each dataset were identified. Two alternative explanations for how mutants defective in embryo development might survive gametophyte development were also discounted because neither genetic redundancy nor residual protein function in weak mutant alleles appeared to explain the different phenotypes observed. Instead, we proposed that residual gene products derived from transcription in heterozygous microsporocytes and megasporocytes often enable mutant gametophytes to survive the loss of an essential gene product and participate in fertilization, after which time the gene disruption eventually limits embryo growth and development (Muralla *et al.*, 2011).

General Features of *EMB* Genes in *Arabidopsis*

The first question about *EMB* genes that needs to be addressed concerns how many such loci are present in the genome. Our best estimate, based on the frequency of seed mutants and duplicate mutant alleles uncovered in mutagenesis experiments, is 750–1000 genes required for seed development in *Arabidopsis* (Meinke *et al.*, 2009b), which corresponds to about 3% of all protein-coding sequences. The current collection of 400 *EMB* genes likely represents at least 40% saturation, sufficient to begin evaluating salient features. Most *EMB* genes are not embryo-specific in their pattern of expression. Embryo development is simply the stage of development when the loss of gene product first becomes critical. Consistent with this idea, weak alleles of many *EMB* genes exhibit phenotypes later in plant development (Muralla *et al.*, 2011). *EMB* genes are widely distributed throughout the five chromosomes and are more likely than the genome as a whole to be present in a single copy. When functionally redundant genes encode a protein required for embryo development, the mutant phenotype is observed only in double or multiple mutants. Examples of such double mutant phenotypes have increased in recent years, reflecting greater emphasis on the use of reverse genetics to study essential cellular functions.

A second question about *EMB* genes concerns the stage of development reached by mutant embryos before seed desiccation. We recently summarized this information for 352 "true" *EMB* genes without evidence of gametophyte defects (Muralla *et al.*, 2011). This analysis updated information published before, using smaller datasets (Tzafrir *et al.*, 2004; Devic, 2008). Based on phenotype data for the strongest allele, 16% of gene disruptions cause embryo development to become arrested at a preglobular stage; 10%, at a preglobular to globular stage; 29%, at the globular stage; and 9% at the transition (heart) stage. Several examples are shown in Figure 1.3. Mutant embryos in 31%

of cases analyzed reach a more advanced, cotyledon stage of development, whereas the remaining 5% have terminal phenotypes that remain to be documented. Several factors may contribute to a delayed onset of developmental arrest in specific mutant embryos, including partially redundant genes, metabolic pathways, or cellular processes; residual protein function in weak mutant alleles; and diffusion of critical metabolites from surrounding maternal tissues. Terminal embryo phenotypes for some mutants are consistently limited to a single stage of development, whereas for other mutants, embryo phenotypes are highly variable and extend through multiple stages of development. In addition, some mutant embryos become arrested shortly after morphological defects are first detected, whereas others continue to develop and produce viable seedlings despite visible defects early in development. Several broad conclusions about mutant phenotypes can nevertheless be made: (1) *EMB* genes clearly differ in how far embryo development can proceed following their disruption; (2) a sizable number of mutants exhibit defects well before the globular stage of embryo development; (3) most of the mutant embryos arrested before the heart stage of development are white or very pale green, consistent with a disruption of chloroplast function, whereas many mutant embryos with defects at the cotyledon stage are green; (4) embryo and endosperm development are affected to a similar extent in some mutants but to different extents in other mutants; and (5) unusual phenotypes do not always result from disruption of "interesting" regulatory genes, and common phenotypes do not exclude the possibility that an important regulatory function has been disrupted.

The role of chloroplast translation in embryo development provides an informative example of how disrupting similar gene products in different plant species can result in different phenotypes. Many *EMB* genes of *Arabidopsis* encode proteins localized to plastids (Hsu *et al.*, 2010; Bryant *et al.*, 2011). Some of these are required for translation of mRNAs encoded by the chloroplast genome. Interfering with chloroplast translation results in embryo lethality in *Arabidopsis*, but in *Brassica* and maize, albino seedlings are produced instead (Zubko and Day, 1998; Asakura and Barkan, 2006). The difference appears to be related to the ability of some plant species to compensate for a disruption of fatty acid biosynthesis in chloroplasts by targeting a modified or duplicated version of the homomeric acetyl-CoA carboxylase to plastids. This homomeric enzyme can substitute for the loss of heteromeric acetyl-CoA carboxylase activity, which depends on production of one subunit (accD) encoded by the chloroplast genome (Bryant *et al.*, 2011). Overall, >20% of chloroplast-localized EMB proteins function in the biosynthesis of amino acids, vitamins, nucleotides, or fatty acids, consistent with the chloroplast localization of these pathways. Major disruptions of chloroplast function, such as interfering with protein import from the cytosol, can also result in embryo lethality, although less severe perturbations often result in reduced embryo pigmentation. By contrast, disruption of essential mitochondrial functions tends to result in gametophyte lethality (Lloyd and Meinke, 2012).

Value of Large Datasets of Essential Genes

Although some authors might argue that establishing large datasets of essential genes in plants is pointless given that all genes must perform an important function because otherwise they would not be maintained by natural selection (Pichersky, 2009), knockouts of essential genes have played a central role in the development of *Arabidopsis* as a model system and in the analysis of a wide range of important biological questions (Meinke *et al.*, 2008, 2009a). Such datasets also contribute a critical plant representative to ongoing comparative and evolutionary studies involving essential genes in microorganisms and multicellular eukaryotes (Liao and Zhang, 2008; Park *et al.*, 2008; Zhang and Lin, 2009; Chen *et al.*, 2010). Because *EMB* genes represent the most common phenotypic marker in *Arabidopsis* and more *emb* mutants have been donated to the *Arabidopsis* stock centers (Meinke and Scholl, 2003) than any other class of mutant, efforts to identify a knockout for every *Arabidopsis* gene (O'Malley and Ecker, 2010) must address the analysis of hundreds of genes known to be required for embryo and gametophyte development.

One objective of recent work in my laboratory has been to establish a comprehensive dataset of *Arabidopsis* genes with a loss-of-function phenotype of any kind to facilitate research in functional genomics, comparative phenomics, and gene discovery relevant to agriculture, bioenergy, and the environment. Our curated dataset of 2400 genes associated with a single mutant phenotype and 400 genes that exhibit a mutant phenotype only when disrupted in combination with a redundant paralog (Lloyd and Meinke, 2012) should provide a foundation for establishing a community-based resource for evaluating genotype-to-phenotype relationships in a model plant.

In addition to enabling comparative studies with essential genes in other organisms, providing a robust collection of genetic markers for ongoing research, facilitating the analysis of essential plant processes, and highlighting *EMB* genes with unknown but essential functions that merit further study, analysis of large collections of mutants defective in embryo development has resolved important questions about the nature of plant auxotrophs; the diversity of interactions between the embryo proper, suspensor, endosperm, and surrounding maternal tissues; and the contributions of specific gene products to cell survival and reproductive development. A noteworthy example of how large datasets of essential genes can contribute to discussions of important biological questions involves the issue of when the transition from maternal to zygotic gene expression occurs during plant embryo development (Muralla *et al.*, 2011). Several reports have claimed that plant embryos are relatively quiescent until the globular stage of development and that in accordance with many animal systems, early plant embryos rely primarily on stored maternal transcripts (Vielle-Calzada *et al.*, 2000; Pillot *et al.*, 2010; Autran *et al.*, 2011). However, this model has been difficult to reconcile with the existence of *emb* mutants with preglobular defects and a zygotic pattern of inheritance. A recent report (Nodine and Bartel, 2012) seems to resolve this conflict by documenting that for most genes, maternal and paternal genomes contribute equally to transcripts found in early embryos, consistent with zygotic gene expression. This study included the analysis of reciprocal crosses and differed from previous studies in that embryos were washed vigorously to remove contaminating RNA derived from the maternal seed coat. When these results are evaluated in the context of long-standing research on embryo-defective mutants – most notably, the identification of 70 *EMB* genes with arrested embryos that fail to reach a globular stage of development (Muralla *et al.*, 2011) – there is compelling evidence for early activation of the zygotic genome in *Arabidopsis* and for the requirement of zygotic gene expression to support cellular functions after fertilization.

Directions for Future Research

Research on large-scale mutant analysis of seed development in *Arabidopsis* has reached a critical stage. Despite remarkable progress in the isolation and characterization of embryo-defective mutants and the use of these mutants to address important topics in plant biology, questions remain about the availability of funding to saturate for this class of mutants and the priority that should be given to identifying additional genes with mutant phenotypes in *Arabidopsis*. For some plant biologists, the time has come to focus instead on applying what has already been learned with *Arabidopsis* to other plant systems. Although this has long been the goal of research on model organisms, there is still much to be learned about how seed development is enabled through the coordinated expression of hundreds of genes with a variety of cellular functions. My hope is that by expanding and coordinating efforts worldwide to characterize essential genes found through reverse genetics with *Arabidopsis*, progress will continue to be made toward reaching the goals that I first set forth in my Ph.D. dissertation – to explore large-scale genetic approaches to seed development in a model plant system and to understand how specific genes, proteins, and cellular processes regulate and support the formation of a mature embryo that contains root and shoot apical meristems, survives desiccation, and germinates to produce a viable plant.

Acknowledgments

Research in the author's laboratory at Oklahoma State University has been supported by grants from the *Arabidopsis* 2010 and Developmental Mechanisms programs at the U.S. National Science Foundation (NSF); Syngenta (Novartis, Ciba-Geigy); the S.R. Noble Foundation in Ardmore, Oklahoma; the U.S. Department of Agriculture (USDA) competitive grants program; and the Oklahoma Center for the Advancement of Science and Technology (OCAST).

Several research collaborators, including most notably David Patton (Syngenta) and Allan Dickerman (VBI), made significant contributions to this work. Over the years, 8 postdoctoral fellows, 7 graduate students, 10 research technicians, several high school students, and more than 65 undergraduate students have contributed to this research. I thank recent members of my laboratory (Rosanna Muralla, Johnny Lloyd, and Nicole Bryant), a former research associate (Iris Tzafrir), and two long-term research technicians (Linda Franzmann and Colleen Sweeney) for major contributions to the SeedGenes project. Rosanna Muralla and Colleen Sweeney developed the figures presented in this chapter.

References

Ajjawi I, Lu Y, Savage LJ, Bell SM, Last RL (2010). Large-scale reverse genetics in *Arabidopsis*: Case studies from the Chloroplast 2010 Project. Plant Physiol 152: 529–540.

Alonso JM, Stepanova AN, Leisse TJ, Kim CJ, Chen H, Shinn P, Stevenson DK, Zimmerman J, Barajas P, Cheuk R, *et al.* (2003). Genome-wide insertional mutagenesis of *Arabidopsis thaliana*. Science 301: 653–657.

Asakura Y, Barkan A (2006). *Arabidopsis* orthologs of maize chloroplast splicing factors promote splicing of orthologous and species-specific group II introns. Plant Physiol 142: 1656–1663.

Autran D, Baroux C, Raissig M, Lenormand T, Wittig M, Grob S, Steimer A, Barann M, Klostermeier UC, Leblanc O, *et al.* (2011). Maternal epigenetic pathways control parental contributions to *Arabidopsis* early embryogenesis. Cell 145: 707–719.

Baus AD, Franzmann L, Meinke DW (1986). Growth *in vitro* of arrested embryos from lethal mutants of *Arabidopsis thaliana*. Theor Appl Genet 72: 577–586.

Becraft PW, Li K, Dey N, Asuncion-Crabb Y (2002). The maize *dek1* gene functions in embryonic pattern formation and cell fate specification. Development 129: 5217–5225.

Berg M, Rogers R, Muralla R, Meinke D (2005). Requirement of aminoacyl-tRNA synthetases for gametogenesis and embryo development in *Arabidopsis*. Plant J 44: 866–878.

Bhattacharyya MK, Smith AM, Ellis TH, Hedley C, Martin C (1990). The wrinkled-seed character of pea described by Mendel is caused by a transposon-like insertion in a gene encoding starch-branching enzyme. Cell 60: 115–122.

Brink RA, Cooper DC (1947). The endosperm in seed development. Bot Rev 13: 423–541.

Bryant N, Lloyd J, Sweeney C, Myouga F, Meinke D (2011). Identification of nuclear genes encoding chloroplast-localized proteins required for embryo development in *Arabidopsis thaliana*. Plant Physiol 155: 1678–1689.

Castle LA, Errampalli D, Atherton TL, Franzmann LH, Yoon EH, Meinke DW (1993). Genetic and molecular characterization of embryonic mutants identified following seed transformation in *Arabidopsis*. Mol Gen Genet 241: 504–514.

Chen S, Zhang YE, Long M (2010). New genes in *Drosophila* quickly become essential. Science 330: 1682–1685.

Clark JK, Sheridan WF (1991). Isolation and characterization of 51 embryo-specific mutations of maize. Plant Cell 3: 935–951.

Demerec M (1923). Heritable characters of maize. XV. Germless seeds. J Hered 14: 297–300.

Devic M (2008). The importance of being essential: *EMBRYO-DEFECTIVE* genes in *Arabidopsis*. CR Biol 331: 726–736.

Emerson RA (1932). A recessive zygotic lethality resulting in 2:1 ratios for normal vs male sterile and colored vs colorless pericarp in F_2 of certain maize hybrids. Science 75: 566.

Errampalli D, Patton D, Castle L, Mickelson L, Hansen K, Schnall J, Feldmann K, Meinke D (1991). Embryonic lethals and T-DNA insertional mutagenesis in *Arabidopsis*. Plant Cell 3: 149–157.

Feldmann KA (1991). T-DNA insertion mutagenesis in *Arabidopsis*: mutational spectrum. Plant J 1: 71–82.

Franzmann L, Patton DA, Meinke DW (1989). *In vitro* morphogenesis of arrested embryos from lethal mutants of *Arabidopsis thaliana*. Theor Appl Genet 77: 609–616.

Franzmann LH, Yoon ES, Meinke DW (1995). Saturating the genetic map of *Arabidopsis thaliana* with embryonic mutations. Plant J 7: 341–350.

Heath JD, Weldon R, Monnot C, Meinke DW (1986). Analysis of storage proteins in normal and aborted seeds from embryo-lethal mutants of *Arabidopsis thaliana*. Planta 169: 304–312.

Hong SK, Kitano H, Satoh H, Nagato Y (1996). How is embryo size genetically regulated in rice? Development 122: 2051–2058.

Hsu SC, Belmonte MF, Harada JJ, Inoue K (2010). Indispensable roles of plastids in *Arabidopsis thaliana* embryogenesis. Curr Genomics 11: 338–349.

José-Estanyol M, López-Ribera I, Bastida M, Jarhmann T, Sánchez-Pons N, Becerra C, Vicient CM, Puigdomènech P (2009). Genetic, molecular and cellular approaches to the analysis of maize embryo development. Int J Dev Biol 53: 1649–1654.

Kamiya N, Nishimura A, Sentoku N, Takabe E, Nagato Y, Kitano H, Matsuoka M (2003). Rice *globular embryo 4* (*gle4*) mutant is defective in radial pattern formation during embryogenesis. Plant Cell Physiol 44: 875–883.

Koornneef M, Meinke D (2010). The development of *Arabidopsis* as a model plant. Plant J 61: 909–921.

Kurata N, Miyoshi K, Nonomura K, Yamazaki Y, Ito Y (2005). Rice mutants and genes related to organ development, morphogenesis and physiological traits. Plant Cell Physiol 46: 48–62.

Kuromori T, Hirayama T, Kiyosue Y, Takabe H, Mizukado S, Sakurai T, Akiyama K, Kamiya A, Ito T, Shinozaki K (2004). A collection of 11800 single-copy *Ds* transposon insertion lines in *Arabidopsis*. Plant J 37: 897–905.

Liao BY, Zhang J (2008). Null mutations in human and mouse orthologs frequently result in different phenotypes. Proc Natl Acad Sci U S A 105: 6987–6992.

Liu CM, McElver J, Tzafrir I, Joosen R, Wittich P, Patton D, Van Lammeren AAM, Meinke D (2002). Condensin and cohesin knockouts in *Arabidopsis* exhibit a *titan* seed phenotype. Plant J 29: 405–415.

Liu CM, Meinke DW (1998). The *titan* mutants of *Arabidopsis* are disrupted in mitosis and cell cycle control during seed development. Plant J 16: 21–31.

Lloyd J, Meinke D (2012). A comprehensive dataset of genes with a loss-of-function mutant phenotype in *Arabidopsis*. Plant Physiol 158: 1115–1129.

Mangelsdorf PC (1923). The inheritance of defective seeds in maize. J Hered 14: 119–125.

Marsden MPF, Meinke DW (1985). Abnormal development of the suspensor in an embryo-lethal mutant of *Arabidopsis thaliana*. Am J Bot 72: 1801–1812.

Mayer U, Torres Ruiz RA, Berleth T, Miséra S, Jürgens G (1991). Mutations affecting body organization in the *Arabidopsis* embryo. Nature 353: 402–407.

McElver J, Tzafrir I, Aux G, Rogers R, Ashby C, Smith K, Thomas C, Schetter A, Zhou Q, Cushman MA, *et al.* (2001). Insertional mutagenesis of genes required for seed development in *Arabidopsis thaliana*. Genetics 159: 1751–1763.

Meinke D, Meinke L, Showalter T, Schissel A, Mueller L, Tzafrir I (2003). A sequence-based map of *Arabidopsis* genes with mutant phenotypes. Plant Physiol 131: 409–418.

Meinke D, Muralla R, Sweeney C, Dickerman A (2008). Identifying essential genes in *Arabidopsis thaliana*. Trends Plant Sci 13: 483–491.

Meinke D, Muralla R, Dickerman A (2009a). An essential line of inquiry. Trends Plant Sci 14: 414–415.

Meinke D, Scholl R (2003). The preservation of plant genetic resources. Experiences with *Arabidopsis*. Plant Physiol 133: 1046–1050.

Meinke D, Sweeney C, Muralla R (2009b). Integrating the genetic and physical maps of *Arabidopsis thaliana*: Identification of mapped alleles of cloned essential (*EMB*) genes. PLoS ONE 4: e7386.

Meinke DW (1982). Embryo-lethal mutants of *Arabidopsis thaliana*: evidence for gametophytic expression of the mutant genes. Theor Appl Genet 63: 381–386.

Meinke DW (1985). Embryo-lethal mutants of *Arabidopsis thaliana*: analysis of mutants with a wide range of lethal phases. Theor Appl Genet 69: 543–552.

Meinke DW (1986). Embryo-lethal mutants and the study of plant embryo development. Oxford Surv Plant Mol Cell Biol 3: 122–165.

Meinke DW (1992). A homoeotic mutant of *Arabidopsis thaliana* with leafy cotyledons. Science 258: 1647–1650.

Meinke DW (1994). Seed development in *Arabidopsis*. In: Meyerowitz EM, Somerville CR (eds). *Arabidopsis thaliana*. Cold Spring Harbor Laboratory Press, Cold Spring Harbor, New York, pp 253–295.

Meinke DW (1995). Molecular genetics of plant embryogenesis. Annu Rev Plant Physiol Plant Mol Biol 46: 369–394.

Meinke DW, Chen J, Beachy RN (1981). Expression of storage-protein genes during soybean seed development. Planta 153: 130–139.

Meinke DW, Cherry JM, Dean C, Rounsley SD, Koornneef M (1998). *Arabidopsis thaliana*: a model plant for genome analysis. Science 282: 662–682.

Meinke DW, Franzmann LH, Nickle TC, Yeung EC (1994). *Leafy cotyledon* mutants of *Arabidopsis*. Plant Cell 6: 1049–1064.

Meinke DW, Sussex IM (1979a). Embryo-lethal mutants of *Arabidopsis thaliana*: a model system for genetic analysis of plant embryo development. Dev Biol 72: 50–61.

Meinke DW, Sussex IM (1979b). Isolation and characterization of embryo-lethal mutants of *Arabidopsis thaliana*. Dev Biol 72: 62–72.

Meyerowitz EM, Somerville CR (1994). *Arabidopsis*. Cold Spring Harbor Laboratory Press, Cold Spring Harbor, New York.

Miséra S, Müller AJ, Weiland-Heidecker U, Jürgens G (1994). The *FUSCA* genes of *Arabidopsis*: negative regulators of light responses. Mol Gen Genet 244: 242–252.

Müller AJ (1963). Embryonentest zum Nachweis rezessiver Letalfaktoren bei *Arabidopsis thaliana*. Biol Zentralbl 82: 133–163.

Muralla R, Chen E, Sweeney C, Gray JA, Dickerman A, Nikolau BJ, Meinke D (2008). A bifunctional locus (*BIO3-BIO1*) required for biotin biosynthesis in *Arabidopsis*. Plant Physiol 146: 60–73.

Muralla R, Lloyd J, Meinke D (2011). Molecular foundations of reproductive lethality in *Arabidopsis thaliana*. PLoS ONE 6: e28398.

Muralla R, Sweeney C, Stepansky A, Leustek T, Meinke D (2007). Genetic dissection of histidine biosynthesis in *Arabidopsis*. Plant Physiol 144: 890–903.

Neuffer MG, Sheridan WF (1980). Defective kernel mutants of maize. I. Genetic and lethality studies. Genetics 95: 929–944.

Nickle TC, Meinke DW (1998). A cytokinesis-defective mutant of *Arabidopsis (cyt1)* characterized by embryonic lethality, incomplete cell walls, and excessive callose. Plant J 15: 321–332.

Nodine MD, Bartel DP (2010). MicroRNAs prevent precocious gene expression and enable pattern formation during plant embryogenesis. Genes Dev 24: 2678–2692.

Nodine MD, Bartel DP (2012). Maternal and paternal genomes contribute equally to the transcriptome of early plant embryos. Nature 482: 94–97.

O'Malley RC, Ecker JR (2010). Linking genotype to phenotype using the *Arabidopsis* unimutant collection. Plant J 61: 928–940.

Park D, Park J, Park SG, Park T, Choi SS (2008). Analysis of human disease genes in the context of gene essentiality. Genomics 92: 414–418.

Patton DA, Franzmann LH, Meinke DW (1991). Mapping genes essential for embryo development in *Arabidopsis thaliana*. Mol Gen Genet 227: 337–347.

Patton DA, Meinke DW (1990). Ultrastructure of arrested embryos from lethal mutants of *Arabidopsis thaliana*. Am J Bot 77: 653–661.

Patton DA, Schetter AL, Franzmann LH, Nelson K, Ward ER, Meinke DW (1998). An embryo-defective mutant of *Arabidopsis* disrupted in the final step of biotin synthesis. Plant Physiol 116: 935–946.

Pichersky E (2009). The validity and utility of classifying genes as "essential." Trends Plant Sci 14: 412–413.

Pillot M, Baroux C, Vazquez MA, Autran D, Leblanc O, Vielle-Calzada JP, Grossniklaus U, Grimanelli D (2010). Embryo and endosperm inherit distinct chromatin and transcriptional states from the female gametes in *Arabidopsis*. Plant Cell 22: 307–320.

Redéi GP (1975). *Arabidopsis* as a genetic tool. Annu Rev Genet 9: 111–127.

Rosso MG, Li Y, Strizhov N, Reiss B, Dekker K, Weisshaar B (2003). An *Arabidopsis thaliana* T-DNA mutagenized population (GABI-Kat) for flanking sequence tag-based reverse genetics. Plant Mol Biol 53: 247–259.

Scanlon MJ, Stinard PS, James MG, Myers AM, Robertson DS (1994). Genetic analysis of 63 mutations affecting maize kernel development isolated from Mutator stocks. Genetics 136: 281–294.

Schauer SE, Jacobsen SE, Meinke DW, Ray A (2002). *DICER-LIKE1*: Blind men and elephants in *Arabidopsis* development. Trends Plant Sci 7: 487–491.

Schneider T, Dinkins R, Robinson K, Shellhammer J, Meinke DW (1989). An embryo-lethal mutant of *Arabidopsis thaliana* is a biotin auxotroph. Dev Biol 131: 161–167.

Schomburg FM, Patton DA, Meinke DW, Amasino RM (2001). *FPA*, a gene involved in floral induction in *Arabidopsis*, encodes a protein containing RNA-recognition motifs. Plant Cell 13: 1427–1436.

Schwartz BW, Yeung EC, Meinke DW (1994). Disruption of morphogenesis and transformation of the suspensor in abnormal *sus*pensor mutants of *Arabidopsis*. Development 120: 3235–3245.

Shellhammer J, Meinke DW (1990). Arrested embryos from the *bio1* auxotroph of *Arabidopsis* contain reduced levels of biotin. Plant Physiol 93: 1162–1167.

Sheridan WF, Clark JK (1993). Mutational analysis of morphogenesis of the maize embryo. Plant J 3: 347–358.

Sheridan WF, Neuffer MG (1980). Defective kernel mutants of maize. II. Morphological and embryo culture studies. Genetics 95: 945–960.

Suzuki M, Ketterling MG, Li QB, McCarty DR (2003). Viviparous1 alters global gene expression patterns through regulation of abscisic acid signaling. Plant Physiol 132: 1664–1677.

Suzuki M, Latshaw S, Sato Y, Settles AM, Koch KE, Hannah LC, Kojima M, Sakakibara H, McCarty DR (2008). The maize *Viviparous8* locus, encoding a putative ALTERED MERISTEM PROGRAM1-like peptidase, regulates abscisic acid accumulation and coordinates embryo and endosperm development. Plant Physiol 146: 1193–1206.

Suzuki M, Settles AM, Tseung CW, Li QB, Latshaw S, Wu S, Porch TG, Schmelz EA, James MG, McCarty DR (2006). The maize *viviparous15* locus encodes the molybdopterin synthase small subunit. Plant J 45: 264–274.

Tian Q, Olsen L, Sun B, Lid SE, Brown RC, Lemmon BE, Fosnes K, Gruis DF, Opsahl-Sorteberg HG, Otegui MS, *et al.* (2007). Subcellular localization and functional domain studies of *DEFECTIVE KERNEL1* in maize and *Arabidopsis* suggest a model for aleurone cell fate specification involving *CRINKLY4* and *SUPERNUMERARY ALEURONE LAYER1*. Plant Cell 19: 3127–3145.

Tzafrir I, Dickerman A, Brazhnik O, Nguyen Q, McElver J, Frye C, Patton D, Meinke D (2003). The *Arabidopsis* SeedGenes Project. Nucleic Acids Res 31: 90–93.

Tzafrir I, McElver JA, Liu CM, Yang LJ, Wu JQ, Martinez A, Patton DA, Meinke DW (2002). Diversity of TITAN functions in *Arabidopsis* seed development. Plant Physiol 128: 38–51.

Tzafrir I, Pena-Muralla R, Dickerman A, Berg M, Rogers R, Hutchens S, Sweeney TC, McElver J, Aux G, Patton D, Meinke D (2004). Identification of genes required for embryo development in *Arabidopsis*. Plant Physiol 135: 1206–1220.

Vernon DM, Meinke DW (1994). Embryogenic transformation of the suspensor in *twin*, a polyembryonic mutant of *Arabidopsis*. Dev Biol 165: 566–573.

Vernon DM, Meinke DW (1995). Late *embryo-defective* mutants of *Arabidopsis*. Dev Genet 16: 311–320.

Vielle-Calzada JP, Baskar R, Grossniklaus U (2000). Delayed activation of the paternal genome during seed development. Nature 404: 91–94.

Yadegari R, Paiva G, Laux T, Koltunow AM, Apuya N, Zimmerman JL, Fischer RL, Harada JJ, Goldberg RB (1994). Cell differentiation and morphogenesis are uncoupled in *Arabidopsis raspberry* embryos. Plant Cell 6: 1713–1729.

Yi G, Lauter AM, Scott MP, Becraft PW (2011). The *thick aleurone1* mutant defines a negative regulation of maize aleurone cell fate that functions downstream of *defective kernel1*. Plant Physiol 156: 1826–1836.

Zhang R, Lin Y (2009). DEG 5.0, a database of essential genes in both prokaryotes and eukaryotes. Nucleic Acids Res 37: D455–D458.

Zubko MK, Day A (1998). Stable albinism induced without mutagenesis: A model for ribosome-free plastid inheritance. Plant J 15: 265–271.

2 Embryogenesis in *Arabidopsis*: Signaling, Genes, and the Control of Identity

D. L. C. Kumari Fonseka, Xiyan Yang, Anna Mudge, Jennifer F. Topping, and Keith Lindsey

Introduction

The embryo of *Arabidopsis thaliana* has become an important reference system in which to study key aspects of plant development. Biologically, the embryo represents the phase of the plant life cycle during which key features of the structural organization of the plant body are established – apical-basal polarity and radial symmetry. In contrast to the animal embryo, the plant embryo is not a miniature adult. It does not contain many organs found in the mature plant. In *Arabidopsis*, the embryo is especially simple, with only a root, cotyledons, and hypocotyl present. Embryos of other species, such as *Zea mays*, have some leaf development in advance of germination, but the *Arabidopsis* embryo is characterized by the superimposition of two developmental axes, the apical-basal and the radial. An unusual feature is that the pattern of cell divisions in the *Arabidopsis* embryo is highly stereotyped and predictable, and this allows the ready identification of genetic mutations that perturb cell patterning mechanisms (Figure 2.1) (Mansfield and Briarty, 1991; Jürgens *et al.*, 1994). *Arabidopsis* embryogenesis is not especially typical of all higher dicotyledonous plants, but its genetic tractability has led to the unraveling of molecular mechanisms that control cell patterning and axis development, and many of these are likely to be conserved between plant species.

These molecular control mechanisms are the focus of this chapter. We consider what is known about the genes necessary for correct embryogenesis and the signaling systems that regulate the temporal and spatial expression of those genes. We do not consider the development of the endosperm, which, although it may influence embryogenesis, is not critical for it, at least at the earliest stages of development. The endosperm is considered in detail in Chapter 3. Also, we are unable to describe the functions of all genes known to be involved in the embryogenic process. Rather, we focus on hormone signaling systems (especially but not exclusively auxin signaling) and how these are used to activate genes in particular tissues as well as how they are themselves influenced by the functions of specific genes to create functional regulatory networks.

Cellular Events

Because mutational screens depend on the identification of abnormal phenotypes, it is appropriate to summarize briefly knowledge of the cellular events of *Arabidopsis* embryogenesis before discussing molecular events. Embryogenesis begins with the fertilization of the haploid egg cell with one

Seed Genomics, First Edition. Edited by Philip W. Becraft.

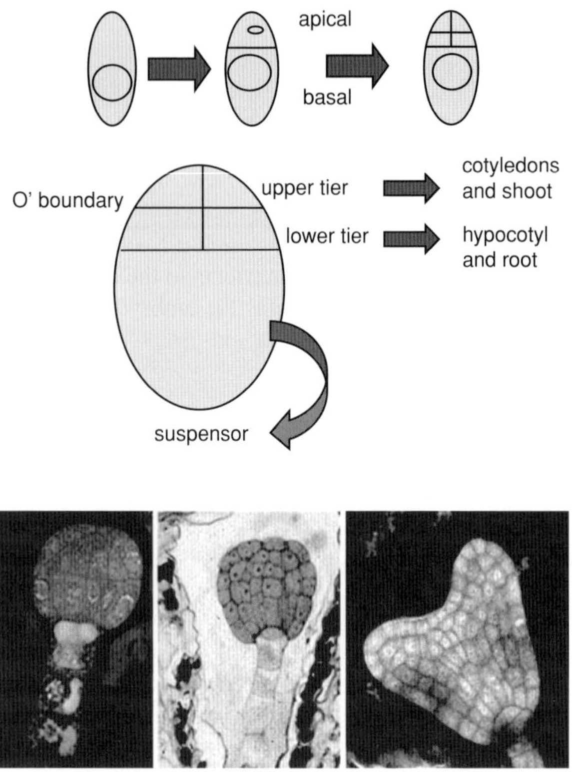

Figure 2.1 *Upper panel,* Diagrammatic representation of early events in *Arabidopsis* embryogenesis. The zygote undergoes an asymmetrical transverse division to form a small apical cell and a larger, more vacuolated basal cell. The basal cell divides to form the suspensor, while the apical cell divides transversely to form upper and lower tiers that develop into the cotyledons and shoot apical meristem and the hypocotyl and root. *Lower panel,* Developing globular *(left, center)* and heart *(right)* stage embryos of *Arabidopsis,* showing the upper cells of the suspensor. (For color detail, see color plate section.)

haploid sperm cell, to create the diploid zygote. As illustrated in Figure 2.1, the zygote divides asymmetrically to form the apical and basal daughter cells with different sizes and cytoplasmic densities (Mansfield and Briarty, 1991). This process defines establishment of the embryonic apical-basal axis, which represents the earliest patterning event in plant embryogenesis. However, it is evident that polarity already exists in the embryo sac, wherein the egg cell and synergids are located at the micropylar end, and the antipodal cells are found at the chalazal end. The egg cell itself also exhibits polarity before fertilization; it has a large vacuole at its micropylar end, whereas the chalazal end is relatively cytoplasmic (Schulz and Jensen, 1968; Mansfield and Briarty, 1991).

The first division of the fertilized egg cell reinforces this polarity. Two daughter cells, an apical cell and a basal cell, adopt different developmental fates in later embryogenesis. The apical cell of the embryo proper divides vigorously and forms most of the embryo. Through perpendicular shifts during three successive rounds of cell division, the apical daughter cell gives rise to an eight-cell embryo proper (defining the octant stage) comprising an apical and central embryo domain. The apical domain, composed of the four uppermost apical cells of the embryo proper, remains largely isodiametric before the initiation of the shoot meristem and most of the cotyledons. The central domain, consisting of the four lower cells of the embryo proper, undergoes divisions that are oriented either parallel or perpendicular to the apical-basal axis, forms the hypocotyl and root, and contributes to part of the cotyledons and the root meristem. By contrast, the basal cell of the zygote divides only horizontally (perpendicular to the apical-basal axis) and to form a filamentous structure: the suspensor and its uppermost cell, the hypophysis. The hypophysis undergoes a sequence of reproducible divisions giving rise to part of the primary root meristem, comprising the quiescent center (QC) and the central (columella) root cap initials (Scheres *et al.*, 1994).The embryonic suspensor pushes the embryo into the lumen of the ovule and provides a connection to the mother tissue, which facilitates nutrition of the embryo proper.

The most basal suspensor cell enlarges dramatically and has abundant contact with surrounding maternal tissues, likely facilitating the supply of nutrients to the embryo. The suspensor appears to have many different functions: it physically projects the embryo into the endosperm and provides both a conduit and a source of hormones and nutrients for the developing embryo. Perhaps the clearest difference in fate between the embryo proper and suspensor is seen as the programmed cell death of the suspensor, which is correspondingly not present in mature seeds (Yeung and Meinke, 1993).

The main elements to be positioned along the apical-basal axis in the *Arabidopsis* embryo are the shoot apical meristem (SAM), cotyledons, hypocotyl, radicle, and root apical meristem. The establishment of the apical-basal axis, in the early stage of embryogenesis, and the radial axis comprising the concentric cell layers of the embryo is crucial for subsequent plant growth and development. The mechanisms regulating these processes are discussed in detail subsequently.

Genes and Signaling – the Global Picture

A range of approaches have been adopted over the past few decades to understand better the genetic basis of embryo development. Most functional analysis of genes has derived from mutational screens and subsequent gene analysis, including more specialized screens such as promoter trapping (Topping *et al.*, 1994). Mutation screens (typically using chemical mutagens such as ethyl methanesulfonate, which induce point mutations, or insertional mutagenesis via T-DNA or transposon insertion into the genome) have proved invaluable for dissecting gene function and form the mainstay of functional analysis. Promoter trapping can identify genes expressed during embryogenesis on the basis of the activation of a promoterless reporter gene in embryonic cell types and can generate mutations that inform gene function. More recently, transcription profiling techniques have been used to study embryo development, aided by the use of laser-capture microdissection (LCM) (Casson *et al.*, 2005; Spencer *et al.*, 2007). This latter technique allows the isolation of small cell clusters, even from specific cellular domains of globular stage embryos of *Arabidopsis*, for RNA extraction and amplification; this overcomes the technical problem faced for many years of not being able to collect sufficient RNA from the youngest embryos for cDNA library construction and analysis. In combination with microarray analysis or RNA sequencing, this method allows high-throughput and high-resolution transcriptional profiling of the developing embryo.

Despite the relative structural simplicity of the *Arabidopsis* embryo, evidence suggests that a large proportion of the genome is active during embryogenesis. The sequencing of the *Arabidopsis* genome has allowed the development of high-throughput gene expression platforms, such as DNA microarray chips, for the transcriptional profiling of the developing embryo. Microarray analysis of RNA samples from apical and basal tissue domains of globular-stage, heart-stage, and torpedo-stage embryos, following LCM (Figure 2.2), revealed that ~75% of the *Arabidopsis* genome on the ATH1 microarray chip is expressed at one or more tissues analyzed (i.e., up to ~17,000 genes) (Casson

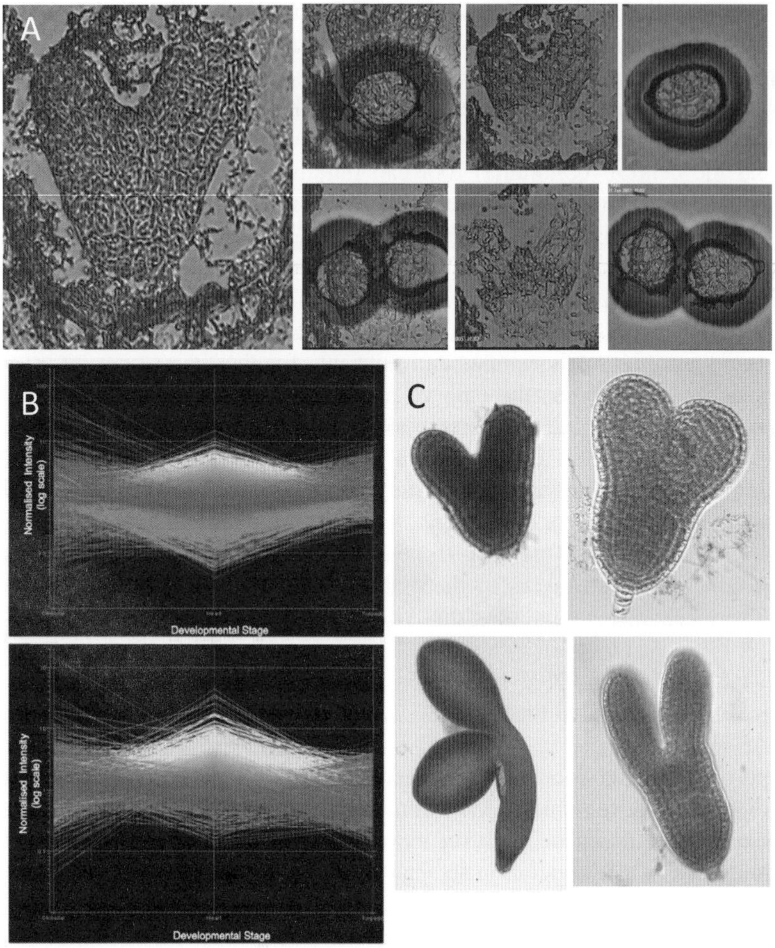

Figure 2.2 *A,* Laser-capture microdissection of cryosectioned heart stage embryo. *B,* Bioinformatics analysis of microarray data from genes expressed in apical *(upper panel)* and basal *(lower panel)* cells of developing embryos. *C,* Promoter-GUS fusion expression patterns of embryonically expressed genes identified by LCM. See Casson *et al.* (2005) and Spencer *et al.* (2007) for further detail. (For color detail, see color plate section.)

et al., 2005; Spencer *et al.*, 2007). This finding is broadly comparable with the proportion of genes found to be expressed in the primary root of *Arabidopsis* (Birnbaum *et al.*, 2003) and illustrates the high level of transcriptional activity during embryo development. The transition from the globular to the heart stage is associated in particular with an upregulation of genes involved in cell cycle control, transcriptional regulation, and energetics and metabolism. The transition from heart to torpedo stage is associated with a repression of cell cycle genes, as cell division slows in favor of cell enlargement, and genes encoding storage proteins are upregulated, together with pathways of cell growth, energy, and cell wall metabolism. The torpedo-stage embryo shows strong functional differentiation in the root and cotyledon, revealed by the distinct classes of genes expressed in these tissues. The transcriptional differences were greater between temporal stages of development than between different tissues of embryos at a given stage, suggesting rapid changes in expression of many genes as development proceeds. Nevertheless, there are clear transcriptional differences between apical and basal regions of embryos at all stages analyzed (Spencer *et al.*, 2007). Le *et al.* (2010) also performed microarray analysis on developing *Arabidopsis* seeds, in part using LCM to isolate RNA from whole developing embryos and other tissues, and found similar numbers of genes expressed (16,000) during seed development; many are stage-specific, in agreement with the data of Casson *et al.* (2005) and Spencer *et al.* (2007).

These results are consistent with older work, based on RNA hybridization studies in tobacco (*Nicotiana tabacum*), which suggested that although populations of transcripts are organ-specific, 60%–77% of plant genes are expressed in heterologous organs (Goldberg, 1988). Goldberg *et al.* (1989) demonstrated that distinct mRNA sets are temporally regulated during embryogenesis, with expression restricted to specific developmental stages, and the LCM data support this global view of large-scale transcriptional changes during embryo development. More recent analysis has shown that, in contrast to animal embryogenesis, in which there is strong maternal control over embryo transcript abundance and development, in *Arabidopsis* the zygotic genome is the principal regulator of early embryogenesis (Nodine and Bartel, 2012). microRNAs (miRNAs) are required to prevent premature expression of transcription factors that, if derepressed, lead to embryo lethality (Nodine and Bartel, 2010).

Coordination of Genes and Cellular Processes: a Role for Hormones

The patterning of cellular organization in the three-dimensional structure of the developing embryo requires the spatial and temporal expression of genes that are revealed by the above-described transcriptomic studies. This control of gene expression is itself presumed to be patterned by signaling systems, in particular (but not necessarily only by), mobile plant hormones, which provide a chemical regulatory framework. The best-characterized hormone system that has been shown to influence embryogenesis, not only in *Arabidopsis* but also in other species, is auxin. The recognition of the importance of auxin in providing positional information in the developing embryo has been instrumental in allowing us to understand better the molecular mechanisms through which specific genes act.

Auxin exists in several forms within the cell, and although its biosynthetic pathway is still incompletely understood, the most abundant biologically active form is free (nonconjugated) indole-3-acetic acid (IAA). There is a tight control over auxin homeostasis to regulate cellular concentration (Woodward and Bartel, 2005), although relatively little is known about specific mechanisms during embryogenesis. It is known, however, that the *YUCCA* gene, which encodes a flavin monooxygenase, is necessary both for auxin biosynthesis and for embryo development (Cheng *et al.*, 2007). It has been found more recently that induction of the SHORT-INTERNODES/STYLISH (SHI/STY) family member *STY1* results in increased transcript levels of the *YUCCA4* gene and higher auxin levels and auxin biosynthesis rates in *Arabidopsis* seedlings (Eklund *et al.*, 2010). Evidence indicates that *STY1*, a likely transcription factor, activates genes involved in auxin biosynthesis through promoter binding, with a role in the formation or maintenance, or both, of the shoot apical meristem, possibly by regulating auxin levels in the embryo.

Synthetic analogues, such as naphthalene-1-acetic acid (NAA) and 2,4-dichorophenoxyacetic acid (2,4-D), have been used experimentally to investigate auxin-mediated effects on development, owing to their different properties; for example, in contrast to IAA or 2,4-D, NAA can readily diffuse into cells without the need for uptake transporters. IAA is a weak acid, and because the extracellular pH is lower than the cytoplasmic pH, it is partially protonated in the extracellular matrix, which increases its membrane permeability, and it diffuses into cells slowly (Blakeslee *et al.*, 2005). However, the AUX1/LAX family of influx carriers promotes rapid auxin uptake into cells (Bennett *et al.*, 1996; Swarup *et al.*, 2008), whereas the PIN-FORMED (PIN) auxin transporters, in association with ATP-binding cassette-like transporters such as AVP1 and PGPs (Li *et al.*, 2005; Mravec *et al.*, 2008), facilitate the efflux of negatively ionized auxin.

Inside the cell, auxin mediates its effects by inducing the derepression of auxin-responsive genes. Such genes contain auxin-responsive *cis*-acting elements in their promoter regions, which bind transcription factors termed auxin response factors (ARFs). *Arabidopsis* contains 23 such proteins, and these typically dimerize and may act as positive or negative regulators of transcription (Reed, 2001). ARF function is repressed by interacting Aux/IAA genes, which, in the absence of auxin, bind them but do not bind DNA. In *Arabidopsis*, there are 29 ARF family members (Reed, 2001; Li *et al.*, 2006). In the presence of auxin, the Aux/IAA proteins are degraded by the proteosome via a mechanism involving the SCF ubiquitin ligase complex, whereby auxin acts as a "molecular glue" to bring together the TIR1 F-box protein of SCF and the target Aux/IAA protein to promote its ubiquitination and subsequent degradation. This mechanism releases the ARF complex to activate (or in some cases block) gene transcription.

The role of auxin as a mediator of cellular patterning in embryogenesis has been suspected for many years now. Liu *et al.* (1993) first demonstrated that auxin transport inhibitors induced developmental defects in embryogenesis, specifically in cultured zygotic embryos of the *Arabidopsis* relative, *Brassica juncea*. Inhibition of polar auxin transport at the globular stage led to the formation of embryos lacking bilateral symmetry at the heart stage. Embryos developed with fused and collar-like cotyledons instead of two distinct lobes, which phenocopied known auxin transport-defective mutants *pin1* (Okada *et al.*, 1991) and *gnom* (Steinmann *et al.*, 1999).

Similarly, Hadfi *et al.* (1998) found that when auxin was supplied to the same *B. juncea* embryos, ball-shaped or cucumber-shaped embryos resulted, possibly because embryos flooded with exogenous auxin are unable to establish the auxin gradients that are essential for normal morphogenesis. Treatment with the antiauxin PCIB inhibited cotyledon growth so that either only one or no cotyledons developed. Correct hypocotyl and radicle growth was also found to require auxin action and movement. When globular-stage embryos were treated with exogenous NPA, a polar auxin transport inhibitor, axis duplication resulted, whereas a later application produced split-collar or collar-like cotyledons. Similar results were found by Fischer *et al.* (1997) for morphogenesis of the embryo of the monocot wheat and for somatic embryos of Norway Spruce (*Picea abies*) (Hakman *et al.*, 2009). Mutations in the *AMINOPEPTIDASE M1* (*APM1*) gene of *Arabidopsis*, encoding a metalloprotease with affinity for and ability to hydrolyze NPA, lead to defects in embryonic and seedling patterning, PIN localization, and auxin transport (Peer *et al.*, 2009).

These results formed the basis for a model in which continuous auxin transport removes auxin from the area between the two emerging cotyledons and supplies the auxin back to the cotyledonary primordia. Auxin removal starts in the central apical region of the globular or early transition embryo, and continues asymmetrically across the apex of the embryo. Inhibition of auxin transport blurs the positional information that is created by its normally precise redistribution, resulting in increased cell division throughout the shoot apex. These findings indicate that auxin translocation is a prerequisite for the radial globular embryo to progress to the bilaterally symmetrical heart stage embryo.

Insight into the molecular mechanisms underpinning these observations came from genetic studies. A key feature of the way by which auxin orchestrates cellular patterning is its directional movement, which leads to the formation of gradients in its concentration. The PIN proteins are encoded by a multigene family, with eight members. Of the eight, PIN1, PIN2, PIN3, PIN4, and

PIN7 are localized to specific faces of the plasma membrane, which provides a mechanism for transport auxin in particular directions. PIN5, PIN6, and PIN8 are located in the endoplasmic reticulum (Mravec *et al.*, 2009), and endoplasmic reticulum localization is likely related to a role for these particular PINs in auxin homeostasis rather than intercellular transport.

The functional significance of PIN proteins in embryogenesis is further revealed by mutational studies. The *pin1* mutant exhibits embryonic defects similar to embryos treated with auxin transport inhibitors (Okada *et al.*, 1991; Liu *et al.*, 1993), and multiple mutants between *pin1*, *pin3*, *pin4*, and *pin7* exhibit a range of embryonic defects (Friml *et al.*, 2003). Immunolocalization of the PIN proteins shows that they have precise patterns in the developing embryo, and their correct function is necessary for correct auxin distribution, as revealed by the pattern of expression of the auxin reporter gene *DR5::GUS* or *DR5::GFP*, and correct development.

The results of this work can be summarized as follows: Once the egg cell has been fertilized and the zygote undergoes a first asymmetrical division, auxin accumulates in the apical cell, leading to the specification of this cell as the founder of the proembryo. This cell acts as an auxin sink, whereby auxin is transported from the adjacent basal cell by PIN7-dependent transport (PIN7 is polarly localized to the apical plasma membrane of the basal cell). This specification is disrupted in *pin7* and multiple *pin* mutants. During the globular stage of development, auxin production may occur in the apical domain of the embryo region. PIN1 basal localization is also established in the provascular cells, which may promote auxin transport basally, toward the developing root promeristem. At this time, the asymmetrical localization of PIN7 is reversed in the basal cells, promoting auxin transport out of the embryo and into the suspensor. PIN4 expression is induced in the basal region of the embryo, moving auxin in association with both PIN1 and PIN7. As a consequence, a new auxin concentration maximum is established in the hypophysis, specifying the founding cells of the root meristem (Figure 2.3).

A different experimental approach to understanding the role of auxin in embryogenesis has been described by Weijers *et al.* (2005). These authors manipulated auxin homeostasis in transgenic *Arabidopsis*, by tissue-specific expression of bacterial genes encoding the indoleacetic acid–tryptophan monooxygenase (*iaaM*) or indoleacetic acid–lysine synthetase (*iaaL*) involved in auxin biosynthesis

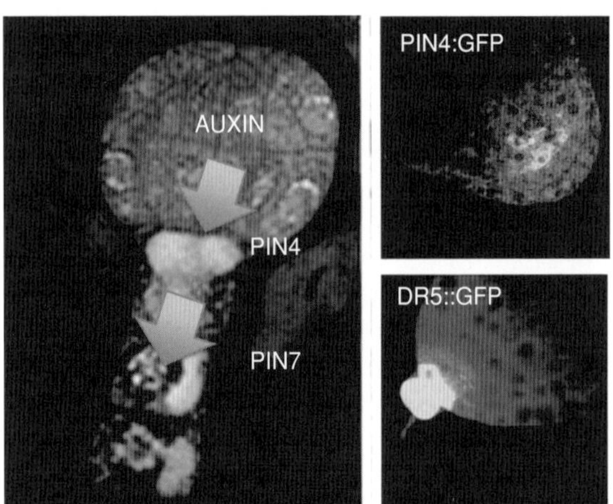

Figure 2.3 Auxin is translocated in the developing globular embryo from the apical region to the hypophysis via PIN4 and then from the hypophysis to the suspensor via PIN4 and PIN7. There is an auxin maximum (seen as DR5::GFP expression) at the basal region of the embryo. (For color detail, see color plate section.)

and conjugation. Despite the demonstrable experimental increases in auxin synthesis and conjugation, there was no discernible effect on either auxin gradients or morphological development of the embryos. This work suggests a strong buffering capacity of the auxin transport system in the developing embryo, and this was shown to be regulated principally by PIN1 and PIN4. Such a robust mechanism probably accounts in part at least for the strongly stereotyped and predictable cellular patterning seen in *Arabidopsis* embryogenesis.

Less information is available for roles for other hormones in the regulation of *Arabidopsis* embryogenesis or at least the early patterning mechanisms. Gibberellins (GA) and abscisic acid (ABA) have been known for many years to regulate later stages of seed development. During the later stages of embryogenesis, from heart stage onward, a biochemical differentiation of the embryo takes place, during which it accumulates chlorophyll (a marker for plastid maturation) and begins synthesis of storage lipids (triacylglycerols) in the plastid and seed storage proteins. In Brassicaceae such as *Arabidopsis*, the embryo is the main storage compartment for these metabolites, and the mature embryo is packed full of oil and protein bodies. GA and ABA play important roles in regulating seed maturation, in concert with sugar metabolism, and links with specific genes are becoming clear, mainly through genetic studies.

Details of seed maturation mechanisms are discussed elsewhere in this book (see Chapter 6), but some key points pertinent to embryogenesis are summarized here. Genes in *Arabidopsis* such as *LEC1*, *LEC2*, *FUS3*, and *ABI3* (part of the so-called LEC1/B3 transcription factor network) are key regulators of aspects of embryo development that include storage product synthesis and seed maturation. LEC1 (LEAFY COTYLEDON 1) was originally identified as a mutant defective in cotyledon identity, showing features of leaflike quality such as trichomes, but also exhibiting morphological defects earlier in embryogenesis, such as abnormal hypocotyl elongation, root pole shape, and shoot meristem organization (Meinke *et al.*, 1994; West *et al.*, 1994). The gene was cloned and found to encode a transcription factor subunit related to the HAP3 subunit of the CCAAT binding factor family (Lotan *et al.*, 1998). *FUS3* and *LEC2* encode B3 domain transcription factors (Luerssen *et al.*, 1998; Stone *et al.*, 2001). *fus3* mutants, similar to *lec1*, show both early and late embryonic phenotypes (Baumlein *et al.*, 1994; Keith *et al.*, 1994). *fus3* (named from the Latin for "purple," reflecting the ectopic anthocyanin accumulation in *fusca* seedlings) forms trichomes on cotyledons, is desiccation-intolerant, and is defective in storage product accumulation. The *FUS3* transcript, similar to *LEC1*, is preferentially localized to the protoderm (the embryonic epidermis) in early embryogenesis, consistent with the ectopic trichome and anthocyanin phenotypes (Tsuchiya *et al.*, 2004). It acts as a negative regulator of the anthocyanin-activating and trichome-activating *TRANSPARENT TESTA 1* (*TTG1*) gene during embryo development. The LEC2 protein is closely related to FUS3 and, similar to FUS3 and LEC1, plays a role in controlling cotyledon identity and storage product accumulation.

Transgenic expression of *LEC1*, *LEC2*, or *FUS3* leads to the activation of embryonic pathways in vegetative tissues, seen as the accumulation of triacylglycerols and seed storage proteins, and even somatic embryogenesis in vegetative tissues (Lotan *et al.*, 1998; Santos Mendoza *et al.*, 2005; Casson and Lindsey, 2006; Stone *et al.*, 2008). LEC1 regulates *LEC2*, *FUS3*, and *ABI3* expression, although LEC2 can also activate *LEC1* and *FUS3*, and *lec2* mutants show reduced *ABI3* expression (To *et al.*, 2006; Baybrook and Harada, 2008). ABI3 is a transcription factor, representing the *Arabidopsis* orthologue of maize VP1, regulating seed maturation and preventing premature germination (vivipary) (Suzuki and McCarty, 2008). The embryo also acquires desiccation tolerance, in part at least through ABA signaling and the accumulation of the protective LATE-EMBRYOGENESIS ABUNDANCE (LEA) proteins (Hundertmark and Hincha, 2008). ABA signaling is now well known as a regulator of embryo maturation and acts via ABI3 (the *abi3* mutant is ABA-insensitive)

(Finkelstein *et al.*, 2002). The link between auxin and storage product accumulation is less clear, but we have shown that LEC1 effects are dependent on auxin and sucrose in *Arabidopsis* (Casson and Lindsey, 2006). Stone *et al.* (2008) also showed that transgenic overexpression of *LEC2* leads to an upregulation of auxin biosynthesis via an induction of *YUCCA* gene expression (*YUC2* and *YUC4*). Auxin upregulates *FUS3* expression, and FUS3 is a positive regulator of ABA levels (Gazzarini *et al.*, 2004). Auxin also represses *LEC2* expression (Casson and Lindsey, 2006), providing further evidence of a link between these transcription factors, auxin, and embryonic identity and morphogenesis.

Gibberellins also play a role in controlling embryogenesis. FUS3 and LEC2 negatively regulate GA synthesis, by repressing the GA synthesis gene *GA3ox2* (Curaba *et al.*, 2004; Gazzarini *et al.*, 2004). LEC2 directly activates *AGAMOUS-LIKE 15* (*AGL15*), which is itself an activator of the catabolic gene *GA2ox6*, leading to a reduction in GA levels (Wang *et al.*, 2004). Overexpression of *AGL15* leads to enhanced somatic embryogenesis in *Arabidopsis* (Harding *et al.*, 2003). Reduced GA levels generally appear to be required for the maintenance of the embryonic development program, whereas increased GA levels (and decreased ABA) promote germination.

Repression of the LEC1/B3 transcription factors in postembryonic (vegetative) tissues is a normal feature of plant development (Figure 2.4). This repression is in part regulated by chromatin remodeling, mediated by the PICKLE (PKL) protein. *pkl* mutants show ectopic embryogenesis and storage oil accumulation, through derepression of *LEC* genes (Ogas *et al.*, 1997, 1999; Dean Rider *et al.*, 2003). The *pkl* phenotype is repressed by exogenous GA and promoted by treatment with GA inhibitors, illustrating the link between *LEC* gene function, GA, and embryogenesis. PKL is also required for the repression of the auxin response factors ARF7 and ARF19 (Fukaki *et al.*, 2006). Three *VP1/ABI3-LIKE* (*VAL*) genes, encoding B3-related transcription factors associated with chromatin factors, also repress *LEC1*, *ABI3*, and *FUS3* expression, as part of a likely chromatin-mediated repression system in vegetative tissues (Suzuki *et al.*, 2007). Similarly, the ectopic activation of embryonic genes by a gain-of-function mutant of *LEC1*, designated the *turnip* mutant, showed an enhanced penetrance of the phenotype in the presence of GA inhibitors (Casson and Lindsey, 2006). A link was also found with auxin – the *tnp* phenotype was enhanced in the

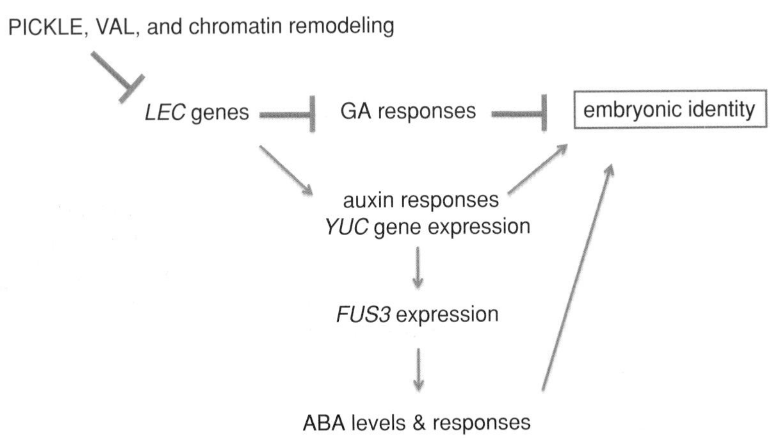

Figure 2.4 *LEC1/B3* genes are repressed in nonembryonic tissues by chromatin-associated factors, including PICKLE and VAL proteins. *LEC* genes appear to promote embryonic identity by repressing GA responses, promoting ABA responses, and may work via auxin.

presence of exogenous auxin, and the *tnp* mutant showed enhanced auxin responses, revealed as an increased level of expression of the auxin-regulated *POLARIS* gene and the *IAA2::GUS* reporter gene. Because *LEC2* expression was not upregulated in *tnp*, the high auxin phenotype was probably not mediated by *LEC2*. However, *LEC1* expression appears not to be regulated by auxin, GA, or cytokinin (Casson and Lindsey, 2006).

A further link between auxin and GAs has been found in relation to embryo development. Willige *et al.* (2010) showed that auxin transport is reduced in GA-deficient mutants, and this is due to reduced PIN protein levels. For PIN2, at least, this reduction is due to vacuolar degradation. Morphologically, this is associated with defective cotyledon development. GA has been shown to influence cotyledon development previously, with the GA signaling-deficient *spindly* mutants having only one cotyledon or fused cotyledons (Hartweck *et al.*, 2002; Silverstone *et al.*, 2007), which is also seen in auxin transport mutants, as described previously.

Relatively little known is about the role of cytokinins during embryogenesis. Cytokinin represses both *LEC2* and *FUS3* transcription (Casson and Lindsey, 2006), and this may be partly due to the negative regulation of auxin transport and responses by cytokinin (Dello Ioio *et al.*, 2008; Ruzicka *et al.*, 2009). Cytokinins are known to regulate cell division, through CYCLIN D3 activation (Riou-Khamlichi *et al.*, 1999). The *ALTERED MERISTEM PROGRAM 1* (*amp1*) mutant, defective in a carboxypeptidase, has enhanced cytokinin levels and associated defects in cotyledon development (Chaudhury *et al.*, 1993). Some alleles also show cell division abnormalities in the basal region of the embryo, and AMP1 plays a role in antagonizing auxin responses necessary for correct embryo and root development, mediated by the ARF5 MONOPTEROS (MP). *amp1* can suppress the *mp* mutant phenotype, by restricting cell number in the embryo (Vidaurre *et al.*, 2007). Repression of cytokinin synthesis by overexpressing genes encoding cytokinin oxidase/dehydrogenease (CKX) enzymes leads to defective embryo size but, surprisingly, larger rather than smaller embryos (Werner *et al.*, 2003). Paradoxically, expression of a cytokinin reporter gene shows activity in the hypophysis, suspensor, and root pole of developing *Arabidopsis* embryos (Müller and Sheen, 2008). This is the area of auxin accumulation in the embryo, as discussed previously, and the data suggest that auxin antagonizes cytokinin responses in this part of the embryo by transcriptional activation of the cytokinin-repressing *ARABIDOPSIS RESPONSE REGULATOR* genes *ARR7* and *ARR15*. Mutation of these genes or ectopic cytokinin signaling in the basal region of the embryo results in a defective root stem-cell system.

Genes and Pattern

So far we have focused the discussion on signaling systems that affect embryogenesis, and an important question relates to the genes that are regulated by these systems and how they exert their cellular effects. Genes essential for correct embryogenesis have been identified predominantly through mutational screens, leading to a classification of mutants as either embryo-lethal or seedling-lethal. The distinction is based on the definition that embryo-lethals arrest at some point during embryogenesis, but before germination, whereas seedling-lethal mutants germinate but arrest at the seedling stage, failing to go on and flower. Meinke's laboratory has been instrumental in characterizing and establishing a database of embryo-lethal mutants, from early work dating back to the 1970s (Meinke and Sussex, 1979), as discussed in Chapter 1. The framework for thinking about the genetic mechanisms regulating pattern formation during *Arabidopsis* embryogenesis gained momentum through screens for and analysis of seedling-lethal mutants and has to a large extent been driven by the laboratory of Jürgens.

A mechanistic understanding of genes required for embryonic and seedling patterning has been made easier by the rapid improvement of knowledge of signaling pathways in plants because many key genes, considered as patterning genes, have an impact on these pathways by acting through them as well as being activated in response to them. Embryonic development, similar to other aspects of development, involves complex gene and signaling networks that are characterized by often counterintuitive feedback loops and interactions, in which each component likely has an effect on all other components, at some level. A "systems biology" approach to understanding embryonic gene and signaling networks is still in its infancy and is expected to be increasingly explored as computational modeling approaches develop further. However, to date, most of our knowledge comes from the more reductionist genetic approach, with all its attendant strengths and weaknesses.

In 1991, Mayer *et al.* published an important article that described an approach to identify patterning genes in *Arabidopsis*, through the screening for seedling-lethals that "lack" specific pattern elements (essentially the shoot and root meristems and hypocotyl that define the apical-basal axis and concentric tissue layers defining the radial axis). Although one possible outcome of such an approach would be to identify sets of transcription factors (e.g., that specify the identity of such pattern elements), genes encoding a wide range of protein functions, from transcription factors to vesicle transport proteins to enzymes, have been elucidated from ensuing studies of such mutants. Perhaps the main value of the approach has been to encourage a consideration of the relevance of the nature of "pattern elements" in plants, given that plants are organisms that exhibit enormous plasticity of development as a consequence of their intercellular communication systems, totipotency, and lack of cell migration.

One gene identified in both embryo-lethal and seedling-lethal mutants screens is *EMB30* (Meinke, 1985), also known as *GNOM* (Mayer *et al.*, 1991, 1993). The mutant embryo can germinate to form an abnormal seedling that may lack apical meristems and in extreme cases develop as a ball of cells lacking polarity and misexpressing genes that are normally expressed in a polar fashion, such as *POLARIS* (Topping and Lindsey, 1997). Sibling seedlings of the same genotype typically show a range of phenotypes, from "golf tee"–like seedlings with fused cotyledons to apolar balls. The embryos show defects from the first division of the zygote, which often lacks the asymmetry typical of wild-type, with subsequent defects in cell division and expansion (Mayer *et al.*, 1993) and in cell wall organization (Shevell *et al.*, 2000).

Cloning of the gene revealed that GNOM is a member of the ARF-GEF family of proteins (ADP-ribosylation factor guanine nucleotide exchange factors), similar to yeast SEC7 (Shevell *et al.*, 1994; Busch *et al.*, 1996). These proteins are reversibly recruited to membranes required, as dimers, for membrane tethering and vesicle fusion (Anders *et al.*, 2008). The fungal toxin Brefeldin A (BFA) is a specific inhibitor of both ARF GEFs and polar auxin transport, and Steinmann *et al.* (1999) showed that BFA-sensitive GNOM is required for the correct polar localization of the PIN1 auxin efflux carrier during embryogenesis. BFA treatment leads to the cytosolic accumulation of PIN1, and this observation has formed the basis of the model for the regulation of PIN localization and dynamics by endocytic recycling, in a pathway dependent on GNOM activity (Geldner *et al.*, 2001, 2003). Auxin transport inhibitors such a triiodobenzoic acid also inhibit PIN localization by disrupting endocytic recycling. There appear also to be distinct ARF-GEF activities translocating vesicle cargoes to the apical versus the basal sides of the cell, and evidence suggests that GNOM has a specific role in vesicle transport to the basal membrane (Kleine-Vehn *et al.*, 2008).

Mutation of the *PINOID* (*PID*) gene, encoding a kinase, also leads to defective PIN localization, and defects in embryogenesis are seen in transgenic plants overexpressing *PID*. In contrast to *gnom* mutants, which show complete disorganization of PIN localization, *PID* overexpressors exhibit a shift from basal to apical PIN1 localization, leading to altered auxin gradients and mis-specification of the hypophysis (Friml *et al.*, 2004). These results suggest that PIN protein phosphorylation is important in determining the fate of the cell to which it is translocated. Quadruple mutants for *PID* and its three most related kinase gene family members (*PID2*, *WAG1*, *WAG2*) exhibit a loss of cotyledon development (Christensen *et al.*, 2000; Friml *et al.*, 2004; Cheng *et al.*, 2008). The PP2A phosphatase anatagonizes *PID* function (Michniewicz *et al.*, 2007).

The asymmetrical division of the embryo defines a critical step in *Arabidopsis* embryogenesis, by creating apical and basal domains with distinct fates. Although suspensor development requires expression of the YODA (YDA) MAP kinase kinase kinase (Lukowitz *et al.*, 2004; Bayer *et al.*, 2009), an early feature is the differential expression between apical and basal regions of the *WUSCHEL-RELATED HOMEOBOX* (*WOX*) transcription factor genes (Haecker *et al.*, 2004). Both *WOX2* and *WOX8* genes are expressed in the zygote, and subsequently *WOX2* expression is found in the apical cell, whereas *WOX8* is expressed in the basal cell. *WOX9* appears to be expressed after the zygotic division, possibly in both apical and basal cells, and *WOX5*, a marker of the QC, is expressed early in the hypophysis, showing its early specification. *WOX8* expression is necessary for *WOX2* expression in the proembryo, and *WOX* genes are also required for correct proembryo development through correct *PIN1* expression and embryonic auxin distribution (Breuninger *et al.*, 2008).

Hypophysis specification also requires auxin signaling and is defective in embryos carrying mutations in *MP*, *PIN* genes (Friml *et al.*, 2003), *AXR6* (Hellman *et al.*, 2003), or genes encoding enzymes in auxin biosynthesis such as *YUC* genes (Cheng *et al.*, 2007) or tryptophan aminotransferases (Stepanova *et al.*, 2008). POLTERGEIST (POL) and the related POLTERGEIST-LIKE1 (PLL1) are phosphatases also necessary for correct hypophysis division and, similar to MP, for vascular development (Song *et al.*, 2008), perhaps suggesting an inductive effect from cells within the embryo.

How is the boundary between apical and basal domains established in the proembryo? One gene with a role in this process is *HANABA TARANU* (*HAN*) (Nawy *et al.*, 2010). *HAN* encodes a GATA factor, in which loss of function causes a shift in gene expression from the basal to the apical region of the proembryo, including *WOX5*; *SUCROSE TRANSPORTER3* (normally expressed in the suspensor); reduced *SHORTROOT* (normally in or adjacent to the hypophysis); and *PIN1* and *PIN7*. These changes in gene expression lead to the ectopic formation of a root meristem in the central region of the globular-shaped *han* embryo. *HAN* activity appears to be necessary for the establishment of correct basal gene expression and auxin distribution required for apical-basal pattern formation.

A second gene to come out of the screen by Mayer *et al.* (1991) and required for the correct development of the *Arabidopsis* apical-basal axis is *MONOPTEROS*. Mutation of *MP* leads to the development of seedlings with no, or a very short, hypocotyl and no primary root, with only a "basal peg" in the position below the cotyledons (Berleth and Jürgens, 1993). Seedlings also typically have abnormal vasculature in the cotyledons, with some strong alleles lacking veins almost completely. These defects can be traced back to early embryogenesis, during which the octant stage has four rather than two tiers of cells, defective cell patterning (especially in the central cells of the vascular primordium) is seen at the heart stage, and asymmetrical cotyledon development occurs. As previously mentioned, *MP* encodes ARF5 required for the activation of auxin response genes, further linking embryo patterning to auxin signaling (Hardtke and Berleth, 1998). MP/ARF5 interacts with the inhibitory Aux/IAA12 protein BODENLOS (BDL) (Hamann *et al.*, 2002), and similar to the loss-of-function *mp* mutant, the gain-of-function *bdl* mutant also exhibits defective apical-basal patterning and root meristem formation in the embryo (Hamann *et al.*, 1999). NONPHOTOTROPIC HYPOCOTYL4 (NPH4) is also an ARF (ARF7), closely related to

MP. NPH4 dimerizes with MP and, similar to MP, homodimerizes and is required for embryonic axis formation (Hardtke *et al.*, 2004). The dimerization of ARFs and interactions with various Aux/IAA proteins likely account for the diversity of auxin responses in different cell types, each of which might be expected to express different combinations of these regulators.

MP and BDL induce auxin transport from a subgroup of embryonic cells in the basal domain into the hypophysis (Weijers *et al.*, 2006). Downstream targets of these transcriptional regulators are beginning to be identified through approaches such as microarray experiments on mutants or overexpressors. For example, following dexamethasone-induction of *MP* expression, Schlereth *et al.* (2010) identified numerous known auxin signaling genes, including basic helix-loop-helix transcription factors *TMO5* and *TMO7* that are expressed in embryonic cells at the root pole. These are required for *MP* function in embryonic root development, and TMO7 is a mobile protein that moves from its site of synthesis in the embryo to its site of action in the hypophysis.

Another target of MP is the AP2 transcription factor DORNRÖSCHEN (DRN) (Cole *et al.*, 2009). Chromatin immunoprecipitation shows direct binding of MP to the *DRN* promoter, controlling its expression in the apical region and cotyledons of the developing embryo. Because DRN and the related DRN-LIKE also interact with the brassinosteroid signaling basic helix-loop-helix protein BIM1, these proteins provide evidence of molecular crosstalk between auxin and brassinosteroids to control embryonic patterning (Chandler *et al.*, 2009).

Further information on the auxin-transcription factor network is revealed by the *topless* (*tpl*) mutant. Severe mutant alleles of *TPL* exhibit a disruption of apical development of the *Arabidopsis* embryo, with homeotic transformation of the shoot into a root (Long *et al.*, 2002, 2006). TPL is a WD-40 repeat-containing protein that interacts with BDL and probably forms part of a protein complex, as suggested by biomolecular fluorescence complementation (BiFC) assays (Szemenyei *et al.*, 2008). *bdl* mutants cannot fully repress auxin responses in a *tpl* background, demonstrating the functional significance of this interaction. In addition, TPL has been found to regulate the *PLT* genes, and ectopic expression of the PLT transcription factors leads to the conversion of shoot to root (Smith and Long, 2010). It was also shown that the class III homeodomain-leucine zipper (HD-ZIP III) transcription factors control embryonic apical fate and are themselves sufficient to drive the conversion of the embryonic root pole into a second shoot pole. There is evidence of an antagonistic relationship between the PLT and HD-ZIP III genes in specifying the root and shoot poles.

The *POPCORN* (*PCN*) gene has been identified more recently that is also required for correct integration of the auxin signaling pathway with embryogenesis and the correct development of shoot and root apical meristems (Xiang *et al.*, 2011). *PCN*, similar to *TPL*, encodes a WD-40 protein, and the *pcn* mutant embryos exhibit abnormal cellular patterning after the two-cell stage of development, with defective bilateral symmetry (cotyledon morphogenesis), hypophysis, and columella development. The mutant also has numerous postembryonic defects suggestive of abnormal meristem function. Consistent with this, the mutant exhibits defective expression of meristem genes, whereby *WUSCHEL* (*WUS*), *CLAVATA3* (*CLV3*), and *SHOOTMERISTEMLESS* (*STM*) are upregulated in the SAM. Auxin responses (as determined by patterns of *DR5::GFP* expression) are altered in the developing embryo and associated with altered localization of PIN1 and expression of *PIN7*, *MP*, *PLT1*, *PLT2*, *PLT3*, *BDL*, and *WOX5*. It is suggested that PCN, similar to TPL, acts as a repressor of auxin responses, although there is no indication of a direct physical interaction with BDL, MP, or TPL. Cotyledon development also requires PIN1 and PID function (Furutani *et al.*, 2004) and *ENHANCER OF PINOID* (Treml *et al.*, 2005).

As discussed earlier, laser-capture microdissection in combination with microarray analysis has led to the identification of thousands of genes that are differentially expressed in apical and basal domains from globular stage onward (Casson *et al.*, 2005; Spencer *et al.*, 2007). One gene that emerged from this screen and has been characterized in some detail is *MERISTEM-DEFECTIVE* (*MDF*) (Casson *et al.*, 2009). Identified on the basis of its expression in the basal domain of the heart-stage embryo, it is required for correct divisions of the basal domain of the globular embryo and subsequent root meristem development and function. The encoded protein is an RS domain protein, with a likely role in regulation of transcription or RNA processing. It is required for correct auxin distribution and *PIN2* expression in the embryo and for *PLT*, *BABY BOOM* (*BBM*), *SCARECROW* (*SCR*), and *SHORTROOT* (*SHR*) expression. It is also required for correct expression of the apical genes *WUS* and *CLV3* and subsequent shoot development. Transgenic overexpression of *MDF* leads to ectopic meristem activity and accumulation of embryonic triacylglycerols, suggesting it acts upstream of multiple embryonic pathways, although it is not itself regulated by auxin. BBM is a member of the PLT protein family, and its overexpression leads also to activation of embryonic pathways in vegetative cells (Boutilier *et al.*, 2002); this may be at least part of the mechanism by which MDF acts.

Once the apical domain of the embryo has been initiated, early activation of the expression of transcription factors leads to the specification of the future SAM. Essential in the process are two types of homeodomain proteins: KNOTTED-related STM (Barton and Poethig, 1993; Long *et al.*, 1996) and WUS (Laux *et al.*, 1996; Mayer *et al.*, 1998). Loss-of-function mutation of either gene leads to a failure of SAM formation; therefore, both genes function as positive regulators of the stem cell population in the developing embryo. In later stages of plant development, the *CLAVATA* genes work in antagonism to WUS, as negative regulators of cell production in the SAM-*clv* mutants are characterized by an excessive number of cells in the SAM (Clark *et al.*, 1993, 1995; Lenhard *et al.*, 2002). CLV1 is a predicted receptor kinase that in association with CLV2 interacts with the CLV3 peptide (Ogawa *et al.*, 2008), and a regulatory loop is established between WUS and the CLV complex to regulate the size of the SAM. It is also possible that WUS has an as yet unclear role in embryogenesis because its overexpression can induce somatic embryo formation on vegetative structures (Zuo *et al.*, 2002). ZWILLE (ZLL) is required to maintain meristem cell identity in the SAM and interacts with STM (Moussian *et al.*, 1998), as does CUP-SHAPED COTYLEDON (CUC) (Aida *et al.*, 1997, 1999), which is also linked to auxin signaling (Aida *et al.*, 2002; Furutani *et al.*, 2004). ZLL is a member of the ARGONAUTE protein family and is required to maintain HD-ZIP III expression levels in the SAM (Liu *et al.*, 2009). Ectopic expression of the HD-ZIP III genes *PHAVOLUTA* (*PHV*) and *PHABULOSA* (*PHB*) in the basal domain of the heart-stage embryo leads to defects in root development, and this is regulated by a SERRATE-miRNA165/166–dependent pathway (Grigg *et al.*, 2009). *HD-ZIP III* genes also interact with the *KANADI* genes (encoding GARP family transcription factors) to regulate *PIN1* expression and embryonic pattern (Izhaki and Bowman, 2007).

These genes mediate their effects through control of hormone responses. STM and the related KNOX family transcription factors repress GA signaling in the SAM. *NTH15* of tobacco encodes a *KNOX* protein, which downregulates the expression of the GA biosynthetic gene *GA20-OXIDASE* in tobacco meristems. It binds directly to the *GA20-OXIDASE* gene promoter, reducing the levels of active GA (Sakamoto *et al.*, 1999). Similarly, KN1 of maize directly binds and activates the promoter of the GA catabolic gene, *GA2OX* (Bolduc and Hake, 2009). Cytokinin has been known for many years to promote shoot meristem formation in plant tissue cultures, and both *STM* and *KNOX* genes are upregulated in cytokinin-overexpressing transgenics (Rupp *et al.*, 1999), as is *WUS* (Bäurle and Laux, 2005; Gordon *et al.*, 2009). *WUS* expression is also induced early in *Arabidopsis* somatic embryogenesis in response to auxin and is necessary for this developmental process (Su *et al.*, 2009). Overexpression of *WUS* can lead to ectopic somatic embryogenesis in vegetative tissues (Zuo *et al.*,

2002). WUS itself has been shown to regulate cytokinin response directly by repressing the *ARR* genes, which themselves negatively regulate cytokinin responses (Leibfried *et al.*, 2005), revealing a cytokinin feedback loop to control stem cell numbers in the SAM. For a more detailed discussion of the molecular development of the SAM, see Simon (2004).

Superimposed on the apical-basal axis during embryogenesis is patterning of the radial axis. In plants, this patterning is seen as concentric rings of tissue, comprising the dermal, ground, and vascular tissues. Radial pattern is established very early in *Arabidopsis* embryogenesis, at the four-cell stage, and this is reinforced by the establishment of the external protoderm in the early globular embryo, followed by further definition of internal cell layers, including the provascular tissues, by heart stage.

Protoderm formation depends at least in part on WOX function. *wox2* mutants are affected in the tangential divisions that define the protoderm at the octant stage of embryogenesis and subsequent proliferative divisions of the protoderm itself, a phenotype exacerbated by additional mutations in *WOX1* and *WOX3* or *WOX8* (Haecker *et al.*, 2004; Breuniger *et al.*, 2008). Establishment of the protoderm is associated with cell layer–specific gene expression domains, such as the *ARABIDOPSIS THALIANA MERISTEM LAYER 1* (*ATML1*) (Lu *et al.*, 1996), *PROTODERMAL FACTOR 2* (*PDF2*) (Abe *et al.*, 2003), and *ARABIDOPSIS CRINKLY 4* (*ACR4*, a receptor-like kinase) (Tanaka *et al.*, 2002; Gifford *et al.*, 2003) genes. *pdf2 atml1* double mutants show defective protoderm development and patterning of other embryo cells (Abe *et al.*, 2003). PDF2 and ATML1, similar to PHV and PHB, are HD-ZIP transcription factors containing a START (StAR-related lipid transfer) domain. In the RNA, this sequence can represent a target site for miRNA-mediated degradation – this occurs for the post-transcriptional processing of *PHB*, *PHV*, and *REVOLUTA* (*REV*) transcripts, as described previously, but may also represent a sterol-binding site in the protein (Lindsey *et al.*, 2003). ACR4 is also essential for root development, both in columella initial divisions and in pericycle divisions leading to the formation of lateral roots (De Smet *et al.*, 2008). Two other kinases are also required for correct development of radial pattern during embryogenesis: RECEPTOR-LIKE PROTEIN KINASE 1 (RPK1) and TOADSTOOL 2 (TOAD2). Double mutants for these genes exhibit cells with incorrect identity, whereby outer cell identities are replaced by inner cell identities (Nodine *et al.*, 2007; Nodine and Tax, 2008). Vascular cell markers are expressed more widely than in wild-type (i.e., in the protoderm), whereas protodermal markers are expressed only transiently. The double mutants typically fail to develop beyond the globular stage.

MP is required for correct vascular patterning during embryogenesis, demonstrating the importance of auxin in this process, as discussed earlier (Berleth and Jürgens, 1993; Hardtke and Berleth, 1998). Mayer *et al.* (1991) also identified the *KNOLLE* and *KEULE* genes as being essential for correct radial pattern. KNOLLE is a syntaxin-like protein (Lukowitz *et al.*, 1996; Lauber *et al.*, 1997) that physically interacts with the Sec1 protein KEULE (Assaad *et al.*, 2001), and together they regulate cytokinesis through roles in controlling vesicle fusion for correct cell plate formation. The KANADI-HD-ZIP III transcription factor system discussed previously also plays a role in radial patterning, as multiple mutants of these genes display defective establishment of the central-peripheral axis (Izhaki and Bowman, 2007).

Mutational screens have also identified regulators of the establishment of cortical and endodermal layers in the root, but which have embryonic expression – SCR and SHR (Benfey *et al.*, 1993; Scheres *et al.*, 1995; Di Laurenzio *et al.*, 1996). Loss-of-function of each gene leads to the formation of a single layer of ground tissue in the *Arabidopsis* root owing to a failure of the formative division in the cortex/endodermis initial cell, which gives rise to the two tissues. Both genes are also coexpressed in the QC of the root, and *SCR* expression is first seen in the hypophysis, before QC formation in the embryo (Wysocka-Diller *et al.*, 2000; Sabatini *et al.*, 2003). These proteins act together as a transcription factor complex to activate a D-type cyclin required for the division of the initial cell (Sozzani *et al.*, 2010). The SHR transcription factor is mobile, moving between cell layers in the root from the site of transcription to site of protein localization in the endodermis nucleus, where it regulates cell fate (Nakajima *et al.*, 2001). It is not yet known whether it is similarly mobile in the embryo, although TMO7 is mobile between the embryo and hypophysis, as indicated earlier (Schlereth *et al.*, 2010).

Other genes required for correct embryonic radial pattern include *HYDRA1* and *FACKEL/HYDRA2*. Each of these genes is required for correct embryogenesis, with mutants showing a range of patterning abnormalities from multiple, disorganized cell layers to abnormal apical meristem function (Mayer *et al.*, 1991; Topping *et al.*, 1997; Jang *et al.*, 2000; Schrick *et al.*, 2000; Souter *et al.*, 2002). Both genes encode enzymes in sterol biosynthesis, and the mutants show defects in hormone signaling (including PIN localization) (Souter *et al.*, 2002; Pullen *et al.*, 2010), cytokinesis, and cell wall construction (Schrick *et al.*, 2004). As vascular and protoderm pattern is disrupted in these mutants, the possibility is raised that lack of specific sterols may lead to defective function of one or more START domain proteins required for radial pattern, such as ATML1, PDF2, PHV, PHB, or REV.

Conclusion and Future Directions

The development of the *Arabidopsis* embryo has provided an outstanding system in which to unravel diverse molecular mechanisms of plant development: cell division control and patterning, gene-signal interactions, cell differentiation, and morphogenesis. The precise patterning of the *Arabidopsis* embryo has greatly facilitated the genetic approach to dissecting mechanisms. We have not touched on epigenetic control in much detail, but this is surely a potentially very fruitful field for future study. Also, we do not know very much about local signalling events, such as the role of cell wall components in regulating cell identity (He *et al.*, 2007), despite some early advances in this area using *Fucus* (Souter and Lindsey, 2000).

Also on the horizon is the more extensive use of systems biology approaches, in which multiple layers of data (transcriptomics, proteomics, metabolomics, protein-protein interactions, and imaging) are integrated using advanced computational techniques to provide a more complete view of embryo development. This information will be increasingly used to inform our understanding of seed development in the major crops, essential if we are to develop a more rational, gene-based approach to improving yield and quality traits, including seedling establishment and plant architecture, to feed a growing world population.

References

Abe M, Katsumata H, Komeda Y, Takahashi T (2003). Regulation of shoot epidermal cell differentiation by a pair of homeodomain proteins in Arabidopsis. Development 130: 635–643.

Aida M, Ishida T, Fukaki H, Fujisawa H, Tasaka M (1997). Genes involved in organ separation in Arabidopsis: an analysis of the cup-shaped cotyledon mutant. Plant Cell 9: 841–857.

Aida M, Ishida T, Tasaka M (1999). Shoot apical meristem and cotyledon formation during Arabidopsis embryogenesis: interaction among the *CUP-SHAPED COTYLEDON* and *SHOOT MERISTEMLESS* genes. Development 126: 1563–1570.

Aida M, Vernoux T, Furutani M, Traas J, Tasaka M (2002). Roles of *PIN-FORMED1* and *MONOPTEROS* in pattern formation of the apical region of the Arabidopsis embryo. Development 129: 3965–3974.

Anders N, Nielsen M, Keicher J, Stierhof Y-D, Furutani M, Tasaka M, Skriver K, Jürgens G (2008). Membrane association of the Arabidopsis ARF Exchange Factor GNOM involves interaction of conserved domains. Plant Cell 20: 142–151.

Assaad FF, Huet Y, Mayer U, Jürgens G (2001). The cytokinesis gene *KEULE* encodes a Sec1 protein that binds the syntaxin KNOLLE. J Cell Biol 152: 531–543.

Barton MK, Poethig RS (1993). Formation of the shoot apical meristem in *Arabidopsis thaliana*: an analysis of development in the wild type and in the *shoot meristemless* mutant. Development 119: 823–831.

Baumlein H, Misera S, Luerssen H, Kolle K, Horstmann C, Wobus U, Muller AJ (1994). The *FUS3* gene of Arabidopsis thaliana is a regulator of gene expression during late embryogenesis. Plant J 6: 367–387.

Bäurle I, Laux T (2005). Regulation of *WUSCHEL* transcription in the stem cell niche of the Arabidopsis shoot meristem. Plant Cell 17: 2271–2280.

Baybrook SA, Harada JJ (2008). LECs go crazy in embryo development. Trends Plant Sci 13: 624–630.

Bayer M, Nawy T, Giglione C, Galli M, Meinnel T, Lukowitz W (2009). Paternal control of embryonic patterning in *Arabidopsis thaliana*. Science 323: 1485–1488.

Benfey PN, Linstead PJ, Roberts K, Schiefelbein JW, Hauser M-T, Aeschbacher RA (1993). Root development in *Arabidopsis:* four mutants with dramatically altered root morphogenesis. Development 119: 57–70.

Bennett MJ, Marchant A, Green HG, May ST, Ward SP, Millner PA, Walker AR, Schultz B, Feldmann KA (1996). *Arabidopsis AUX1* gene: a permease-like regulator of root gravitropism. Science 273: 948–950.

Berleth T, Jürgens G (1993). The role of the *monopteros* gene in organising basal development in the Arabidopsis embryo. Development 118: 575–587.

Birnbaum K, Shasha DE, Wang JY, Jung JW, Lambert GM, Galbraith DW, Benfey PN (2003). A gene expression map of the *Arabidopsis* root. Science 302: 1956–1960.

Blakeslee JJ, Peer WA, Murphy AS (2005). Auxin transport. Curr Opin Plant Biol 8: 494–500.

Bolduc N, Hake S (2009). The maize transcription factor KNOTTED1 directly regulates the gibberellin catabolism gene *ga2ox1*. Plant Cell 21: 1647–1658.

Boutilier K, Offringa R, Sharma VK, Kieft H, Ouellet T, Zhang LM, Hattori J, Liu CM, van Lammeren AAM, Miki BLA, Custers JBM, Campagne MMV (2002). Ectopic expression of *BABY BOOM* triggers a conversion from vegetative to embryonic growth. Plant Cell 14: 1737–1749.

Breuninger H, Rikirsch E, Hermann M, Ueda M, Laux T (2008). Differential expression of *WOX* genes mediates apical-basal axis formation in the Arabidopsis embryo. Dev Cell 14: 867–876.

Busch M, Mayer U, Jürgens G (1996). Molecular analysis of the Arabidopsis pattern formation of gene GNOM: gene structure and intragenic complementation. Mol Gen Genet 250: 681–691.

Casson SA, Lindsey K (2006). The *turnip* mutant of Arabidopsis reveals that *LEAFY COTYLEDON1* expression mediates the effects of auxin and sugars to promote embryonic cell identity. Plant Physiol 142: 526–541.

Casson S, Spencer M, Walker K, Lindsey K (2005). Laser capture microdissection for the analysis of gene expression during embryogenesis of Arabidopsis. Plant J 42: 111–123.

Casson SA, Topping JF, Lindsey K (2009). MERISTEM-DEFECTIVE, an RS domain protein, is required for the correct meristem patterning and function in Arabidopsis. Plant J 57: 857–869.

Chandler JW, Cole M, Flier A, Werr W (2009). BIM1, a bHLH protein involved in brassinosteroid signalling, controls Arabidopsis embryonic patterning via interaction with DORNRÖSCHEN and DORNRÖSCHEN-LIKE. Plant Mol Biol 69: 57–68.

Chaudhury AM, Letham S, Craig S, Dennis ES (1993). *amp1* – a mutant with high cytokinin levels and altered embryonic pattern, faster vegetative growth, constitutive photomorphogenesis and precocious flowering. Plant J 4: 907–916.

Cheng Y, Dai X, Zhao Y (2007). Auxin synthesized by the YUCCA flavin monooxygenases is essential for embryogenesis and leaf formation in Arabidopsis. Plant Cell 19: 2430–2439.

Cheng Y, Qin G, Dai X, Zhao Y (2008). *NPY* genes and AGC kinases define two key steps in auxin-mediated organogenesis in Arabidopsis. Proc Natl Acad Sci U S A 105: 21017–21022.

Christensen SK, Dagenais N, Chory J, Weigel D (2000). Regulation of auxin response by the protein kinase PINOID. Cell 100: 469–478.

Clark SE, Running MP, Meyerowitz EM (1993). *CLAVATA1*, a regulator of meristem and flower development in Arabidopsis. Development 119: 397–418.

Clark SE, Running MP, Meyerowitz EM (1995). *CLAVATA3* is a specific regulator of shoot and floral meristem development affecting the same processes as *CLAVATA1*. Development 121: 2057–2067.

Cole M, Chandler J, Weijers D, Jacobs B, Comelli P, Werr W (2009). DORNRÖSCHEN is a direct target of the auxin response factor MONOPTEROS in the *Arabidopsis* embryo. Development 136: 1643–1651.

Curaba J, Moritz T, Blervaque R, Parcy F, Raz V, Herzog M, Vachon G (2004). *AtGA3ox2*, a key gene responsible for bioactive gibberellin biosynthesis, is regulated during embryogenesis by *LEAFY COTYLEDON2* and *FUSCA3* in *Arabidopsis*. Plant Physiol 136: 3660–3669.

De Smet I, Vassileva V, De Rybel B, Levesque MP, Grunewald W, Van Damme D, Van Noorden G, Naudts M, Van Isterdael G, De Clercq R, Wang JY, Meuli N, Vanneste S, Friml J, Hilson P, Jürgens G, Ingram GC, Inze D, Benfey P, Beeckman T (2008). Receptor-like kinase ACR4 restricts formative cell divisions in the *Arabidopsis* root. Science 322: 595–597.

Dean Rider S Jr, Henderson JT, Jerome RE, Edenberg HJ, Romero-Severson J, Ogas J (2003). Coordinate repression of regulators of embryonic identity by PICKLE during germination in Arabidopsis. Plant J 35: 33–43.

Dello Ioio R, Nakamura K, Moubayidin L, Perilli S, Taniguchi M, Morita MT, Aoyama T, Costantino P, Sabatini S (2008). A genetic framework for the control of cell division and differentiation in the root meristem. Science 322: 1380–1384.

Di Laurenzio L, Wysocka-Diller J, Malamy JE, Pysh L, Helariutta Y, Freshour G, Hahn MG, Feldmann KA, Benfey PN (1996). The *SCARECROW* gene regulates an asymmetric cell division that is essential for generating the radial organization of the *Arabidopsis* root. Cell 8: 423–433.

Eklund DM, Staldal, V, Valsecchi I, Cierlik I, Eriksson C, Hiratsu K, Ohme-Takagi M, Sundstrom JF, Thelander M, Ezcurra I, Sundberg E (2010). The *Arabidopsis thaliana* STYLISH1 protein acts as a transcriptional activator regulating auxin biosynthesis. Plant Cell 22: 349–363.

Finkelstein RR, Gampala SS, Rock CD (2002). Abscisic acid signaling in seeds and seedlings. Plant Cell 14: S15–S45.

Fischer C, Speth V, Fleig Eberenz S, Neuhaus G (1997). Induction of zygotic polyembryos in wheat: influence of auxin polar transport. Plant Cell 9: 1767–1780.

Friml J, Vieten A, Sauer M, Weijers D, Schwarz H, Hamann T, Offringa R, Jürgens G (2003). Efflux-dependent auxin gradients establish the apical-basal axis of *Arabidopsis*. Nature 426:147–153.

Friml J, Yang X, Michniewicz M, *et al.* (2004). A PINOID-dependent binary switch in apical-basal PIN polar targeting directs auxin efflux. Science 306: 862–865.

Fukaki H, Taniguchi N, Tasaka M (2006). PICKLE is required for SOLITARY-ROOT/IAA14-mediated repression of ARF7 and ARF19 activity during Arabidopsis lateral root initiation. Plant J 48: 380–389.

Furutani M, Vernoux T, Traas J, Kato T, Tasaka M, Aida M (2004). PIN-FORMED1 and PINOID regulate boundary formation and cotyledon development in *Arabidopsis* embryogenesis. Development 131: 5021–5030.

Gazzarrini S, Tsuchiya Y, Lumba S, Okamoto M, McCourt P (2004). The transcription factor FUSCA3 controls developmental timing in Arabidopsis through the hormones gibberellin and abscisic acid. Dev Cell 7: 373–378.

Geldner N, Anders N, Wolters H, Keicher J, Kornberger W, Muller P, Delbarre A, Ueda T, Nakano A, Jürgens G (2003). The Arabidopsis GNOM ARF-GEF mediates endosomal recycling, auxin transport, and auxin-dependent plant growth. Cell 112: 219–230.

Geldner N, Friml J, Stierhof Y-D, Jürgens G, Palme K (2001). Auxin transport inhibitors block PIN1 cycling and vesicle trafficking. Nature 413: 425–428.

Gifford ML, Dean S, Ingram GC (2003). The Arabidopsis *ACR4* gene plays a role in cell layer organisation during ovule integument and sepal margin development. Development 130: 4249–4258.

Goldberg RB (1988). Plants—novel developmental processes. Science 240: 1460–1467.

Goldberg RB, Barker SJ, Perezgrau L (1989). Regulation of gene expression during pant embryogenesis. Cell 56: 149–160.

Gordon SP, Chickarmane VS, Ohno C, Meyerowitz EM (2009). Multiple feedback loops through cytokinin signaling control stem cell number within the Arabidopsis shoot meristem. Proc Natl Acad Sci U S A 106: 16529–16534.

Grigg SP, Galinha C, Kornet N, Canales C, Scheres B, Tsiantis M (2009). Repression of apical homeobox genes is required for embryonic root development in Arabidopsis. Curr Biol 19: 1485–1490.

Hadfi K, Speth V, Neuhaus G (1998). Auxin-induced developmental patterns in Brassica juncea embryos. Development 125: 879–887.

Haecker A, Gross-Hardt R, Geiges B, Sarkar A, Breuninger H, Herrmann M, Laux T (2004). Expression dynamics of *WOX* genes mark cell fate decisions during early embryonic patterning in *Arabidopsis thaliana*. Development 131: 657–668.

Hakman I, Hallberg H, Palovaara J (2009). The polar auxin transport inhibitor NPA impairs embryo morphology and increases the expression of an auxin efflux facilitator protein PIN during Picea abies somatic embryo development. Tree Physiol 29: 483–496.

Hamann T, Benková E, Baurle I, Kientz M, Jürgens G (2002). The Arabidopsis *BODENLOS* gene encodes an auxin response protein inhibiting MONOPTEROS-mediated embryo patterning. Genes Dev 16: 1610–1615.

Hamann T, Mayer U, Jürgens G (1999). The auxin-insensitive *bodenlos* mutation affects primary root formation and apical-basal patterning in the *Arabidopsis* embryo. Development 126: 1387–1395.

Harding EW, Tang W, Nichols KW, Fernandez DE, Perry SE (2003). Expression and maintenance of embryogenic potential is enhanced through constitutive expression of *AGAMOUS-LIKE 15*. Plant Physiol 133: 653–663.

Hardtke CS, Berleth T (1998). The Arabidopsis gene *MONOPTEROS* encodes a transcription factor mediating embryo axis formation and vascular development. EMBO J 17: 1405–1411.

Hardtke CS, Ckurshumova W, Vidaurre DP, Singh SA, Stamatiou G, Tiwari SB, Hagen G, Guilfoyle TJ, Berleth T (2004). Overlapping and non-redundant functions of the *Arabidopsis* auxin response factors MONOPTEROS and NONPHOTOTROPIC HYPOCOTYL 4. Development 131: 1089–1100.

Hartweck LM, Scott CL, Olszewski NE (2002). Two O-linked N-acetylglucosamine transferase genes of *Arabidopsis thaliana* L. Heynh. have overlapping functions necessary for gamete and seed development. Genetics 161: 1279–1291.

He YC, He YQ, Qu LH, Sun MX, Yang H (2007). Tobacco zygotic embryogenesis in vitro: the original cell wall of the zygote is essential for maintenance of cell polarity, the apical-basal axis and typical suspensor formation. Plant J 49: 515–527.

Hellmann H, Hobbie L, Chapman A, Dharmasiri S, Dharmasiri N, del Pozo C, Reinhardt D, Estelle M (2003). Arabidopsis *AXR6* encodes CUL1 implicating SCF E3 ligases in auxin regulation of embryogenesis. EMBO J 22: 3314–3325.

Hundertmark M, Hincha D (2008). LEA (Late Embryogenesis Abundant) proteins and their encoding genes in *Arabidopsis thaliana*. BMC Genomics 9: 118.

Izhaki A, Bowman JL (2007). *KANADI* and class III HD-Zip gene families regulate embryo patterning and modulate auxin flow during embryogenesis in *Arabidopsis*. Plant Cell 19: 495–508.

Jang JC, Fujioka S, Tasaka M, Seto H, Takatsuto S, Ishii A, Aida M, Yoshida S, Sheen J (2000). A critical role of sterols in embryonic patterning and meristem programming revealed by the fackel mutants of *Arabidopsis thaliana*. Genes Dev 14: 1485–1497.

Jürgens G, Ruiz RAT, Berleth T (1994). Embryonic pattern formation in flowering plants. Ann Rev Genet 28: 351–371.

Keith KA, Kraml M, Dengler NG, McCourt P (1994). *fusca3*: a heterochronic mutation affecting late development in Arabidopsis. Plant Cell 6: 589–600.

Kleine-Vehn J, Dhonukshe P, Sauer M, Brewer PB, Wisniewska J, Paciorek T, Benkova E, Friml J (2008). ARF GEF-dependent transcytosis and polar delivery of PIN auxin carriers in Arabidopsis. Curr Biol 18: 526–531.

Lauber MH, Waizenegger I, Steinmann T, Schwartz H, Mayer U, Hwang I, Lukowitz W, Jürgens G (1997). The Arabidopsis KNOLLE protein is a cytokinesis-specific syntaxin. J Cell Biol 139: 1485–1493.

Laux T, Mayer KF, Berger J, Jürgens G (1996). The *WUSCHEL* gene is required for shoot and floral meristem integrity in Arabidopsis. Development 122: 87–96.

Le BH, Cheng C, Bui AQ, Wagmaister JA, Henry KF, Pelletier J, Kwong L, Belmonte M, Kirkbride R, Horvath S, Drews GN, Fischer RL, Okamuro JK, Harada JJ, Goldberg RB (2010). Global analysis of gene activity during *Arabidopsis* seed development and identification of seed-specific transcription factors. Proc Natl Acad Sci U S A 107: 8063–8070.

Leibfried A, To JP, Busch W, Stehling S, Kehle A, Demar M, Kieber JJ, Lohmann JU (2005). WUSCHEL controls meristem function by direct regulation of cytokinin-inducible response regulators. Nature 438: 1172–1175.

Lenhard M, Jürgens G, Laux T (2002). The *WUSCHEL* and *SHOOTMERISTEMLESS* genes fulfil complementary roles in Arabidopsis shoot meristem regulation. Development 129: 3195–3206.

Li J, Dai X, Zhao Y (2006). A role for auxin response factor 19 in auxin and ethylene signaling in Arabidopsis. Plant Physiol 140: 899–908.

Li J, Yang H, Peer WA, Richter G, Blakeslee J, Bandyopadhyay A, Titapiwantakun B, Richards EL, Krizek B, Murphy AS, Gilroy S and Gaxiola R (2005) *Arabidopsis* H + -PPase AVP1 regulates auxin-mediated organ development. Science 310: 121–125.

Lindsey K, Pullen ML, Topping JF (2003). Importance of plant sterols in pattern formation and hormone signalling. Trends Plant Sci 8: 521–525.

Liu CM, Xu ZH, Chua NH (1993). Auxin polar transport is essential for the establishment of bilateral symmetry during early plant embryogenesis. Plant Cell 5: 621–630.

Liu Q, Yao X, Pi L, Wang H, Cui X, Huang H (2009). The *ARGONAUTE10* gene modulates shoot apical meristem maintenance and leaf polarity establishment by repressing miR165/166 in Arabidopsis. Plant J 58: 27–40.

Long JA, Moan EI, Medford JI, Barton MK (1996). A member of the KNOTTED class of homeodomain proteins encoded by the *STM* gene Arabidopsis. Nature 379: 66–69.

Long JA, Ohno C, Smith ZR, Meyerowitz EM (2006). TOPLESS regulates apical embryonic fate in Arabidopsis. Science 312: 1520–1523.

Long JA, Woody S, Poethig S, Meyerowitz EM, Barton K (2002). Transformation of shoots into roots in Arabidopsis embryos mutant at the *TOPLESS* locus. Development 129: 2797–2806.

Lotan T, Ohto M, Yee KM, West MA, Lo R, Kwong RW, Yamagishi K, Fischer RL, Goldberg RB, Harada JJ (1998). *Arabidopsis LEAFY COTYLEDON1* is sufficient to induce embryo development in vegetative cells. Cell 93: 1195–1205.

Lu P, Porat R, Nadeau JA, O'Neill SD (1996). Identification of a meristem L1 layer-specific gene in Arabidopsis that is expressed during embryonic pattern formation and defines a new class of homeobox genes. Plant Cell 8: 2155–2168.

Luerssen H, Kirik V, Herrmann P, Misera S (1998). *FUSCA3* encodes a protein with a conserved VP1/AB13-like B3 domain which is of functional importance for the regulation of seed maturation in *Arabidopsis thaliana*. Plant J 15: 755–764.

Lukowitz W, Mayer U, Jürgens G (1996). Cytokinesis in the Arabidopsis embryo involves the syntaxin-related KNOLLE gene product. Cell 84: 61–71.

Lukowitz W, Roeder A, Parmenter D, Somerville CR (2004). A MAPKK kinase gene regulates extra-embryonic cell fate in Arabidopsis. Cell 116: 109–119.

Mansfield SG, Briarty LG (1991). Early embryogenesis in *Arabidopsis thaliana*. II. The developing embryo. Can J Bot 69: 461–476.

Mayer KFX, Schoof H, Haecker A, Lenhard M, Jürgens G, Laux T (1998). Role of *WUSCHEL* in regulating stem cell fate in the *Arabidopsis* shoot meristem. Cell 95: 805–815.

Mayer U, Buttner G, Jürgens G (1993). Apical-basal pattern formation in the Arabidopsis embryo: studies on the role of the *gnom* gene. Development 117: 149–162.

Mayer U, Torres-Ruiz RA, Berleth T, Misera S, Jürgens G (1991). Mutations affecting body organization in the *Arabidopsis* embryo. Nature 353: 402–407.

Meinke DW (1985). Embryo-lethal mutants of *Arabidopsis thaliana*: analysis of mutants with a wide range of lethal phenotypes. Theor Appl Genet 69: 543–552.

Meinke D, Franzmann LH, Nickle TC, Yeung EC (1994). *Leafy cotyledon* mutants of *Arabidopsis*. Plant Cell 6: 1049–1064.

Meinke DW, Sussex IM (1979). Embryo-lethal mutants of *Arabidopsis thaliana*: a model system for genetic analysis of plant embryo development. Dev Biol 72: 50–61.

Michniewicz M, Zago MK, Abas L, Weijers D, Schweighofer A, Meskiene I, Heisler MG, Ohno C, Zhang J, Huang F, Schwab R, Weigel D, Meyerowitz EM, Luschnig C, Offringa R, Friml J (2007). Antagonistic regulation of PIN phosphorylation by PP2A and PINOID directs auxin flux. Cell 130: 1044–1056.

Moussian B, Schoof H, Haecker A, Jürgens G, Laux T (1998). Role of the *ZWILLE* gene in the regulation of central shoot meristem cell fate during Arabidopsis embryogenesis. EMBO J 17: 1799–1809.

Mravec J, Kubes M, Bielach A, Gaykova V, Petrásek J, Skupa P, Chand S, Benková E, Zazimalová E, Friml J (2008). Interaction of PIN and PGP transport mechanisms in auxin distribution-dependent development. Development 135: 3345–3354.

Mravec J, Skupa P, Bailly A, Hoyerova K, Krecek P, Bielach A, Petrasek J, Zhang J, Gaykova V, Stierho Y-D, Dobrev PI, Schwarzerova K, Rolcik J, Seifertova D, Luschnig C, Benkova E, Zazimalova E, Geisler M, Friml J (2009), Subcellular homeostasis of phytohormone auxin is mediated by the ER-localized PIN5 transporter. Nature 459: 1136–1140.

Müller B, Sheen J (2008). Cytokinin and auxin interaction in root stem-cell specification during early embryogenesis. Nature 453: 1094–1097.

Nakajima K, Sena G, Nawy T, Benfey PN (2001). Intercellular movement of the putative transcription factor SHR in root patterning. Nature 413: 307–311.

Nawy T, Bayer M, Mravec J, Friml J, Birnbaum KD, Lukowitz W (2010). The GATA factor HANABA TARANU is required to position the proembryo boundary in the early Arabidopsis embryo. Dev Cell 19: 103–113.

Nodine MD, Bartel DP (2010). MicroRNAs prevent precocious gene expression and enable pattern formation during plant embryogenesis. Genes Dev 24: 2678–2692.

Nodine MD, Bartel DP (2012). Maternal and paternal genomes contribute equally to the transcriptome of early plant embryos. Nature 482: 94–97.

Nodine MD, Tax FE (2008). Two receptor-like kinases required together for the establishment of *Arabidopsis* cotyledon primordia. Dev Biol 314: 161–170.

Nodine MD, Yadegari R, Tax FE (2007). RPK1 and TOAD2 are two receptor-like kinases redundantly required for *Arabidopsis* embryonic pattern formation. Dev Cell 12: 943–956.

Ogas J, Cheng JC, Sung ZR, Somerville C (1997). Cellular differentiation regulated by gibberellin in the Arabidopsis thaliana pickle mutant. Science 277: 91–94.

Ogas J, Kaufmann S, Henderson J, Somerville C (1999). PICKLE is a CHD3 chromatin-remodeling factor that regulates the transition from embryonic to vegetative development in *Arabidopsis*. Proc Natl Acad Sci U S A 96: 13839–13844.

Ogawa M, Shinohawa H, Sakagami Y, Matsubayashi Y (2008). *Arabidopsis* CLV3 directly binds CLV1 ectodomain. Science 319: 294.

Okada K, Ueda J, Komaki MK, Bell CJ, Simura Y (1991). Requirement of the auxin polar transport system in early stages of *Arabidopsis* floral bud formation. Plant Cell 3: 677–684.

Peer WA, Hosein FN, Bandyopadhyay A, Makam SN, Otegui MS, Lee G-J, Blakeslee JJ, Cheng Y, Titapiwatanakun, B, Yakubov, B, Bangari B, Murphy AS (2009). Mutation of the membrane-associated M1 protease APM1 results in distinct embryonic and seedling developmental defects in Arabidopsis. Plant Cell 21: 1693–1721.

Pullen M, Clark N, Zarinkamar F, Topping J, Lindsey K (2010). Analysis of vascular development in the hydra sterol biosynthetic mutants of Arabidopsis. PLoS ONE 5: e12227.

Reed JW (2001). Roles and activities of AUX/IAA proteins in *Arabidopsis*. Trends Plant Sci 6: 420–425.

Riou-Khamlichi C, Huntley R, Jacqmard A, Murray JA (1999). Cytokinin activation of *Arabidopsis* cell division through a D-type cyclin. Science 283: 1541–1544.

Rupp HM, Frank M, Werner T, Strnad M, Schmulling T (1999). Increased steady state mRNA levels of the *STM* and *KNAT1* homeobox genes in cytokinin overproducing Arabidopsis thaliana indicate a role for cytokinins in the shoot apical meristem. Plant J 18: 557–563.

Ruzicka K, Simasková M, Duclercq J, Petrasek J, Zazımalova E, Simon S, Friml J, Van Montagu MC, Benkova E (2009). Cytokinin regulates root meristem activity via modulation of the polar auxin transport. Proc Natl Acad Sci U S A 106: 4284–4289.

Sabatini S, Heidstra R, Wildwater M, Scheres B (2003). SCARECROW is involved in positioning the stem cell niche in the *Arabidopsis* root meristem. Genes Dev 17: 354–358.

Sakamoto T, Nishimura A, Tamaoki M, Kuba M, Tanaka H, Iwahori S, Matsuoka M (1999). The conserved KNOX domain mediates specificity of tobacco KNOTTED1-type homeodomain proteins. Plant Cell 11: 1419–1432.

Santos Mendoza M, Dubreucq B, Miquel M, Caboche M, Lepiniec L (2005). *LEAFY COTYLEDON 2* activation is sufficient to trigger the accumulation of oil and seed specific mRNAs in Arabidopsis leaves. FEBS Lett 579: 4666–4670.

Scheres B, Di Laurenzio L, Willemsen V, Hauser M-T, Janmaat K, Weisbeek P, Benfey PN (1995). Mutations affecting the radial organisation of the *Arabidopsis* root display specific defects throughout the embryonic axis. Development 121: 53–62.

Scheres B, Wolkenfelt H, Willemsen V, Terlouw M, Lawson E, Dean C, Weisbeek P (1994). Embryonic origin of the Arabidopsis root and root-meristem initials. Development 120: 2475–2487.

Schlereth A, Moeller B, Liu W, Kientz M, Flipse J, Rademacher EH, Schmid M, Jürgens G, Weijers D (2010). MONOPTEROS controls embryonic root initiation by regulating a mobile transcription factor. Nature 464: 913–916.

Schrick K, Fujioka S, Takatsuto S, Stierhof YD, Stransky H, Yoshida S, Jürgens G (2004). A link between sterol biosynthesis, the cell wall, and cellulose in *Arabidopsis*. Plant J 38: 227–243.

Schrick K, Mayer U, Horrichs A, Kuhnt C, Bellini C, Dangl J, Schmidt J, Jürgens G (2000). FACKEL is a sterol C-14 reductase required for organized cell division and expansion in *Arabidopsis* embryogenesis. Genes Dev 14: 1471–1484.

Schulz R, Jensen WA (1968). *Capsella* embryogenesis: the egg, zygote and young embryo. Am J Bot 55: 807–819.

Shevell D, Kunkel T, Chua NH (2000). Cell wall alterations in the Arabidopsis *emb30* mutant. Plant Cell 12: 2047–2059.

Shevell DE, Leu WM, Gillmor CS, Xia G, Feldmann KA, Chua NH (1994). EMB30 is essential for normal cell division, cell expansion, and cell adhesion in Arabidopsis and encodes a protein that has similarity to Sec7. Cell 77: 1051–1062.

Silverstone AL, Tseng T-S, Swain SM, Dill A, Jeong SY, Olszewski NE, Sun T-P (2007). Functional analysis of SPINDLY in gibberellin signaling in Arabidopsis. Plant Physiol 143: 987–1000.

Simon R (2004). Development of the shoot apical meristem. In: Lindsey K (ed). *Polarity in Plants*. Blackwell, Oxford, UK.

Smith ZR, Long JA (2010). Control of Arabidopsis apical-basal embryo polarity by antagonistic transcription factors. Nature 464: 423–426.

Song SK, Hofhuis H, Lee MM, Clark SE (2008). Key divisions in the early *Arabidopsis* embryo require POL and PLL1 phosphatases to establish the root stem cell organizer and vascular axis. Dev Cell 15: 98–109.

Souter M, Lindsey K (2000). Polarity and signalling in plant embryogenesis. J Exp Bot 347: 971–983.

Souter M, Topping J, Pullen M, Friml J, Palme K, Hackett R, Grierson D, Lindsey K (2002). *hydra* mutants of *Arabidopsis* are defective in sterol profiles and auxin and ethylene signaling. Plant Cell 14: 1017–1031.

Sozzani R, Cui H, Moreno-Risueno MA, Busch W, Van Norman JM, Vernoux T, Brady SM, Dewitte W, Murrau JAH, Benfey PN (2010). Spatiotemporal regulation of cell-cycle genes by SHORTROOT links patterning and growth. Nature 466: 128–132.

Spencer MW, Casson SA, Lindsey K (2007). Transcriptional profiling of the Arabidopsis embryo. Plant Physiol 143: 924–940.

Steinmann T, Geldner N, Grebe M, Mangold S, Jackson CL, Paris S, Gälweiler L, Palme K, Jürgens G (1999). Coordinated polar localization of auxin efflux carrier PIN1 by GNOM ARF GEF. Science 286: 316–318.

Stepanova AN, Robertson-Hoyt J, Yun J, Benavente LM, Xie DY, Dolezal K, Schlereth A, Jürgens G, Alonso JM (2008). TAA1-mediated auxin biosynthesis is essential for hormone crosstalk and plant development. Cell 133: 177–191.

Stone SL, Braybrook SA, Paula SL, Kwong LW, Meuser J, Pelletier J, Hsieh TF, Fischer RL, Goldberg RB, Harada JJ (2008). Arabidopsis *LEAFY COTYLEDON2* induces maturation traits and auxin activity: implications for somatic embryogenesis. Proc Natl Acad Sci U S A 105: 3151–3156.

Stone SL, Kwong LW, Yee KM, Pelletier J, Lepiniec L, Fischer RL, Goldberg RB, Harada JJ (2001). *LEAFY COTYLEDON2* encodes a B3 domain transcription factor that induces embryo development. Proc Natl Acad Sci U S A 98: 11806–11811.

Su YH, Zhao XY, Liu Y, Zhang CL, O'Neill SD, Zhang XS (2009). Auxin-induced *WUS* expression is essential for embryonic stem cell renewal during somatic embryogenesis in Arabidopsis. Plant J 59: 448–460.

Suzuki M, McCarty DR (2008). Functional symmetry of the B3 network controlling seed development. Curr Opin Plant Biol 11: 548–553.

Suzuki M, Wang HH-Y, McCarthy DR (2007). Repression of the *LEAFY COTYLEDON 1/B3* regulatory network in plant embryo development by *VP1/ABSCISIC ACID INSENSITIVE 3-LIKE* B3 genes. Plant Physiol 143: 902–911.

Swarup K, Benkova E, Swarup R, *et al.* (2008). The auxin influx carrier LAX3 promotes lateral root emergence. Nat Cell Biol 10: 946–954.

Szemenyei H, Hannon M, Long JA (2008). TOPLESS mediates auxin-dependent transcriptional repression during *Arabidopsis* embryogenesis. Science 319: 1384–1386.

Tanaka H, Watanabe M, Watanabe D, Tanaka T, Machida C, Machida Y (2002). *ACR4*, a putative receptor kinase gene of *Arabidopsis thaliana*, that is expressed in the outer cell layers of embryos and plants, is involved in proper embryogenesis. Plant Cell Physiol 43: 419–428.

To A, Valon C, Savino G, Guilleminot J, Devic M, Giraudat J, Parcy F (2006). A network of local and redundant gene regulation governs Arabidopsis seed maturation. Plant Cell 18: 1642–1651.

Topping JF, Agyeman F, Henricot B, Lindsey K (1994). Identification of molecular markers of embryogenesis in *Arabidopsis thaliana* by promoter trapping. Plant J 5: 895–903.

Topping JF, Lindsey K (1997). Promoter trap markers differentiate structural and positional components of polar development in *Arabidopsis*. Plant Cell 9: 1713–1725.

Topping JF, May VJ, Muskett PR, Lindsey K (1997). Mutations in the *HYDRA* genes of Arabidopsis perturb cell shape and disrupt embryonic and seedling morphogenesis. Development 124: 4415–4424.

Treml BS, Winderl S, Radykewicz R, Herz M, Schweizer G, Hutzler P, Glawischnig E, Torres Ruiz RA (2005). The gene *ENHANCER OF PINOID* controls cotyledon development in the Arabidopsis embryo. Development 132: 4063–4074.

Tsuchiya Y, Nambara E, Naito S, McCourt P (2004). The *FUS3* transcription factor functions through the epidermal regulator *TTG1* during embryogenesis in *Arabidopsis*. Plant J 37: 73–81.

Vidaurre DP, Ploense S, Krogan NT, Berleth T (2007). *AMP1* and *MP* antagonistically regulate embryo and meristem development in *Arabidopsis*. Development 134: 2561–2567.

Wang H, Caruso LV, Downie AB, Perry SE (2004). The embryo MADS domain protein AGAMOUS-LIKE 15 directly regulates expression of a gene encoding an enzyme involved in gibberellin metabolism. Plant Cell 16: 1206–1219.

Weijers D, Sauer M, Meurette O, Friml J, Ljung K, Sandberg G, Hooykaas P, Offringa R (2005). Maintenance of embryonic auxin distribution for apical-basal patterning by PIN-FORMED-dependent auxin transport in Arabidopsis. Plant Cell 17: 2517–2526.

Weijers D, Schlereth A, Ehrismann JS, Schwank G, Kientz M, Jürgens G (2006). Auxin triggers transient local signalling for cell specification in *Arabidopsis* embryogenesis. Dev Cell 10: 265–270.

Werner T, Motyka V, Laucou V, Smets R, Van Onckelen H, Schmülling T (2003). Cytokinin-deficient transgenic Arabidopsis plants show multiple developmental alterations indicating opposite functions of cytokinins in regulating shoot and root meristem activity. Plant Cell 15: 2532–2550.

West M, Yee KM, Danao J, Zimmerman JL, Fischer RL, Goldberg RB, Harada JJ (1994). *LEAFY COTYLEDON1* is an essential regulator of late embryogenesis and cotyledon identity in Arabidopsis. Plant Cell 6: 1731–1745.

Willige B, Isono E, Richter R, Zourelidou M, Schwechheimer C (2010). Gibberellin regulates PIN-FORMED abundance and is required for auxin transport-dependent growth and development in *Arabidopsis thaliana*. Plant Cell 23: 2184–2195.

Woodward AW, Bartel B (2005). Auxin: regulation, action, and interaction. Ann Bot 95: 707–735.

Wysocka-Diller JW, Helariutta Y, Fukaki H, Malamy JE, Benfey PN (2000). Molecular analysis of SCARECROW function reveals a radial patterning mechanism common to root and shoot. Development 127: 595–603.

Xiang D, Yang H, Venglat P, Cao Y, We R, Ren M, Stone S, Wang E, Wang H, Xiao W, Weijers D, Berleth T, Selvaraj G, Datla R (2011). *POPCORN* functions in the auxin pathway to regulate embryonic body plan and meristem organization in *Arabidopsis*. Plant Cell 23: 4348–4367.

Yeung EC, Meinke DW (1993). Embryogenesis in angiosperms: development of the suspensor. Plant Cell 5: 1371–1381.

Zuo J, Niu QW, Frugis G, Chua NH (2002). The *WUSCHEL* gene promotes vegetative-to-embryonic transition in Arabidopsis. Plant J 30: 349–359.

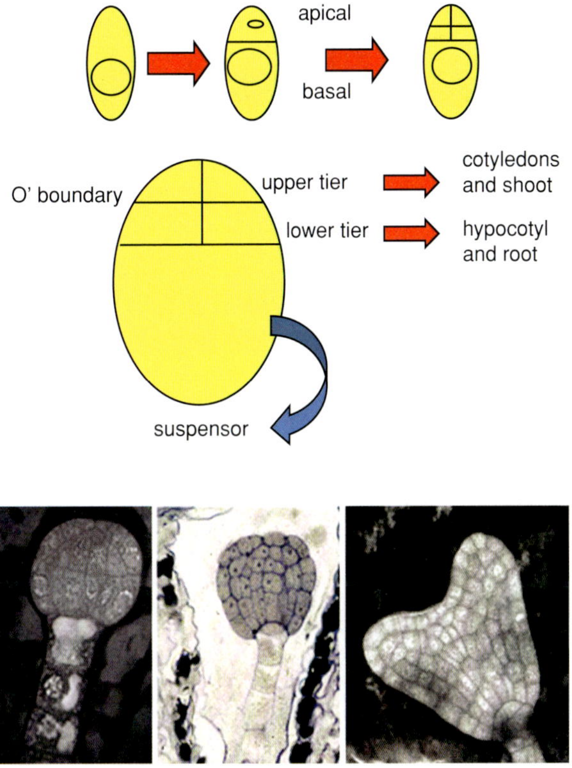

Plate 2.1 *Upper panel,* Diagrammatic representation of early events in *Arabidopsis* embryogenesis. The zygote undergoes an asymmetrical transverse division to form a small apical cell and a larger, more vacuolated basal cell. The basal cell divides to form the suspensor, which the apical cell divides transversely to form upper and lower tiers that develop into the cotyledons and shoot apical meristem and the hypocotyl and root. *Lower panel,* Developing globular *(left, center)* and heart *(right)* stage embryos of *Arabidopsis,* showing the upper cells of the suspensor.

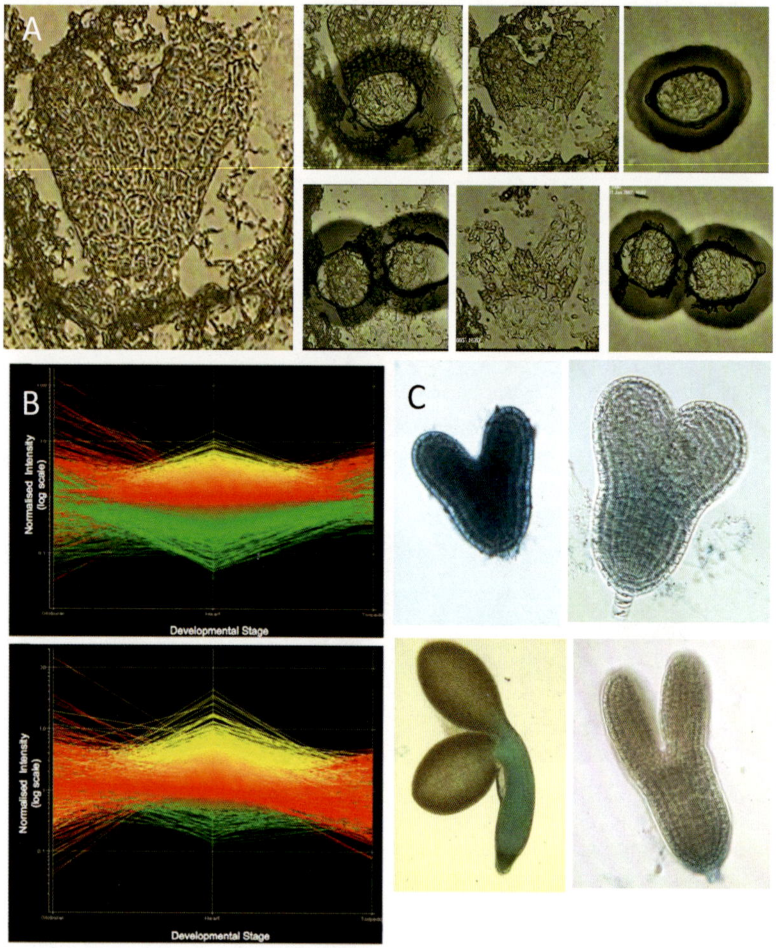

Plate 2.2 *A,* Laser-capture microdissection of cryosectioned heart stage embryo. *B,* Bioinformatics analysis of microarray data from genes expressed in apical *(upper panel)* and basal *(lower panel)* cells of developing embryos. *C,* Promoter-GUS fusion expression patterns of embryonically expressed genes identified by LCM. See Casson *et al.* (2005) and Spencer *et al.* (2007) for further detail.

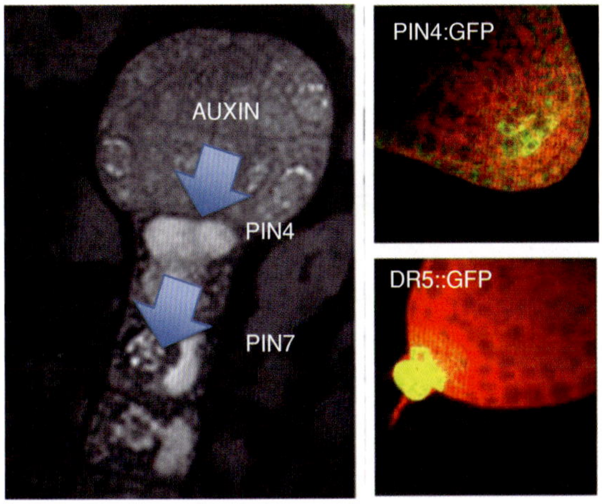

Plate 2.3 Auxin is translocated in the developing globular embryo from the apical region to the hypophysis via PIN4 and then from the hypophysis to the suspensor via PIN4 and PIN7. There is an auxin maximum (seen as DR5::GFP expression) at the basal region of the embryo.

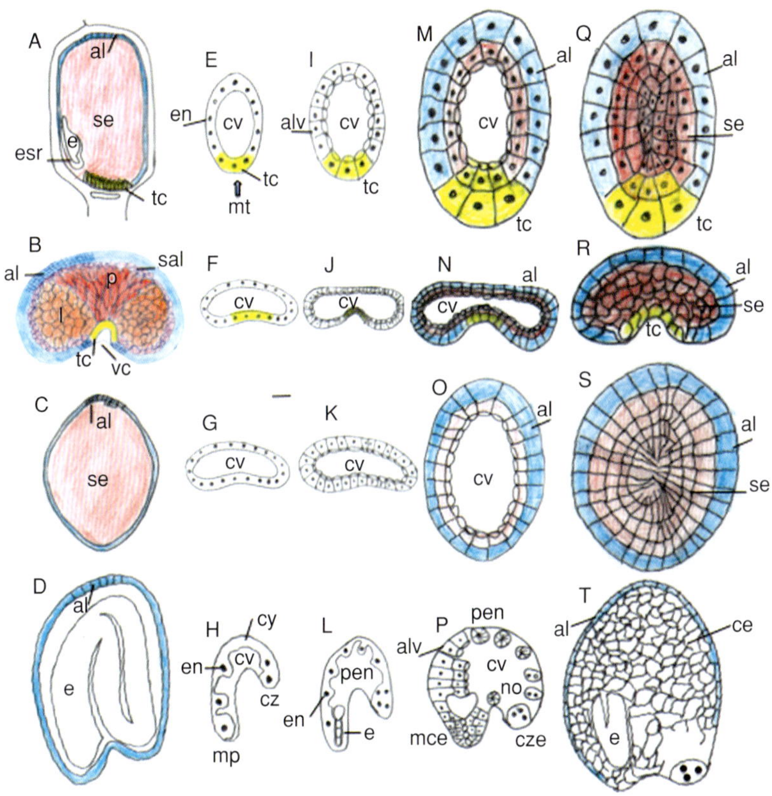

Plate 3.1 Overview of endosperm structure. *A–D*, Mature endosperm of maize *(A)*, barley *(B)*, rice *(C)*, and *Arabidopsis thaliana (D)*. al, aleurone; e, embryo; esr, embryo surrounding region; i, irregular starchy endosperm cells; ma, modified aleurone layer (transfer cells); p, prismatic cells; sal, subaleurone layer; se, starchy endosperm; tc, transfer cells; vc, ventral crease. *E–H*, Coenocytic stage endosperm of maize *(E)*, barley *(F)*, rice *(G)*, and *Arabidopsis thaliana (H)*. cv, central vacuole; tc, transfer cell region; *Meg1*, maternal transcript involved in induction of transfer cell fate specification; Mp, micropylar end; cz, chalazal end; en, endosperm nucleus; cy, endosperm cytoplasm. *I–L*, Alveolar stage of endosperm development of maize *(I)*, barley *(J)*, rice *(K)*, and *Arabidopsis thaliana (L)*. *M–P*, Cellularizing endosperm with one peripheral cell layer and inner alveoli in maize *(M)*, barley *(N)*, rice *(O)*, and *Arabidopsis thaliana (P)*. *Q–T*, Fully cellular endosperm of maize *(Q)*, barley *(R)*, rice *(S)*, and *Arabidopsis thaliana (T)*.

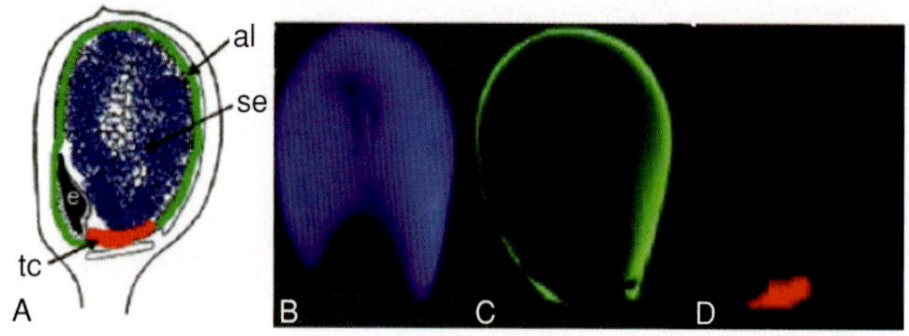

Plate 3.3 Kernel of "triple marked" maize endosperm. *A*, Diagram of the three major endosperm tissues. al, aleurone; e, embryo; se, starchy endosperm; tc, transfer cells. *B–D*, Expression of fluorescent cell-type markers in 12 DAP endosperm. Various cell-specific promoters are used to regulate expression of fluorescent proteins with differing spectral properties.B, γ-Zein:AmCyan expression fluoresces blue in the starchy endosperm. *C*, Ltp2:ZsYellow expression is specific for the aleurone layer. *D*, End1:DsRed expression marks transfer cells. (Modified from Gruis *et al.* [2006].)

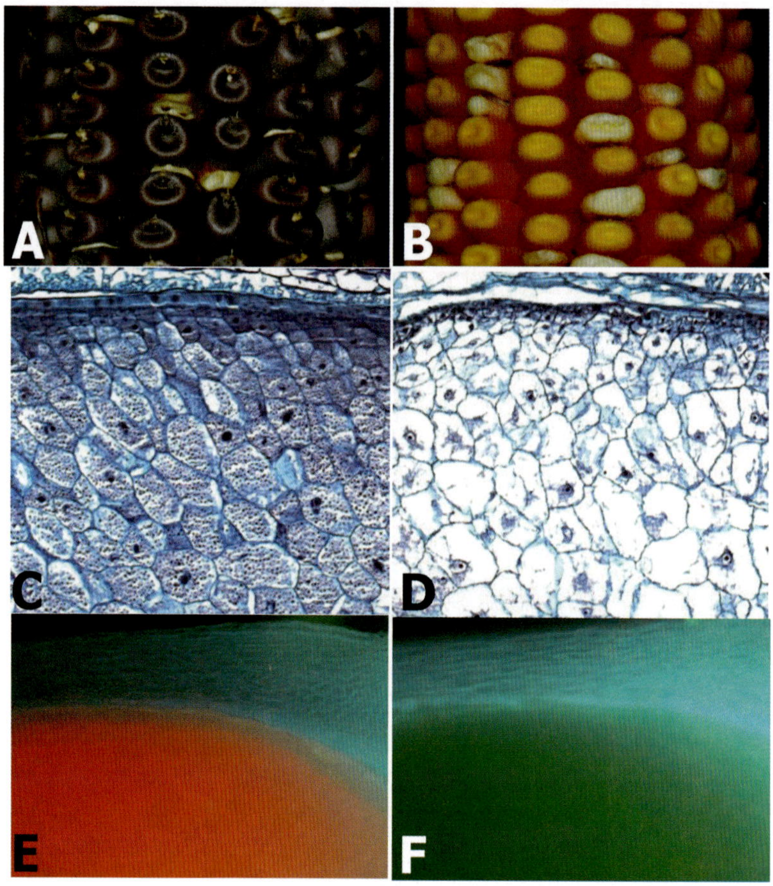

Plate 3.4 Examples of maize kernel mutants. *A*, Ear segregating an *emp* mutant. *B*, Ear segregating a *dek* mutant. *C*, Histological section of a wild-type kernel with cells packed with starch grains. *D*, Section of a mutant endosperm defective in starch accumulation. *E*, Expression of an FL2-RFP fluorescent marker for α-zein accumulation in wild-type (Mohanty *et al.*, 2009). *F*, Lack of FL2-RFP expression in a mutant kernel.

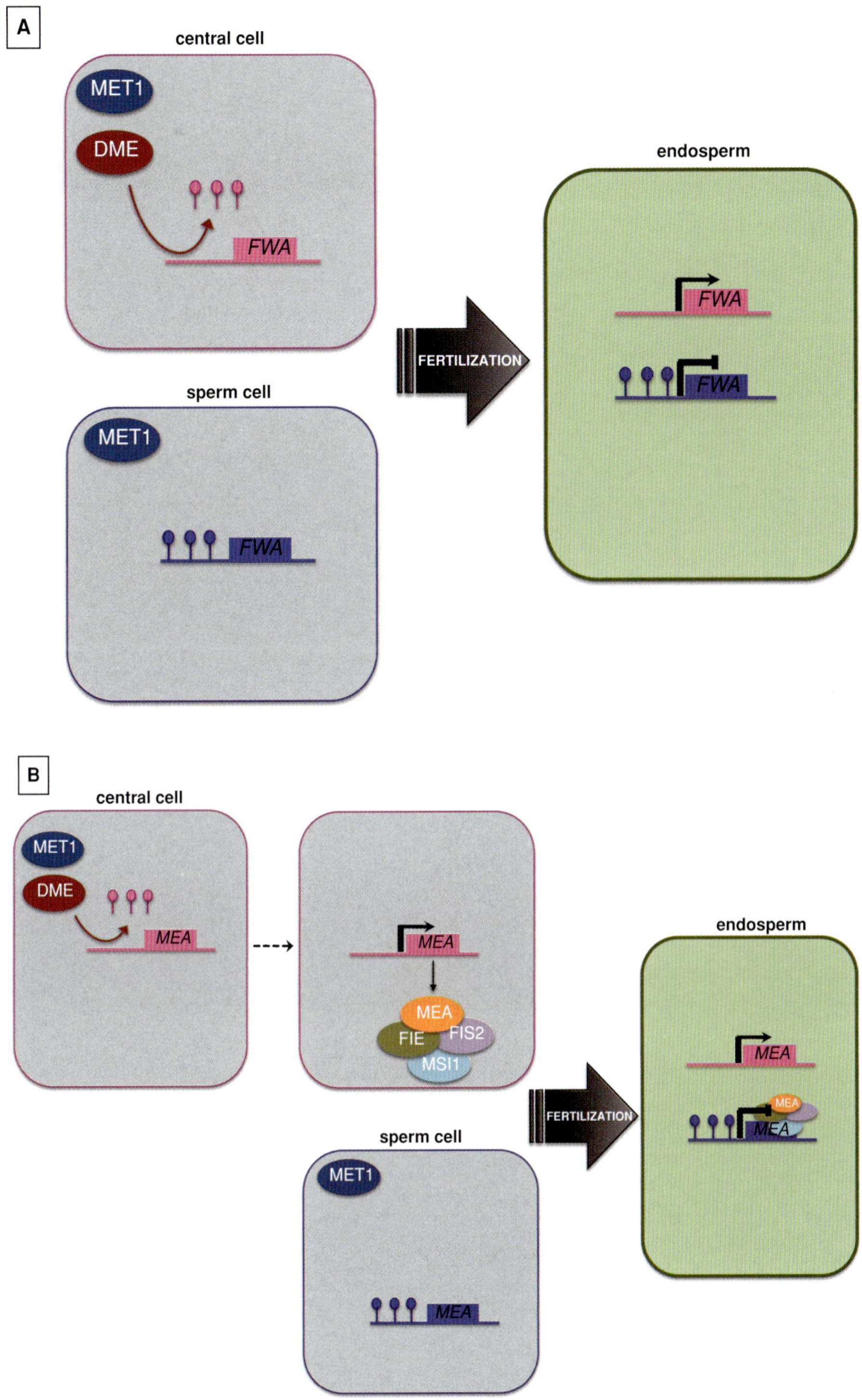

Plate 4.1 *A–C*, Models for *Arabidopsis* genomic imprinting: *FWA* model *(A)*, *MEA* model *(B)*, and *PHE1*-model *(C)*. "Lollipops" represent 5-methyl cytosines. The red bar on the maternal *PHE1* allele in represents the location of the 3′-tandem repeats targeted for DME-mediated DNA demethylation.

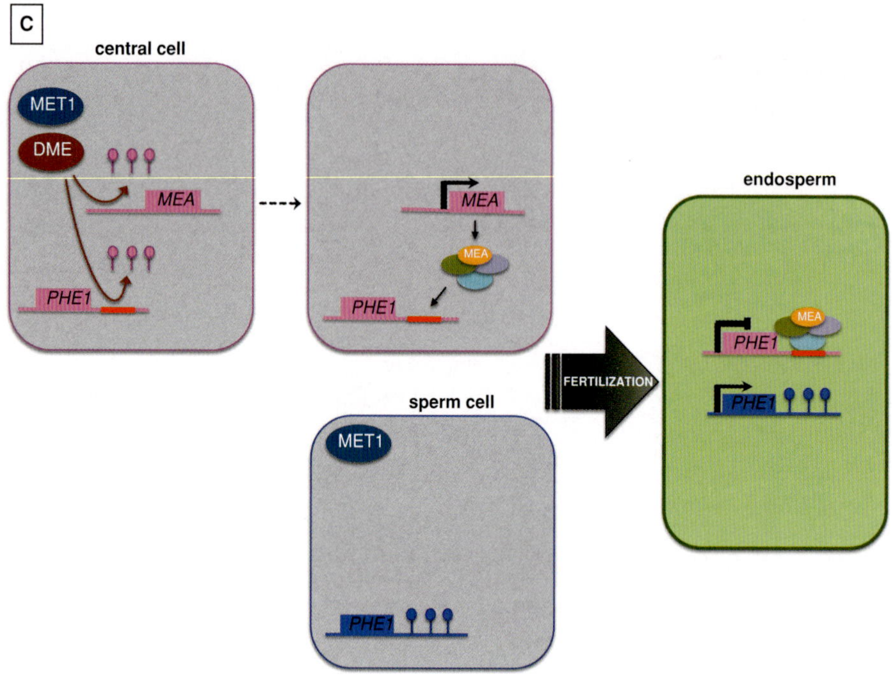

Plate 4.1 (*Continued*)

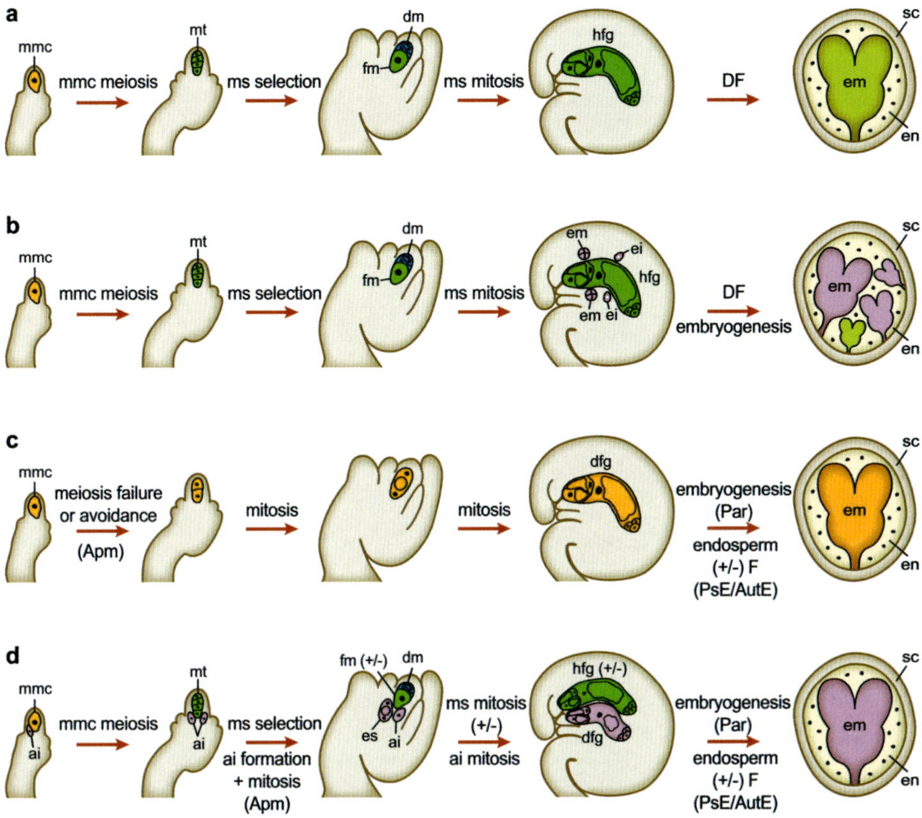

Plate 5.1 Seed development in sexual and apomictic species. *a,* Events of sexual seed formation. *b,* Events of sporophytic apomixis. *c,* Events of diplosporous apomixis and, *d,* aposporous apomixis, where endosperm formation may or may not require fertilization. In the case of apospory, haploid female gametophytes may coexist with the aposporous gametophyte, or degeneration of the haploid gametophyte may occur during the events of aposporous gametophyte development. See text for further details. ai, aposporous initial cells; Apm, apomeiosis; AutE; autonomous endosperm formation; DF, double fertilization; dfg, diploid female gametophyte; dm, degenerating megaspores; ei, embryo initial cell; em, embryo; en, endosperm; F, fertilization; fm, functional megaspore; hfg, haploid female gametophyte; mmc, megaspore mother cell; ms, megaspore; mt, meiotic tetrad; Par, parthenogenesis; PsE, pseudogamous endosperm formation which requires fertilization; sc, seed coat; (+/−), present or absent. (Modified from Drews and Koltunow, 2011.)

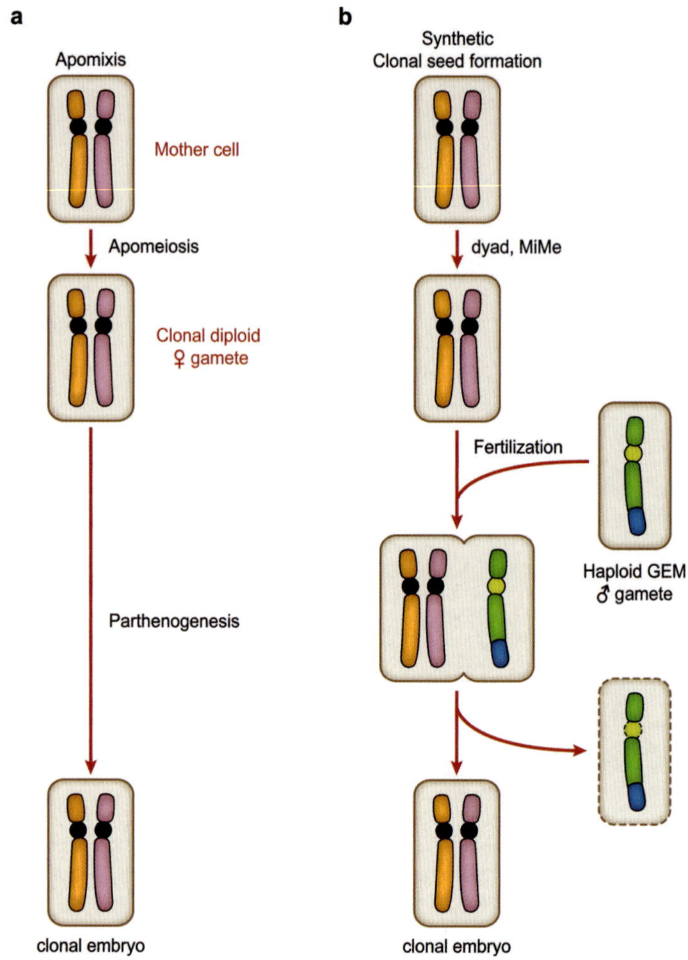

Plate 5.2 Natural and synthetic clonal reproduction through seed. *a,* In natural apomicts, clonal embryos with a maternal genotype develop without fertilization. *b,* Induction of clonal embryos in sexual plants as described by Marimuthu *et al.* (2011) involves combining mutants so that gametogenesis occurs without meiosis and subsequent fertilization with a parent whose chromosomes are modified to be eliminated after fertilization. (Modified from Marimuthu *et al.*, 2011.)

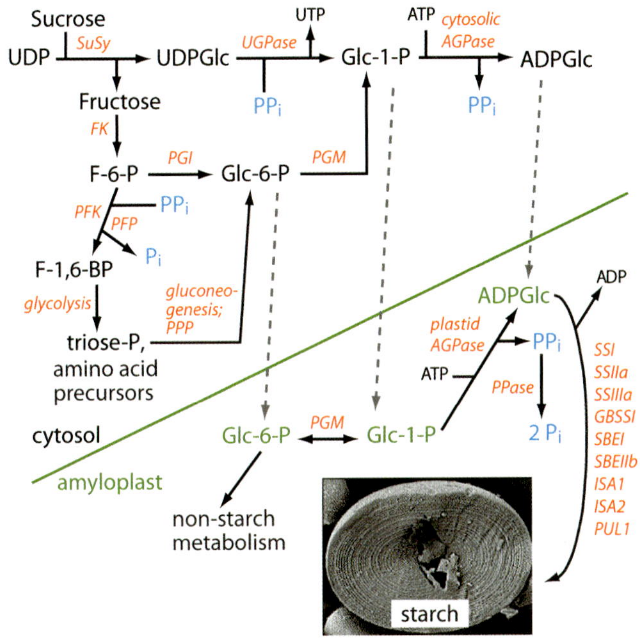

Plate 7.1 Outline of endosperm metabolism from sucrose to starch. *Gray lines* indicate transport from the cytosol into the amyloplast stroma. Glc-1-P is supplied either by UGPase in the "direct" path from sucrose or from metabolic conversion to triose phosphates and resynthesis back into hexose phosphate. Enzymes noted as being involved in starch biosynthesis have been identified by genetic analyses; this does not preclude the involvement of additional isoforms. SuSy, sucrose synthase; UGPase, UDP glucose pyrophosphorylase; FK, fructokinase; PGI, phosphoglucoisomerase; PGM, phosphoglucomutase; PFK, phosphofructokinase; PFP, pyrophosphate-dependent phosphofructokinase; PPP, pentose phosphate pathway; PPase, pyrophosphatase; F-6-P, fructose-6-phosphate; F-1,6-BP, fructose-1,6-bisphosphate.

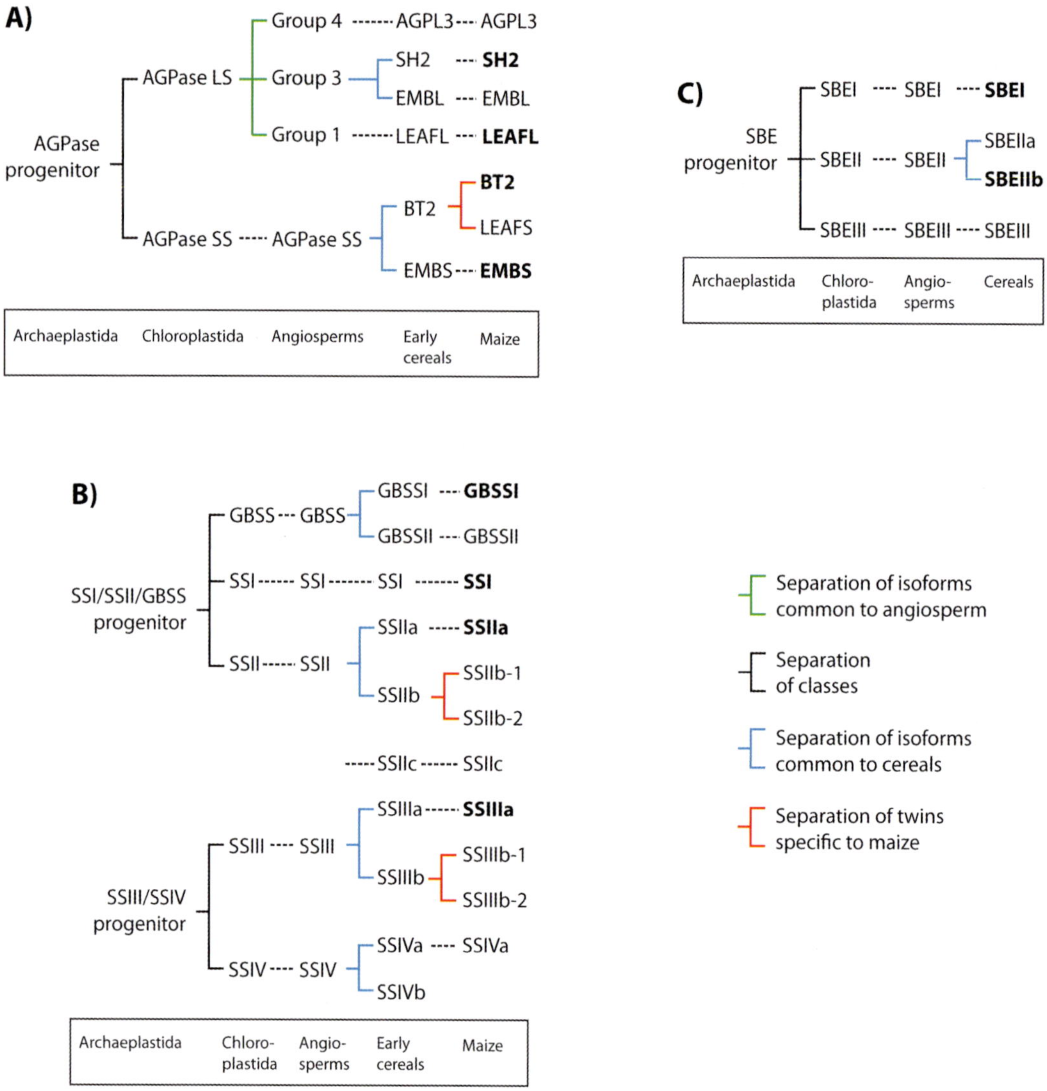

Plate 7.2 Proposed evolution of maize starch biosynthetic genes. Enzymes in bold text have been shown by genetic analysis to be involved in endosperm starch biosynthesis. Only lineages that lead to genes extant in maize are shown. *A)* AGPase. A single gene in the Archaeplastida progenitor, derived from a cyanobacterium, duplicated to form AGPase LS and AGPase SS that are conserved in all chloroplast-containing species. An angiosperm progenitor underwent further duplications to generate a minimum of three isoforms of AGPase LS and one of AGPase SS before separation of monocots and eudicots. In the Graminacea, a whole-genome duplication resulted in two isoforms of the group 3 AGPase LS and two isoforms of AGPase SS. A whole-genome duplication generated the duplicate gene encoding LEAFS that is present in maize but not in rice. *B)* SS. Two genes were present in the primordial Archaeplastida progenitor before divergence of plastid-containing organisms. At the establishment of the Chloroplastida, these had duplicated to form five SS classes, each of which is extant in all green algae and land plants. In the Graminacea, a whole-genome duplication resulted in isoforms of GBSS, SSII, and SSIII, each of which is present in maize and rice. SSIV was also duplicated at this stage, but one copy was not retained in maize. SSIIc could have originated either in an ancestor common to both monocots and eudicots or after that division. A whole-genome duplication generated twin genes for SSIIb and SSIIIb that are present in maize but not in rice. *C)* SBE. Three classes of SBE had been formed before divergence of green algae and land plants by duplication of an ancestral gene. In general, all three classes are conserved in Chloroplastida species, although at least one exception is known. The ancestral whole-genome duplication in the Graminaceae gave rise to two isoforms of the SBE II class, both of which are present in maize and rice. (Adapted from Rosti and Denyer [2007]; Deschamps *et al.* [2008]; Georgelis *et al.* [2008]; and Yan *et al.* [2009a, 2009b].)

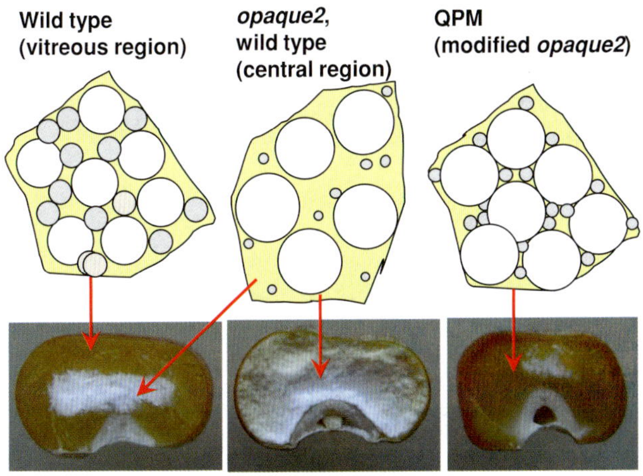

Plate 8.2 Diagram of protein body and starch grain interaction in vitreous endosperm formation. Individual cells of developing endosperm are represented with the relative size and abundance of starch grains (white spheres) and zein protein bodies (gray spheres) that are thought to result in vitreous or opaque endosperm in normal, *opaque2,* and modified *opaque2* (QPM) kernels.

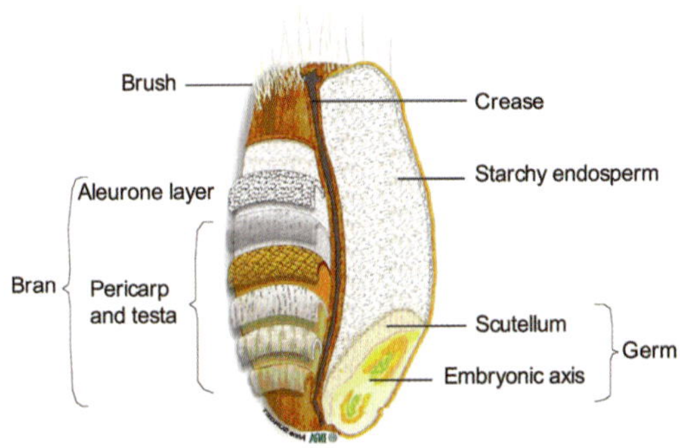

Plate 9.1 Component tissues of wheat grain. (From Surget and Barron [2005], with permission.)

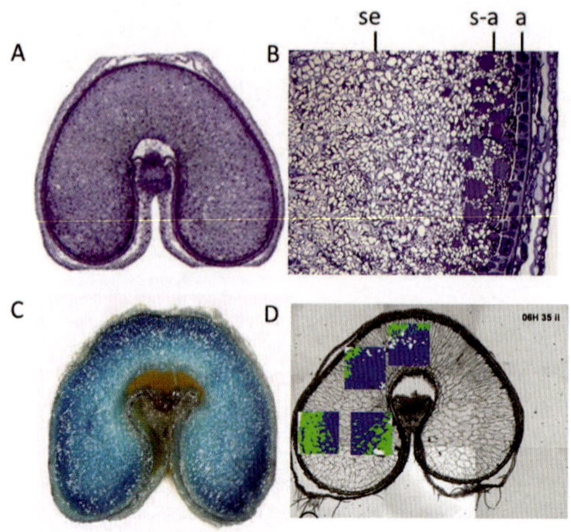

Plate 9.2 Transverse sections of developing caryopses of wheat, showing gradients in composition in the endosperm. *A* and *B,* Low-magnification and high-magnification images of a transverse section of the whole caryopsis at about 28 days after anthesis, stained with toluidine blue to show protein. Aleurone *(a)* and subaleurone *(s-a)* cells are rich in protein, and central starchy endosperm *(se)* cells are rich in starch granules (unstained). *C,* Expression of low-molecular-weight glutein subunit-promoter-GUS fusion construct in a developing endosperm, showing high expression in the subaleurone and outer starchy endosperm cells. *D,* Overlaying of Fourier-transform infra-red microspectroscopy images onto a section of cell wall only at 35 days after anthesis. The images are colored to show the distributions of high-substitution (shown in blue) and low-substitution (shown in green) forms of arabinoxylan in the cell wall. (*A* and *B,* Courtesy Cristina Sanchez-Gritsch and Paola Tosi, Rothamsted Research; *C,* courtesy Caroline Sparks and Huw Jones, Rothamsted Research; *D,* from Toole *et al.* [2010] with permission.)

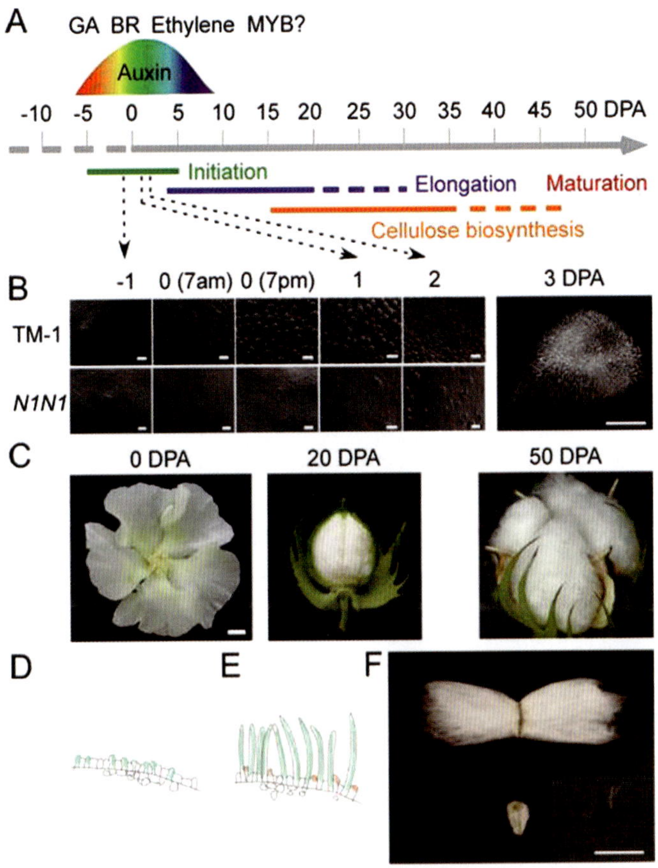

Plate 11.1 Cotton fiber development. *A,* Continuous and overlapping stages of cotton fiber development, including fiber cell initiation, elongation, cellulose (secondary cell wall) biosynthesis, and maturation. During elongation and maturation stages, cellulose biosynthesis is the dominant cellular metabolic process. (Secondary cell wall accumulation starts by depositing the cellulose under the primary cell wall after 15 DPA.) Regulation of phytohormones, including auxin, brassinosteroid (BR), and ethylene and transcription regulators, including MYB proteins, is implicated in cotton fiber cell initiation. *B,* Scanning electron microscope images of cotton fiber initiation processes. TM-1, *G. hirsutum* L. TM-1; N1N1, naked seed mutant. The ovules were collected at −1, 0 (7 a.m. and 7 p.m.), and 1 and 2 days after anthesis (0 = on the day of anthesis). Bar = 20 μm; bar = 100 μm for TM-1 ovule at 3 DPA. *C,* TM-1 cotton flower on 0 DPA, boll on 20 DPA and 50 DPA. *D,* Diagram of cotton fiber lint cell initiation on 0 DPA (green). *E,* Diagram of fuzz fiber cell initiation on 5–7 DPA (orange) along with lint fiber (green). F, Mature cotton lint and fuzz fiber on TM-1 seed. *Inset,* Mature seed of the fiberless mutant *XuFL.*

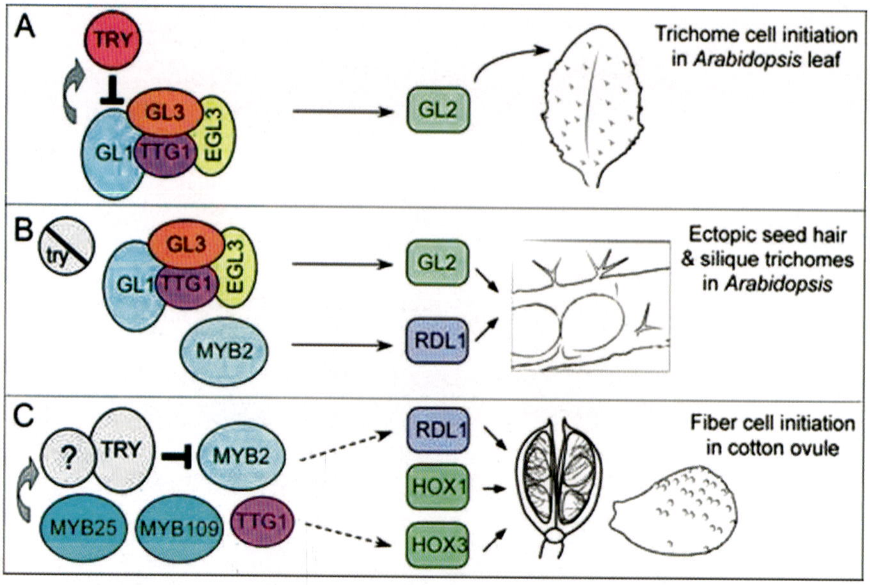

Plate 11.2 Models for leaf trichome and seed hair development. *A,* The trichome initiation complex consisting of GL1, GL3, EGL3, TTG1, and a negative regulator TRY promotes GL2 expression, leading to initiation of leaf trichomes in *Arabidopsis thaliana. B,* Cotton fiber genes, including *MYB2* and *RDL1,* stimulate the development of ectopic seed hairs and silique trichomes in *A. thaliana. C,* Multiple transcription factors identified in cotton fibers are shown to play roles in the development of seed hair and fiber in cotton. (From Guan *et al.* [2011].)

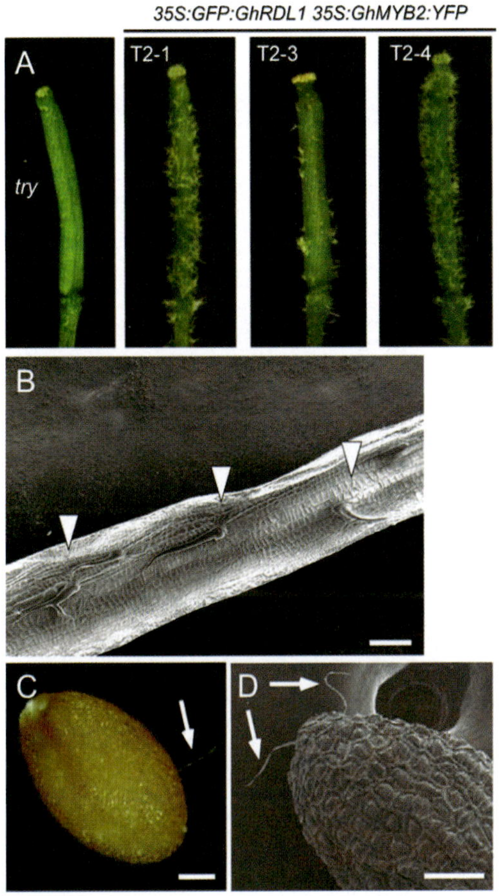

Plate 11.3 Development of ectopic trichomes inside and outside of siliques in transgenic *Arabidopsis thaliana* plants overexpressing *GhMYB2* and *GhRDL1*. *A,* Ectopic trichomes outside of siliques in transgenic *A. thaliana* plants. *B,* Ectopic trichomes inside of siliques *(arrows)* in transgenic *A. thaliana* plants. *C* and *D,* Seed hair observed in transgenic *A. thaliana* seeds under light microscope *(C)* and Scanning electron microscope *(D)*. (From Guan *et al.*, [2011].)

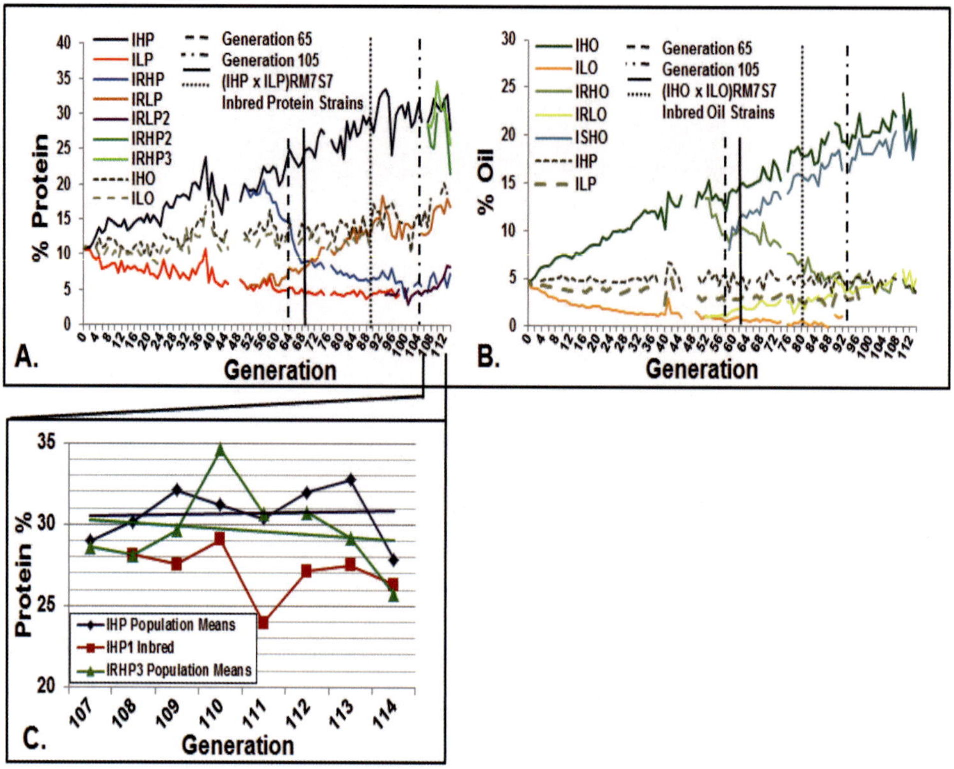

Plate 12.1 *A* and *B*, Selection responses in the Illinois Protein Strains *(A)* and Illinois Oil Strains *(B)*. Protein concentrations for IHO and ILO are plotted as *dashed lines* in *(A)*, and oil concentrations for IHP and ILP are plotted as *dashed lines* in *(B)*. The vertical lines highlight important generations of the experiment: generation 65, for which oldest seed is still available; generation 70, from which QTL mapping populations have been derived from crosses of selected strains; generation 90, as a source of inbred lines derived from the Illinois Protein Strains; and generation 105 from which inbred lines for the Illinois Oil strains are being produced. *C,* Expanded view of the last eight cycles of selection response in IHP and Illinois Reverse High Protein 3 compared with IHP1 inbred line.

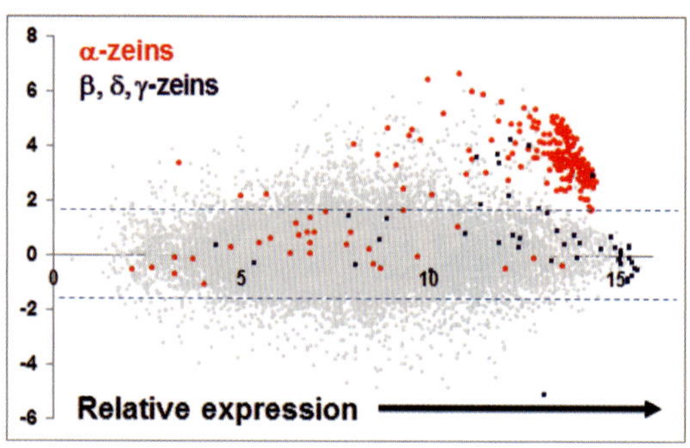

Plate 12.3 Relative zein gene expression levels in 16 DAP seeds as determined by microarray. Plotted is the expression ratio of IHP1 relative to ILP1 versus average probe signal intensity. *Dashed lines* represent cutoffs for features exhibiting statistically significant differential intensity (*t*-test, *P* value $< .05$; FDR adjusted *q* value < 0.05). The features annotated as α-zeins are plotted in red, and β-zeins, δ-zeins, and γ-zeins are plotted in blue. All other features are plotted in gray.

Plate 12.4 Phenotypes of *opaque2* introgressions into IHP1 and B73 inbred lines. *A*, Wild-type IHP1 ear. *B*, IHP1 ear homozygous for the *o2* mutation. *C*, SDS-PAGE of total seed proteins from IHP1, B73, and o2 introgressions into each of these genetic backgrounds. The relative migration of the 22-kDa α-zeins is indicated.

Plate 12.6 Photographs of kernels produced following backcrosses of *FL2-mRFP* transgenic events to the Illinois Protein Strains inbreds (TG event 52) and B73 (TG event 47). Photographs were taken at regular intervals during the period of zein accumulation from 12–24 days after pollination (DAP).

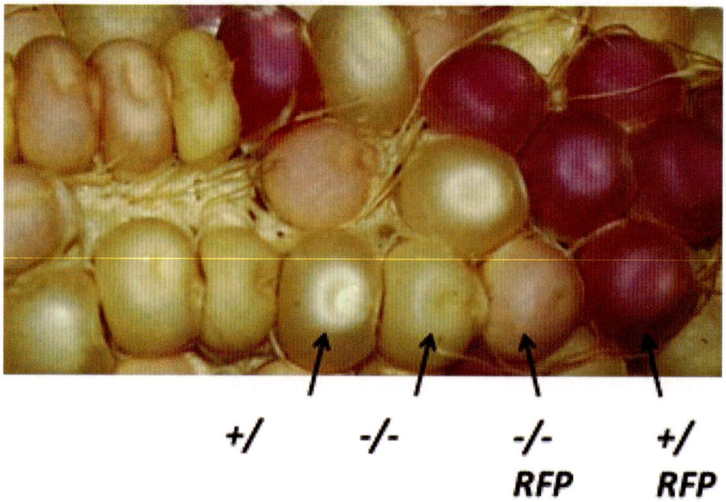

Plate 12.8 Photograph of an IHP1 ear segregating for the *o2* mutation and the *FL2-mRFP* transgene. The *o2* mutant phenotype is chalky and opaque in appearance (−/−) compared with wild-type kernels (+/). Kernels containing both the *o2* mutation and the *FL2-mRFP* transgene (−/−; RFP) illustrate reduced expression of *FL2-mRFP*, as detected by a significant decrease in pink coloration of the kernels compared with kernels containing only the *FL2-mRFP* transgene, which are much darker (+/; RFP).

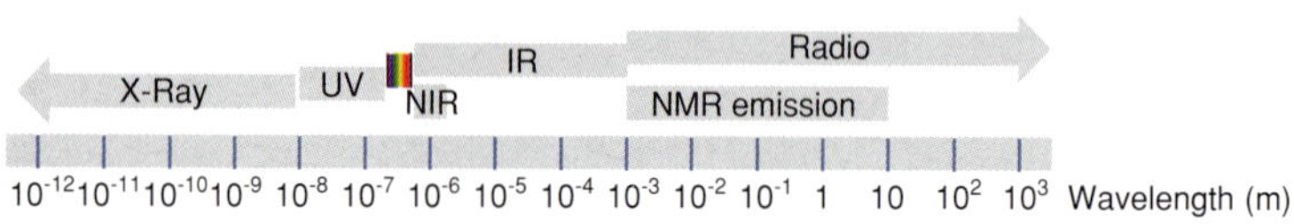

Plate 13.1 Electromagnetic spectrum showing wavelength ranges for machine vision technologies.

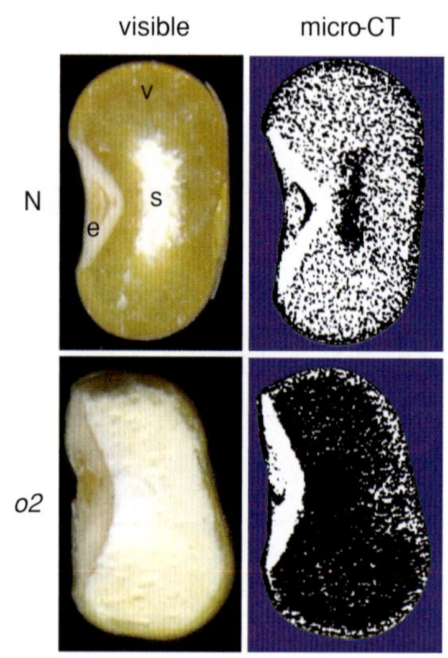

Plate 13.2 Comparison of seed cross sections from normal (N) and *opaque2* (*o2*) mutant. Visible panels are hand sections made after completing CT scans. Micro-CT scans have an identical threshold for attenuation with white pixels indicating regions above the density threshold and black pixels indicating regions below. *v,* vitreous endosperm; *s,* starchy endosperm; *e,* embryo.

3 Endosperm Development

Odd-Arne Olsen and Philip W. Becraft

Introduction

The endosperm of cereal grains is an important source of food, feed, and industrial raw material. Biologically, the endosperm is considered a key factor in the evolution of land plants, helping seeds to survive under harsh conditions. The endosperm functions to nurture the embryo during early stages of development, and in species such as cereals that have persistent endosperms, it also functions as a food reserve to nourish the growing seedling during germination. These same food reserves provide nutrition to humans and livestock as well as industrial feedstock. In 1900, an unusual feature of the angiosperm seed was discovered – the embryo and the endosperm originate in two separate fertilization events, termed "double fertilization" (Nawaschin, 1898; Guignard, 1901). In this process, the embryo forms from the fusion of one haploid male gamete (sperm) and a haploid egg cell. The endosperm is derived from the fusion of a second sperm with two haploid central cell nuclei to form a triploid primary endosperm nucleus. This mode of development is called the *Polygonum* type and is typical of most familiar seeds including crops such as cereal grains and soy.

New genomics tools and resources are allowing the experimental and comparative analysis of endosperm biology at levels and scales unimaginable just a few years ago. Genome level data are rapidly accumulating on the developmental biology, molecular biology, biochemistry, physiology, and evolution of an increasing number of species. Computational tools are being developed that integrate these various levels of information to provide a systems view of endosperm function. It is hoped that such a systems-level understanding will lay the foundation for novel strategies to improve grain quality and quantity for food, feed, and industrial purposes.

Overview of Endosperm Structure and Development

An overview of the structure of the endosperm of maize, barley, rice and *Arabidopsis thaliana* is given in Figure 3.1. Cereal endosperm contains four major cell types, each with specialized functions. The starchy endosperm (central endosperm) is the major storage cell type, transfer cells transport nutrient solutes from maternal tissues into the seed, aleurone functions in mineral storage and to digest storage products to feed the germinating seedling, and the embryo surrounding cells are hypothesized to function in communication and nutrient transfer between the endosperm

Seed Genomics, First Edition. Edited by Philip W. Becraft.

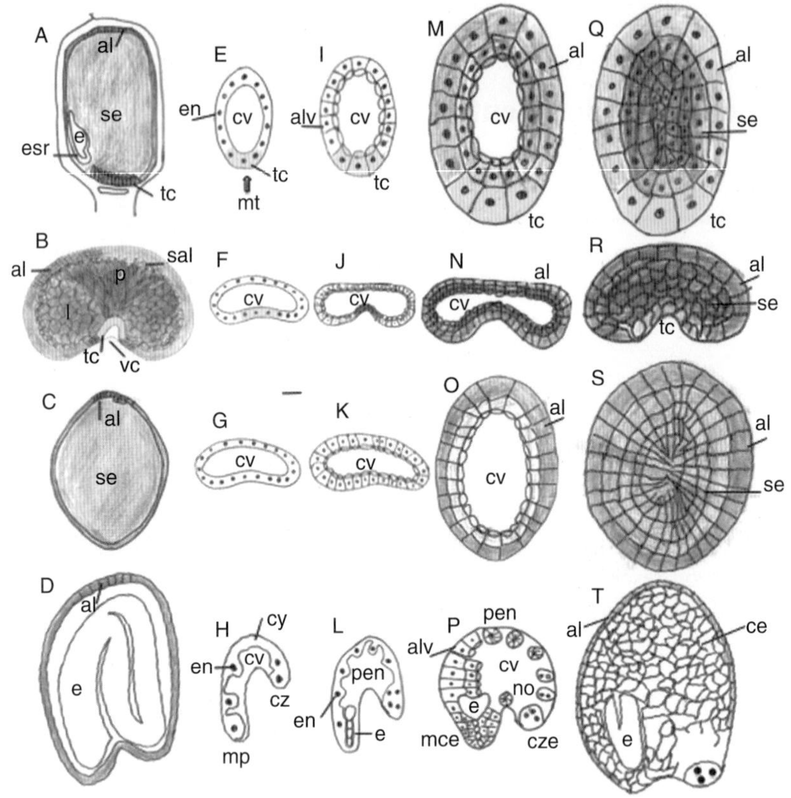

Figure 3.1 Overview of endosperm structure. *A–D*, Mature endosperm of maize *(A)*, barley *(B)*, rice *(C)*, and *Arabidopsis thaliana (D)*. al, aleurone; e, embryo; esr, embryo surrounding region; i, irregular starchy endosperm cells; p, prismatic cells; sal, subaleurone layer; se, starchy endosperm; tc, transfer cells; vc, ventral crease. *E–H*, Coenocytic stage endosperm of maize *(E)*, barley *(F)*, rice *(G)*, and *Arabidopsis thaliana (H)*. cv, central vacuole; tc, transfer cell region; mt, *Meg1* maternal transcript involved in induction of transfer cell fate specification; Mp, micropylar end; cz, chalazal end; en, endosperm nucleus; cy, endosperm cytoplasm. *I–L*, Alveolar stage of endosperm development of maize *(I)*, barley *(J)*, rice *(K)*, and *Arabidopsis thaliana (L)*; alv, alveoli; pen, peripheral endosperm. *M–P*, Cellularizing endosperm with one peripheral cell layer and inner alveoli in maize *(M)*, barley *(N)*, rice *(O)*, and *Arabidopsis thaliana (P)*. *Q–T*, Fully cellular endosperm of maize *(Q)*, barley *(R)*, rice *(S)*, and *Arabidopsis thaliana (T)*. (For color detail, see color plate section.)

and embryo. The transient endosperm of *Arabidopsis* is not a significant storage tissue, with the cotyledons of the embryo filling that role.

These and most other species of economic interest display a nuclear pattern of endosperm development, whereon nuclei proliferate in a coenocytic phase before undergoing cellularization (Brown et al. 1994; Berger, 2003; Olsen, 2004b; Sabelli and Larkins, 2009; Becraft and Gutierrez-Marcos, 2012). In contrast, cellular endosperm development lacks a coenocytic phase. Beginning at fertilization, cell proliferation involves typical cell divisions in which cytokinesis accompanies nuclear division. Although striking similarities exist in nuclear endosperm development between monocot species such as the cereals and eudicot species such as *Arabidopsis*, the nuclear mode of development is believed to have evolved independently in the two groups from a cellular ancestral state (Geeta, 2003). Consequently, it cannot be assumed that similar features of the two types of endosperms have identical genetic bases. Current phylogenetic data suggest that monocots represent a monophyletic group that shares a common ancestor, although, as shown subsequently, significant morphological differences exist between the endosperms of the main cereals. Molecular data also support this conclusion, in that substantial portions of wheat and maize endosperm transcripts lack clear orthologues in rice (Lai *et al.*, 2004; Drea *et al.*, 2005).

At maturity, the maize endosperm (Figure 3.1*A*) consists of an internal mass of starchy endosperm cells, the major storage cell type; these cells are packed with starch granules and storage proteins. Starchy endosperm is surrounded by a single layer of aleurone cells (Becraft, 2007). Aleurone cells have thick cell walls rich in arabinoxylan (Dornez *et al.*, 2011) and a dense granular cytoplasm containing aleurone grains and other inclusions (Jones, 1969; Morrison *et al.*, 1975; Bechtel and Pomeranz, 1977). In particular, genotypes of maize, anthocyanin pigments, accumulate at maturity (Cone, 2007). Aleurone cells become desiccation tolerant and are the only living cell type in the endosperm of mature seed. When dry grains are imbibed in water, gibberellin (GA) signaling from the embryo induces the aleurone to produce and secrete hydrolases, particularly proteases and glucanases. These enzymes break down starch and storage proteins in the dead starchy endosperm cells, liberating sugars and amino acids to be absorbed by the germinating embryo. In the periphery of the starchy endosperm, one or two layers, often referred to as the subaleurone layer, contain less starch and more proteins than the central part. The basal transfer cell layer forms at the interface between the endosperm cavity and the maternal chalazal tissue. The transfer layer facilitates transfer of solutes, including amino acids, sucrose, and monosaccharides, across the plasmalemma from the maternal pedicel tissue to the endosperm compartment (Thompson *et al.*, 2001). The transfer cells also express several proteins such as BAPs (BASAL LAYER-TYPE ANTIFUNGAL PROTEINs) that have antimicrobial properties suggesting a role in defense against invading pathogens (Serna *et al.*, 2002). BETL1 is synthesized and located in basal endosperm cells, where it is tightly bound to the cell wall (Hueros *et al.*, 1995), whereas BAP2 is secreted into the intercellular matrix of the basal endosperm and accumulates predominantly in the adjacent thick-walled cell layer of the pedicel (Serna *et al.*, 2002). At early developmental stages, in the part of the starchy endosperm facing the embryo, the embryo surrounding region (ESR) forms and persists until approximately 12 days after pollination (DAP) (not shown) (Cosségal *et al.*, 2007). The function of this tissue is unknown but is hypothesized to be important for communication between the endosperm and embryo.

The barley endosperm (Figure 3.1*B*) has a similar structure, with the exception of the aleurone layer, which is three cell layers thick (Olsen and Weschke, 2012). Barley also has a subaleurone layer. In contrast to maize, the barley endosperm consists of wings with so-called irregular starchy endosperm cells and a body of prismatic starchy endosperm cells over the transfer cells toward the dorsal side. The transfer cells form in the crease of the grain and are known as modified aleurone. Although not as distinct as in maize, the barley endosperm contains cells similar to those in the ESR region. Wheat endosperm is similar in structure to barley except that the aleurone layer is typically singular but occasionally has up to three cell layers in some regions of the grain (not shown) (Drea *et al.*, 2005). In rice, the endosperm displays a radial symmetry in transverse sections giving it a tubelike appearance (Figure 3.1*C*). The aleurone layer varies in thickness from one cell layer in most regions to three cell layers. Notably, the rice endosperm lacks a distinct transfer cell layer. In mature seeds of *Arabidopsis thaliana*, the nonpersistent endosperm consists only of an aleurone layer, and the rest of the seed cavity is filled by the embryo (Vaughn and Whitehouse, 1971; Chamberlain and Horner, 1990; Groot and van Caeseele, 1993). The endosperm that initially forms after fertilization (see later) is gradually depleted as the embryo grows (Brown *et al.*, 1999).

The endosperm of maize, barley, wheat, rice, and *Arabidopsis* all have an initial coenocytic endosperm stage consisting of a thin layer of cytoplasm with nuclei lining the wall of the embryo sac (Figure 3.1*E–H*). The cereal endosperm display radial symmetries (Figure 3.1*E–G*), whereas the central cell of *Arabidopsis* has three regions that become distinct as the seed grows: the embryo-surrounding region or micropylar endosperm (MCE), the peripheral endosperm (PEN) in the central chamber, and the chalazal endosperm (CZE) (Figure 3.1*H*) (Brown and Lemmon, 1999; Boisnard-Lorig *et al.*, 2001; Sorensen *et al.*, 2002). The coenocytic stage results from repeated nuclear divisions of the primary endosperm nucleus after fertilization without the formation of cell walls between daughter nuclei. The mechanisms for suppression of phragmoplast formation in early endosperm are unknown. The process for endosperm cellularization was first described in barley and is illustrated in Figure 3.2. A transverse section of a coenocytic stage barley endosperm and

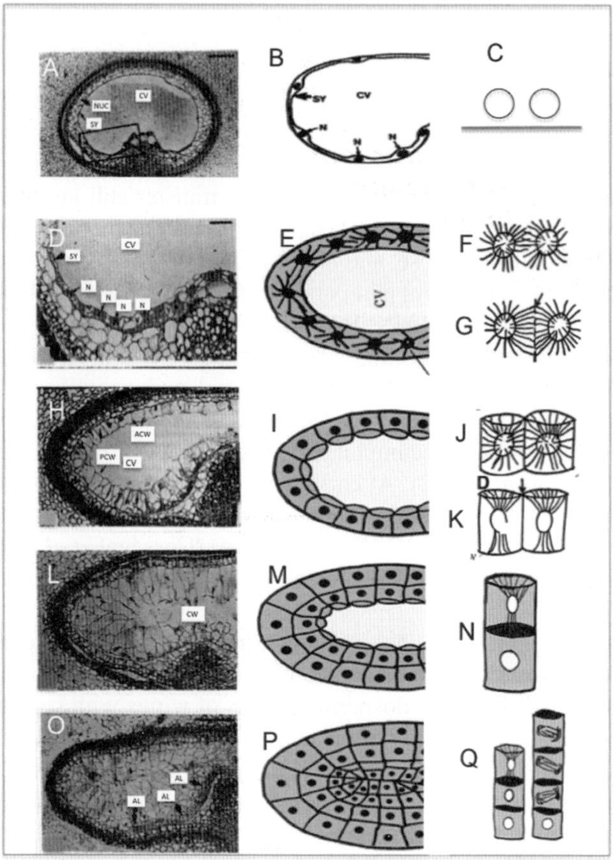

Figure 3.2 Barley endosperm cellularization process. *Left column* shows micrographs of transverse sections (2 μm) of plastic embedded barley endosperm at various developmental stages during the cellularization process. *Middle column* shows line drawings of the stage shown in the first column. *Right column* shows line drawings of endosperm nuclei and the associated microtubular arrays participating in the cellularization process. *A–C*, Coenocytic stage. Individual nuclei embedded in a thin layer of peripheral cytoplasm surround a large central vacuole. SY, coenocyte; CV, central vacuole; N, nucleus. *D–G*, Initiation of cellularization process. Radial microtubular systems (RMSs) on neighboring nuclei *(F)* form interzones acting as primitive phragmoplasts to form the alveolar cell walls *(G)*. *H–K*, Alveolar stage. Cell walls have formed alveoli around each endosperm nucleus *(J)*. A canopy of microtubules extends the alveoli toward the center of the endosperm cavity *(K)*. ACW, anticlinal cell wall; CV, central vacuole; PCW, periclinal cell wall. *L–N*, Early cellularization stage with one peripheral layer of cells and internal layer of alveoli. A periclinal division in an alveolus produces a peripheral daughter cell and an internal alveolus with its opening toward the central vacuole. *O–Q*, Fully cellular endosperm. Repetitions of the initial cellularization process in the internal alveolar layers fill in the endosperm cavity, completing the cellularization process. AL, aleurone.

line drawing representations of the same and two coenocytic nuclei are shown in Figure 3.2*A–C*. In the next phase of endosperm development, each nucleus is encased in tubelike cell wall structures (alveoli) with the openings pointing inward toward the center of the endosperm (Figure 3.2*H–K*). The formation of alveoli occurs via the so-called radial microtubular system (RMS) that forms on the membrane of coenocytic nuclei after the coenocyte has developed numerous vacuoles (Figure 3.2*D–F*). Cell walls form in the interzones between RMS from neighboring nuclei (Figure 3.2*G, arrow*). These interzones function as primitive phragmoplasts depositing the cell wall material initially consisting of a high proportion of callose. This process leads to the first layer of alveoli (Figure 3.2*H–K*). In *Arabidopsis* endosperm at the globular embryo stage, the coenocytic cytoplasm of the MCE surrounds the developing embryo, and the multinucleate PEN coenocyte is a thin peripheral layer with evenly spaced nuclei (Fig. 3.1*L*).

After the initial formation of the alveolar layer, nuclei divide periclinally with an accompanying cytokinesis, producing a peripheral layer of cells and an internal layer of alveoli (Figure 3.1*M–P*, Figure 3.2*L–N*). This process is repeated in the new alveolar layer until after several rounds the endosperm becomes fully cellular (Figure 3.1*Q–T*, Figure 3.2*O–Q*). In *Arabidopsis*, cellularization is first completed in the MCE around the embryo, whereas the PEN remains coenocytic (Figure 3.1*P*). The CZE remains coenocytic until late stages of seed maturation (Sorensen *et al.*, 2002). In contrast to cereals, the central cell mass of the *Arabidopsis* endosperm is digested during embryo growth, leaving only the peripheral layer of aleurone cells in mature seeds (Figure 3.1*D*).

Knowledge about the molecular mechanisms directing the cell cycle in coenocytic endosperm is still scarce. The key event in forming the coenocyte, phragmoplast suppression, appears similar in the species described in Figure 3.1, although in eudicot and monocot lineages, nuclear endosperm development appears to have independently evolved from the ancestral cellular mode of development (Geeta, 2003). Are the molecular controls for suppression identical? Both types of endosperm recruited the ancient program for cellularization via RMS. The basis for the halt in cell cycle progression is also unknown, as are the initiation mechanisms for the full typical plant cell cycle when the nuclei of the alveoli start to divide.

So far, little is known about the molecular biology of endosperm cellularization. As expected, the genes involved in alveolar cell wall formation are similar to the default plant mitotic cell cycle. The functions of other key genes are likely to be revealed in mutants that are arrested at the coenocytic stage. One such barley mutant, B15, suppresses phragmoplast formation to give a coenocytic endosperm that fails to cellularize (Bosnes *et al.*, 1987, 1992). Further candidates for coenocytic and cellularization genes are to be found among other barley mutants arrested at the coenocytic stage described by Ramage and Crandall (1981). Molecular cloning of barley mutant genes are now feasible as a result of the development of genomics resources, including the availability of increasingly denser genetic maps (Mayer *et al.*, 2011). Similar to barley mutants, the *ede1* mutant in *Arabidopsis* fails to cellularize, featuring an aberrant microtubule cytoskeleton where the RMS and cytokinetic phragmoplasts are absent (Pignocchi *et al.*, 2009). The EDE1 protein accumulates in nuclear caps in premitotic cells, colocalizes along microtubules of the spindle and phragmoplast, and interacts with 14-3-3 upsilon in a yeast two-hybrid assay (Pignocchi and Doonan, 2011). Another interesting approach to understanding molecular events in the coenocytic endosperm is transcriptome analysis using laser capture microdissection as reported for 4-day-old *Arabidopsis* coenocytic endosperm (Day *et al.*, 2008). These authors identified coenocytic endosperm-expressed genes that are similar to genes involved in conventional somatic mitosis with cytokinesis as well as cytoskeleton associated genes that may act to facilitate coenocytic development. Transcriptome data such as these, together with systematic mutant screens, are likely to contribute important insights into early endosperm development.

Endosperm Cell Fate Specification and Differentiation

Transfer Cells

The first molecular indication of cell fate specification reported for cereal endosperm is the differential expression of transcripts localized to the region of coenocytic endosperm above the maternal vasculature that later develops into transfer cells (modified aleurone); this implicates maternal signaling as a possible positional cue to induce transfer cell fate. One example of such a transcript is *End1* from barley (Doan et al., 1996) and wheat, which is strongly expressed in the ventral or adaxial endosperm overlying the crease at 3 DAA. The transcript continues to accumulate as the modified aleurone cellularizes and remains at a high level in the outer two cell layers through 9 DAA (Drea *et al.*, 2005). END1 is of unknown function but is similar to a family of Cys-rich proteins in *Arabidopsis* classified as proteinase inhibitors, storage proteins, and lipid transfer proteins (Drea *et al.*, 2005). In addition to *End1*, several other genes with a similar pattern of expression, including β-expansin, are detectable (Drea *et al.*, 2005).

A similar pattern of expression as for *End1* is seen in the transcript for the maize MYB-related protein-1 (MRP1, originally designated ZmMRP-1), which encodes a single Myb-repeat protein (Gomez *et al.*, 2002). This protein was shown to be necessary and sufficient to induce transfer cell differentiation in maize endosperm (Gomez *et al.*, 2009). MRP1 transcriptionally activates numerous transfer cell–specific genes in the maize endosperm (Gomez *et al.*, 2002), including *Meg1* (Gutierrez-Marcos *et al.*, 2004). Costa *et al.* (2012) showed that the *meg1* gene also acts upstream of MRP1, suggesting a positive feedback loop that establishes transfer cell fate and maintains its cellular differentiation throughout development. Strong support for this comes from decreased MRP1 expression in *meg1-RNAi* lines as well as ectopic MRP1 expression and transfer cell formation resulting from misexpression of *meg1*.

Another source of evidence for the role of maternal signaling in transfer cell development in maize comes from endosperm *in vitro* organ cultures. In such cultures, derived from isolated young endosperm at 4 DAP grown either in liquid cultures or on agar medium without maternal tissue influence, starchy endosperm and aleurone cells develop but not transfer cells (Gruis *et al.*, 2006).

Although not yet investigated, it is reasonable to speculate that transfer cell specification occurs by a similar mechanism in maize and in other cereal species with transfer cells (e.g., barley and wheat). In their survey of wheat endosperm gene expression using in situ hybridization, Drea *et al.* (2005) reported 44 transcripts that were predominant and 32 that were specific to the modified aleurone cell layer. This group included the largest number without clear orthologues in rice, a result well in line with the fact that rice lacks transfer cells.

The transfer cells are the earliest endosperm cell type to differentiate. Drea *et al.* (2005) found that the modified aleurone layer and the adjacent central endosperm region have reduced levels of histone expression by 6–7 DAA compared with the rest of the endosperm. This reduction coincides with increased calcofluor staining of cellulose demonstrating that this part of the endosperm undergoes differentiation and cell cycle exit (Drea *et al.*, 2005).

Little is known about transfer cell function in *Arabidopsis*, although the specialized pad of tissue known as the chalazal proliferating tissue positioned in the tip of the chalazal chamber has been suggested to serve a role similar to the transfer cells in cereal endosperm (Brown *et al.*, 1999). The mechanism for cell fate specification of this tissue is unknown.

The last few years have brought important insights into transfer cell developmental biology. In maize, the maternal *Meg1* transcript was identified as a long suspected maternal factor to initiate transfer cell fate specification. Strong indications that a similar system operates in barley and wheat comes from the differential expression of the *End1* and similar transcripts in the endosperm coenocyte over the nucellar projection. In contrast, the lack of localized transfer cells in rice endosperm suggests that a similar system of *Meg1*-like and *End1*-like genes may not be present in rice. The contrast between rice and the other cereals opens up further investigations of molecular mechanisms of transfer cell specification and possibly also strategies to increase grain yield by directed changes in transfer cell function or morphology.

Starchy Endosperm Cells

Starchy endosperm cells represent the largest body of cells in the cereal endosperm (Figure 3.1*A–C*). Starchy endosperm cells accumulate starch and storage proteins. Prolamins and glutelins are encoded by transcripts that are expressed differentially in these cells (see Chapters 8 and 9). Starchy endosperm cells are derived from two sources. The first source is the inner cells of cell files that are present at the completion of endosperm cellularization (Figure 3.1*Q–S* and Figure 3.2*P*). Soon after cellularization culminates, cell division resumes in the inner cells of these files (Figure 3.2*Q*). In contrast to the alveolar divisions, which are strictly periclinal, the division planes are oriented randomly, and the cell file pattern is soon lost. The second source of starchy endosperm cells is the inner daughter cells of aleurone cells that divide periclinally giving rise to the subaleurone cells (Morrison *et al.*, 1975; Becraft *et al.*, 2000).

Cell cycle regulation in cereal endosperm has been studied most intensively in maize, where, similar to barley, a phase of mitotic cell division occurring after cellularization (Figure 3.2*Q*) is largely responsible for generating the final population of endosperm cells. This period in maize lasts 8–12 DAP in the central endosperm (Sabelli and Larkins, 2009). In maize, cell division occurs in waves, stopping first at the base of the endosperm and then in the central region. Other growth parameters, such as increases in the size of nuclei, starch granules, and cells, suggest that differentiation follows the cell division gradients (Kowles and Phillips, 1988). The mitotic index peaks around 8–10 DAP and then declines sharply. During the period from 8–é12 DAP, the maize endosperm grows rapidly to fill the seed cavity. This growth appears to be correlated with endoreduplication because the mean volume of centrally located nuclei increases roughly 10-fold (Kowles and Phillips, 1985). Endoreduplication seems to be a common feature of cereal endosperms (Chojecki *et al.*, 1986; Ramachandran and Raghavan, 1989; Giese, 1992; Kladnik *et al.*, 2006; Sabelli and Larkins, 2009). From around 9 DAP, maize starchy endosperm cells transition from a mitotic to an endoreduplication cell cycle in which repeated rounds of DNA replication without cytokinesis lead to DNA contents of as much as 96C (Sabelli and Larkins, 2009). Similar to animals, plant cell cycle progression is regulated by cyclin-dependent kinases (CDKs) and their cyclin partners, CDK inhibitors, and retinoblastoma-related (RBR) proteins (Sabelli and Larkins, 2009). In maize endosperm, CDK activity peaks at the time of the onset of endoreduplication (10–12 DAP), convincingly supporting the view that the switch from the mitotic to the endoreduplication cell cycle occurs via downregulation of mitotic CDKs and upregulation of S-phase CDKs (Grafi and Larkins, 1995; Sun *et al.*, 1999; Leiva-Neto *et al.*, 2004; Coelho *et al.*, 2005; Sabelli and Larkins, 2009). Increasing evidence also implicates RBRs in endosperm development (Sabelli and Larkins, 2009). The function of endosperm endoreduplication remains unresolved because impairing endoreduplication had no obvious effect on grain size or yield (Sabelli and Larkins, 2009).

In their analysis of wheat endosperm transcription, Drea *et al.* (2005) detected 26 transcripts that were central endosperm predominant and 46 that were central endosperm specific. Among transcripts in the first group, most were shared between starchy endosperm, transfer cells, and aleurone (e.g., the formate dehydrogenase transcript, which was expressed in all cell types until at least 13 DAA). Other transcripts continued expressing in the starchy endosperm while being downregulated in the aleurone (e.g., α-thionin in a pattern similar to the storage protein gliadin that starts to accumulate in the starchy endosperm at 9 DAA). Among the investigated wheat starchy endosperm transcripts, the cereal with the largest divergence was rice.

Around the midpoint of seed development, cereal endosperms undergo programmed cell death (PCD), which was suggested by Nguyen *et al.* (2007) to facilitate nutrient hydrolysis and uptake by the embryo at germination. PCD in maize starchy endosperm initiates at approximately 16 DAP in the central starchy endosperm cells and in apical cells near the silk scar and eventually leads to the death of the half-top of the endosperm at 28 DAP (Young and Gallie, 1999). In wheat, PCD initiates stochastically and leads to the death of all starchy endosperm cells by 30 DAP (Young and Gallie, 1999). PCD in plants bears typical characteristics of some PCD in animals, including DNA fragmentation, but the effector caspases have not been clearly identified. Instead, evidence points to a role of proteases with caspase-like activity (Hatsugai *et al.*, 2004; Nguyen *et al.*, 2007). Endosperm PCD is novel compared with typical PCD programs where cellular corpse processing occurs concurrently with cell death and is controlled by an endogenous cellular program. In endosperm by contrast, cellular corpse processing is temporally delayed until germination and is controlled exogenously by the activities of aleurone cells (van Doorn *et al.*, 2011).

Ethylene is a clear candidate as a regulator of PCD (Young *et al.*, 1997). In addition, ABA biosynthesis affects PCD indirectly, via ethylene biosynthesis, as shown by increased ethylene levels and PCD in maize viviparous mutants in which the ABA biosynthetic pathway is compromised (Young *et al.*, 1997; Sabelli and Larkins, 2009; Becraft and Gutierrez-Marcos, 2012).

Aleurone Cells

Cell fate specification of aleurone cells in maize follows a simple pattern described by the so-called surface rule; all cells positioned on the surface of the endosperm assume aleurone cell fate, and interior cells become starchy endosperm cells (Olsen, 2004a). An illustration of this principle is easily observable with *in vitro* maize endosperm cultures in which aleurone cells and starchy endosperm cells are marked with cell-specific green and cyan fluorescent proteins (Figure 3.3) (Gruis *et al.*, 2006). In such cultures, the endosperm becomes fragmented and irregular. Regardless

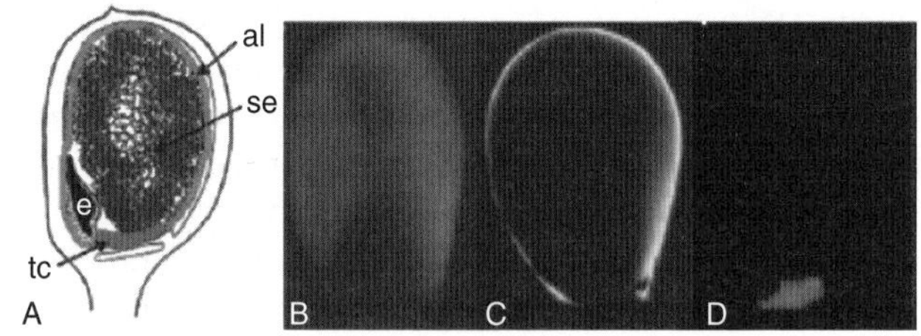

Figure 3.3 Kernel of "triple marked" maize endosperm. *A*, Diagram of the three major endosperm tissues. al, aleurone; e, embryo; se, starchy endosperm; tc, transfer cells. *B–D*, Expression of fluorescent cell-type markers in 12 DAP endosperm. Various cell-specific promoters are used to regulate expression of fluorescent proteins with differing spectral properties.B, γ-Zein:AmCyan expression fluoresces blue in the starchy endosperm. *C*, Ltp2:ZsYellow expression is specific for the aleurone layer. *D*, End1:DsRed expression marks transfer cells. (Modified from Gruis *et al.* [2006].) (For color detail, see color plate section.)

of shape, all surfaces are covered by aleurone cells (green); internal cells have starchy endosperm identity and fluoresce blue.

The fate of aleurone cells appears to be specified early in endosperm development. As described previously, the surface layer of endosperm displays a distinct behavior from the internal cells at the outset of cellularization, suggesting that independent developmental trajectories for aleurone and starchy endosperm are already established. Transcriptomic analysis in wheat further shows that by 6 DAA the surface layer has acquired a molecular signature of aleurone, very different from starchy endosperm (Gillies *et al.*, 2012). Although aleurone cells have not fully differentiated by this point, the transcriptional profile is more similar to later stage aleurone than to internal cells.

Despite the early establishment of distinct position-dependent cell identities, cell fate in the endosperm periphery remains flexible throughout endosperm development. In culture, starchy endosperm cells that become positioned on the surface transdifferentiate to aleurone cells, and aleurone cells that become internalized convert to starchy endosperm (Gruis *et al.*, 2006; Reyes *et al.*, 2010, 2011). Also *in planta*, aleurone cell fate remains plastic throughout late stages of endosperm development, up to the last cell division, even after they have acquired the typical aleurone cytological features (Becraft and Asuncion-Crabb, 2000; Becraft and Yi, 2011). A signaling system thus functions throughout endosperm development to specify and maintain aleurone cell identity.

Genetic studies in maize have identified several components of a signaling system required for aleurone cell differentiation. Mutations in the *crinkly4* (*cr4*) and *defective kernel1* (*dek1*) genes disrupt aleurone cell fate specification indicating these genes are positively acting factors required for this process (Becraft *et al.*, 1996, 2002; Lid *et al.*, 2002). CR4 is a receptor kinase (Becraft *et al.*, 1996; Jin *et al.*, 2000), and DEK1 is a novel plasmamembrane–localized protein with 21 transmembrane helices, an extracellular loop domain, and a cytoplasmic domain containing a calpain-like protease (Lid *et al.*, 2002; Wang *et al.*, 2003). Genetic studies suggest CR4 and DEK1 function together in a signaling system (Becraft *et al.*, 2002), and the colocalization of these proteins in endocytic vesicles supports that hypothesis (Tian *et al.*, 2007). Negatively acting factors also have been identified in maize. The *supernumerary aleurone layers1* (*sal1*) mutant produces multiple aleurone layers instead of the typical single layer (Shen *et al.*, 2003). SAL1 has homology to class E vacuolar sorting proteins, suggesting that it might regulate aleurone cell number by functioning in retrograde cycling of the DEK1 or CR4 proteins from the plasmamembrane. This would serve to dampen the amplitude of signaling and restrict aleurone cell fate to just the outer layer. In the *sal1* mutant, DEK1 or CR4, or both, would accumulate and provide a higher level of signaling to promote aleurone differentiation. The colocalization of SAL1 with DEK1 and CR4 in endocytic vesicles is consistent with this hypothesis (Tian *et al.*, 2007). This would place SAL1 upstream of DEK1 and CR4 in the regulatory pathway. The *thk1* mutant also causes multiple aleurone layers; however, this mutant is epistatic to *dek1*, suggesting that THK1 functions downstream of DEK1 (Yi *et al.*, 2011). The molecular identity of THK1 is unknown. Other factors have been identified in rice and barley. A double knockdown of RISBZ1 and RPBF transcription factors caused a multilayered, disordered aleurone in rice (Kawakatsu *et al.*, 2009). How or if these functions relate to the DEK1 signaling system is unknown. The *defective seed1* (*des5/B5*) mutant of barley creates a single aleurone layer, as opposed to the normal three layers, suggesting the normal gene functions to promote multilayering (Olsen *et al.*, 2008). *HvCr4* transcript levels were significantly decreased in *des5*, consistent with the hypothesis that the level of CR4 signaling might modulate aleurone cell layer number. However, the inheritance of aleurone layer number as a quantitative trait argues that the regulation could be complex and involve additional factors yet to be discovered (Jestin *et al.*, 2008).

There have also been indications of hormonal functions in aleurone specification. Expression of isopentenyl transferase under the control of a senescence inducible SAG-12 promoter induced aleurone mosaicism in the crowns of maize kernels suggesting that cytokinin can inhibit aleurone differentiation (Geisler-Lee and Gallie, 2005). Treating plants with the auxin transport inhibitor NPA caused a multiple aleurone cell layer suggesting that auxins also function in regulating aleurone fate (Forestan *et al.*, 2010).

Mutant analyses have also identified a collection of genes that appear to function downstream of the cell fate decision and to be required for aleurone differentiation (Gavazzi *et al.*, 1997; Becraft and Asuncion-Crabb, 2000; Lid *et al.*, 2004). As of yet, the molecular identification of these genes have not been reported.

At late stages of development, during seed maturation, the aleurone cells are the only endosperm cells not to undergo cell death. The aleurone layer undergoes ABA-dependent maturation, which involves the expression of dehydrins and the acquisition of desiccation tolerance, similar to the embryo maturation program. The VIVIPAROUS1 (VP1) transcription factor and ABA biosynthetic enzymes are required for this process. In viviparous mutants, the aleurone layer prematurely enters the germination program and begins secreting hydrolytic enzymes to digest the contents of starchy endosperm cells.

In contrast to the starchy endosperm, programmed cell death occurs during germination in aleurone cells and is induced by GA, not ethylene (Nguyen *et al.*, 2007). A transcriptome analysis of barley showed that proteases and the ethylene and ABA pathways are implicated in endosperm PCD and maturation during barley grain development (Sreenivasulu *et al.*, 2008).

In summary, aleurone cell fate specification in maize endosperm follows a simple rule; cells on the surface become aleurone cells, whereas interior cells are specified as starchy endosperm cells. The *Dek1* and *Cr4* genes are known to act in aleurone cell specification. The molecular basis for three aleurone cell layers in barley, instead of one layer as in maize and wheat, is currently unknown. Cloning of the mutant genes that give multiple layers in maize and a single layer in barley may reveal the mechanism regulating aleurone cell layer numbers in cereals, opening up strategies to increase the number of oil-containing, nutritious aleurone cells in species with only a single cell layer.

Embryo-Surrounding Region

The embryo-surrounding region (ESR) has been most intensively studied in maize where it lines the cavity of the endosperm in which the embryo develops. These cells have dense cytoplasmic contents (Schel *et al.*, 1984; Schel and Kieft, 1986; Kowles and Phillips, 1988) and are characterized by the expression of the *Esr1*, *Esr2*, *Esr3* (Opsahl-Ferstad *et al.*, 1997), *ZmAE1* (*Zea mays androgenic embryo1*), and *ZmAE3* (Magnard *et al.*, 2000) genes between 5 and 20 DAP. The ESR3 protein is part of a family of small hydrophilic proteins that share a conserved motif with CLAVATA3 (CLV3), a ligand of the receptor-like kinases CLV1 and CLV2 functioning to regulate meristem size in *Arabidopsis* (Fletcher *et al.*, 1999; Ogawa *et al.*, 2008).

Future studies are needed to clarify the role of ESR proteins in ESR signaling. The function of the ESR in maize is unknown but may include a role in embryo nutrition or in communication between the embryo and endosperm, or both. The ESR cells disintegrate and disappear at 12 DAP (Kieselbach and Walker, 1952). Barley lacks a prominent ESR region, but a small part of the endosperm situated close to the embryo that appears to be similar to the maize ESR was termed "zone 1" by Engell (1989).

Genomic Resources

Among the cereals, the most extensive genomic resources are available for rice and maize. Each has a sequenced genome, microarrays, genetic transformation, and sequence indexed insertional mutant collections (Settles *et al.*, 2007; Jung *et al.*, 2008; Vollbrecht *et al.*, 2010). In addition, other resources such as fluorescent protein markers to identify various cell types or subcellular compartments are being developed (Mohanty *et al.*, 2009). Such markers greatly facilitate developmental and genetic studies as exemplified by the work of Gruis *et al.* (2006). These investigators used a "triple marked" line containing transgenes specifically labeling aleurone, starchy endosperm, and transfer cells (Figure 3.3) and demonstrated the ability of *in vitro* cultured endosperm to differentiate aleurone and starchy endosperm but not transfer cells. Figure 3.4 illustrates the use of such markers to analyze the effects of mutations on the differentiation of various cell types. Among the small grain cereals, wheat genome sequencing is currently underway, and a physical map for barley has been published, both using single chromosome arm isolation, BAC library construction, and sequencing

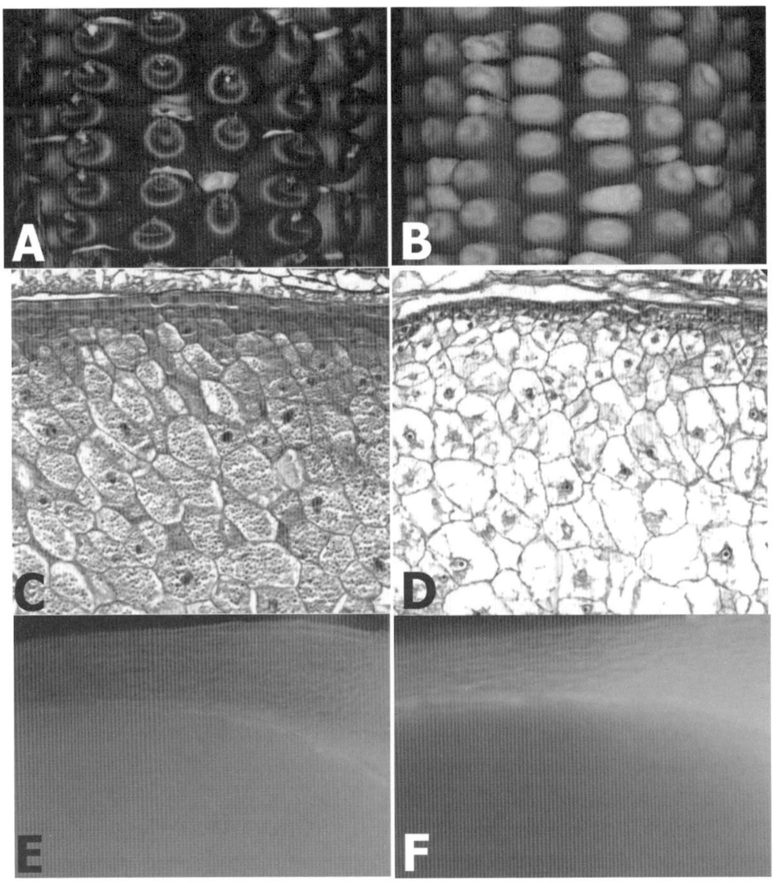

Figure 3.4 Examples of maize kernel mutants. *A*, Ear segregating an *emp* mutant. *B*, Ear segregating a *dek* mutant. *C*, Histological section of a wild-type kernel with cells packed with starch grains. *D*, Section of a mutant endosperm defective in starch accumulation. *E*, Expression of an FL2-RFP fluorescent marker for α-zein accumulation in wild-type (Mohanty *et al.*, 2009). *F*, Lack of FL2-RFP expression in a mutant kernel. (For color detail, see color plate section.)

of individual chromosome arms (Mayer *et al.*, 2011). These technological advances are expected to faciltitate the identification of mutant genes in wheat and barley in the near future.

Maize has particularly rich collections of seed mutants, many of which are available through the Maize Genetics Cooperation Stock Center (Sachs, 2009). The first systematic collection of Mu-tagged maize lines for reverse genetics screening, TUSC (Trait Utility System for Corn), facilitated the cloning of several maize genes, including *An1* (Bensen *et al.*,1995), *Dek1*, and *Sal1* (Table 3.1) (Lid *et al.*, 2002; Shen *et al.*, 2003). A large-scale mutagenesis with *Mutator* transposons produced a collection of nearly 2000 independent maize mutants with kernel phenotypes (McCarty *et al.*, 2005; Settles *et al.*, 2007). More than half of these mutants showed mutant phenotypes known as *empty pericarp* (*emp*) or *defective kernel* (*dek*) (Figure 3.4), depending on the severity of the phenotype; more filial tissue is apparent in *dek* than *emp* mutant kernels (Neuffer and Sheridan, 1980; Sheridan and Neuffer, 1980; Scanlon *et al.*, 1994). There is overlap in *emp* and *dek* mutant classes and some mutants with variable expressivity often display both phenotypes. Genes producing *emp* or *dek* mutant phenotypes are essential for endosperm or endosperm and embryo development. Genes required specifically for embryo development produce embryo-specific (*emb*) mutant phenotypes (Clark and Sheridan, 1991; Sheridan and Clark, 1993). Although *emb* mutants produce nearly normal endosperm, *emp* and *dek* mutants produce neither normal endosperm nor embryos. Embryos can be rescued in culture for some *dek* mutants (Sheridan and Neuffer, 1980), demonstrating that embryo viability is dependent on endosperm functions but not vice versa.

An early and conservative estimate suggested there are >285 loci essential for maize seed viability, based on allele frequencies in EMS-induced seed mutants (Neuffer and Sheridan, 1980). Subsequent analyses identified multiple essential genes not identified in the earlier study, verifying that the estimate is low (Clark and Sheridan, 1991; Scanlon *et al.*, 1994; Costa *et al.*, 2003; Gutierrez-Marcos *et al.*, 2006; Yi *et al.*, 2009). *Arabidopsis* is estimated to have 500–750 genes essential for seed development (see Chapter 1) (McElver *et al.*, 2001). Analysis of >300 such genes revealed that virtually every class of cellular function was represented (McElver *et al.*, 2001; Tzafrir *et al.*, 2004; Drea *et al.*, 2005; Devic, 2008; Meinke *et al.*, 2008). A unifying theme was that genes essential for seed development are less likely than other genes to contain paralogs in the genome.

Available information on the molecular identity of seed-lethal mutants in maize (Table 3.1) is consistent with the array of functions seen in *Arabidopsis*, although the data are currently too sparse for general conclusions. The large number, indistinct macroscopic phenotypes, and lethality of *dek/emp* mutants make them difficult to study using traditional approaches. Although essential genes are likely to be involved in all aspects of seed biology, including development, this important class of mutants has been poorly studied. These mutants, including use of the high-throughput maize *in vitro* endosperm culture system for transformation of aleurone and starchy endosperm cells (Reyes *et al.*, 2010) are promising tools for enhancing understanding of cereal endosperm development.

Transcriptional Profiling of Endosperm Development

Transcriptomic analyses have been performed for most of the major grain cereals, including maize, rice, wheat, and barley (Laudencia-Chingcuanco *et al.*, 2007; Hansen *et al.*, 2009; Liu *et al.*, 2010; Sekhon *et al.*, 2011; Gillies *et al.*, 2012; Xue *et al.*, 2012). As would be expected for cereals, genes involved in sucrose and starch metabolism and protein synthesis were enriched in the endosperm. Additionally, genes involved in defense, signal transduction, and transcription are highly represented.

Table 3.1 Known Maize *dek/emp* Mutant Genes

Mutant	Gene ID	Molecular Function	References
defective kernel1 (dek1)	GRMZM2G321753	21-transmembrane calpain protease	Lid *et al.*, 2002; Wang *et al.*, 2003
Supernumerary aleurone layer 1 (sal1)	GRMZM2G117935	Endosome trafficking	Shen *et al.*, 2003
opaque5 (o5)	GRMZM2G142873	Monogalactosyldiacylglycerol synthase	Myers *et al.*, 2011
discolored1 (dsc1)	GRMZM2G117329	ADP–ribosylation factor–GTPase activating protein	Takacs *et al.*, 2012
empty pericarp2 (emp2)	GRMZM2G039155	Heat shock factor binding protein	Fu *et al.*, 2002
empty pericarp4 (emp4)	GRMZM2G092198	Mitochondrion-targeted pentatricopeptide repeat protein	Gutierrez-Marcos *et al.*, 2007
indeterminate gametophyte1 (ig1)	GRMZM2G118250	LOB transcription factor	Evans, 2007
zmk1	GRMZM2G171279 (ambiguous – could be paralog)	Auxin-induced K^+ uptake channel	Philippar *et al.*, 2006
etched1 (et1)	GRMZM2G157574	Plastidal transcription factor	da Costa e Silva *et al.*, 2004

In rice, transcripts from 18,401 of the 46,857 gene models were expressed in developing endosperm, analyzed at 3, 6, 9, and 16 DAP. Hence, slightly more than a third of the genome is involved in endosperm development and function (Xue *et al.*, 2012). Given the unique aspects of endosperm function and development, surprisingly few genes are endosperm-specific. Microarray analysis of developing rice caryopses identified 474 genes expressed in endosperm but not embryo (Xue *et al.*, 2012). Microarray analysis during a time course of maize seed development identified 168 endosperm-specific genes (Sekhon *et al.*, 2011). In contrast, deep sequencing of maize transcriptomes identified no genes specific to the 14 DAP endosperm, although embryo-specific genes were found (Waters *et al.*, 2011). The contrasting results likely reflect the different sensitivities of deep sequencing versus microarray methods. Additionally, the number of genes expressed in 16 DAP rice endosperm was significantly less than at earlier stages (Xue *et al.*, 2012), suggesting that a fuller examination of different developmental stages and more precise dissection of specific regions or cell types would likely reveal genes expressed uniquely in the endosperm. Consistent with this, Gillies *et al.* (2012) reported in wheat that the transcriptomes of the starchy endosperm and aleurone were substantially different from one another and that both changed dynamically during development. Nonetheless, the paucity of endosperm-specific genes is a surprising result given the unique and highly specialized aspects of endosperm development and function.

A coexpression meta-analysis of rice microarray data suggested that endosperm-specific transcription factors were largely involved in nutrient storage and response to abscisic acid (Xue *et al.*, 2012). Numerous protein kinase genes were found to be predominantly expressed in the endosperm, including two genes for SnRKs, involved in regulating carbohydrate metabolism, and many receptor-like kinases, calcium-dependent kinases, and casein kinases. These were suggested to function in combination with the transcription factors to regulate cellular and metabolic networks involved with endosperm development (Xue *et al.*, 2012).

In a time course of maize grain development, principle component analysis of global gene expression showed that whole seeds from 2–8 DAP clustered together, possibly reflecting high mitotic activity in the endosperm during that period (Sekhon *et al.*, 2011). Transcription profiles in the endosperm underwent a transition from 12–20 DAP, after which little change occurred up through 24 DAP. At this point, the endosperm reaches a steady state where cell division and differentiation occur around the periphery of the endosperm, with a gradient of storage product accumulation and cell maturation progressing inward until finally internal cells undergo PCD. This steady state would be maintained until grains underwent desiccation, which would likely involve expression of new genes. Similar transitions were observed in barley (Hansen *et al.*, 2009), and reprogramming occurred in wheat between 3 and 7 DAA, between 7 and 14 DAA, and between 21 and 28 DAA (Laudencia-Chingcuanco *et al.*, 2007). These periods correspond to endosperm cellularization and proliferation, the onset of grain filling, and seed maturation and desiccation.

Gillies *et al.* (2012) similarly found a dynamic profile with RNAseq analysis. They analyzed wheat aleurone and starchy endosperm at 6, 9, and 14 DAA. Gene expression differences reflected the different metabolic activities of the two cell types (e.g., lipid versus starch accumulation). Less predictable were differences in classes of regulatory activities, such as a higher level of receptor signaling functions in aleurone. Also, the two tissues appeared to be under differential stress that changed over time. Although both tissues showed dynamic expression profiles as expected for a developmental progression, the major changes occurred at different times. In starchy endosperm, there was a major reprogramming between 6 and 9 DAA as the tissue transitioned from proliferation to grain filling, whereas in aleurone the major changes occurred at 14 DAA as the cells become fully differentiated. These findings reinforce the need to examine subpopulations of endosperm cells over a time course to understand fully the developmental progression of cellular activities.

Considering the high levels of conservation among genomes of cereals and in the functions of cereal endosperms, there appears to be surprising diversity in endosperm transcriptomes. In a comparative study of maize and rice, 78% of maize endosperm cDNAs mapped to homologous sequences of the rice genome, leaving 22% of the endosperm transcriptome unique to maize (Lai *et al.*, 2004); 4139 unique maize cDNAs mapped to 3108 rice loci suggesting that many of these genes were tandemly duplicated in rice. This highlights two reasons why genomic analyses must be performed for each individual species: (1) not all genes are present in every species and can be studied only where they reside; (2) because duplicated genes are less likely to show mutant phenotypes, the array of genes amenable to mutant analysis likely differs substantially among cereal species.

Gene Imprinting in Cereal Endosperm

Gene imprinting refers to the differential expression of a gene depending on whether it was inherited from the female or male parent. Imprinting occurs in plants and mammals and is most prevalent in extraembryonic tissues that provide nutritional support to the developing embryo: the placenta in mammals and the endosperm in plants. The function of imprinting is generally believed to relate to controlling resource allocation among offspring (Haig, 2004). Endosperm gene imprinting has been extensively studied in *Arabidopsis* and is discussed in detail in Chapter 4. In cereals, more recent global transcript profiling studies using deep sequencing approaches on seeds from reciprocal crosses have begun to shed additional light on the extent of imprinting, have confirmed the prevalence of imprinting in the endosperm compared with embryos, and have provided new insights into mechanisms and the potential functional significance of imprinting. Such analyses identified 165 maternally and 43 paternally expressed genes in *Arabidopsis* endosperm (Gehring *et al.*, 2011), 93 maternally and 72 paternally biased genes in rice (Luo *et al.*, 2011), and 68 maternally and 111 paternally preferred genes in maize (Zhang *et al.*, 2011). Imprinted genes appear to be enriched for regulatory functions, including nucleic acid binding factors, kinases, and chromatin regulatory factors (Waters *et al.*, 2011). Additionally, imprinted noncoding transcripts were identified in rice and maize (Luo *et al.*, 2011; Zhang *et al.*, 2011). In maize, several paternally expressed noncoding RNAs mapped to intronic regions of maternal-specific protein-coding genes (Zhang *et al.*, 2011), reminiscent of an imprinting mechanism known in mammals (Frost and Moore, 2010). Other observations hint at previously unknown imprinting mechanisms. DNA methylation was examined in maize, and all imprinted loci showed maternal hypomethylation, regardless of whether the gene was maternally or paternally expressed (Waters *et al.*, 2011; Zhang *et al.*, 2011). Such genes are likely subject to regulation by a polycomb group repressive complex, as described in Chapter 4. In rice, parent-specific alternative transcript splicing was also observed (Luo *et al.*, 2011). Among the functional classes of imprinted genes were a prevalence of genes involved in chromatin modeling and regulatory genes such as signal transduction or transcription factors (Gehring *et al.*, 2011; Zhang *et al.*, 2011).

The function of imprinting is generally hypothesized to regulate nutrient allocation among offspring, as has been demonstrated in mammals. Functional studies lend strong support to this hypothesis. The *meg1* gene, described earlier, is important for transfer cell differentiation and is an imprinted gene with only the maternal allele expressed in the early endosperm. Replacing the promoter with a nonimprinted transfer cell promoter resulted in more extensive transfer layer formation and an increase in seed size (Costa *et al.*, 2012). Imprinting the male copy serves to place nutrient allocation under maternal control.

Several lines of evidence suggest that nutrient allocation might not be the only purpose for imprinting. A generally low level of conservation in imprinted genes among *Arabidopsis*, rice, and maize (Gehring *et al.*, 2011; Hsieh *et al.*, 2011; Luo *et al.*, 2011; Zhang *et al.*, 2011) could be an argument that most imprinting is unlikely to control genes of fundamental importance for seed development. However, low conservation levels among imprinted placental genes are observed between mouse and human where important developmental functions are well established (Frost and Moore, 2010), and a limited overlap was observed among different *Arabidopsis* datasets (Gehring *et al.*, 2011; Hsieh *et al.*, 2011), so these comparisons must be viewed with caution. There are some examples of conserved imprinted genes among maize, rice, and *Arabidopsis* (Waters *et al.*, 2011; Zhang *et al.*, 2011). However, the facts that imprinting in some genes is allele-specific (Kermicle, 1970; Waters *et al.*, 2011; Zhang *et al.*, 2011) and that transposon insertions are often associated with imprinted loci (Wolff *et al.*, 2011) suggest that some imprinting may be a manifestation of transposon silencing mechanisms that maintain genome integrity during reproductive development (Wollmann and Berger, 2011).

Conclusion

Endosperm development displays several features that are novel relative to the rest of plant development. Early stages of endosperm development are highly conserved among plant species, whereas later stages differ substantially. Cereal endosperm has received particular attention because of its persistence in mature grains and because of its significance for food, feed, and industrial uses. Genomic analyses have validated the high degree of genetic conservation among cereal species but have also highlighted differences. Although much information will translate among cereal species, a substantial amount will not, and analyses must be conducted for each system. New genomic tools and resources being rapidly developed for functional and comparative studies will quickly increase our understanding of endosperm development to a high level of sophistication. With that, our ability to improve and engineer endosperm properties for human needs will be greatly enhanced.

Acknowledgments

The authors thank members of the Becraft laboratory for helpful reading of the manuscript. Research in the author's laboratory is supported by grant IOS-1121738 from the U.S. National Science Foundation to P.W.B. and grants from the Norwegian Research Council and the E. U. to O.-A.O.

References

Bechtel DB, Pomeranz Y (1977). Ultrastructure of the mature ungerminated rice (Oryza sativa) caryopsis: the caryopsis coat and the aleurone cells. Am J Bot 64: 966–973.

Becraft PW (2007). Aleurone cell development. In: Olsen OA (ed). *Endosperm: Developmental and Molecular Biology*. Springer, Berlin, Germany, pp 45–56.

Becraft PW, Asuncion-Crabb Y (2000). Positional cues specify and maintain aleurone cell fate in maize endosperm development. Development 127: 4039–4048.

Becraft PW, Gutierrez-Marcos J (2012). Endosperm development: dynamic processes and cellular innovations underlying sibling altruism. WIREs Dev Biol 1: 579–593.

Becraft PW, Li K, Dey N, Asuncion-Crabb YT (2002). The maize dek1 gene functions in embryonic pattern formation and in cell fate specification. Development 129: 5217–5225.

Becraft PW, Stinard PS, McCarty D (1996). CRINKLY4: a TNFR-like receptor kinase involved in maize epidermal differentiation. Science 273: 1406–1409.

Becraft PW, Yi G. (2011). Regulation of aleurone development in cereal grains. J Exp Bot 62: 1669–1675.

Bensen RJ, Johal GS, Crane VC, Tossberg JT, Schnable PS, Meeley RB, Briggs SP (1995). Cloning and characterization of the maize an1 gene. Plant Cell 7: 75–84.

Berger F (2003). Endosperm: the crossroad of seed development. Curr Opin Plant Biol 6: 42–50.

Boisnard-Lorig C, Colon-Carmona A, Bauch M, Hodge S, Doerner P, Bancharel E, Dumas C, Haseloff J, Berger F (2001). Dynamic analyses of the expression of the HISTONE::YFP fusion protein in Arabidopsis show that syncytial endosperm is divided in mitotic domains. Plant Cell 13: 495–509.

Bosnes M, Harris E, Ailgeltiger L, Olsen OA (1987). Morphology and ultrastructure of 11 barley shrunken endosperm mutants. Theor Appl Genet 74: 177–187.

Bosnes M, Weideman F, Olsen OA (1992). Endosperm differentiation in barley wild-type and sex mutants. Plant J 2: 661–674.

Brown RC, Lemmon BE, Olsen O-A (1994) Endosperm development in barley: microtubule involvement in the morphogenetic pathway. Plant Cell 6: 1241–1252.

Brown R, Lemmon BE, Nguyen H, Olsen OA (1999). Development of endosperm in Arabidopsis thaliana. Sex Plant Reprod 12: 32–42.

Chamberlain MA, Horner HT (1990). An ultrastructural study of endosperm wall formation and degeneration in Glycine max (L.). Merr Am J Bot 77: 33.

Chojecki AJS, Bayliss MW, Gale MD (1986). Cell production and DNA accumulation in the wheat endosperm, and their association with grain weight. Ann Bot 58: 809–818.

Clark J, Sheridan W (1991). Isolation and characterization of 51 embryo-specific mutations of maize. Plant Cell 3: 935–951.

Coelho CM, Dante RA, Sabelli PA, Sun Y, Dilkes BP, Gordon-Kamm WJ, Larkins BA (2005). Cyclin-dependent kinase inhibitors in maize endosperm and their potential role in endoreduplication. Plant Physiol 138: 2323–2336.

Cone KC (2007). Anthocyanin synthesis in maize aleurone tissue. In: Olsen OA (ed). *Endosperm: Developmental and Molecular Biology*. Springer, Berlin, Germany, pp 121–140.

Cosségal M, Vernoud V, Depège N, Rogowsky PM (2007). The embryo surrounding region. In: Olsen OA (ed). *Endosperm: Developmental and Molecular Biology*. Springer, Berlin, Germany, pp 52–72.

Costa LM, Gutierrez-Marcos JF, Brutnell TP, Greenland AJ, Dickinson HG (2003). The globby1-1 (glo1-1) mutation disrupts nuclear and cell division in the developing maize seed causing alterations in endosperm cell fate and tissue differentiation. Development 130: 5009–5017.

Costa LM, Yuan J, Rouster J, Paul W, Dickinson H, Gutierrez-Marcos JF (2012). Maternal control of nutrient allocation in plant seeds by genomic imprinting. Curr Biol 22: 160–165.

Day RC, Herridge RP, Ambrose BA, Macknight RC (2008). Transcriptome analysis of proliferating Arabidopsis endosperm reveals biological implications for the control of syncytial division, cytokinin signaling, and gene expression regulation. Plant Physiol 148: 1964–1984.

Day RC, Mueller S, Macknight RC (2009). Identification of cytoskeleton-associated genes expressed during Arabidopsis syncytial endosperm development. Plant Signal Behav 4: 883–886.

da Costa e Silva O, Lorbiecke R, Garg P, Muller L, Wassmann M, Lauert P, Scanlon M, Hsia AP, Schnable PS, Krupinska K, Wienand U (2004). The Etched1 gene of Zea mays (L.) encodes a zinc ribbon protein that belongs to the transcriptionally active chromosome (TAC) of plastids and is similar to the transcription factor TFIIS. Plant J 38: 923–939.

Devic M (2008). The importance of being essential: EMBRYO-DEFECTIVE genes in Arabidopsis. wComptes Rendus Biol 331: 726.

Doan DNP, Linnestad C, Olsen OA (1996) Isolation of molecular markers from the barley endosperm coenocyte and the surrounding nucellus cell layers. Plant Mol. Biol. 31: 877–886.

Dornez E, Cuyvers S, Holopainen U, Nordlund E, Poutanen K, Delcour JA, Courtin CM (2011). Inactive fluorescently labeled xylanase as a novel probe for microscopic analysis of arabinoxylan containing cereal cell walls. J Agric Food Chem 59: 6369–6375.

Drea S, Leader DJ, Arnold BC, Shaw P, Dolan L, Doonan JH (2005). Systematic spatial analysis of gene expression during wheat caryopsis development. Plant Cell 17: 2172–2185.

Engell K (1989). Embryology of barley: time coarse and analysis of controlled fertilization and early embryo formation based on serial sections. Nord J Bot 9: 265–280.

Evans MMS (2007). The indeterminate gametophyte1 gene of maize encodes a LOB domain protein required for embryo sac and leaf development. Plant Cell 19: 46–62.

Fletcher JC, Brand U, Running MP, Simon R, Meyerowitz EM (1999). Signaling of cell fate decisions by CLAVATA3 in Arabidpsis shoot meristems. Science 283: 1911–1914.

Forestan C, Meda S, Varotto S (2010). ZmPIN1-mediated auxin transport is related to cellular differentiation during maize embryogenesis and endosperm development. Plant Physiol 152: 1373–1390.

Frost JM, Moore GE (2010). The importance of imprinting in the human placenta. PLoS Genet 6: e1001015.

Fu S, Meeley R, Scanlon MJ (2002). Empty pericarp2 encodes a negative regulator of the heat shock response and is required for maize embryogenesis. Plant Cell 14: 3119–3132.

Gavazzi G, Dolfini S, Allegra D, Castiglioni P, Todesco G, Hoxha M (1997). Dap (Defective aleurone pigmentation) mutations affect maize aleurone development. Mol Gen Genet 256: 223–230.

Geeta R (2003). The origin and maintenace of nucelar endosperm: viewing development through a phylogenetic lens. Proc R Soc Lond B 270: 29–35.

Gehring M, Missirian V, Henikoff S (2011). Genomic analysis of parent-of-origin allelic expression in Arabidopsis thaliana seeds. PLoS One 6: e23687.

Geisler-Lee J, Gallie DR (2005). Aleurone cell identity is suppressed following connation in maize kernels. Plant Physiol 139: 204–212.

Giese H (1992). Replication of DNA during barley endosperm development. Can J Bot 70: 313–318.

Gillies SA, Futardo A, Henry RJ (2012). Gene expression in the developing aleurone and starchy endosperm of wheat. Plant Biotechnol J DOI: 10.1111/j.1467-7652.2012.00705.x.

Gomez E, Royo J, Guo Y, Thompson R, Hueros G (2002). Establishment of cereal endosperm expression domains: identification and properties of a maize transfer cell-specific transcription factor, ZmMRP-1. Plant Cell 14: 599–610.

Gomez E, Royo J, Muniz LM, Sellam O, Paul W, Gerentes D, Barrero C, Lopez M, Perez P, Hueros G (2009). The maize transcription factor myb-related protein-1 is a key regulator of the differentiation of transfer cells. Plant Cell 21: 2022–2035.

Grafi G, Larkins BA (1995). Endoreduplication in maize endosperm: involvement of M phase-promoting factor inhibition and induction of S phase-related kinases. Science 269: 1262–1264.

Groot EP, van Caeseele LA (1993). The development of the aleurone layer in canola (Brassica napus). Can J Bot 71: 1193–1201

Gruis DF, Guo H, Selinger D, Tian Q, Olsen OA (2006). Surface position, not signaling from surrounding maternal tissues, specifies aleurone epidermal cell fate in maize. Plant Physiol 141: 898–909.

Guignard L (1901). La double fècondation dans le mais. J Bot 15: 37–50.

Gutierrez-Marcos JF, Costa LM, Biderre-Petit C, Khbaya B, O'Sullivan DM, Wormald M, Perez P, Dickinson HG (2004). Maternally expressed gene1 is a novel maize endosperm transfer cell-specific gene with a maternal parent-of-origin pattern of expression. Plant Cell 16: 1288–1301.

Gutierrez-Marcos JF, Costa LM, Evans MMS (2006). Maternal gametophytic *baseless1* is required for development of the central cell and early endosperm patterning in maize (Zea mays). Genetics 174: 317–329.

Gutierrez-Marcos JF, Dal Pra M, Giulini A, Costa LM, Gavazzi G, Cordelier S, Sellam O, Tatout C, Paul W, Perez P, Dickinson HG, Consonni G (2007). Empty pericarp4 encodes a mitochondrion-targeted pentatricopeptide repeat protein necessary for seed development and plant growth in maize. Plant Cell 19: 196–210.

Haig D (2004). Genomic imprinting and kinship: how good is the evidence? Annu Rev Genet 38: 553–585.

Hansen M, Friis C, Bowra S, Holm PB, Vincze E (2009). A pathway-specific microarray analysis highlights the complex and co-ordinated transcriptional networks of the developing grain of field-grown barley. J Exp Bot 60: 153–167.

Hatsugai N, Kuroyanagi M, Yamada K, Meshi T, Tsuda S, Kondo M, Nishimura M, Hara-Nishimura I (2004). A plant vacuolar protease, VPE, mediates virus-induced hypersensitive cell death. Science 305: 855–858.

Hsieh TF, Shin J, Uzawa R, Silva P, Cohen S, Bauer MJ, Hashimoto M, Kirkbride RC, Harada JJ, Zilberman D, *et al.* (2011). Regulation of imprinted gene expression in Arabidopsis endosperm. Proc Natl Acad Sci U S A 108: 1755–1762.

Hueros G, Varotto S, Salamini F, Thompson RD (1995). Molecular characterization of BET1, a gene expressed in the endosperm transfer cells of maize. Plant Cell 7: 747–757.

Jestin L, Ravel C, Auroy S, Laubin B, Perretant M-R, Pont C, Charmet G (2008). Inheritance of the number and thickness of cell layers in barley aleurone tissue (Hordeum vulgare L.): an approach using F2–F3 progeny. Theor Appl Genet 116: 991–1002.

Jin P, Guo T, Becraft PW (2000). The maize CR4 receptor-like kinase mediates a growth factor-like differentiation response. Genesis 27: 104–116.

Jones RL (1969). The fine structure of barley aleurone cells. Planta 85: 359–375.

Jung KH, An G, Ronald PC (2008). Towards a better bowl of rice: assigning function to tens of thousands of rice genes. Nat Rev Genet 9: 91–101.

Kawakatsu T, Yamamoto MP, Touno SM, Yasuda H, Takaiwa F (2009). Compensation and interaction between RISBZ1 and RPBF during grain filling in rice. Plant J 59: 908–920.

Kermicle JL (1970). Dependence of the R-mottled aleurone phenotype in maize on mode of sexual transmission. Genetics 66: 69–85.

Kieselbach TA, Walker ER (1952). Structure of certain specialized tissues in the kernel of corn. Am J Bot 39: 561–569.

Kladnik A, Chourey Prem S, Pring DR, Dermastia M (2006). Development of the endosperm of Sorghum bicolor during the endoreduplication-associated growth phase. J Cereal Sci 43: 209–215.

Kowles RV, Phillips RL (1988). Endosperm development in maize. Int Rev Cytol 112: 97–136.

Kowles RV, Phillips RL (1985). DNA amplification patterns in maize (Zea mays) endosperm nuclei during kernel development. Proc Natl Acad Sci U S A 82: 7010–7014.

Lai J, Dey N, Kim CS, Bharti AK, Rudd S, Mayer KFX, Larkins BA, Becraft P, Messing J (2004). Characterization of the maize endosperm transcriptome and its comparison to the rice genome. Genome Res 14: 1932–1937.

Laudencia-Chingcuanco DL, Stamova BS, You FM, Lazo GR, Beckles DM, Anderson OD (2007). Transcriptional profiling of wheat caryopsis development using cDNA microarrays. Plant Mol Biol 63: 651–668.

Leiva-Neto JT, Grafi G, Sabelli PA, Woo YM, Dante RA, Maddock S, Gordon-Kamm WJ, Larkins BA (2004). A dominant negative mutant of cyclin-dependent kinase A reduces endoreduplication but not cell size or gene expression in maize endosperm. Plant Cell 16: 1854–1869.

Lid SE, Al RH, Krekling T, Meeley RB, Ranch J, Opsahl-Ferstad HG, Olsen OA (2004). The maize disorganized aleurone layer 1 and 2 (dil1, dil2) mutants lack control of the mitotic division plane in the aleurone layer of developing endosperm. Planta 218: 370–378.

Lid SE, Gruis D, Jung R, Lorentzen JA, Ananiev E, Chamberlin M, Niu XM, Meeley R, Nichols S, Olsen OA (2002). The defective kernel 1 (dek1) gene required for aleurone cell development in the endosperm of maize grains encodes a membrane protein of the calpain gene superfamily. Proc Natl Acad Sci U S A 99: 5460–5465.

Liu X, Guo T, Wan X, Wang H, Zhu M, Li A, Su N, Shen Y, Mao B, Zhai H, *et al.* (2010). Transcriptome analysis of grain-filling caryopses reveals involvement of multiple regulatory pathways in chalky grain formation in rice. BMC Genomics 11: 730.

Luo M, Taylor JM, Spriggs A, Zhang H, Wu X, Russell S, Singh M, Koltunow A (2011). A genome-wide survey of imprinted genes in rice seeds reveals imprinting primarily occurs in the endosperm. PLoS Genet 7: e1002125.

Magnard JL, Le Deunff E, Domenech J, Rogowsky PM, Testillano PS, Rougier M, Risueno MC, Vergne P, *et al.* (2000). Endosperm-specific features at the onset of microspore embryogenesis in maize. Plant Mol Biol 44: 559–574.

Mayer KFX, Martis M, Hedley PE, Šimková H, Liu H, Morris JA, Steuernagel B, Taudien S, Roessner S, Gundlach H, Kubaláková M, Suchánková P, Murat F, Felder M, Nussbaumer T, Graner A, Salse J, Endo T, Sakai H, Tanaka T, Itoh T, Sato K, Platzer M, Matsumoto T, Scholz U, Doležel J, Waugh R, Stein N (2011). Unlocking the barley genome by chromosomal and comparative genomics. Plant Cell 23: 1249–1263.

McCarty DR, Settles M, Suzuki M, Tan B, Latshaw S, Porch T, Robin K, Baier J, Avigne W, Lai J, *et al.* (2005). Steady-state transposon mutagenesis in inbred maize. Plant J 44: 52–61.

McElver J, Tzafrir I, Aux G, Rogers R, Ashby C, Smith K, Thomas C, Schetter A, Zhou Q, Cushman MA, *et al.* (2001). Insertional mutagenesis of genes required for seed development in Arabidopsis thaliana. Genetics 159: 1751–1763.

Meinke D, Muralla R, Sweeney C, Dickerman A (2008). Identifying essential genes in Arabidopsis thaliana. Trends Plant Sci 13: 483–491.

Mohanty A, Luo A, DeBlasio S, Ling X, Yang Y, Tuthill DE, Williams KE, Hill D, Zadrozny T, Chan A, Sylvester AW, Jackson D (2009). Advancing cell biology and functional genomics in maize using fluorescent protein-tagged lines. Plant Physiol 149: 601–605.

Morrison IN, Kuo J, O'Brien TP (1975). Histochemistry and fine structure of developing wheat aleurone cells. Planta 123: 105–116.

Myers AM, James MG, Lin Q, Yi G, Stinard PS, Hennen-Bierwagen TA, Becraft PW (2011). Maize opaque5 encodes monogalactosyldiacylglycerol synthase and specifically affects galactolipids necessary for amyloplast and chloroplast function. Plant Cell 23: 2331–2347.

Nawaschin S (1898). Resulte eine revison der Befruchtungsvorgange bei Lilium martagon und Ritillaria tenella. Bull Acad Imperial Sci, StPeterburg 9: 377–382.

Neuffer MG, Sheridan WF (1980). Defective kernel mutants of maize. 1. Genetic and lethality studies. Genetics 95: 929–944.

Nguyen HN, Sabelli PA, Larkins BA (2007). Endoreduplication and programmed cell death in the cereal endosperm. In: Olsen OA (ed). *Endosperm: Developmental and Molecular Biology*. Springer, Berlin, Germany, pp 21–43.

Ogawa M, Shinohara H, Sakagami Y, Matsubayashi Y (2008) Arabidopsis CLV3 Peptide Directly Binds CLV1 Ectodomain. Science 319: 294.

Olsen LT, Divon HH, Al R, Fosnes K, Lid SE, Opsahl-Sorteberg HG (2008). The defective seed5 (des5) mutant: effects on barley seed development and HvDek1, HvCr4, and HvSal1 gene regulation. J Exp Bot 59: 3753–3765.

Olsen OA (2004a). Nuclear endosperm development in cereals and Arabidopsis thaliana. Plant Cell 16: S214–S227.

Olsen OA (2004b). Dynamics of maize aleurone cell formation: The "surface-"rule. Maydica 49: 37–40.

Olsen OA, Weschke W (2012). Genetic and molecular aspects of barley grain development. Barley Chemistry and Technology, AACC series of monographs on grain chemistry and technology. AACC International PRESS, St. Paul, Minnesota.

Opsahl-Ferstad HG, Le Deunff E, Dumas C, Rogowsky PM (1997). ZmEsr, a novel endosperm-specific gene expressed in a restricted region around the maize embryo. Plant J 12: 235–246.

Philippar K, Buchsenschutz K, Edwards D, Loffler J, Luthen H, Kranz E, Edwards KJ, Hedrich R (2006). The auxin-induced K(+) channel gene Zmk1 in maize functions in coleoptile growth and is required for embryo development. Plant Mol Biol 61: 757–768.

Pignocchi C, Doonan JH (2011). Interaction of a 14-3-3 protein with the plant microtubule-associated protein EDE1. Ann Bot 107: 1103–1109.

Pignocchi C, Minns GE, Nesi N, Koumproglou R, Kitsios G, Benning C, Lloyd CW, Doonan JH, Hills MJ (2009). ENDOSPERM DEFECTIVE1 is a novel microtubule-associated protein essential for seed development in arabidopsis. Plant Cell 21: 90–105.

Ramachandran C, Raghavan V (1989). Changes in nuclear DNA content of endosperm cells during grain development in rice (Oryza sativa). Ann Bot 64: 459–468.

Ramage RT, Crandall CL (1981). Defective endosperm xenia (dex) mutants. Barley Genet Newslett 11: 32–33.

Reyes FC, Chung T, Holding D, Jung R, Vierstra R, Otegui MS (2011). Delivery of prolamins to the protein storage vacuole in maize aleurone cells. Plant Cell 23: 769–784.

Reyes FC, Sun B, Guo H, Gruis D, Otegui MS (2010). Agrobacterium tumefaciens-mediated transformation of maize endosperm as a tool to study endosperm cell biology. Plant Physiol 153: 624–631.

Sabelli PA, Larkins BA (2009). The development of endosperm in grasses. Plant Physiol 149: 14–26.

Sachs MM (2009). Cereal germplasm resources. Plant Physiol 149: 148–151.

Scanlon MC, James MG, Stinard PS, Myers AM, Roberston DS (1994). Characterization of ten new mutations of the maize Etched-1 locus. Maydica 39: 301–308.

Schel JHN, Kieft H (1986). An ultrastructural study of embryo and endosperm development during in vitro culture of maize ovaries (Zea mays). Can J Bot 64: 2227–2238.

Schel JHN, Kieft H, van Lammeren AAM (1984). Interactions between embryo and endosperm during early developmental stages of maize caryopses (Zea mays). Can J Bot 62: 2842–2853.

Sekhon RS, Lin H, Childs KL, Hansey CN, Buell CR, de Leon N, Kaeppler SM (2011). Genome-wide atlas of transcription during maize development. Plant J 66: 553–563.

Serna A, Maitz M, O'Connell T, Santandrea G, Thevissen K, Tienens K, Hueros G, Faleri C, Cai G, Lottspeich F, *et al.* (2001). Maize endosperm secretes a novel antifungal protein into adjacent maternal tissue. Plant J 25: 687–698.

Settles AM, Holding DR, Tan BC, Latshaw SP, Liu J, Suzuki M, Li L, O'Brien BA, Fajardo DS, Wroclawska E, Tseung CW, Lai J, Hunter CT 3rd, Avigne WT, Baier J, Messing J, Hannah LC, Koch KE, Becraft PW, Larkins BA, McCarty

DR (2007). Sequence-indexed mutations in maize using the UniformMu transposon-tagging population. BMC Genomics 8: 116.

Shen B, Li C, Min Z, Meeley RB, Tarczynski MC, Olsen OA (2003). sal1 determines the number of aleurone cell layers in maize endosperm and encodes a class E vacuolar sorting protein. Proc Natl Acad Sci U S A 100: 6552–6557.

Sheridan WF, Clark JK (1993). Mutational analysis of morphogenesis of the maize embryo. Plant J 3: 347–358.

Sheridan WF, Neuffer MG (1980). Defective kernel mutants of maize II. Morphological and embryo culture studies. Genetics 95: 945–960.

Sorensen MB, Mayer U, Lukowitz W, Robert H, Chambrier P, Jurgens G, Somerville C, Lepiniec L, Berger F (2002). Cellularisation in the endosperm of Arabidopsis thaliana is coupled to mitosis and shares multiple components with cytokinesis. Development 129: 5567–5576.

Sreenivasulu N, Usadel B, Winter A, Radchuk V, Scholz U, Stein N, Weschke W, Strickert M, Close TJ, Stitt M, *et al.* (2008). Barley grain maturation and germination: metabolic pathway and regulatory network commonalities and differences highlighted by new MapMan/PageMan profiling tools. Plant Physiol 146: 1738–1758.

Sun YJ, Dilkes BP, Zhang CS, Dante RA, Carneiro NP, Lowe KS, Jung R, Gordon-Kamm WJ, Larkins BA (1999). Characterization of maize (Zea mays L.) Wee1 and its activity in developing endosperm. Proc Natl Acad Sci U S A 96: 4180–4185.

Takacs EM, Suzuki M, Scanlon MJ (2012). Discolored1 (DSC1) is an ADP-ribosylation factor-GTPase activating protein required to maintain differentiation of maize kernel structures. Front Plant Sci 3: 115.

Thompson RD, Hueros G, Becker HA, Maitz M (2001). Development and functions of seed transfer cells. Plant Sci 160: 775–783.

Tian Q, Olsen L, Sun B, Lid SE, Brown RC, Lemmon BE, Fosnes K, Gruis DF, Opsahl-Sorteberg HG, Otegui MS, *et al.* (2007). Subcellular localization and functional domain studies of DEFECTIVE KERNEL1 in maize and Arabidopsis suggest a model for aleurone cell fate specification involving CRINKLY4 and SUPERNUMERARY ALEURONE LAYER1. Plant Cell 19: 3127–3145.

Tzafrir I, Pena-Muralla R, Dickerman A, Berg M, Rogers R, Hutchens S, Sweeney TC, McElver J, Aux G, Patton D, *et al.* (2004). Identification of genes required for embryo development in Arabidopsis. Plant Physiol 135: 1206–1220.

van Doorn WG, Beers EP, Dangl JL, Franklin-Tong VE, Gallois P, Hara-Nishimura I, Jones AM, Kawai-Yamada M, Lam E, Mundy J, Mur LAJ, Petersen M, Smertenko A, Taliansky M, Van Breusegem F, Wolpert T, Woltering E, Zhivotovsky B, Bozhkov PV (2011). Morphological classification of plant cell deaths. Cell Death Differ 18: 1241–1246.

Vaughn JC, Whitehouse JM (1971). Seed structure and the taxonomy of the Cruciferae. Bot J Linn Soc 64: 383–409.

Vollbrecht E, Duvick J, Schares JP, Ahern KR, Deewatthanawong P, Xu L, Conrad LJ, Kikuchi K, Kubinec TA, Hall BD, *et al.* (2010). Genome-wide distribution of transposed dissociation elements in maize. Plant Cell 22: 1667–1685.

Wang C, Barry JK, Min Z, Tordsen G, Rao AG, Olsen OA (2003). The calpain domain of the maize DEK1 protein contains the conserved catalytic triad and functions as a cysteine proteinase. J Biol Chem 278: 34467–34474.

Waters AJ, Makarevitch I, Eichten SR, Swanson-Wagner RA, Yeh C-T, Xu W, Schnable PS, Vaughn MW, Gehring M, Springer NM (2011). Parent-of-origin effects on gene expression and DNA methylation in the maize endosperm. Plant Cell 23: 4221–4233.

Wolff P, Weinhofer I, Seguin J, Roszak P, Beisel C, Donoghue MTA, Spillane C, Nordborg M, Rehmsmeier M, Köhler C (2011). High-resolution analysis of parent-of-origin allelic expression in the arabidopsis endosperm. PLoS Genet 7: e1002126.

Wollmann H, Berger F (2011). Epigenetic reprogramming during plant reproduction and seed development. Curr Opin Plant Biol 15: 63–69.

Xue LJ, Zhang JJ, Xue HW (2012). Genome-wide analysis of the complex transcriptional networks of rice developing seeds. PLoS One 7: e31081.

Yi G, Lauter AM, Scott AM, Becraft PW (2011). The thick aleurone1 mutant defines a negative regulation of maize aleurone cell fate that functions downstream of defective kernel. Plant Physiol 156: 1826–1836.

Yi G, Luth D, Goodman TD, Lawrence CJ, Becraft PW (2009). High-throughput linkage analysis of Mutator insertion sites in maize. Plant J 58: 883–892.

Young TE, Gallie DR (1999). Analysis of programmed cell death in wheat endosperm reveals differences in endosperm development between cereals. Plant Mol Biol 39: 915–926.

Young TE, Gallie DR, DeMason DA (1997). Ethylene-mediated programmed cell death during maize endosperm development of wild-type and shrunken2 genotypes. Plant Physiol 115: 737–751.

Zhang M, Zhao H, Xie S, Chen J, Xu Y, Wang K, Zhao H, Guan H, Hu X, Jiao Y, Song W, Lai J (2011). Extensive, clustered parental imprinting of protein-coding and noncoding RNAs in developing maize endosperm. Proc Natl Acad Sci U S A 108: 20042–20047.

4 Epigenetic Control of Seed Gene Imprinting

Christian A. Ibarra, Jennifer M. Frost, Juhyun Shin, Tzung-Fu Hsieh, and Robert L. Fischer

Introduction

Genomic imprinting is a process found in mammals and flowering plants and is defined as the monoallelic expression of a gene in a parent-of-origin dependent manner. Imprinting is established epigenetically because the two alleles of an imprinted gene are identical in DNA sequence but differ in expression state. Epigenetic marks, mainly DNA methylation and histone modifications, are often associated with repressed chromatin and reside on specific parental alleles resulting in transcriptional silencing (Allis *et al.*, 2007). In both plants and mammals, most genomic imprints are established after meiosis in the gametes and, for the most part, remain after fertilization. To date, the number of imprinted genes in mammals is ~120, representing <1% of the genes in the genome (Glaser *et al.*, 2006).

In the case of the flowering plant *Arabidopsis thaliana*, the current number of imprinted genes is ~50, a finding resulting from several more recent high-throughput allele-specific transcriptome studies (Tiwari *et al.*, 2010; Gehring *et al.*, 2011; Hsieh *et al.*, 2011; McKeown *et al.*, 2011; Wolff *et al.*, 2011). Similar genome-wide imprinting screens in maize (Waters *et al.*, 2011) and rice (Luo *et al.*, 2011) have also greatly increased the number of known imprinted genes in these agriculturally significant monocots. As described subsequently, the plant-imprinting field has graduated to the "omics" era, as a considerable number of transcriptome, methylome, and epigenome studies have appeared in the literature in the last few years.

Genomic Imprinting and Parental Conflict Theory

Parental Conflict Theory

Genomic imprinting is a process that results in only a single allele of a gene being expressed. As such, it seems a precarious evolutionary choice, reducing the robustness of the genome and potentially reducing the fitness of the organism. This is especially so in the case of imprinted genes, which have been shown to be indispensible for mammalian and plant development (Huh *et al.*, 2008; Bartolomei and Ferguson-Smith, 2011; Wilkins and Ubeda, 2011).

Several theories have been proposed attempting to explain the evolutionary origins of genomic imprinting (Haig, 2000; Day and Bonduriansky, 2004; Wolf and Hager, 2006). However, the most

Seed Genomics, First Edition. Edited by Philip W. Becraft.

widely accepted theory is the kinship or parental conflict theory (Haig, 2000). This theory suggests that imprinting arose from a conflict between parents regarding how much of the mother's nutrients should support her offspring from potentially multiple fathers (Moore and Haig, 1991). From an evolutionary perspective, there is no drive for the father to favor maternal nutrient distribution to the offspring of other broods. His main aim is to promote the growth of the fetus and the placenta, possibly at the cost of other fetuses, which may not be his, both in the current and in future pregnancies, maximizing the propagation of his genes. Conversely, the mother wishes to curb the growth of the fetus and the placenta, preserving her fecundity and the viability of other fetuses she may be carrying or wishes to carry in the future, maximizing the propagation of her genes.

This conflict is potentially powerful enough to promote the evolution of unequal gene expression between selected parental alleles over time. Genes promoting the increased allocation of maternal resources to the offspring would be expressed from the paternally inherited genome (paternally expressed genes [PEGs]) and repressed on the maternally inherited genome. Genes promoting a more equal distribution of maternal resources to the offspring would be expressed on the maternally inherited genome (maternally expressed genes [MEGs]) and repressed on the paternally inherited genome.

Parental Conflict Occurs in Tissues That Support the Embryo

The mammalian placenta and flowering plant endosperm are similar in function, being directly responsible for bringing maternal and embryonic circulation into contact, facilitating nutrient exchange and determining resource allocation to the embryo. They both serve as important foci for potential parental conflict (Frost and Moore, 2010; Li and Dickinson, 2010). The evolution of the placenta has been significantly advantageous to mammals, providing the fetus with a longer gestation period resulting in a more developed offspring at birth (Constancia *et al.*, 2004). Of the ~120 imprinted genes currently known in mammals, most are expressed in both the fetus and the placenta. A subset of imprinted genes are expressed only in the placenta, and some genes are more widely expressed but imprinted only in the placenta and not the fetus (Lewis *et al.*, 2004; Coan *et al.*, 2005; Monk *et al.*, 2009). The ontology of many mammalian imprinted genes shows them to be important in regulating the placenta's ability to transfer nutrients and to regulate embryo development (Constancia *et al.*, 2004; Gregg *et al.*, 2010).

The flowering plant endosperm acts as a conduit between the sporophyte and the embryo, working to transfer maternal nutrients to the embryo. In addition, the endosperm has been shown to synthesize extensive amounts of lipids, proteins, and starch (Olsen, 2004; Huh *et al.*, 2008). In dicots, such as *Arabidopsis*, the endosperm is mostly consumed during seed development by the embryo, whereas in monocots it persists in the mature seed and is consumed by the embryo during germination (Olsen, 2004; Brown and Lemmon, 2007). Persistent endosperm provides a selective advantage for monocots (i.e., cereals) allowing for better seed dispersal, long-term dormancy, and effective seedling establishment.

The endosperm has long been recognized for its nutritive value. The cereal endosperm has been highly advantageous to humans, being the primary source of nutrition worldwide, providing at least ≥60% of daily caloric intake (http://www.who.int/nutrition/topics/3_foodconsumption/en/index1.html). Almost all genomic imprinting in plants occurs solely in the endosperm, a situation reminiscent of mammals, where the placenta is the focal organ for imprinting. This highlights the potential importance of the flowering plant endosperm as a tissue that promotes parental conflict.

Parental Conflict Predicts That Imprinted Genes Will Have Embryo-Supporting Functions

In mammals, the physiology of several imprinted genes is strongly supportive of the parental conflict theory. A key example is *insulin-like growth factor 2 (Igf2)*, a highly potent growth factor that is paternally expressed in the mammalian fetus and placenta. In mice, a reduction in *Igf2* expression leads to a 40% reduction in fetal growth (DeChiara *et al.*, 1990). Manipulation of the *Igf2/H19* imprinting control region to promote biallelic *Igf2* transcription results in fetal and placental overgrowth (Leighton *et al.*, 1995), a molecular etiology and phenotype mirrored by Beckwith-Wiedemann syndrome in humans (Cooper *et al.*, 2005). This overgrowth is also true globally, demonstrated by the phenotypes resulting from the creation of uniparental mice. Concepti carrying paternal-only genetic material consist of vestigial embryonic material, with overgrowth of placental trophoblast cells, and concepti with maternal-only genetic material consist of a disorganized embryo, with little or no placental material (McGrath and Solter, 1984; Surani *et al.*, 1984).

Some imprinted genes found in the *Arabidopsis* endosperm appear also to have functions in accordance with the parental conflict theory. Two of the best examples are the well-studied MEGs *MEDEA* (*MEA*) (Grossniklaus *et al.*, 1998; Kiyosue *et al.*, 1999) and *FERTILIZATION INDEPENDENT SEED 2* (*FIS2*) (Luo *et al.*, 1999). *MEA* encodes as SET-domain protein that is homologous to *Drosophila Enhancer of Zeste [E(z)]*, whereas *FIS2* encodes a zinc-finger transcription factor homologous to *Drosophila Suppressor of Zeste 12 [Su(z)12]* (Grossniklaus *et al.*, 1998; Luo *et al.*, 1999). SET domains in *Arabidopsis* are responsible for the post-translational methylation of lysine residues on histone tails. In *Drosophila*, both E(z) and Su(z)12 along with other proteins form the polycomb group (PcG) complex, which carries out transcriptional gene silencing through trimethylation (me3) of the 27th lysine residue on the tail of histone H3 (H3K27). Remarkably, loss-of-function mutations in both *MEA* and *FIS2* produce autonomous endosperm without ever undergoing fertilization. This endosperm is severely compromised as a consequence of failing to undergo the distinct phases of endosperm development (i.e., nuclear proliferation, cellularization, and maturation) and is marked by excessive nuclear proliferation, a loss of anterior and posterior polarity, and an aborted embryo. Both *mea* and *fis2* phenotypes suggest the normal function of these genes is to suppress central cell proliferation until fertilization, a function that is reflective of the parental conflict theory because both *MEA* and *FIS2* are MEGs (Huh *et al.*, 2008).

More recently, high-throughput genome-wide imprinting studies in *Arabidopsis* resulted in the discovery of additional MEGs and PEGs (Tiwari *et al.*, 2010; Gehring *et al.*, 2011; Hsieh *et al.*, 2011; McKeown *et al.*, 2011; Wolff *et al.*, 2011). These experiments revealed imprinted genes that potentially function to regulate endosperm nutrient flow, embryo growth, and development. Many of these imprinted genes encoded transcription factors, auxin and ethylene signaling related proteins, and components of the ubiquitin-26S proteasome pathway. Epigenetic regulatory proteins important for histone and DNA methylation and small RNA pathway proteins were also found to be imprinted (Hsieh *et al.*, 2011). These findings suggest the parental conflict between maternal nutrient allocation to the embryo manifests at different regulatory levels, including the chromatin level, post-translational level, and at a protein-protein interaction level (Tiwari *et al.*, 2010; Gehring *et al.*, 2011; Hsieh *et al.*, 2011; McKeown *et al.*, 2011; Wolff *et al.*, 2011).

Epigenetic Regulators of *Arabidopsis* Imprinting

DNA Methylation

Eukaryotic DNA methylation is the addition of a methyl group to the 5-carbon of cytosine. The process is enzymatic, carried out by DNA methyltransferases that function to transfer a methyl group from the universal methyl donor S-adenosyl methionine (SAM) and covalently attach it to the cytosine base, resulting in the formation of 5-methyl cytosine (5mC). This slight modification to the DNA molecule can have a profound effect, sufficient to cause gene silencing. Various models have been proposed to explain the repressive nature of this DNA methylation mark (Cedar and Bergman, 2012). One potential mechanism in which DNA methylation may lead to transcriptional repression is by acting as a blocking agent that physically inhibits the interaction of DNA with transcription factors and other DNA binding proteins essential for transcription (Watt and Molloy, 1988). Alternatively, DNA methylation has also been shown to be strongly associated with the formation of repressive heterochromatin (Klose and Bird, 2006). In this case, the methylation mark is recognized by proteins possessing methyl binding domains (MBDs), which bind to the methylated site and further recruit other proteins, many of which are chromatin remodelers that change the local chromatin architecture into a more heterochromatic state (Cedar and Bergman, 2012).

Eukaryotic DNA methylation is involved in diverse biological processes (He *et al.*, 2011). In mammals, DNA methylation has been shown to be involved in cellular differentiation, X-chromosome inactivation, transposon silencing, promoter silencing, and imprinting (Reik, 2007). In plants, the repressive effects of DNA methylation are used routinely to silence transposable elements, functioning as a genomic defense mechanism (Zemach *et al.*, 2010b). DNA methylation is critical to organism development because null mutations in some of the DNA methyltransferases display embryonic lethality in mammals (Li *et al.*, 1992) and aberrant morphological effects in plants (Zemach *et al.*, 2010b). In both plants and mammals, DNA methylation is critical to genomic imprinting, carrying out allele-specific silencing (Li *et al.*, 1993).

DNA methylation in plants occurs in the contexts of CG, CHG, and CHH (H = A, C, or T) (Cokus *et al.*, 2008). This is strikingly different to the situation in mammals, where DNA methylation occurs overwhelmingly in the CG context, having only a small amount of CHH methylation present in the germline (Ramsahoye *et al.*, 2000; Lister *et al.*, 2009). Studies over the last decade have revealed the varied methylation contexts to be the result of disparate underlying molecular mechanisms, involving distinct DNA methyltransferases, small RNAs (sRNAs), and protein factors (Law and Jacobsen, 2010).

DNA methylation in the CG context is carried out by methyltransferase-1 (MET1), which is a homologue of the mammalian DNA methyltransferase-1 (DNMT1) (Vongs *et al.*, 1993). Similar to its mammalian homologue, MET1 functions to maintain CG methylation in the genome from one mitotic replication to the next. An important component of CG maintenance methylation is the Set or Ring-Associated domain (SRA) containing protein Variation In Methylation 1 (VIM1) in plants, and Ubiquitin-like plant Homeodomain and Ring Finger Domain-1 (UHRF1) in mammals (Law and Jacobsen, 2010). SRA domains are known to bind hemimethylated DNA, and it is believed that UHRF1 recruits DNMT1 to the replication fork as it binds to the nascent hemimethylated CG sites formed after replication (Bostick *et al.*, 2007). The UHRF1 recruitment of DNMT1 to the replication fork restores the two hemimethylated daughter DNA strands to a fully methylated state (Bostick *et al.*, 2007). Null mutations in *Arabidopsis MET1* generate plants with very little CG methylation that display defects in morphology and fertility (Law and Jacobsen, 2010; Zemach *et al.*, 2010b). A key aspect of the establishment of *Arabidopsis* gene imprinting is the MET1-mediated DNA methylation of specific parental alleles during gametogenesis (Huh *et al.*, 2008).

Maintaining CHG methylation in *Arabidopsis* is achieved via the plant-specific DNA methyltransferase CMT3 and its ability to associate with specific histone modifications (Bartee *et al.*, 2001; Lindroth *et al.*, 2001; Johnson *et al.*, 2007). In addition to a methyltransferase domain, CMT3 possesses a chromodomain that binds specifically to histone 3 lysine 9 dimethylation (H3K9Me2) marks (Bartee *et al.*, 2001; Lindroth *et al.*, 2001) suggesting H3K9Me2 may recruit CMT3 for DNA methylation. The H3K9Me2 mark is established by the histone methyltransferase Supressor of Variegation 3-9 Homologue 4 (SUV4), which contains both a SET and an SRA domain. The SET domain in SUV4 is responsible for dimethylation of H3K9, whereas the SRA domain is believed to bind to methylated DNA. A mechanism for CHG maintenance methylation has been proposed suggesting a reinforcing "loop" between DNA methylation and histone modification, where H3K9me2 marks guide CMT3 to loci to promote DNA methylation, and the DNA methylation marks themselves guide SUV4 to carry out H3K9 dimethylation (Law and Jacobsen, 2010).

The maintenance of CHH methylation is achieved through an elaborate RNAi-mediated process known as RNA-directed DNA methylation (RdDM) (Wassenegger *et al.*, 1994; Henderson and Jacobsen, 2007; Matzke *et al.*, 2009). RdDM is initiated by a plant-specific DNA-dependent RNA polymerase, RNA-Polymerase IV (POLIV), which targets transposons and repetitive elements for transcription, resulting in the production of aberrant single-stranded RNA (ssRNA). The ssRNA serves as a template for RNA replication, carried out by RNA-dependent RNA Polymerase 2 (RDR2) leading to the formation of double-stranded RNA (dsRNA), which is processed into small 24-nucleotide (nt) primary small interfering RNAs (siRNAs) via the Dicer-Like 3 (DCL3) endoribonuclease. The primary siRNAs are further processed by the covalent attachment of a methyl-group onto the 2′-OH group on the 3′-terminal nucleotide of the siRNA, carried out by the SAM-dependent RNA methyltransferase, HUA Enhancer 1 (HEN1). The mature siRNAs are then loaded onto protein complexes that contain Argonaute 4 (AGO4), a member of the Argonaute family of small RNA (sRNA) binding proteins. Argonaute (AGO) proteins are highly conserved across eukaryotes and associate with different classes of sRNAs, such as microRNAs (miRNAs), siRNAs, and the animal related piwi-interacting RNAs (piRNAs) (He *et al.*, 2011). Most AGOs assemble into the RNA-Induced Silencing Complex (RISC), which functions to carry out post-transcriptional gene silencing (Czech and Hannon, 2011). However, in the case of RdDM, the siRNA-loaded AGO4 (siRNA/AGO) further recruits an additional component, RNA Polymerase V (POLV). POLV, similar to POLIV, is a plant-specific DNA-dependent RNA polymerase; however, the role of POLV in RdDM is to transcribe noncoding RNA transcripts from intergenic regions. The function of the POLV-transcript is to tether the siRNA/AGO complex near the DNA, enabled by base-pair complementarity between the AGO-associated siRNA and POLV transcript. The RNA-binding protein Kow Domain-Containing Transcription Factor 1 (KTF1) is then recruited to the nascent complex, further strengthening this base-pair complementarity (Schwartz and Pirrotta, 2007) and finalizing the formation of the mature RdDM-effector complex. The RdDM-effector complex stimulates the recruitment of the *de novo* methyltransferase Domains Rearranged Methyltransferase 2 (DRM2), which establishes and maintains DNA maintenance methylation in the CHH context (Law and Jacobsen, 2010).

Histone Modifications

The histone tails emanating from the core nucleosomes of chromatin are subject to numerous reversible chemical modifications, including methylation, demethylation, acetylation, deacetylation, phosphorylation, ubiquitinylation, sumolyation, and ADP-ribosylation (Kouzarides, 2007). These specific modifications cause structural changes to chromatin architecture and to the genes residing within, leading to transcriptional activation or repression depending on the type of modification (Kouzarides, 2007).

The histone modification most critical to plant imprinting is the trimethylation of H3K27, which leads to transcriptional repression (Pirrotta, 1998; Schwartz and Pirrotta, 2007, 2008). H3K27me3 is carried out by the Polycomb group complex (PcG), a 600-kDa multiprotein complex originally discovered as a major regulator of homeotic *Hox* genes in *Drosophila*. *Hox* genes are responsible for setting up the fundamental body plan and segmentation of the developing fruit fly, and the repressive function of the PcG complex ensures their correct temporospatial expression. Null mutations in members of the PcG protein complex cause striking developmental abnormalities in *Drosophila*, such as the ectopic placement of sex combs on the second and third legs of the adult male fly (Grimaud *et al.*, 2006). *Drosophila* PcG proteins are found in two distinct complexes: PcG Repressive Complex 1 (PRC1) and PcG Repressive Complex 2 (PRC2). However, only the PRC2 has been found in *Arabidopsis* (Gehring *et al.*, 2006; Huh *et al.*, 2008).

In *Arabidopsis*, there are several PRC2 variants that are key players in numerous developmentally related processes, such as the change from vegetative phase to reproductive phase, vernalization, flowering time, endosperm development, and imprinting (Hennig and Derkacheva, 2009). The *Arabidopsis* PRC2 involved in endosperm imprinting (also called the FIE-MEA complex) consists of the SET-domain containing protein Medea (MEA), a C2H2 zinc-finger protein Fertilization Independent Seed 2 (FIS2), and two WD-40 proteins Fertilization Independent Endosperm (FIE) and Multicopy Supressor of IRA 1 (MSI1) (Hennig and Derkacheva, 2009). Similar to its *Drosophila* homologue, PRC2 in *Arabidopsis* appears indispensable for development because null mutations in the PRC2 component genes *MEA, FIE, FIS2*, and *MSI1* all cause dramatic developmental abnormalities (Huh *et al.*, 2008; Hennig and Derkacheva, 2009). As described subsequently, PRC2 in *Arabidopsis* acts to establish the imprinting of many genes.

Active DNA Demethylation

DNA demethylation occurs in many organisms both as a passive and as an active process. Passive DNA demethylation is the removal of 5mC by a failure of the DNA methylation maintenance machinery to methylate a particular cytosine during replication. Continual replication in the absence of DNA methylation gradually dilutes the methylation mark, until eventually methylation is absent. The process of active DNA demethylation involves the enzymatic removal of 5mC (He *et al.*, 2011). Because DNA methylation functions as a repressive epigenetic mark, one could surmise that demethylation would lead to transcriptional derepression (activation). Active DNA demethylation is functionally antagonistic to DNA methylation, generally resulting in gene expression (Ooi and Bestor, 2008). Over the last decade, a considerable amount of work has gone into elucidating the mechanisms of active DNA demethylation.

DNA Glycosylases and Base Excision Repair Pathway

DNA glycosylases function to remove damaged or mispaired bases from genomic DNA and are the first enzymes in the the base excision repair (BER) pathway (Jacobs and Schar, 2012). A damaged base is initially recognized by a DNA glycosylase, which flips the base out from the double helix and hydrolyzes the N-glycosidic bond (ribose-base), resulting in removal of the base from the double helix. This produces an apurinic/apyrimidinic (AP) site, which is further processed by an AP lyase that nicks the sugar-phosphate backbone. The single-strand break in the sugar-phosphate backbone activates an AP endonuclease to create a 3′-hydroxyl, allowing a DNA polymerase to add the correct base. A DNA ligase is then used to seal the nick, completing the BER pathway.

Active DNA Demethylation Is Carried out by 5mC-Specific DNA Glycosylases and Components of BER Pathway

In *Arabidopsis*, active DNA demethylation is carried out by the DEMETER (DME) family of DNA glycosylases and components of the base excision repair (BER) pathway (He *et al.*, 2011). In contrast to other DNA glycosylases, the primary function of the DME glycosylases is to remove 5mC, instead of damaged or mispaired bases (Choi *et al.*, 2002; Gehring *et al.*, 2006). The methylated cytosine is flipped out of the double helix, and its N-glycosidic bond is hydrolyzed by the DME DNA glycosylase enzyme, resulting in the expected AP site and a free methylated base (5mC). DME further nicks the sugar-phosphate backbone at the AP site, as a consequence of being a bifunctional helix-hairpin-helix DNA glycosylase, which has both DNA glycosylase and AP lyase activity (Gehring *et al.*, 2006). At this point, downstream components of the BER pathway are activated (i.e., AP endonuclease, DNA polymerase, and DNA ligase), resulting in replacement of 5mC with unmethylated cytosine.

The DME family of DNA glycosylases consists of Demeter (DME), Repressor Of Silencing 1 (ROS1), Demeter-Like 2 (DML2), and Demeter-Like 3 (DML3). Each has different functions in *Arabidopsis*, and *ROS1*, *DML2*, and *DML3* are expressed broadly throughout the plant, mainly in adult vegetative tissue. A genome-wide methylation analysis of whole plant tissue possessing a triple mutation in *ros1 dme2 dml3* found several hundred hypermethylated loci compared with wild-type tissue, suggesting they are targeted by these three DNA glycosylases in wild-type plants (Penterman *et al.*, 2007; Lister *et al.*, 2009). Most of these demethylated loci appeared at the 5′ and 3′ ends of genes, a pattern opposite to that of DNA methylation. A model was proposed suggesting the three DNA glycosylases worked to "prune" away DNA methylation from the ends of genes. In essence, this 5′ and 3′ gene-adjacent demethylation is proposed to act to remove misdirected DNA methylation that may have made its way into transcriptionally sensitive areas of genes (Penterman *et al.*, 2007; Zhu *et al.*, 2007).

DME Is a Reproductive Companion Cell Demethylase Critical to Imprinting

In contrast to ROS1, DML2, and DML3, which are broadly expressed throughout the plant, DME has been shown to be primarily expressed in reproductive tissues, specifically the companion cells of the developing female and male gametophytes (Choi *et al.*, 2002; Schoft *et al.*, 2011). DME-mediated DNA demethylation occurs in the central cell of the female gametophyte before fertilization, resulting in the maternal expression of several genes (Gehring *et al.*, 2006; Huh *et al.*, 2008). Activity of DME in the central cell is essential for setting up the allele-specific methylation patterns that ultimately establish genomic imprinting in the *Arabidopsis* endosperm (Choi *et al.*, 2002; Gehring *et al.*, 2006). Central cell DME-mediated DNA demethylation is required for subsequent seed development because seeds inheriting a maternal *dme*-mutant allele are nonviable. Thus, *dme*-mutant heterozygous plants generate seeds in which half of the progeny abort, and the other half are normal (Choi *et al.*, 2002).

DME is also expressed in the vegetative cell of the male gametophyte (Schoft *et al.*, 2011). Similar to the *dme*-mutant female gametophyte, a male pollen inheriting a *dme*-mutant allele also displays a mutant phenotype in the form of an inability to germinate properly. Locus-specific methylation analysis revealed that the vegetative cell genome has several loci targeted by DME for DNA demethylation. These loci were the same as the loci in the central cell, suggesting DME may be targeting similar loci in both the central cell and the vegetative cell for active DNA demethylation (Schoft *et al.*, 2011).

Mechanisms Establishing *Arabidopsis* Gene Imprinting

Until recently, there were only ten known imprinted genes in *Arabidopsis* (Bauer and Fischer, 2011). Many genes have been studied in great detail, leading to the elucidation of three key epigenetic strategies that are used to promote parent-of-origin monoallelic silencing: (1) DME-mediated DNA demethylation, (2) MET1-mediated DNA methylation maintenance, and (3) PRC2-mediated transcriptional repression (Bauer and Fischer, 2011). The recent expansion in the number of known *Arabidopsis* imprinted genes provides many more genes whose imprinted expression regulation may be studied (Tiwari *et al.*, 2010; Gehring *et al.*, 2011; Hsieh *et al.*, 2011; McKeown *et al.*, 2011; Wolff *et al.*, 2011). Data from a genome-wide imprinting study show many of these genes fall into one of the three established regulatory groups; however, numerous genes use apparently novel silencing mechanisms (Hsieh *et al.*, 2011).

MEGs Established by DME-Mediated DNA Demethylation

FWA is maternally expressed and paternally silenced in the endosperm, and its imprinted status in *Arabidopsis* has been known for some time. Using expression of a *pFWA::FWA-GFP* transgene, it was shown that *FWA* is expressed only in the central cell and endosperm (Kinoshita *et al.*, 2004). The ectopic expression of FWA protein, a homeodomain-containing transcription factor, causes a late flowering phenotype (Soppe *et al.*, 2000). In the endosperm, an upstream region of the promoter for the maternal *FWA* allele is hypomethylated compared with the paternal *FWA* allele and maternal and paternal *FWA* alleles in other tissues (embryo, seed coat, leaf, and pollen). The hypomethylated maternal *FWA* allele is expressed whereas the hypermethylated paternal *FWA* allele is not, resulting in monoallelic expression in the endosperm. As expected, a *MET1* mutation in the pollen donor results in biallelic *FWA* expression in the endosperm, whereas mutations in *DME* in the female gametophyte result in silencing of the maternal *pFWA::GFP* allele (Kinoshita *et al.*, 2004). Another imprinted gene in *Arabidopsis*, Polycomb-group protein *FIS2*, is regulated in a similar way. An imprinting regulation model can be made from such data; *MET1* maintains 5′ CG methylation in the promoter region of the paternal allele, whereas *DME*, a gene that is expressed in the central cell and not in sperm, demethylates and activates the maternal allele (Figure 4.1*A*). DNA methylation–mediated repression of the paternal allele results, whereas DME-mediated demethylation of the maternal allele results in expression of the maternal allele after fertilization (Jullien *et al.*, 2006).

One can predict that other maternally expressed genes whose regulation is consistent with the above-described model will display biallelic expression in a paternal *met1* background, whereas its expression would be downregulated in a maternal *dme* background. Genome-wide allele-specific transcriptome studies identified nine new imprinted genes that are in accord with these predictions (Table 4.1). Although the number of new imprinted genes is small, they potentially play an important role in endosperm development because their functions indicate they may act to regulate the expression of other genes (Hsieh *et al.*, 2011).

MEGs Established by DME-Mediated DNA Demethylation and PRC2 Repression

The maternally expressed gene *MEA* uses a more complex imprinting expression mechanism. DME in the central cell demethylates CG sites flanking *MEA*, which is required for *MEA* maternal expression. Because paternal *met1* did not release silencing of the *MEA* paternal allele, it is known that *MEA* regulation is not simply dependent on DNA methylation. Instead, mutation of the maternal *FIE* allele (or other components of PRC2) resulted in biallelic expression of *MEA* in the endosperm (Gehring *et al.*, 2006). Another maternally expressed gene, *AfH5*, exhibits similar behavior; its promoter is a target for PRC2, and mutation of maternal *MEA* releases *AfH5* paternal silencing

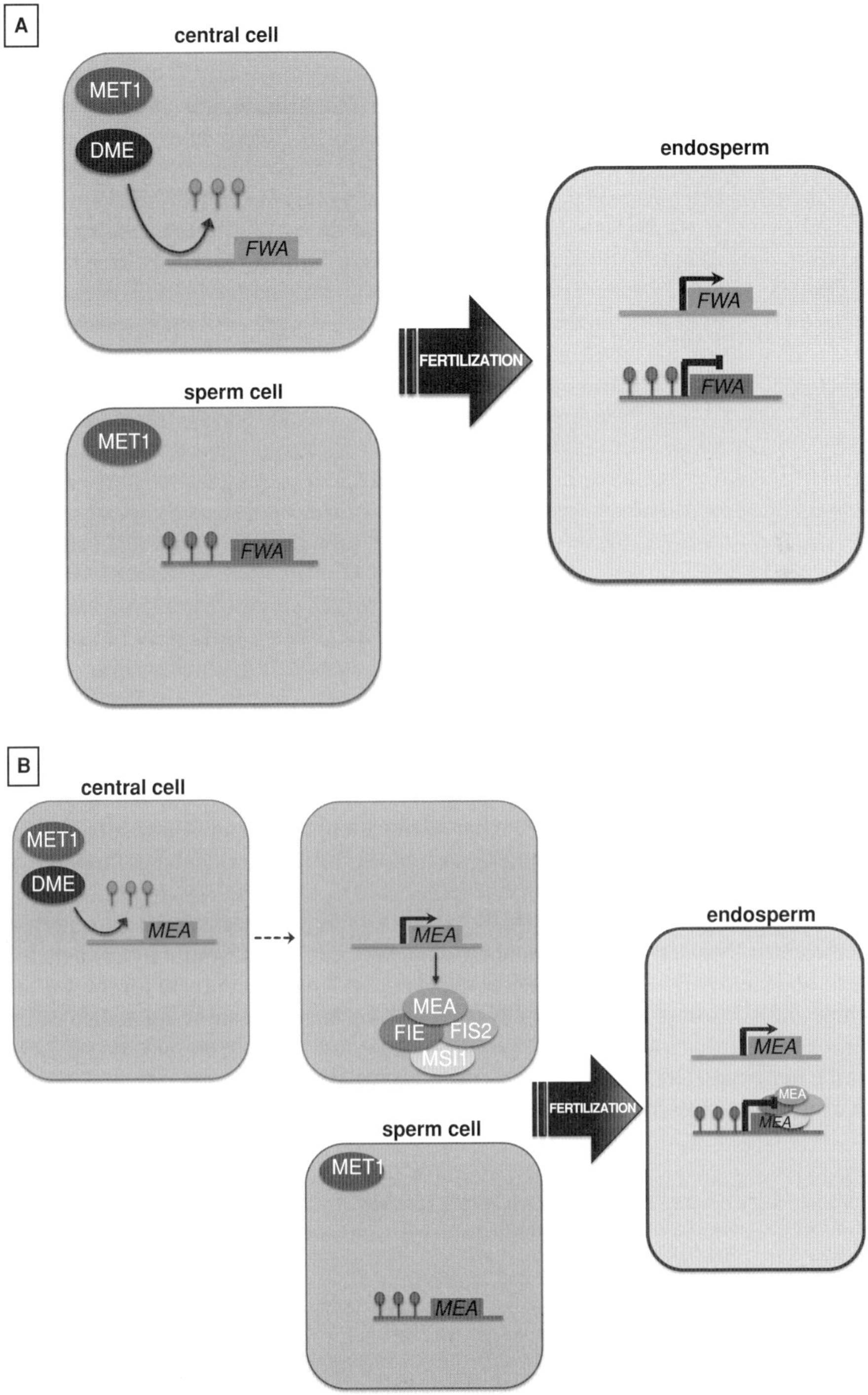

Figure 4.1 *A–C,* Models for *Arabidopsis* genomic imprinting: *FWA* model *(A)*, *MEA* model *(B)*, and *PHE1*-model *(C)*. "Lollipops" represent 5-methyl cytosines. The red bar on the maternal *PHE1* allele in represents the location of the 3′-tandem repeats targeted for DME-mediated DNA demethylation. (For color detail, see color plate section.)

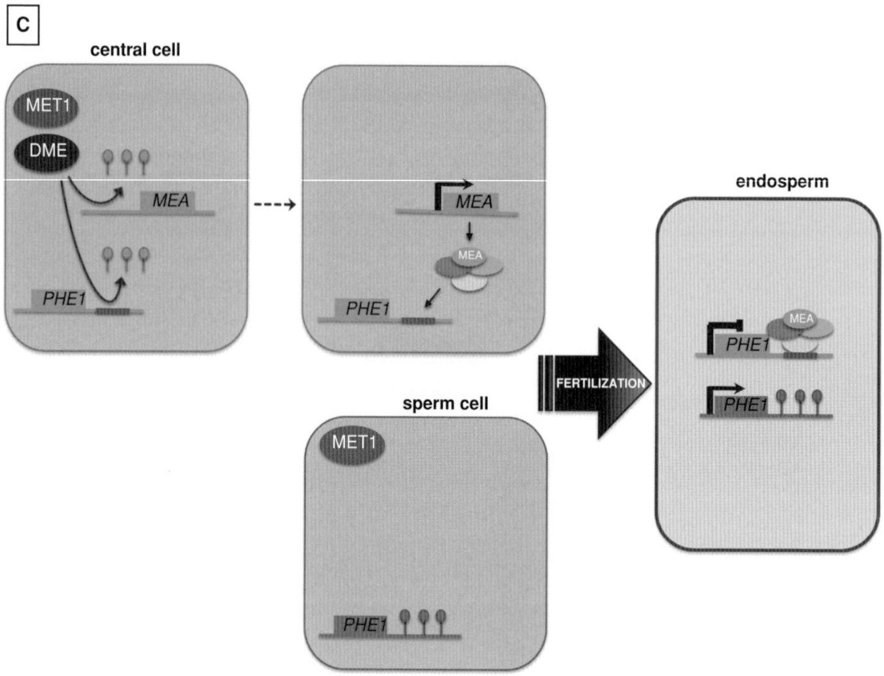

Figure 4.1 (*Continued*)

(Gerald *et al.*, 2009). The model proposed to account for the regulation of these two genes is that before fertilization, DME demethylates and activates *MEA* in the central cell. The *MEA* protein then forms part of the PRC2 complex with *FIE*. After fertilization, in the endosperm, the PRC2 complex silences the paternal allele of imprinted genes, presumably through enrichment of repressive histone modifications on this allele (Figure 4.1*B*).

For imprinted genes that follow the DME/PRC2 model, it is expected that paternal allele silencing will not be disrupted by paternal *MET1* mutations but will be affected by mutations in maternal *FIE*. In addition to *MEA* and *AfH5,* 20 new genes have been found more recently to follow this expression pattern (Table 4.2). Most of these genes were found to have functions related to intermediary metabolism or signaling. However, only two of these genes, *At1g69900* and *At5g47770*, became biallelic in the maternal *DME* mutant, as was the case for *MEA*. Because DME is needed for *MEA* and *FIS2* (maternal) expression, this means that 18 of the new FIE-related MEGs are paternally

Table 4.1 New MEGs Established by DME-Mediated DNA Demethylation

ATMYB3R2	at4g00540	Transcription factor
ERF/AP2	At4g31060	Transcription factor
–	At1g59930	PHERES-like imprinted transcription factor-related (truncated)
JLO	At4g00220	Transcription factor, regulates PIN expression for auxin transport
EIN2	At5g03280	ethylene hormone signaling
DRB2	At2g28380	DS RNA-binding, Component of sRNA pathway
SUVH8	At2g24740	SET-domain protein (H3K9 methyltransferase)
SDC	At2g17690	F-box (targets protein for ubiquitylation, sending it to the 26S proteasome)
MRU1	At5g35490	Unknown function. Overexpressed in non-CG met mutant

Table 4.2 New MEGs Regulated by DME-Mediated DNA Demethylation and PRC2 Complex

–	At1g08050	Zinc finger family (C3HC4-type RING finger)
–	At1g21790	TRAM, LAG1 and CLN8 lipid-sensing domain containing protein
–	At1g69900	Actin cross-linking protein
–	At1g76250	–
–	At1g76820	eukaryotic translation initiation factor 2 family binding to GTP, have GTPase activity
AtSKP2	At1g77000	Homolog to human SKP2, may be involved in degradation of a CDK inhibitor
–	At2g17990	–
ADS2	At2g31360	Homolog to acyl desaturase in cyanobacteria, yeast and mammals.
–	At3g17250	PP2 (Protein phosphastase 2C)-related
–	At3g22810	PI (Phosphoinositide-binding) protein involved in signal transduction
AtTKP5	At4g01840	K^+ channel family protein
ACX1	At4g16760	Acyl-CoA oxidase involved in jasmonate biosynthesis
CAND6	At5g02630	Lung seven transmembrane receptor family
–	At5g03370	Acylphosphatase
–	At5g22920	(CHY-type) Zinc finger protein
–	At5g24460	–
–	At5g42235	Defensin-like (DEFL) family protein
FPS1	At5g47770	Farnesyl diphosphate synthase1 involved in isopropenoid biosynthesis
AGP22	At5g53250	Arabinogalactan protein22
ENODL1	At5g53870	Early nodulin-like protein1 that carry electron in plasma membrane

silenced in the absence of MEA and FIS2 components of PRC2. One possible explanation for the more limited effect of the *dme* mutation compared with the *fie* mutation is that *FIE* is a single-copy gene required for all PRC2 complex formation (Ohad *et al.*, 1999), whereas MEA and FIS2 are both interchangeable with other PcG family members (i.e., Swinger and Curly Leaf for MEA; Vernalization 2, and Embryonic Flower 2 for FIS2), which might provide redundant PRC2 functionality in a *dme* mutant background.

PEGs Established by DME-Mediated DNA Demethylation and PRC2 Repression

PHERES1 (PHE1) was the first paternally expressed imprinted gene to be discovered in *Arabidopsis*, and its analysis resulted in a paradigm for paternal-specific expression (Kohler *et al.*, 2012). *PHE1* is a type I MADS-box transcription factor that is expressed after fertilization in the chalazal endosperm and repressed in other parts of the seed by PRC2. Mutation of *PHE1* does not result in seed abortion, but *phe1* partially rescues *mea*, suggesting that the seed abortion phenotype of *mea* is partially due to the failure of PRC2 to reduce the dosage of *PHE1* through transcription of only one allele (Kohler *et al.*, 2012). Disruption of the DNA methylation pattern on the maternal *PHE1* allele did not reactivate the silenced maternal allele. On the contrary, paternal *MET1* mutations decreased paternal *PHE1* expression after fertilization. The silent maternal *PHE1* allele is maintained by PRC2, and loss of maternal PRC2 results in loss of *PHE1* imprinting (Kohler *et al.*, 2005). DNA methylation of tandem repeats located 2 kb downstream (3′ side) of the *PHE1* coding region is required for *PHE1* expression (Villar *et al.*, 2009). A model was proposed suggesting the downstream *PHE1* repeats on the maternal allele are demethylated by DME in the central cell, allowing the binding of PRC2 and repression of the *PHE1* maternal allele. In contrast, the paternal allele is expressed because its 3′-tandem repeats remained methylated, preventing PRC2 binding (Figure 4.1*C*) (Makarevich *et al.*, 2006).

Table 4.3 New PEGs Regulated by DME-Mediated DNA Demethylation and PRC2 Complex

–	At2g36560	Contain DNA binding domain
YUC10	At1g48910	Flavin monooxygenase related to auxin biosynthesis
VIM5	At1g57800	maintenance of DNA methylation
–	At4g11940	–
–	At5g63740	–
SUVH7	At1g1770	SET-domain protein (H3K9 methyltransferase)

PEGs that follow the *PHE1* model would be expected to become biallelically expressed in endosperm inheriting a maternal *fie* and *dme* and downregulated in endosperm inheriting a paternal *met1* mutant allele. Among nine recently identified PEGs, all but one (VIM5) became biallelically expressed in *fie* endosperm and downregulated in *met1* (Table 4.3). Five genes showed biallelic expression in *dme,* whereas three genes, *At1g31640 (AGL92), At1g660410* (encodes an F-box protein), and *At2g21930* (encodes an F-box protein), are still imprinted in *dme* mutant endosperm. This finding suggests that most PEGs are regulated in a manner similar to *PHE1. VIM5*, related to *VIM1* and expressed only in the endosperm, seems to be strictly regulated by PRC2, whereas *At1g660410* and *At2g21930* are not regulated by *DME.*

Imprinting in the Embryo

Although genes with imprinted expression in the endosperm are plentiful, there are only a few examples of imprinted genes in the plant embryo. A study by Jahnke and Scholten (2009) showed that in reciprocal crosses of different maize inbred lines, only the maternal transcripts of *maternally expressed in embryo 1 (mee1)* are detected in both embryo and endosperm. The expressed maternal alleles of *mee1* correlate with hypomethylation in the promoter and coding sequences of the *mee1* gene. The silent paternal alleles are hypermethylated at both CG and CHG contexts suggesting that allele-specific differential methylation plays a role in the imprinted expression of *mee1* in the embryo and endosperm. The *mee1* gene is demethylated and expressed in the maize central cell but methylated and silent in both the sperm and the egg cell. On fertilization, the maternal *mee1* allele rapidly becomes demethylated in the embryo, whereas the paternal allele remains methylated. This observation suggests the presence of epigenetic marks other than DNA methylation that allow the zygote to differentiate between the parental alleles of this gene, selecting only the maternal *mee1* allele for active demethylation. In contrast to the situation in the endosperm, differential methylation of the two parental *mee1* alleles in the embryo does not persist throughout embryogenesis, and by 16 days after pollination, the maternal *mee1* allele has become methylated and silent.

The occurrence of imprinted expression during early embryo development, although rare, might not be specific to maize. Nodine and Bartel (2012) used high-throughput RNA-sequencing techniques to sequence the transcripts of hybrid embryos from crosses between two polymorphic *Arabidopsis* accessions. They found that despite equal contribution from both parental genomes for 1- to 32-cell stage embryos, a small group of genes clearly display parentally biased expression patterns (77 maternally enriched genes and 45 paternally enriched genes) (Nodine and Bartel, 2012). It is thought that the imprinted expression of these genes in early *Arabidopsis* embryogenesis does not persist in later seed development because similar genome-wide high-throughput sequencing approaches assessing imprinted genes at later stages of seed development identified only very few parentally biased transcripts in the embryo (Gehring *et al.*, 2011; Hsieh *et al.*, 2011).

Imprinting in Monocots

More recent high-throughput imprinting studies in *Zea mays* (Waters *et al.*, 2011; Zhang *et al.*, 2011) and *Oryza sativa* (Luo *et al.*, 2011) have greatly increased the number of imprinted genes in these agriculturally important monocots. Similar to *Arabidopsis*, gene imprinting in cereals almost exclusively occurs in the endosperm (Jahnke and Scholten, 2009).

Only a few genes were found to be mutually imprinted in *Arabidopsis*, maize, and rice (Waters *et al.*, 2011). However, because only one third to one half of the genes assessed in each species were analyzed for imprinting (a consequence of SNP availability and read coverage (Waters *et al.*, 2011)), conclusions made from cross-species comparisons derived from gene homology are limited. Nonetheless, one difference between the seeds of monocots and dicots is the development of their endosperm: monocots have a persistent endosperm, which remains surrounding the embryo until germination, whereas dicots have a nonpersistent endosperm that is almost completely gone by the time of germination (Olsen, 2004). Potential differences in the types of imprinted genes between *Arabidopsis*, rice, and maize may be a reflection of the developmental pathway of their respective endosperm.

An association between DNA methylation and imprinting has been observed in both rice and maize (Springer and Gutierrez-Marcos, 2009). Additionally, in maize, strong correlations have been made between the epigenetic marks of DNA methylation and H3K27me3 and gene imprinting (Springer and Gutierrez-Marcos, 2009; Raissig *et al.*, 2011). Finally, maize may also use active chromatin modification in the form of H3 and H4 acetylation (H3/H4Ac) to regulate imprinting (Haun and Springer, 2008). Active DNA demethylation has yet to be genetically identified in monocots, although *in silico* investigation has identified orthologs of the DME family of DNA glycosylases in rice (Zemach *et al.*, 2010a) and likely maize. Rice plants with a null mutation in a rice DME-ortholog, *ROS1a*, display reproductive phenotypes similar to phenotypes observed in *Arabidopsis dme* mutant plants (Ono *et al.*, 2012). Taken together, these findings suggest the mechanisms used to establish plant gene imprinting may be common between monocots and dicots.

Imprinting in Rice

Previous studies in rice endosperm led to the discovery of two imprinted genes, *OsFIE1* (Luo *et al.*, 2009) and *OsMADs87* (Ishikawa *et al.*, 2011). *OsFIE1* is a MEG specifically expressed in the endosperm that is related to the imprinted maize *FIE1* gene (Luo *et al.*, 2009). In *Arabidopsis*, null mutations in *FIE* result in an overproliferated endosperm. However, a T-DNA insertion in *OsFIE1* showed no such phenotype, suggesting imprinted *OsFIE1* in rice is not involved in the repression of central cell development (Luo *et al.*, 2009). The *OsMADs87* gene in rice was found to be a homologue of the paternally expressed *Arabidopsis* imprinted gene, *PHE1*. However, in rice, *OsMADs87* is a maternally expressed imprinted gene (Ishikawa *et al.*, 2011), as opposed to the paternal expression of *PHE1* in *Arabidopsis*. The function of *OsMADs87* in the endosperm is unknown.

A high-throughput transcriptome study conducted in rice brought the current number of known imprinted genes to 121, of which 62 are MEGs and 59 are PEGs (Luo *et al.*, 2011). All gene imprinting was found in the endosperm except for one gene, *Os10g05750*, which was a MEG in both the endosperm and the embryo. *Os10g05750* encodes an allergenic protein in olive (*Olea europaea* L.) that is thought to control pollen tube emergence and guidance (de Dios Alche *et al.*, 2004). Ontological analysis of imprinted rice genes show 30% of them to be of unknown function, whereas the remaining genes were suggested to be involved in regulatory processes such as signal transduction, cellular component organization, and DNA and RNA binding (Luo *et al.*, 2011).

The imprinted rice dataset was further compared with some more recent *Arabidopsis* high-throughput imprinting studies (Gehring *et al.*, 2009, 2011; Hsieh *et al.*, 2011; Wolff *et al.*, 2011) to find genes with similar homology. Overall, when using the filtered data presented in the literature, a low-degree of conservation was detected between the rice and *Arabidopsis* imprinted gene datasets. A low level of conservation could suggest that plant gene imprinting evolved rapidly and independently in monocots and dicots. Of the 27 candidate genes that showed significant homology between the rice and *Arabidopsis* datasets, several have key functions in the sRNA pathway (*AGO* and *DsRNA BINDING*), chromatin remodeling (*SUVH* and *PICKLE-LIKE*), and DNA methylation (*VIM5*) (Luo *et al.*, 2011). If imprinting did evolve independently, the dosage attenuation of these common, key regulatory genes must be an important process. Alternatively, as previously mentioned, the lack of conservation between the imprinted gene datasets of rice and *Arabidopsis* may be due to the technical limitation of available SNPs or inappropriate data filtering.

Imprinting in Maize

Before more recent high-throughput approaches, earlier studies in *Zea mays* resulted in identification of a few imprinted genes. *fie1* and *fie2* are MEGs found in the endosperm and are homologues of *Arabidopsis FIE* (Springer *et al.*, 2002; Danilevskaya *et al.*, 2003; Gutierrez-Marcos *et al.*, 2003). *fie1* is expressed solely in the endosperm, displaying "binary" imprinting throughout seed development. Binary imprinting is defined to be when one parental allele is completely silenced, contrasting with "differential" imprinting when only a relative difference between parental allele expression is observed (Gutierrez-Marcos and Springer, 2009). In the case of *fie2*, expression is more ubiquitous, occurring in both reproductive and vegetative tissues, and is imprinted only during the early stages of endosperm development (Danilevskaya *et al.*, 2003; Gutierrez-Marcos *et al.*, 2003, 2006). Null-mutations in *fie1* and *fie2* have not yet been identified, and consequently the function of these two genes has not been determined.

meg1 is a MEG that displays imprinting during the early stages of endosperm development (Gutierrez-Marcos *et al.*, 2004). Notably, *meg1* imprinting occurs in the basal transfer region of the endosperm, in the transfer cells that function to facilitate the movement of nutrients from the maternal tissues of the sporophyte to the endosperm (Olsen, 2004). *meg1* encodes a novel protein that was shown to be glycosylated and localized to the cell wall ingrowths of the basal transfer cells (Gutierrez-Marcos *et al.*, 2004). A more recent study by Costa *et al.* (2012) has further demonstrated *meg1* to be required for transfer cell formation and differentiation. Maternally expressed imprinted *meg1* was also shown to be essential in regulating maternal nutrient uptake, sucrose partitioning, and increasing seed biomass yield, as a *meg1*-null mutant produced smaller seeds (Costa *et al.*, 2012). These results indicate that maternally expressed *meg1* promotes nutrient allocation to the embryo instead of suppressing it – a finding that is not in accord with the parental conflict theory (Haig, 2000). Costa *et al.* (2012) suggested imprinting may not have arisen from parental conflicts over nutrient amounts to offspring (Moore and Haig, 1991) but instead from maternal-offspring influences over placental-like functions. In this case, the evolutionary drive causing imprinting seems more conducive to a coadaptation theory of genomic imprinting (Wolf and Hager, 2006).

The *nrp1* gene is another MEG in maize, exclusively expressed in the endosperm (Guo *et al.*, 2003). The function of *nrp1* is unknown, but it encodes a putative transcription factor of the NAM gene family.

A genome-wide transcriptome study conducted by Waters *et al.* (2011) has increased the number of known potentially imprinted genes in maize to ~100, of which 54 are MEGs and 46 are PEGs. Gene ontology analysis of the imprinted genes showed an enrichment for orthologs of chromatin-related proteins and transcription factors, such as *VIM1* (Woo *et al.*, 2008), *VERNALIZATION 5* (Greb *et al.*, 2007), *FIE* (Ohad *et al.*, 1999), and *CURLY LEAF* (Goodrich *et al.*, 1997).

Evolution of Plant Imprinting

Similarities between mutant phenotypes, imprinting mechanisms, and sites of gene imprinting (placenta and endosperm) support the idea of convergent evolution of genomic imprinting in plants and mammals (Feil and Berger, 2007) and provide persuasive evidence that the seminal theory of Moore and Haig (1991) may be correct. The facts that angiosperms contain an organ synonymous to the placenta and that they also imprint genes are striking. That the blunt method of altering genomic ploidy results in a change in seed size is consistent with the parental conflict theory in *Arabidopsis* and maize. The generation of maize endosperm that departs from the normal 2:1 ratio between female and male genomes exerts a dramatic effect on seed development. Paternal genomic excess prolongs early endosperm development and delays cellularization, resulting in larger seeds, whereas maternal genome excess results in the opposite phenotype, accelerating the onset of cellularization and resulting in smaller seeds (Lin, 1984; Lin and Dickinson, 2010). Likewise, in *Arabidopsis*, doubling the maternal genome contribution results in a smaller endosperm and embryo, and doubling the paternal contribution results in a larger endosperm and embryo (Scott *et al.*, 1998). These results support the idea that gene imprinting in monocots and dicots is intimately involved with resource allocation to the embryo.

Seed size itself is not a barrier to maternal fitness in the same way as it is to placental mammals. More likely, the nutrient transfer capacity of the endosperm is key in the balance of parental resources. In maize, the previously mentioned maternally expressed imprinted *meg1* gene was shown to be fundamental for the development of the endosperm nutrient transfer cells that are at the interface between the mother and the seed (Gutierrez-Marcos *et al.*, 2004). *meg1* was found to regulate maternal nutrient uptake, sucrose partitioning, and seed biomass; however, its maternal expression is at odds with the predictions of parental conflict.

An alternative, although not mutually exclusive, hypothesis is that plant gene imprinting is a by-product of transposable element (TE) silencing. Most imprinted genes in *Arabidopsis* and rice are flanked by transposable elements (TEs). DNA demethylation appears to occur at TEs that flank *Arabidopsis* imprinted genes (Gehring *et al.*, 2009; Hsieh *et al.*, 2009; Schoft *et al.*, 2011). It is possible that there is an underlying requirement for DNA demethylation in companion cells (central cell and vegetative cell) to ensure the RdDM-mediated methylation of TEs in the gametes (Schoft *et al.*, 2011). The fact that DNA demethylation of central and vegetative cells may be similar, despite their dramatically different functions in plant reproduction, argues that TE silencing, rather than gene imprinting, may be the basal function of DNA demethylation in plant reproduction.

Kohler *et al.* (2012) examined unfiltered next generation sequencing data for imprinted genes and found that a higher level of conservation potentially exists in both the genes and the function of genes imprinted in rice, maize, and *Arabidopsis*. These clades are separated by ~150 million years, similar to the divergence of humans and marsupials. Therefore, it remains plausible that imprinting has a specific purpose (Kohler *et al.*, 2012). Although DNA demethylation may have initially evolved to ensure TE silencing, its persistence may also be due to the benefit conferred through dosage attenuation of particular genes involved in seed development and nutrient allocation.

Conclusion

Plant genomic imprinting has been demonstrated in both dicots and monocots, occurring almost exclusively in the endosperm. The parental conflict theory suggests that gene imprinting arose to mediate the competition between parents regarding the level of maternal nutrients and resources allocated to the offspring. This conflict is predicted to occur in the tissues that deliver nutrients to the offspring. However, the findings of maternally expressed *meg1* in maize functioning to promote resource allocation to the embryo suggests a maternal-offspring coadaptation model for imprinting may be a driving force for imprinting as well. Whether or not these different selective forces for imprinting overlap or are mutually exclusive remains to be determined. Nevertheless, gene imprinting in both mammals and plants occurs in structures at the interface between the maternal circulation and the embryo – in mammals, it is the placenta, whereas in plants, it is the endosperm, and both organs act to regulate nutrient exchange.

Epigenetic mechanisms in the form of DNA methylation and histone modifications are the molecular factors responsible for carrying out imprinting in both mammals and plants. However, plants have an additional mechanism in the form of active DNA-demethylation, which is performed by the DME family of plant-specific DNA glycosylases. DME-mediated DNA demethylation involves a co-opting of the BER pathway that results in the removal of 5mC. Notably, DME-mediated DNA demethylation effectively causes the transcription of genes and appears to be an integral part of imprinting regulatory strategies for genes.

High-throughput next generation sequencing has advanced the field of plant gene imprinting, leading to the discovery of more imprinted genes not only in the model plant *Arabidopsis* but also in economically relevant crop plants, rice and maize. The discovery of new plant imprinted genes reveals these genes to be from a range of ontological classes, and the function of dosage attenuation in these proteins provides exciting new avenues of research. Current data suggest that monocot species use DNA methylation and histone modifications to establish genomic imprinting in the same way as *Arabidopsis*, indicating that imprinting and its regulatory strategies are conserved in these highly diverged organisms. Further work must be done to investigate imprinting regulation in these species.

Acknowledgments

The authors gratefully acknowledge the support of the National Institutes of Health (R01-GM069415) and the National Science Foundation (MCB-0918821 and IOS-1025890) to R.L.F. and a Ruth L. Kirschstein NIH Predoctoral Fellowship (GM093633) to C.A.I.

References

Allis CD, Jenuwein T, Reinberg D (2007). *Epigenetics*. Cold Spring Harbor Laboratory Press, Cold Spring Harbor, New York, New York.

Bartee L, Malagnac F, Bender J (2001). Arabidopsis cmt3 chromomethylase mutations block non-CG methylation and silencing of an endogenous gene. Genes Dev 15: 1753–1758.

Bartolomei MS, Ferguson-Smith AC (2011). Mammalian genomic imprinting. Cold Spring Harb Perspect Biol 3: a002592.

Bauer MJ, Fischer RL (2011). Genome demethylation and imprinting in the endosperm. Curr Opin Plant Biol 14: 162–167.

Bostick M, Kim JK, Esteve PO, Clark A, Pradhan S, Jacobsen SE (2007). UHRF1 plays a role in maintaining DNA methylation in mammalian cells. Science 317: 1760–1764.

Brown RC, Lemmon BE (2007). The developmental biology of cereal endosperm. In: Olsen OA (ed). *Endosperm: Developmental and Molecular Biology*. Springer, Berlin, Germany, pp 1–20.

Cedar H, Bergman Y (2012). Programming of DNA Methylation Patterns. Annual Review of Biochemistry 81.

Choi Y, Gehring M, Johnson L, Hannon M, Harada JJ, Goldberg RB, Jacobsen SE, Fischer RL (2002). DEMETER, a DNA glycosylase domain protein, is required for endosperm gene imprinting and seed viability in *Arabidopsis*. Cell 110: 33–42.

Coan PM, Burton GJ, Ferguson-Smith AC (2005). Imprinted genes in the placenta: a review. Placenta 26(Suppl): S10–S20.

Cokus SJ, Fing S, Zhang X, Chen Z, Merriman B, Haudenschild CD, Pradhan S, Nelson SF, Pellegrini M, Jacobsen SE (2008). Shotgun bisulphite sequencing of the Arabidopsis genome reveals DNA methylation patterning. Nature 452: 215–219.

Constancia M, Kelsey G, Reik W (2004). Resourceful imprinting. Nature 432: 53–57.

Cooper WN, Luharia A, Evans GA, Raza H, Haire AC, Grundy R, Bowdin SC, Riccio A, Sebastio G, Bliek J, *et al.* (2005). Molecular subtypes and phenotypic expression of Beckwith-Wiedemann syndrome. Eur J Hum Genet 13: 1025–1032.

Costa LM, Yuan J, Rouster J, Paul W, Dickinson H, Gutierrez-Marcos JF (2012). Maternal control of nutrient allocation in plant seeds by genomic imprinting. Curr Biol 22: 160–165.

Czech B, Hannon GJ (2011). Small RNA sorting: matchmaking for Argonautes. Nat Rev Genet 12: 19–31.

Danilevskaya ON, Hermon P, Hantke S, Muszynski MG, Kollipara K, Ananiev EV (2003). Duplicated fie genes in maize: expression pattern and imprinting suggest distinct functions. Plant Cell 15: 425–438.

Day T, Bonduriansky R (2004). Intralocus sexual conflict can drive the evolution of genomic imprinting. Genetics 167: 1537–1546.

DeChiara TM, Efstratiadis A, Robertsen EJ (1990). A growth-deficiency phenotype in heterozygous mice carrying an insulin-like growth factor II gene disrupted by targeting. Nature 345: 78–80.

de Dios Alche J, M'rani-Alaoui M, Castro AJ, Rodriquez-Garcia MI (2004). Ole e 1, the major allergen from olive (Olea europaea L.) pollen, increases its expression and is released to the culture medium during in vitro germination. Plant Cell Physiol 45: 1149–1157.

Feil R, Berger F (2007). Convergent evolution of genomic imprinting in plants and mammals. Trends Genet 23: 192–199.

Frost JM, Moore GE (2010). The importance of imprinting in the human placenta. PLoS Genet 6: e1001015.

Gehring M, Bubb KL, Henikoff S (2009). Extensive demethylation of repetitive elements during seed development underlies gene imprinting. Science 324: 1447–1451.

Gehring M, Bubb KL, Henikoff S (2009). Extensive demethylation of repetitive elements during seed development underlies gene imprinting. Science 324: 1447–1451.

Gehring M, Huh JH, Hsieh TF, Penterman J, Choi Y, Harada JJ, Goldberg RB, Fischer RL (2006). DEMETER DNA glycosylase establishes MEDEA polycomb gene self-imprinting by allele-specific demethylation. Cell 124: 495–506.

Gehring M, Missirian V, Henikoff S (2011). Genomic analysis of parent-of-origin allelic expression in Arabidopsis thaliana seeds. PLoS One 6: e23687.

Gerald JNF, Hui PS, Berger F (2009). Polycomb group-dependent imprinting of the actin regulator AtFH5 regulates morphogenesis in Arabidopsis thaliana. Development 136: 3399–3404.

Glaser RL, Ramsay JP, Morison IM (2006). The imprinted gene and parent-of-origin effect database now includes parental origin of de novo mutations. Nucl Acids Res 34: D29–D31.

Goodrich J, Puangsomlee P, Martin M, Long D, Meyerowitz EM, Coupland G (1997). A polycomb-group gene regulates homeotic gene expression in *Arabidopsis*. Nature 386: 44–51.

Greb T, Mylne JS, Crevillen P, Geraldo N, An H, Gendall AR, Dean C (2007). The PHD finger protein VRN5 functions in the epigenetic silencing of Arabidopsis FLC. Curr Biol 17: 73–78.

Gregg C, Zhang J, Weissbourd B, Luo S, Schroth GP, Haig D, Dulac C (2010). High-resolution analysis of parent-of-origin allelic expression in the mouse brain. Science 329: 643–648.

Grimaud C, Nègre N, Cavalli G (2006). From genetics to epigenetics: the tale of Polycomb group and trithorax group genes. Chromosome Res 14: 363–375.

Grossniklaus U, Vielle-Calzada J-P, Hoeppner MA, Gagliano WB (1998). Maternal control of embryogenesis by *MEDEA*, a polycomb-group gene in *Arabidopsis*. Science 280: 446–450.

Guo M, Rupe MA, Danilevskaya ON, Yang X, Hu Z (2003). Genome-wide mRNA profiling reveals heterochronic allelic variation and a new imprinted gene in hybrid maize endosperm. Plant J 36: 30–44.

Gutierrez-Marcos JF, Costa LM, Biderre-Petit C, Khbaya B, O'Sullivan DM, Wormald M, Perez P, Dickinson HG (2004). maternally expressed gene1 is a novel maize endosperm transfer cell-specific gene with a maternal parent-of-origin pattern of expression. Plant Cell 16 1288–1301.

Gutierrez-Marcos JF, Costa LM, Dal Pra M, Scholten S, Kranz E, Perez P, Dickinson HG (2006). Epigenetic asymmetry of imprinted genes in plant gametes. Nat Genet 38: 876–878.

Gutierrez-Marcos JF, Pennington PD, Costa LM, Dickinson HG (2003). Imprinting in the endosperm: a possible role in preventing wide hybridization. Philos Trans R Soc Lond B Biol Sci 358: 1105–1111.

Haig D (2000). The kinship theory of genomic imprinting. Annu Rev Ecol Syst 31: 9–32.

Haun WJ, Springer NM (2008). Maternal and paternal alleles exhibit differential histone methylation and acetylation at maize imprinted genes. Plant J 56: 903–912.

He X-J, Chen T, Zhu J-K (2011). Regulation and function of DNA methylation in plants and animals. Cell Res 21: 442–465.

Henderson IR, Jacobsen SE (2007). Epigenetic inheritance in plants. Nature 447: 418–424.

Hennig L, Derkacheva M (2009). Diversity of Polycomb group complexes in plants: same rules, different players? Trends Genet 25: 414–423.

Hsieh TF, Ibarra CA, Silva P, Zemach A, Eshed-Williams L, Fischer RL, Zilberman D (2009). Genome-wide demethylation of Arabidopsis endosperm. Science 324: 1451–1454.

Hsieh TF, Shin J, Uzawa R, Silva P, Cohen S, Bauer MJ, Hashimoto M, Kirkbride RC, Harada JJ, Zilberman D, *et al.* (2011). Regulation of imprinted gene expression in Arabidopsis endosperm. Proc Natl Acad Sci U S A 108: 1755–1762.

Huh JH, Bauer MJ, Hsieh TF, Fischer RL (2008). Cellular programming of plant gene imprinting. Cell 132: 735–744.

Ishikawa R, Ohnishi T, Kinoshita Y, Eiguchi M, Kurata N, Kinoshita T (2011). Rice interspecies hybrids show precocious or delayed developmental transitions in the endosperm without change to the rate of syncytial nuclear division. Plant J 65: 798–806.

Jacobs AL, Schar P (2012). DNA glycosylases: in DNA repair and beyond. Chromosoma 121: 1–20.

Jahnke S, Scholten S (2009). Epigenetic resetting of a gene imprinted in plant embryos. Curr Biol 19: 1677–1681.

Johnson LM, Bostick M, Zhang X, Kraft E, Henderson I, Callis J, Jacobsen SE (2007). The SRA methyl-cytosine-binding domain links DNA and histone methylation. Curr Biol 17: 379–384.

Jullien PE, Kinoshita T, Ohad N, Berger F (2006). Maintenance of DNA methylation during the Arabidopsis life cycle is essential for parental imprinting. Plant Cell 18: 1360–1372.

Kinoshita T, Miura A, Choi Y, Kinoshita Y, Cao X, Jacobsen SE, Fischer RL, Kakutani T (2004). One-way control of FWA imprinting in Arabidopsis endosperm by DNA methylation. Science 303: 521–523.

Kiyosue T, Ohad N, Yadegari R, Hannon M, Dinneny J, Wells D, Katz A, Margossian L, Harada JJ, Goldberg RB, Fischer RLF (1999). Control of fertilization-independent endosperm development by the MEDEA Polycomb gene in Arabidopsis. Proc Natl Acad Sci U S A 96: 4186–4191.

Klose RJ, Bird AP (2006). Genomic DNA methylation: the mark and its mediators. Trends Biochem Sci 31: 89–97.

Kohler C, Page DR, Gagliardini V, Grossniklaus U (2005). The Arabidopsis thaliana MEDEA Polycomb group protein controls expression of PHERES1 by parental imprinting. Nat Genet 37: 28–30.

Kohler C, Wolff P, Spillane C (2012). Epigenetic mechanisms underlying genomic imprinting in plants. Annu Rev Plant Biol 63.

Kouzarides T (2007). Chromatin modifications and their function. Cell 128: 693–705.

Law JA, Jacobsen SE (2010). Establishing, maintaining and modifying DNA methylation patterns in plants and animals. Nat Rev Genet 11: 204–220.

Leighton PA, Ingram RS, Eggenschwiler J, Efstratiadis A, Tilghman SM (1995). Disruption of imprinting caused by deletion of the H19 gene region in mice. Nature 375: 34–39.

Lewis A, Mitsuya K, Umlauf D, Smith P, Dean W, Walter J, Higgins M, Feil R, Reik W (2004). Imprinting on distal chromosome 7 in the placenta involves repressive histone methylation independent of DNA methylation. Nat Genet 36: 1291–1295.

Li E, Beard C, Jaenisch R (1993). Role for DNA methylation in genomic imprinting. Nature 366: 362–365.

Li E, Bestor TH, Jaenisch R (1992). Targeted mutation of the DNA methyltransferase gene results in embryonic lethality. Cell 69: 915–926.

Li N, Dickinson HG (2010). Balance between maternal and paternal alleles sets the timing of resource accumulation in the maize endosperm. Proc R Soc B Biol Sci 277: 3–10.

Lin B-Y (1984). Ploidy barrier to endosperm development in maize. Genetics 107: 103–115.

Lin N, Dickinson HG (2010). Balance between maternal and paternal alleles sets the timingof resource accumulation in the maize endosperm. Proc Biol Sci 277: 3–10.

Lindroth AM, Cao X, Jackson JP, Zilberman D, McCallum CM, Henikoff S, Jacobsen SE (2001). Requirement of CHROMOMETHYLASE3 for maintenance of CpXpG methylation. Science 292: 2077–2080.

Lister R, Pelizzola M, Dowen RH, Hawkins RD, Hon G, Tonti-Filippini J, Nery JR, Lee L, Ye Z, No QM, *et al.* (2009). Human DNA methylomes at base resolution show widespread epigenomic differences. Nature 462: 315–322.

Luo M, Bilodeau P, Koltunow A, Dennis ES, Peacock WJ, Chaudhury AM (1999). Genes controlling fertilization-independent seed development in *Arabidopsis thaliana*. Proc Natl Acad Sci U S A 96: 296–301.

Luo M, Platten D, Chaudhury A, Peacock WJ, Dennis ES (2009). Expression, imprinting, and evolution of rice homologs of the polycomb group genes. Mol Plant 2: 711–723.

Luo M, Taylor JM, Spriggs A, Zhang H, Wu X, Russell S, Singh M, Koltunow A (2011). A genome-wide survey of imprinted genes in rice seeds reveals imprinting primarily occurs in the endosperm. PLoS Genet 7: e1002125.

Makarevich G, Leroy O, Akinci U, Schubert D, Clarenz O, Goodrich J, Grossniklaus U, Kohler C (2006). Different Polycomb group complexes regulate common target genes in Arabidopsis. EMBO Rep 7: 947–952.

Matzke M, Kanno T, Daxinger L, Huettel B, Matzke AJ (2009). RNA-mediated chromatin-based silencing in plants. Curr Opin Cell Biol 21: 367–376.

McGrath J, Solter D (1984). Completion of mouse embryogenesis requires both the maternal and paternal genomes. Cell 37: 179–183.

McKeown PC, Laouielle-Duprat S, Prins P, Wolff P, Schmid MW, Donoghue MT, Fort A, Duszynska D, Comte A, Lao NT, *et al.* (2011). Identification of imprinted genes subject to parent-of-origin specific expression in Arabidopsis thaliana seeds. BMC Plant Biol 11: 113.

Monk D, Arnaud P, Frost J, Hills FA, Stanier P, Feil R, Moore GE (2009). Reciprocal imprinting of human GRB10 in placental trophoblast and brain: evolutionary conservation of reversed allelic expression. Hum Mol Genet 18: 3066–3074.

Moore T, Haig D (1991). Genomic imprinting in mammalian development: a parental tug-of-war. Trends Genet 7: 45–48.

Nodine MD, Bartel DP (2012). Maternal and paternal genomes contribute equally to the transcriptome of early plant embryos. Nature 482: 94–97.

Ohad N, Yadegari R, Margossian L, Hannon M, Michaeli D, Harada JJ, Goldberg RB, Fischer RL (1999). Mutations in *FIE*, a WD polycomb group gene, allow endosperm development without fertilization. Plant Cell 11: 407–415.

Olsen OA (2004). Nuclear endosperm development in cereals and Arabidopsis thaliana. Plant Cell 16: S214–S227.

Ono A, Yamaguchi K, Fukada-Tanaka S, Terada R, Mitsui T, Iida S (2012). A null mutation of ROS1a for DNA demethylation in rice is not transmittable to progeny. Plant J DOI: 10.1111/j.1365-313X.2012.05009.x.

Ooi SK, Bestor TH (2008). The colorful history of active DNA demethylation. Cell 133: 1145–1148.

Penterman J, Zilberman D, Huh JH, Ballinger T, Henikoff S, Fischer RL (2007). DNA demethylation in the Arabidopsis genome. Proc Natl Acad Sci U S A 104: 6752–6757.

Pirrotta V (1998). Polycombing the genome: PcG, trxG, and chromatin silencing. Cell 93: 333–336.

Raissig MT, Baroux C, Grossniklaus U (2011). Regulation and flexibility of genomic imprinting during seed development. Plant Cell 23: 16–26.

Ramsahoye BH, Biniszkiewicz D, Lyko F, Clark V, Bird AP, Jaenisch R (2000). Non-CpG methylation is prevalent in embryonic stem cells and may be mediated by DNA methyltransferase 3a. Proc Natl Acad Sci U S A 97: 5237–5242.

Reik W (2007). Stability and flexibility of epigenetic gene regulation in mammalian development. Nature 447: 425–432.

Schoft VK, Chumak N, Choi Y, Hannon M, Garcia-Aguilar M, Machlicova A, Slusarz L, Mosiolek M, Park JS, Park GT, *et al.* (2011). Function of the DEMETER DNA glycosylase in the Arabidopsis thaliana male gametophyte. Proc Natl Acad Sci U S A 108: 8042–8047.

Schwartz YB, Pirrotta V (2007). Polycomb silencing mechanisms and the management of genomic programmes. Nat Rev Genet 8: 9–22.

Schwartz YB, Pirrotta V (2008). Polycomb complexes and epigenetic states. Curr Opin Cell Biol 20: 266–273.

Scott RJ, Spielman M, Bailey J, Dickinson HG (1998). Parent-of-origin effects on seed development in *Arabidopsis thaliana*. Development 125: 3329–3341.

Soppe WJJ, Jacobsen SE, Alonso-Blanco C, Jackson JP, Kakutani T, Koornneef M, Peeters AJM (2000). The late flowering phenotype of *fwa* mutants is caused by gain-of-function epigenetic alleles of a homeodomain gene. Mol Cell 6: 791–802.

Springer NM, Danilevskaya ON, Hermon P, Helentjaris TG, Phillips RL, Kaeppler HF, Kaeppler SM (2002). Sequence relationships, conserved domains, and expression patterns for maize homologs of the polycomb group genes E(z), esc, and E(Pc). Plant Physiol 128: 1332–1345.

Springer NM, Gutierrez-Marcos JF (2009). Imprinting in maize. In: Bennetzen JL, Hake S (eds). *Maize Handbook, volume II: Genetics and Genomics*. Springer, New York, pp 428–440.

Surani MA, Barton SC, Norris ML (1984). Development of reconstituted mouse eggs suggests imprinting of the genome during gametogenesis. Nature 308: 548–550.

Tiwari S, Spielman M, Schulz R, Oakey R, Kelsey G, Salazar A, Zhang K, Pennell R, Scott R (2010). Transcriptional profiles underlying parent-of-origin effects in seeds of Arabidopsis thaliana. BMC Plant Biol 10: 72.

Villar CBR, Erilova A, Makarevich G, Trosch R, Kohler C (2009). Control of PHERES1 imprinting in Arabidopsis by direct tandem repeats. Mol Plant 2: 654–660.

Vongs A, Kakutani T, Martienssen RA, Richards EJ (1993). *Arabidopsis thaliana* DNA methylation mutants. Science 260: 1926–1928.

Wassenegger M, Heimes S, Riedel L, Sanger HL (1994). RNA-directed de novo methylation of genomic sequences in plants. Cell 76: 567–576.

Waters AJ, Makarevitch I, Eichten SR, Swanson-Wagner RA, Yeh CT, Xu W, Schnable PS, Vaughn MW, Gehring M, Springer NM (2011). Parent-of-origin effects on gene expression and DNA methylation in the maize endosperm. Plant Cell 23: 4221–4233.

Watt F, Molloy PL (1988). Cytosine methylation prevents binding to DNA of a HeLa cell transcription factor required for optimal expression of the adenovirus major late promoter. Genes Dev 2: 1136–1143.

Wilkins JF, Ubeda F (2011). Diseases associated with genomic imprinting. In: Xiaodong C, Robert MB (eds). *Progress in Molecular Biology and Translational Science*. Academic Press, Wilkins, Massachusetts, pp 401–445.

Wolf JB, Hager R (2006). A maternal-offspring coadaptation theory for the evolution of genomic imprinting. PLoS Biol 4: e380.

Wolff P, Weinhofer I, Seguin J, Roszak P, Beisel C, Donoghue MT, Spillane C, Nordborg M, Rehmsmeier M, Kohler C (2011). High-resolution analysis of parent-of-origin allelic expression in the Arabidopsis endosperm. PLoS Genet 7: e1002126.

Woo HR, Dittmer TA, Richards EJ (2008). Three SRA-domain methylcytosine-binding proteins cooperate to maintain global CpG methylation and epigenetic silencing in Arabidopsis. PLoS Genet 4: e1000156.

Zemach A, Kim MY, Silva P, Rodrigues JA, Dotson B, Brooks MD, Zilberman D (2010a). Local DNA hypomethylation activates genes in rice endosperm. Proc Natl Acad Sci U S A 107: 18729–18734.

Zemach A, McDaniel IE, Silva P, Zilberman D (2010b): Genome-wide evolutionary analysis of eukaryotic DNA methylation. Science 328, 916–919.

Zhang M, Zhao H, Xie S, Chen J, Xu Y, Wang K, Zhao H, Guan H, Hu X, Jiao Y, *et al.* (2011). Extensive, clustered parental imprinting of protein-coding and noncoding RNAs in developing maize endosperm. Proc Natl Acad Sci U S A 108: 20042–20047.

Zhu J, Kapoor A, Sridhar VV, Agius F, Zhu JK (2007). The DNA glycosylas/lyase ROS1 functions in pruning DNA methylation patterns in Arabidopsis. Curr Biol 17: 54–59.

5 Apomixis

Anna M. G. Koltunow, Peggy Ozias-Akins, and Imran Siddiqi

Introduction

Food security is a challenge for the future of our planet with global population expected to reach 9 billion by 2050. Food production must increase on diminishing arable land, regardless of changing climate pressures. Societal expectations of modern agribusiness also demand decreased dependence on agrichemicals and increased stewardship of land and water resources, conserving them for future generations (Tester and Langridge, 2010).

Hybrid crops exhibit superior yields (15%–35%), attributed to the little understood phenomenon of heterosis or hybrid vigor (Kempe and Gils, 2011). In addition to increased yield, hybrids display improved physical stability, higher responses to fertilizers, better root penetration, and improved tolerance to drought and heat (Kempe and Gils, 2011). However, the production and use of hybrids in world food production are restricted to a few crops, such as maize, sorghum, sunflower, cucumber, onion, and some varieties of rice. Hybrid yield advantages, increased genetic diversity, and accompanying agronomic trait gains from other significant crops including wheat and soybean, currently generated from homozygous inbred varieties, are yet to be realized. The multiple benefits of hybrids come at the cost of expensive and often technically demanding commercial hybrid seed production methodologies. For example, the cross may need to be carried out in a particular direction, and means to prevent seeds forming from the parents can be technically difficult and labor intensive. Parental lines must be maintained so that hybrid seed can be continuously remade, because seeds from high yielding hybrids cannot simply be resown to generate improved crops in subsequent generations; this is due, in part, to the intrinsic nature of sexual reproduction that drives genetic diversity. The combination of meiosis during male and female gamete formation and subsequent gamete fusion at fertilization to generate the embryo lead to segregation of alleles in the progeny, and a coincident decline in heterosis is observed in subsequent seed generations. Economic hybrid seed production coupled with heterotic stability and, preferably, preservation of the hybrid seed genotype would accelerate wider use of hybrids in agriculture.

Some plants naturally produce seeds asexually by a suite of mechanisms under the collective term of apomixis. Apomixis influences reproductive development within the ovule; male gamete development is generally not affected. In contrast to sexual seed formation, apomictic reproduction avoids meiosis and fertilization to generate an embryo that is solely maternal in genotype. The maternal genotype remains fixed and is stably maintained, barring mutation, from one seed generation to the next. Application of apomixis in hybrid seed production would enable hybrid genotype fixation

Seed Genomics, First Edition. Edited by Philip W. Becraft.

and inexpensive, large-scale hybrid seed multiplication. The economic benefits of the application of apomixis to agriculture are immense and have been described elsewhere (Hanna, 1995; Koltunow *et al.*, 1995a; Vielle-Calzada *et al.*, 1996; Ramulu *et al.*, 1999). Apomixis is not evident in current major cereal, grain, and vegetable crops. It is primarily found in non-agronomic plants, and only a few of these are sufficiently related to cross with crops for introgression of the trait by hybridization.

Apomictic species are being used as research models for the identification and isolation of the genes conferring apomixis and the downstream pathways they control, to provide tools for the development of apomixis as a technology in hybrid seed breeding. Few species are obligate apomicts; most are facultative apomicts, meaning that they retain the capacity to form seeds via the sexual pathway to varying degrees (Asker and Jerling, 1992). The analysis of evolved apomictic species with various mechanisms is informative for understanding molecular, biochemical, and ecological factors associated with high rates of apomictic seed set. In addition to elucidating the molecular basis for apomixis in evolved species, a parallel research approach is aimed at identifying mutations in genes or novel alleles that modify aspects of sexual seed development giving apomixis pathway–like phenotypes, and combining them to synthesize a pathway resembling apomixis that enables formation of seeds containing embryos with a maternal genotype (Chaudhury and Peacock, 1993).

In this chapter, we review progress in the understanding of the developmental biology of apomixis and its evolution, distribution, genetic regulation, and progress toward isolating apomixis genes from functional apomictic species. We describe genes involved in sexual reproduction that, when mutated, stimulate apomixis-like phenotypes, and we highlight more recent success in the generation of asexual seeds in *Arabidopsis* following the combination of specific mutants. We also consider some issues for the installation of apomixis in hybrid crops in light of current knowledge.

Biology of Apomixis in Natural Systems

Overview of Sexual Seed Formation Pathway

Apomixis is broadly divided into two types: gametophytic and sporophytic. Both types omit steps found in the sexual pathway. The diversity of apomictic pathways has been covered in many reviews (Nogler, 1984a; Koltunow, 1993; Koltunow and Grossniklaus, 2003). The distinctions between sporophytic and gametophytic apomixis are best understood if compared with the sequence of events during sexual seed formation (Figure 5.1). The sexual mode of seed formation is retained in most apomicts owing to their facultative nature. In angiosperms, sexual reproduction is characterized by an alternation of generations life cycle involving the transition between a diploid (2n) sporophytic generation and a haploid (n) gametophytic generation that gives rise to a diploid sporophyte following double fertilization. Diploid (2n) male and female gamete precursor cells termed microspore and megaspore mother cells undergo meiosis in the anther and ovule to produce haploid (n) female megaspores and male microspores. They subsequently undergo mitosis to give rise to multicellular gametophytes, the male pollen grain and female embryo sac containing the sperm and egg cells. Figure 5.1*a* shows the events of female gametophyte development in the *Polygonum*-type embryo sac, the most common type formed in angiosperms. A single megaspore mother cell differentiates and undergoes megasporogenesis, which involves the events of meiotic reduction to produce four haploid megaspores, and subsequent megaspore selection where three of the four megaspores formed degenerate. The surviving megaspore undergoes megagametogenesis, which involves three rounds of mitosis without cytokinesis generating a syncytium of eight nuclei. Subsequent cellularization

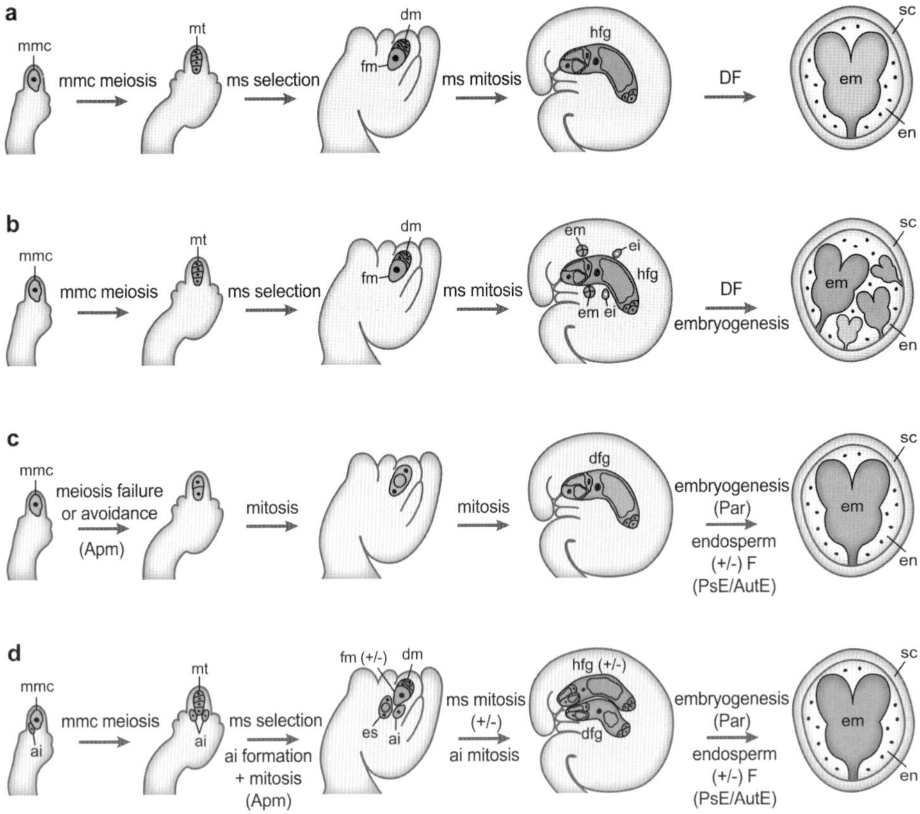

Figure 5.1 Seed development in sexual and apomictic species. *a,* Events of sexual seed formation. *b,* Events of sporophytic apomixis. *c,* Events of diplosporous apomixis and, *d,* aposporous apomixis, where endosperm formation may or may not require fertilization. In the case of apospory, haploid female gametophytes may coexist with the aposporous gametophyte, or degeneration of the haploid gametophyte may occur during the events of aposporous gametophyte development. See text for further details. ai, aposporous initial cells; Apm, apomeiosis; AutE; autonomous endosperm formation; DF, double fertilization; dfg, diploid female gametophyte; dm, degenerating megaspores; ei, embryo initial cell; em, embryo; en, endosperm; F, fertilization; fm, functional megaspore; hfg, haploid female gametophyte; mmc, megaspore mother cell; ms, megaspore; mt, meiotic tetrad; Par, parthenogenesis; PsE, pseudogamous endosperm formation which requires fertilization; sc, seed coat; (+/−), present or absent. (Modified from Drews and Koltunow, 2011.) (For color detail, see color plate section.)

and cell differentiation events result in the formation of a mature female gametophyte containing an egg cell, two synergids, three antipodals, and a central cell with two haploid nuclei.

Subsequent seed formation requires double fertilization. A haploid sperm cell fuses with the haploid egg nucleus to generate a diploid zygote, with a 1 maternal (m):1 paternal (p) genome ratio, which differentiates into the diploid embryo. Fusion of a second sperm cell with the two central cell nuclei generates a triploid endosperm precursor cell with a 2m:1p genome ratio that proliferates to form nutritive endosperm. In cereals such as maize, the 2m:1p genome ratio is critical for endosperm viability (Koltunow and Grossniklaus, 2003). In many grasses and cereals, the antipodals proliferate after fertilization, whereas they degenerate before or during fertilization in other species. The ovule tissues surrounding the fertilized female gametophyte develop into the seed coat (Maheshwari, 1950; Willemse and van Went, 1984; Drews and Koltunow, 2011).

Sporophytic Apomixis

Sporophytic apomixis involves the direct formation of one or more embryos from diploid ovule cells, termed embryo initial (ei) cells, which differentiate adjacent to a developing female gametophyte (Figure 5.1*b*). It is common in mango and *Citrus* and is often termed adventitious embryony. Analyses in *Citrus* show that ei cells are specified early in ovule development, during megaspore mitosis. If the adjacent haploid embryo sac is fertilized, the zygotic embryo may or may not survive the competition for resources associated with multiple embryo growth in the viable polyembryonic seeds formed. If the embryo sac is not fertilized, multiple embryos derived from ei cells apparently grow by using nutrient derived from digestion of the massive nucellus in the ovule, and they arrest at the globular and heart stages of embryogenesis. Tissue culture can be used to rescue these embryos, which develop into viable seedlings (Koltunow *et al.*, 1995b). Polyembryonic seeds are commonly used to generate rootstocks for citrus propagation (Kepiro and Roose, 2010). Sporophytic apomixis is of minor focus in this chapter. It has not been extensively examined, although independent genetic studies indicate dominant control and simple inheritance of polyembryony in citrus (Garcia *et al.*, 1999; Nakano *et al.*, 2008a, 2008b; Kepiro and Roose, 2010).

Gametophytic Apomixis

Gametophytic apomixis results in the formation of a diploid or chromosomally unreduced female gametophyte because meiosis is bypassed by processes grouped under the term apomeiosis (Apm) (Figure 5.1*c* and *d*). An egg cell differentiates in the unreduced embryo sac and develops into an embryo in the absence of fertilization in a process termed parthenogenesis (Par). Endosperm formation in gametophytic apomicts may be fertilization independent (autonomous endosperm [AutE]) or may require fertilization (pseudogamous endosperm [PsE]). Autonomous endosperm formation is rare in apomicts and is predominantly found in the Asteraceae. Table 5.1 provides examples of numerous monocot and eudicot gametophytic apomicts and their modes of unreduced embryo sac and endosperm formation.

Apomeiosis occurs by two major routes, diplospory (Dip) (Figure 5.1*c*) and apospory (Apo) (Figure 5.1*d*), distinguished by the origin of the precursor cell giving rise to the unreduced embryo sac. In diplospory, the megaspore mother cell remains unreduced because either meiosis never initiates (mitotic diplospory, *Antennaria*-type) or it fails (meiotic diplospory), and the cell undergoes restitution at an early stage. Recombination has been observed in some forms of meiotic diplospory (van Baarlen *et al.*, 2000). Mitotic divisions ensue to generate an embryo sac (Figure 5.1*c*). In apospory, the megaspore mother cell differentiates, and sexual female gametophyte development initiates. Additional cells, termed aposporous initial (ai) cells, differentiate in close proximity to cells undergoing sexual female gametophyte development. One or more ai cells undergo mitosis to produce unreduced embryo sacs. The timing and frequency of ai cell appearance in relation to the events of the sexual pathway vary in different species. Typically, ai cells appear soon after megaspore mother cell enlargement or during megaspore tetrad formation (Figure 5.1*d*). In aposporous *Pennisetum* and *Hieracium* species, the sexual pathway terminates soon after ai cells begin to enlarge or undergo mitosis (Snyder, 1955; Peel *et al.*, 1997; Koltunow *et al.*, 1998, 2000, 2011a), whereas in other species both reduced and aposporous embryo sacs may coexist in ovules (Figure 5.1*d*).

Many unreduced embryo sacs contain eight nuclei before cellularization occurs and structurally resemble the sexual *Polygonum*-type embryo sac described earlier. These "eight-nucleate" unreduced embryo sacs are termed *Hieracium*-type in the apomixis literature. The *Panicum*-type, "four

Table 5.1 Features of Gametophytic Apomictic Model Systems under Study

Species	Apomixis type[a]	Traits Mapped[b]	Loci	References for Inheritance and Molecular Markers	Chromosomal Location	Transformable	Genome Sequence/No. Nucleotide Sequences in GenBank[c]
Taraxacum officinale (e)[d]	Dip AutE	Apm/Par/AutE	2–3	Tas and Van Dijk, 1999; Van Dijk *et al.*, 1999; Vijverberg *et al.*, 2004, 2010	nd[e]	nd	No/264 nucleotide; 41296 EST; 0 GSS; 0 SRA
Tripsacum dactyloides (m)	Dip PsE	Apm	1	Leblanc *et al.*, 1995; Grimanelli *et al.*, 1998b	nd	nd	No/548 nucleotide; 5 EST; 48 GSS; 0 SRA
Erigeron annuus (e)	Dip AutE	Apm/Par-AutE	2	Noyes and Rieseberg, 2000; Noyes *et al.*, 2007	nd	nd	No/51 nucleotide; 0 EST; 0 GSS; 0 SRA
Boechera divaricarpa (e)	Dip PsE	Clonal progeny	nd	Schranz *et al.*, 2006	nd	nd	*B. holboellii* pending/597 nucleotide *for B. divaricarpa*
Pennisetum squamulatum (m)	Apo (4-nucl)[f] PsE	Apm	1	Ozias-Akins *et al.*, 1993, 1998; Goel *et al.*, 2006; Huo *et al.*, 2009; Zeng *et al.*, 2011	Apm: telomeric (Akiyama *et al.*, 2004)	nd[g]	No/84 nucleotide; 0 EST; 2389 GSS; SRA Accn. SRX047367
Cenchrus ciliaris (m)	Apo (4-nucl) PsE	Apm or Apm-Par	1	Sherwood *et al.*, 1994; Gustine *et al.*, 1997; Roche *et al.*, 1999; Jessup *et al.*, 2002; Dwivedi *et al.*, 2007	Apm: interstitial/ pericentromeric (Akiyama *et al.*, 2005)	No stable	No/47 nucleotide; 21733 EST; 2145 GSS; 0 SRA
Paspalum simplex (m)	Apo (8-nucl)[h] PsE	Apm-Par	1	Pupilli *et al.*, 2001; Labombarda *et al.*, 2002	Distal (Calderini *et al.*, 2006)	nd	No/7 nucleotide; 0 EST; 0 GSS; 0 SRA
Hieracium spp (e)	Apo AutE	Par; Apm/Par-AutE	1 (Par); 2 (Apm/Par-AutE)	Bicknell *et al.*, 2000; Catanach *et al.*, 2006; Koltunow *et al.*, 2011a; Okada *et al.*, 2011	Apm (*LOA*) subtelomeric (Okada *et al.*, 2011)	Yes (Bicknell and Borst, 1994)	No/840 nucleotide; 132 EST; 0 GSS; SRA Accn. SRX082257
Hypericum perforatum (e)	Apo PsE	Apm/Par	2	Schallau *et al.*, 2010	nd	Yes (Franklin *et al.*, 2007)	No/63 nucleotide; 3 EST; 0 GSS; 22 SRA[i]

(*continued*)

Table 5.1 *(Continued)*

Species	Apomixis type[a]	Traits Mapped[b]	Loci	References for Inheritance and Molecular Markers	Chromosomal Location	Transformable	Genome Sequence/No. Nucleotide Sequences in GenBank[c]
Brachiaria brizantha (m)	Apo (4-nucl) PsE	Apm	1	do Valle and Savidan, 1996; Pessino *et al.*, 1997, 1998	nd	No stable	No/16 nucleotide; 2203 EST; 0 GSS; 0 SRA
Brachiaria humidicola (m)	Apo (4-nucl) PsE	Apm	1	Zorzatto *et al.*, 2010	nd		No/31 nucleotide; 0 EST; 0 GSS; 0 SRA
Panicum maximum (m)	Apo (4-nucl) PsE	Apm	1	Hanna *et al.*, 1973; Savidan, 1980; Ebina *et al.*, 2005	nd	No stably transformed plants[g]	No[j]/116 nucleotide; 196 EST; 0 GSS; 0 SRA
Paspalum notatum (m)	Apo (4-nucl) PsE	Apm	1	Martinez *et al.*, 2001, 2003; Stein *et al.*, 2007; Acuna *et al.*, 2009, 2011	nd	Yes (Smith *et al.*, 2002; Altpeter and James, 2005)	No/80 nucleotide; 0 EST; 0 GSS; 0 SRA
Poa pratensis (m)	Apo (8-nucl) PsE	Par; Apm/Par; Apm/Par	1 (Par); 2 (Apm/Par); 5 (Apm/Par)	Barcaccia *et al.*, 1998; Albertini *et al.*, 2001; Matzk *et al.*, 2005	nd	Yes (Ha *et al.*, 2001; Gao *et al.*, 2006)	No/239 nucleotide; 0 EST; 0 GSS; 0 SRA
Ranunculus auricomus (e)	Apo PsE	Apm/Par	1–2	Nogler, 1984b, 1995	nd	nd	No/8 nucleotide; 0 EST; 0 GSS; 0 SRA

[a]The mode of gametophytic apomixis in each species is denoted by the mode of unreduced embryo sac formation, either diplosporous (Dip) or aposporous (Apo). All form unreduced embryo sacs and undergo autonomous embryo development from unreduced eggs. The mode of endosperm formation is autonomous (AutE) or pseudogamous (PsE), which requires fertilization.

[b]Apm, apomeiosis; Par, parthenogenesis; AutE, autonomous endosperm.

[c]http://www.ncbi.nlm.nih.gov/ entries as of 06 Dec 2011. EST, expressed sequence tag; GSS, genome survey sequence; SRA, sequence read archive.

[d](e), eudicot; (m), monocot.

[e]nd, not determined (no physical data).

[f]Unreduced 4-nucleate *Panicum*-type aposporous embryo sacs are formed.

[g]Sexual *Pennisetum glaucum* has been transformed (Goldman *et al.*, 2003); sexual *Panicum virgatum* (switchgrass) has been transformed (Somleva *et al.*, 2002; Li and Qu, 2011).

[h]Unreduced 8-nucleate aposporous embryo sacs are formed.

[i]*Hypericum perforatum* SRA accessions: SRX062061-SRX062062, SRX062064-SRX062066, SRX062068-SRX062069, SRX062072-SRX062073, SRX062133-SRX062141, SRX062143-SRX062146.

[j]There is no genome sequence for *Panicum maximum,* although switchgrass (*P. virgatum*) and a diploid relative (*P. hallii*) are being sequenced by DOE-JGI; most *Panicum* sequences in GenBank (>700,000 ESTs; ∼200,000 GSS; 114 SRA) are from switchgrass.

nucleate" embryo sac found in many aposporous grasses is a notable exception. Only two rounds of mitosis occur, generating four nuclei that typically form an egg, two synergids, and a uninucleate central cell after cellularization. This diploid, uninucleate central cell in four-nucleate *Panicum*-type embryo sacs may be an adaptation to maintain a viable 2m:1p genome ratio in the pseudogamous endosperm, following fertilization (Koltunow and Grossniklaus, 2003).

Variations in mature unreduced embryo sac structure have been observed within the same genus. *Paspalum notatum* is characterized by four-nucleate *Panicum*-type aposporous embryo sacs, whereas *P. simplex* forms eight-nucleate aposoporous embryo sacs that are essentially indistinguishable from reduced embryo sacs of the *Polygonum*-type. Cytological analyses in *Hieracium* subgenus *Pilosella* species also indicate that the unreduced eight-nucleate *Hieracium*-type embryo sac is not always the stable outcome in aposporous species described by Rosenberg (1907) when apomixis was discovered in the genus. Abnormality in both the number of mitotic divisions and the way embryo sacs form have been described that lead to variations in embryo sac structure. For example, multiple developing embryo sacs may amalgamate to form a single structure of varying nuclear content in some aposporous *Hieracium* species (Koltunow *et al.*, 1998, 2000, 2011a).

Phylogenetic and Geographical Distribution of Apomixis

Perhaps the most recent comprehensive analysis of the phylogenetic distribution of apomixis was undertaken by Carman (1997). Of the 460 angiosperm families according to Judd *et al.* (1994), 33 contain gametophytic apomicts and are distributed across the more primitive lineages of Ranunculales and Piperales to the more advanced Poales and Asterales (Carman, 1997). More genera in the Poaceae (grass) and Asteraceae (sunflower) families contain apomicts than those of other plant families, with at least 38 and 35 genera (Ozias-Akins, 2006). Asteraceae and Poaceae are the first and fourth largest plant families in terms of number of genera and species, separated by Orchidaceae (orchids) and Fabaceae (legumes) (www.mobot.org). Gametophytic apomicts have not been reported in the Fabaceae even though it is a member of the rosids where other families demonstrating apomixis occur.

Apomixis is not the prevailing reproductive mode in evolution (Rollins, 1967). Given its broad distribution among angiosperms, yet infrequent occurrence, gametophytic apomixis is regarded as polyphyletic, and apomicts that either are more recently derived or confer some adaptive advantage are those that are extant. Asexual reproduction may be favored in geographically marginal habitats where sexual reproduction has a low rate of success (Silvertown, 2008), according to the survivalist "escape from sterility" yet "escape into a blind alley of evolution" theories described several decades ago (Darlington, 1939). However, geographical apomixis cannot be explained by a single predominant evolutionary force such as hybridization, polyploidy, male function, niche differentiation, or biotic interactions (Hoerandl, 2006). The "twiggy" phylogenetic distribution of asexuals, usually interpreted as an evolutionary dead end pattern, can also be predicted by three neutral models, only one involving extinction (Schwander and Crespi, 2009). The first posits that transition from sexual to asexual reproduction is rare, the second that origin and neutral loss are in equilibrium (Janko *et al.*, 2008), and the third that hybrid origin of asexuals is more frequent in high latitudes where extinction rates also are higher. These models were tested empirically with data only from animals but were proposed as alternatives to selective models used to explain "twigginess" of both animal and plant asexuals (i.e., mutation accumulation and ecological adaptation). Most investigations of distribution do not take into account the facultativeness of apomixis where the capacity for sexual reproduction within a clone enables recombination and gene flow to varying degrees, a phenomenon well documented in *Ranunculus*, *Taraxacum* (Hoerandl and Paun, 2007), and *Hieracium* (Fehrer *et al.*, 2007).

Part of the difficulty in phylogenetic comparison of sexuals and asexuals for testing of these models in plants originates with defining the species; asexuals often are considered members of polyploid agamic complexes (de Wet, 1968; Hoerandl, 1998). To study asexuals empirically compared with sexuals and to avoid the confounding effects of polyploidy and changes in ploidy invariably encountered among apomicts, Johnson *et al.* (2010) took advantage of a reproductive anomaly in evening primrose (*Oenothera*), permanent translocation heterozygosity or PTH, that parallels apomixis in its clonal reproductive outcome but occurs in a diploid organism. Diversification rates in asexual lineages were significantly faster (eightfold) than in sexual lineages where recombination and segregation can be homogenizing forces (Johnson *et al.*, 2011). This study provides evidence that asexual lineages are not evolutionary dead ends despite a constraint for adaptive evolution.

Inheritance of Apomixis

As discussed previously, sporophytic and gametophytic apomixis can be broken down into three developmental components. These include the avoidance of meiosis (Apm) to generate an unreduced embryo sac containing a cell capable of fertilization-independent embryogenesis (Par) and the generation of viable endosperm, which may occur by fertilization-dependent or fertilization-independent means. One to five dominant genetic loci have been proposed to control one or more of the three aforementioned components of apomixis in different species. However, the apparent simple Mendelian inheritance observed in many of these species (Table 5.1) is, most likely, an artifact of genes residing in a region of the genome suppressed for recombination (Ozias-Akins *et al.*, 1998; Roche *et al.*, 2001a; Pupilli *et al.*, 2004; Ozias-Akins and van Dijk, 2007). Recombination in some species has led to separation of traits for unreduced embryo sac formation and parthenogenesis, suggesting that at least two genes are required for apomixis. Even in these cases, however, one of the two loci may be in a region of reduced recombination (e.g., parthenogenesis in *Taraxacum* (van Dijk *et al.*, 2009) and diplospory in *Erigeron* (Noyes and Rieseberg, 2000)). Low recombination around apomixis loci may result from local chromosomal rearrangements that become fixed because of the asexual mode of reproduction. These recombination-deficient regions may ensure linkage disequilibrium for coding and regulatory regions that maximize the penetrance of apomixis (Vijverberg *et al.*, 2010).

Of the genetic mapping studies that have been conducted (Ozias-Akins and van Dijk, 2007), the most informative are in monocots *Pennisetum*, *Paspalum*, *Poa*, and *Tripsacum*, all members of the grass family (Poaceae), and in eudicots *Hieracium*, *Erigeron*, and *Taraxacum* (Table 5.1). No in-depth genetic studies have been undertaken on the near-diploid apomictic *Boechera holboellii* that is closely related to the model sexual plant species *Arabidopsis thaliana* and whose genome is presently being sequenced by the DOE Joint Genome Institute.

Simple dominant Mendelian inheritance of apospory or apomixis has been observed in *Pennisetum squamulatum*, *Cenchrus ciliaris*, *Panicum maximum*, *Tripsacum dactyloides*, multiple *Brachiaria* and *Paspalum* species, and *Ranunculus auricomus* where crosses between apomictic males and sexual females have been conducted (Table 5.1). Intraspecific crosses are possible when a sexual polyploid ecotype occurs, or the chromosome number of a sexual diploid has been artificially doubled to make a ploidy compatible cross with an apomictic tetraploid. In other species where no sexual ecotypes have been identified, interspecific crosses have been necessary to study inheritance. Given the heterozygosity of polyploid apomicts, genetic mapping can be conducted in the F_1

generation where sexual and apomictic modes of reproduction segregate 1:1. Significant deviation from 1:1 segregation has been observed most prominently in *Paspalum notatum* and interpreted as selection against male gametes carrying an apomixis allele either because of pleiotropy, incompletely penetrant lethality or a linked deleterious allele (Martinez *et al.*, 2001). Transmission of the *LOA* apomeiosis locus in aposporous *Hieracium* varies among species with exceptionally poor transmission of *LOA*-linked sequences when apomictic *H. caespitosum* (C36) is the pollen donor (Okada *et al.*, 2011). Less extreme segregation distortion has been observed in other *Hieracium* species (Catanach *et al.*, 2006; Okada *et al.*, 2011) and in other apomicts (Ozias-Akins *et al.*, 1998; Grimanelli *et al.*, 1998a, 1998b; Ozias-Akins and van Dijk, 2007). There is also evidence that transmission can vary between male and female gametes when sufficient numbers of reduced female gametes can be analyzed (Roche *et al.*, 2001b).

Although in most grass species, the events of unreduced embryo sac formation and parthenogenesis have not been genetically separated, there are some possible exceptions, most notably in *Poa pratensis* (Albertini *et al.*, 2001). Of 38 hybrids from a sexual × apomict cross, 22 were identified as non-parthenogenetic, and 20 of those showed only sexual development. In the two exceptional cases, aposporous embryo sacs were formed, but no embryos developed after auxin treatment, the typical test for parthenogenesis in *Poa* (Matzk, 1991), suggesting that unreduced embryo sac formation and parthenogenesis were independently controlled.

Kaushal *et al.* (2008) also described an "uncoupling" of apomixis components in *Panicum maximum* based on single seed flow cytometric seed screening (FCSS) (Matzk *et al.*, 2000). Reproductive versatility was demonstrated in some lines, for example, apospory with varying levels of parthenogenesis and occasional autonomous instead of the expected pseudogamous endosperm formation. None of these developmentally variable lines exclusively lacked parthenogenesis (i.e., formed fertilized unreduced eggs) or formed reduced eggs that underwent autonomous embryogenesis (i.e., lacked nonreduction). These results are better interpreted perhaps as possible examples of early failure of parthenogenesis in the case of putative autonomous endosperm detection or incomplete penetrance or expressivity of apomictic components.

Similar observations that occurrence of apomeiosis and parthenogenesis were highly correlated but that parthenogenesis frequency never exceeded apomeiosis frequency were made from FCSS of *Boechera spp.* accessions (Aliyu *et al.*, 2010). In *Boechera*, few controlled crosses to examine inheritance of apomeiosis and parthenogenesis have been made, and they have not supported a simple model of inheritance for apomixis (Schranz *et al.*, 2005, 2006). Many progeny of controlled crosses have been sterile (Schranz *et al.*, 2005), interpreted as lack of transmission of apomixis, and encumber genetic analysis of progeny. Cytological observation of embryo sac development was not conducted in these studies where inferences for genetic control of apomixis were made from assessment of progeny uniformity, fertility, and karyotype analyses. The transmission of apomixis was not associated with a single dominant factor on the heterochromatic (*Het*) chromosome that typically is found in apomicts because triploid progeny derived from a cross of diploid sexual *B. stricta* with diploid *B. divaricarpa* contained the full chromosome complement from the diploid apomictic parent, including the *Het* chromosome, but did not provide evidence for apomictic reproduction on progeny analysis (Schranz *et al.*, 2006). An apomixis factor on *Het* was not completely ruled out given that recombination could have occurred during the formation of the unreduced gametes through meiotic diplospory.

A complex five-locus model was invoked by Matzk *et al.* (2005) to explain variable penetrance of apomixis components in *Poa* that involves apospory initiator and preventer genes, parthenogenesis initiator and preventer genes, and a megaspore development gene. Segregation of apomixis components in progeny of several crosses was interpreted according to the model. Their analyses confirmed that aposporous plants lacking parthenogenetic capacity can be recovered in *Poa*, but the converse was not found. In these examples, it appears that parthenogenesis is largely dependent on apomeiosis, leaving open the question of control by an independent genetic element or pleiotropy.

Genetic studies in species of the family Asteraceae resolve parthenogenesis and apomeiosis as genetically distinct. Unlinked loci controlling apomeiosis and parthenogenesis have been clearly demonstrated in *Erigeron annuus*, *Taraxacum officinale*, and *Hieracium caespitosum* (reclassified as *H. praealtum*, genotype R35). All of these species form autonomous endosperm (Table 5.1). Genetic mapping of structured populations was used to determine the unlinked nature of these loci in *Erigeron* (Noyes and Rieseberg, 2000) and *Taraxacum* (Tas and van Dijk, 1999; van Dijk and Bakx-Schotman, 2004; van Dijk *et al.*, 1999), whereas mapping of mutations was used in *Hieracium* (Catanach *et al.*, 2006).

In *Taraxacum*, hybrid progeny were recovered at low frequency from a cross of a sexual diploid with apomictic triploids, and ploidy levels ranged from diploid to triploid and tetraploid. All tetraploids and one third of the triploids set seeds after emasculation establishing their apomictic mode of reproduction (Tas and van Dijk, 1999). Of the two thirds "nonapomictic" triploids and diploids, no seeds were set after emasculation. Subsequent analysis determined that diploids were sexual, but some of the "nonapomictic" triploids formed unreduced female gametes that developed only if fertilized indicating that apomeiosis and parthenogenesis were uncoupled in *Taraxacum* (van Dijk *et al.*, 1999). In one of the triploids, endosperm did initiate in the absence of pollination, but the egg cell remained undivided, suggesting that autonomous endosperm formation and parthenogenesis may also be uncoupled (van Dijk *et al.*, 2003).

Further genetic analysis in *Taraxacum* has focused on the *DIP* (diplospory) locus in genetic backgrounds where parthenogenesis is absent, simplifying genetic analysis and demonstrating that *DIP* behaves as a dominant, simplex allele (van Dijk and Bakx-Schotman, 2004; Vijverberg *et al.*, 2004). Yet this apparent simplicity may, as with several grasses, be complicated by duplication and local chromosomal rearrangements, complexity that was introduced to explain recombinants with reduced penetrance of diplospory (Vijverberg *et al.*, 2010).

Although no large-scale repression of recombination has been observed around *DIP* in *Taraxacum*, another diplosporous member of the same plant family, *Erigeron*, has shown recombination reduction associated with diplospory (Noyes and Rieseberg, 2000). As with *Taraxacum*, uncoupling of *DIP* and parthenogenesis allowed the recovery of *DIP*-only genotypes that showed an increase in chromosome number owing to fertilization of the unreduced egg cell (Noyes, 2005). The converse, parthenogenesis of reduced eggs, was not recovered, although functional apomixis was restored in progeny of *DIP*-only genotypes crossed with a female sterile plant that contained all of the molecular markers linked with parthenogenesis (Noyes, 2006). Parthenogenesis was restored more frequently after fertilization with diploid rather than haploid male gametes, although gametophytic versus zygotic lethality was not determined. Three apomicts with restored parthenogenesis did not contain any of the parthenogenesis-linked markers showing that linkage could be broken. No evidence for independent control of parthenogenesis and autonomous endosperm formation has been observed in *Erigeron,* and autonomous seed development was considered possibly due to pleiotropy of a single gene (Noyes *et al.*, 2007).

Mutation mapping in *Hieracium praealtum* and supporting genetic crosses have established that apomeiosis and autonomous seed initiation are controlled by two dominant, independent loci (Catanach *et al.*, 2006; Koltunow *et al.*, 2011a). Deletion of the dominant apomeiosis locus, *LOSS OF APOMEIOSIS* (LOA), recovered *loa* mutants, characterized by their inability to form ai cells. Deletion of the *LOSS OF PARTHENOGENESIS* (*LOP*) locus in *H. praealtum* recovered *LOAlop* mutants that formed unreduced embryo sacs but were unable to undergo autonomous seed formation. Pollination of *LOAlop* mutants resulted in progeny with increased ploidy level indicating fertilization of unreduced eggs. On analysis of unpollinated *loa* mutants carrying a functional *LOP* locus (*loaLOP*), 50% of seeds (*loaLOP*) developed as a consequence of autonomous embryo and endosperm formation, and progeny had half the chromosome number of the parent. This finding indicated that autonomous seed development occurred from a meiotically reduced egg and that *LOP* was segregating gametophytically in reduced embryo sacs owing to loss of *LOA* function (Koltunow *et al.*, 2011a). The identified *loalop* mutants exhibited a complete reversion to sexual reproduction, confirming that sexual reproduction is intact in the apomict *H. praealtum*, and it is the default reproductive pathway once *LOA* and *LOP* are eliminated from the genome. Further evidence in *Hieracium praealtum* suggests that *LOP* acts pleiotropically to induce autonomous embryo and endosperm development, or, if multiple genes are required, they are tightly linked (Koltunow *et al.*, 2011a).

An aposporous eudicot outside of the sunflower family with *Hieracium*-type unreduced embryo sac development but primarily pseudogamous endosperm formation is *Hypericum perforatum* (Martonfi *et al.*, 1996). *H. perforatum* displays considerable heterogeneity for reproductive outcomes where 11 routes have been observed (Matzk *et al.*, 2001). Seeds derived through apomeiosis without parthenogenesis and parthenogenesis without apomeiosis were identified suggesting that genetic components for these processes could be separated (Matzk *et al.*, 2001; Barcaccia *et al.*, 2006). Genetic analysis confirmed that apospory and parthenogenesis were under independent genetic control with each segregating as dominant, simplex alleles in $4x$ sexual $\times$ $4x$ apomict crosses, although linkage in *cis* at 20 cM was estimated (Schallau *et al.*, 2010).

Apomeiosis and parthenogenesis are likely under independent genetic control, but this independence cannot be easily demonstrated in some species because of their tight linkage. At the present time, there is little evidence for independent genetic control of autonomous endosperm formation. These straightforward conclusions are based on the treatment of apomeiosis, autonomous embryo, and autonomous endosperm development as qualitative traits. Quantitative analysis of these same traits has established that additional loci and genes influence penetrance or expressivity (Matzk *et al.*, 2005).

Genetic Diversity in Natural Apomictic Populations

Most apomicts are polyploid and heterozygous, implicating hybrid origin (Carman, 1997). Polyploidy and heterozygosity function as reservoirs of allelic diversity. True clonality in nature (i.e., obligate apomixis) is rare (Asker and Jerling, 1992), even though that would be the goal of engineered apomixis in crops. Among natural populations, greater than expected variation has been observed (Ellstrand and Roose, 1987; Campbell and Dickinson, 1990; Asker and Jerling, 1992; Hoerandl and Paun, 2007). Simulation of genetic variation in populations for which apomixis has become fixed showed that young populations reflect remnant variability of the sexual genotypes in the same geographic region, except where low pollen fertility affects the probability of fixation, and is strongly dependent on population size (Adolfsson and Bengtsson, 2007). As a population fixed for apomixis ages, genetic variability is expected to decrease because of random genetic drift or selection, although mutation acts to introduce new variation. This model for variation in an asexual population is independent of potential reproductive versatility related to the facultative nature of apomixis.

Facultative apomixis gives rise to varying frequencies of sexually derived progeny including those derived from fertilization of a meiotic egg $(n+n)$, along with ploidy changes related to progeny derived from autonomous development of a meiotic egg (n + 0) and progeny derived from fertilization of an unreduced egg (2n + n) (Asker and Jerling, 1992; Hoerandl and Paun, 2007). These arise as a result of uncoupling of apomeiosis and parthenogenesis either genetically or secondary to incomplete penetrance of apomixis in the ovule (Koltunow *et al.*, 2011a).

Male meiosis in many apomicts is relatively normal leading to viable male gametes that contribute to gene flow between apomicts and sexuals in overlapping geographic regions (van Dijk, 2003). Gene flow is more successful in taxa where interploidy crosses are tolerated, or apomictic and sexual genotypes of the same ploidy coexist. In *Taraxacum,* most apomicts found in nature are triploid, but rare tetraploids also exist in mixed polyploid asexual-diploid sexual populations (Verduijn *et al.*, 2004). Although rare, tetraploids were shown to be the primary source of pollen for the formation of new triploids from sexual, diploid mothers. In the aposporous subgenus *Hieracium Pilosella*, hybridization is common between sexual and apomictic species, and complex population structures with varying ploidy states exist (Fehrer *et al.*, 2007).

Molecular Relationships between Sexual and Apomictic Pathways

Developmentally, gametophytic apomixis appears to omit key steps present in the sexual pathway, including meiosis during gametophyte formation and the requirement for fertilization for embryo and, in some species, endosperm formation. Apomixis and sexuality in gametophytic apomicts are not mutually exclusive developmental programs because they coexist in facultative apomicts. Given these features, it has been proposed that gametophytic apomixis is not a new pathway requiring its own unique infrastructure. Instead, apomixis is thought to result from alterations in control of the sexual pathway, leading to a truncated pathway characterized by the omission of key steps or changes in the timing of events (Koltunow, 1993; Grimanelli *et al.*, 2001; 2003; Koltunow and Grossniklaus, 2003; Tucker *et al.*, 2003). Because most apomicts are polyploid, it has also been proposed that apomixis arose from heterochronic expression of the sexual pathway genes as a consequence of hybridization (Carman, 1997).

Cytological studies in *Tripsacum* showing alteration in timing of developmental events (Grimanelli *et al.*, 2003) and global analysis of gene expression in ovules of *Boechera* (Sharbel *et al.*, 2009, 2010) support the idea of a change in timing of the sexual program in apomixis. Developmental studies in *Hieracium* based on examining marker gene expression have provided support at the molecular level to the view that apomixis results from deregulated expression of the sexual program (Tucker *et al.*, 2003). Marker genes that are expressed in the female gametophyte, in the embryo, and during endosperm development show a similar expression pattern between sexual and asexual *Hieracium*; however, an early marker of the megaspore mother cell (MMC) is not expressed in aposporous initial cells (Tucker *et al.*, 2003). These findings are consistent with the view that apospory in *Hieracium* results in the initiation of gametogenesis directly from a somatic cell, bypassing the normal sequence of MMC specification, meiosis, and gametogenesis that is followed in the sexual pathway. Transcriptome analyses of *Hieracium* aposporous initial cells captured by laser microdissection should provide further information on gene expression programs in this enlarging cell (Koltunow *et al.*, 2011b).

The observation that deletion of *LOA*, the apomeiosis locus, and *LOP*, the locus required for autonomous seed initiation in aposporous *Hieracium praealtum*, results in a reversion to sexual reproduction indicates these loci are not essential for sexual reproduction and further supports their likely involvement in deregulating the default sexual program (Koltunow *et al.*, 2011a). Comparative transcriptional analyses of the changes in gene expression programs in aposporous *Hieracium* and derived mutants may provide information on the downstream processes influenced by the *LOA* and *LOP* apomixis loci (Koltunow *et al.*, 2011b).

The initiation of the sexual program in aposporous *Hieracium* is required to activate *LOA* function, which enables aposporous initial cell formation. Prevention of MMC meiosis by transgenic ablation, using a cytotoxic marker targeted to the enlarging MMC, inhibits aposporous initial cell formation in *H. piloselloides* (Koltunow *et al.*, 2011a). This provides initial evidence for signaling between sexual and apomictic pathways in *Hieracium*.

The locus controlling apomixis in facultative apomictic maize-*Tripsacum* hybrids containing one *Tripsacum* and two maize genomes behaves differently according to context (Leblanc *et al.*, 2009). When present in unreduced female gametes, it is functional, but when present in reduced gametes, it is nonfunctional and gives rise to progeny that reproduce sexually. These results suggest that the *Tripsacum* apomixis control region may interact with other *Tripsacum*-specific genes or alleles to confer apomixis, which is a barrier to transmission of the trait to sexual maize. The absence of these *Tripsacum*-specific genes in reduced gametes, together with apomixis, and the presence of a sexual phenotype (Leblanc *et al.*, 2009) also suggest that apomixis in diplosporous *Tripsacum* may be imposed on sexuality as found for aposporous *Hieracium* so that loss of apomixis results in reversion to sexuality.

Given the facultative nature of apomixis and outcomes of genetic analyses in other apomictic species, discussed previously, the default reproductive mode is also likely to be sexual reproduction in species other than *Hieracium* and *Tripsacum*. The efficiency of apomictic seed set in facultative apomicts is likely to be dependent on the degree of dominance and the penetrance of the apomictic pathway over the sexual, and potential signaling between the two pathways that may lead to the demise of sexual gametophyte formation in some species.

Features of Chromosomes Carrying Apomixis Loci and Implications for Regulation of Apomixis

Relationships between chromosome structure and the genetic control of apomixis have been proposed in numerous studies. The *DIP* locus of *Taraxacum* is thought to be located on a satellite chromosome (van Dijk and Bakx-Schotman, 2004). All diplosporous *Boechera* apomicts studied contain one or two aberrant chromosomes, *Het* and *Del* (Kantama *et al.*, 2007), although *Het* does not carry a single dominant locus for apomixis (Schranz *et al.*, 2006), and *Del* is not always present. Association of diplospory loci with chromosomal features in these species has not been confirmed.

The chromosome carrying the apomeiotic *LOA* locus in three *Hieracium* subgenus *Pilosella* species shares features with the Apospory Specific Genomic Region (ASGR = Apm + Par) conferring aposporous apomixis in *Pennisetum squamulatum* (Akiyama *et al.*, 2004; Okada *et al.*, 2011). Both are located at the distal subtelomeric end of an individual chromosome in a hemizygous chromosomal region containing repetitive sequences (Table 5.1). The AGSR of aposporous *C. ciliaris*, a *Pennisetum* relative, is pericentromeric in location and also surrounded by complex repeats (Akiyama *et al.*, 2005). This repeat structure is not universally detected at the ASGR of all 12 species of apomictic *Pennisetum* examined (Akiyama *et al.*, 2011). By contrast, the apospory locus in *Paspalum notatum*, a member of the same tribe of the Poaceae family, also has a hemizygous chromosomal location, but there is currently no association of the region with complex repeats (Calderini *et al.*, 2006). Similarly, *LOA*-specific linked markers and associated chromosomal structures are not conserved in aposporous *Hieracium aurantiacum*, which is found in a different haplotype network group to the three other species where chromosomal features are conserved (Okada *et al.*, 2011). This may reflect polyphyletic evolution of the apospory locus within the grass family as well as subgenus *Pilosella*.

Partial sequencing of genomic DNA at the ASGR (Conner *et al.*, 2008) in *Pennisetum* and *Cenchrus* and DNA at the *LOA* locus in *H. praealtum* tightly linked to apospory shows the presence of abundant Ty-copia and Ty-gypsy transposons in addition to complex locus associated repeats that show no obvious sequence conservation (Okada *et al.*, 2011). These monocot and eudicot species diverged 42–47 million years ago (Barreda *et al.*, 2010), and comparisons of sequences at apomixis loci indicate the ASGR and *LOA* loci do not share a common ancestral origin. Instead, the sequence comparisons indicate convergent evolution of these repeat-rich hemizygous chromosomal regions containing apospory loci.

Current sequence information indicates these apomixis associated genomic regions are gene poor. The prediction is that genes present in these repetitive regions would be under strong selection, presumably with an important functional role, to have survived invasion by transposons and other structural chromosome rearrangements. The hemizygous chromosomal repeat-rich apomeiosis loci would be predicted to be suppressed for recombination, which is the case for *Pennisetum* (Goel *et al.*, 2006). Some limited recombination occurs at the *LOA* locus in *Hieracium praealtum* (Okada *et al.*, 2011) and tetraploid *H. piloselloides* (unpublished), which may facilitate identification of the critical sequences associated with apospory.

The conservation of chromosomal structures in these divergent monocot and eudicot species also suggests that such complex repeat regions associated with these loci may be required for function and maintenance of the apomixis trait. Epigenetic pathways may influence gene expression at apomixis loci surrounded by transposons and complex repeat structures because such regions are often targets of small RNAs that induce RNA-directed DNA methylation that suppresses or alters gene expression (Martienssen *et al.*, 2004; Vaucheret, 2008). Disruptions in epigenetic small RNA and DNA methylation pathways in sexual *Arabidopsis* and maize have led to gametophytic apomixis-like phenotypes in these sexual plants (discussed subsequently). There is currently no direct evidence for a role or the involvement of these pathways in natural apomicts.

Observations of apomixis mutants in *Hieracium* where deletions of *LOA* and *LOP* loci lead to a reversion to sexual reproduction implies that it is unlikely that these loci contain genes essential for the progression of the sexual pathway. These apomixis loci might contain genes that interfere with genetic and epigenetic programs (described subsequently) that function to restrict the numbers of female gamete precursor cells to a single cell and regulate the orderly progression of meiosis and double fertilization in the sexual pathway.

Genes Associated with Apomixis

The genes responsible for the manifestation of apomixis have not yet been identified. The genetic analysis of apomixis, development of markers, partial sequencing of genomic regions containing apomixis loci, and analysis of genes differentially expressed in apomictic versus sexual plants have identified numerous candidate "apomixis" genes. Some of these have functional counterparts in sexual reproduction (discussed subsequently).

The partially sequenced *P. squamulatum* ASGR contains a *BABYBOOM* (BBM)-like gene (Conner *et al.*, 2008). *BBM* was identified in *Arabidopsis* and is an AP2-domain transcription factor. Over-expression in *Arabidopsis* leads to the development of embryos and cotyledons from vegetative tissues (Boutilier *et al.*, 2002). Differential screening of sexual and apomictic ovules in

Poa pratensis has identified a *SOMATIC EMBRYOGENESIS RECEPTOR (SERK)-LIKE KINASE, PpSERK*, and *APOSTART*. *PpSERK* is thought to be the switch that channels embryo sac development and embryo induction. *APOSTART*, a gene with homology to the *Arabidopsis* lipid-binding START-domain containing protein, is thought to be involved in meiosis and programmed cell death (Albertini *et al.*, 2005). Analysis of the *Hypericum* apospory locus (*HAPPY*) has identified a CAPS marker cosegregating with apospory but not with parthenogenesis, which is independently controlled in *Hypericum* (Schallau *et al.*, 2010). The marker is part of a gene showing high similarity to the ARIADNE (ARI) family of proteins, a class of ring finger proteins that putatively function as E3 ubiquitin protein ligases and has high similarity to *Arabidopsis ARI7*. Comparison of sexual and apomictic alleles showed that the allele identified by the CAPS marker in aposporous plants was truncated and surrounded by copia-like transposons and pseudogenes. The tetraploid apomict also carries three other intact alleles, whereas the sexual plant has four intact copies of the gene (Schallau, 2010). The role of all of these genes in apomixis is unclear. In some cases, progress to clarify their functional roles is hampered by the inability to transform sexual and apomictic species under study (Table 5.1), and surrogate sexual models are being used.

Mutations in members of the *Arabidopsis* FERTILIZATION INDEPENDENT SEED (FIS) complex initiate endosperm proliferation in the absence of fertilization (see later) (Rodrigues *et al.*, 2010a). *Hieracium* orthologues of two *Arabidopsis FIS*-complex genes (*FIE* and *MSI1*), whose functions are elaborated on further subsequently, are not linked to the autonomous seed initiation (*LOP*) locus in *Hieracium praealtum* (Rodrigues *et al.*, 2010b; Koltunow *et al.*, 2011a, 2011b). Downregulation of *Hieracium FIE* (*HFIE*) does not lead to autonomous seed formation in sexual *Hieracium*. However, *HFIE* function is required for the formation of viable fertilization-dependent embryo development in sexual *Hieracium* and for viable autonomous embryo formation in apomictic *Hieracium* (Rodrigues *et al.*, 2008).

Transferring Apomixis to Sexual Plants: Clues from Apomicts

To what extent does the evolution of apomixis in natural systems affect attempts to transfer apomixis or engineer apomixis in crops for which true-breeding hybrids provide a performance advantage? Transfer has been unsuccessful where traditional breeding has taken advantage of crop species with close apomictic relatives (Savidan, 2000). In the case of *Tripsacum*, we discussed previously evidence suggesting the presence of barriers in the transfer of apomixis to maize (Leblanc *et al.*, 2009). The greatest progress to date has been made with introgression of the trait into pearl millet where a single alien chromosome persists in the most advanced apomictic lines (Singh *et al.*, 2010; Zeng *et al.*, 2011).

In retrospect, the slow progress can be explained by the necessity to work at the polyploid level and the genetic consequence of chromosomal repatterning that has occurred in the absence of recombination owing to apomeiosis. Similarly, segregation distortion against apomixis through male (Martinez *et al.*, 2001; Nogler, 1984b; Ozias-Akins *et al.*, 1998) or female (Roche *et al.*, 2001b) gametes may be due to linked deleterious alleles. These properties discourage the use of natural apomicts to develop diploid apomictic crops through hybridization.

If the identified apomixis genes prove nonfunctional in sexual plants because they require additional species-specific factors, strategies for the transfer of apomixis to sexual species may require identification and analysis of the downstream target genes that are regulated by the apomixis control region. Did polyploidy (through hybridization) play a central role in the origin of apomixis, or did it arise as a by-product of enhanced unreduced gamete formation? Will polyploidy be required for apomixis to function? Derived apomictic polyhaploids ($n + 0$ progeny) suggest that the latter is not the case even though polyhaploids usually are less fit presumably because of the unmasking of deleterious mutations from the genome of the maternal apomict (Dujardin and Hanna, 1986; Leblanc *et al.*, 1996; Bicknell, 1997).

Synthetic Approach to Building Apomixis

The idea that apomixis results from temporal and spatial alterations in the action of the sexual program suggests the possibility that variant alleles of genes that act in the normal sexual pathway can result in components of apomixis, and that it ought to be possible to identify such genes by mutational analysis in sexual plants (Chaudhury and Peacock, 1993). By identifying and combining the components, it should also be possible to build apomixis in a sexual plant. Molecular genetic analysis in model sexual plants (mainly *Arabidopsis* and rice) has provided support for the validity of this approach. A selected set of genes that are relevant in the context of apomixis is presented in Table 5.2.

Specification of MMC

Analysis of genes controlling sporogenous cell identity and fate in *Arabidopsis*, rice, and maize has identified genes involved in specification of the MMC. Mutations in *Arabidopsis SPOROCYTELESS/NOZZLE* result in failure to form a MMC and defects in nucellar cell identity (Schiefthaler *et al.*, 1999; Yang *et al.*, 1999). Mutations in *Arabidopsis WUSCHEL*, a regulator of stem cell identity in the shoot apical meristem, also result in defects in MMC specification (Gross-Hardt *et al.*, 2002). *WUSCHEL* acts by controlling the expression of *WINDHOSE1* and *WINDHOSE2,* which encode small peptides and act in conjunction with *TORNADO2* in controlling megasporogenesis (Lieber *et al.*, 2011). In addition, the *MULTIPLE ARCHESPORIAL CELLS1* (*MAC1*) gene of maize (Sheridan *et al.*, 1996), and the *MULTIPLE SPOROCYTES1* (*MSP1*) and *TAPETUM DETERMINANT LIKE1a* (*TDL1a*) genes in rice (Nonomura *et al.*, 2003; Zhao *et al.*, 2008) are required to limit the number of MMCs to one per ovule. Loss of gene function results in multiple MMCs indicating that these genes negatively regulate sporogenous cell fate in the ovule. The phenotype bears some relationship to apospory wherein multiple cells in the ovule are capable of initiating gametogenesis, although in apospory this occurs without initiating cells undergoing meiosis. *MSP1* encodes a leucine-rich repeat receptor kinase, and *TLD1a* encodes its ligand. Analogous phenotypes to *msp1* mutants occur in male but not female sporogenous development in the *extra sporogenous cells/excess microsporocytes1* (*exs/ems1*) as well as *tapetum determinant1* (*tpd1*) mutants of *Arabidopsis* (Canales *et al.*, 2002; Zhao *et al.*, 2002; Yang *et al.*, 2003). *EXS/EMS1* and *TPD1* are orthologous to *MSP1* and *TDL1a*.

Loss of *ARGONAUTE9* function as well as other genes in the small RNA pathway, *RNA-DEPENDENT RNA POLYMERASE6* (*RDR6*) and *SUPPRESSOR OF GENE SILENCING3* (*SGS3*), in *Arabidopsis* has been shown to result in dominant loss of restriction in gametic cell identity in the ovule leading to expression of gametic markers in multiple cells of the nucellus that have not undergone meiosis (Olmedo-Monfil *et al.*, 2010). The implication of these results is that gametic cell fate can be uncoupled from meiosis in *ago9* mutants; however, it is unclear whether some of these cells can go on to form functional apomeiotic gametes. A phenotype similar to *ago9* is observed in mutant alleles of *MNEME* (*MEM*), which encodes a DEAD/DEAH box RNA-helicase that shows increased expression in the MMC (Schmidt *et al.*, 2011). In addition to loss of restriction of gametic identity, *mem* mutants also show defects in female gametogenesis and embryogenesis.

Table 5.2 Selected Mutants and Genes from Sexual Plants that Show Features of Apomixis

Mutant/Gene	Mutant Phenotype	Molecule	Reference
Apomeiosis			
SWI1	Diplospory-like	Meiosis-specific chromatin-associated protein	Ravi *et al.*, 2008
AM1	Meiotic non-reduction	SWI1 homolog	Pawlowski *et al.*, 2009
mac1	Multiple archesporial cells	nd	Sheridan *et al.*, 1996
MSP1	Multiple archesporial cells	Leucine-rich repeat receptor kinase	Nonomura *et al.*, 2003
TDL1a	Multiple archesporial cells	Small extracellular protein	Zhao *et al.*, 2008
MiMe (*Atspo11-1 Atrec8 osd1*)	Diplospory-like	Atspo11-1: double-strand endonuclease; Atrec8: meiosis-specific cohesion; osd1: cell-cycle progression protein	d'Erfurth *et al.*, 2009
MiMe-2 (Atspo11-1 Atrec8 tam1)	Diplospory-like	tam1: CYCLIN A1;2	d'Erfurth *et al.*, 2010
AGO9	Apospory-like	ARGONAUTE protein	Olmedo-Monfil *et al.*, 2010
AGO104	Diplospory-like	ARGONAUTE protein	Singh *et al.*, 2011
RDR6	Apospory-like	RNA-dependent RNA polymerase	Olmedo-Monfil *et al.*, 2010
SGS3	Apospory-like	RNA binding protein	Olmedo-Monfil *et al.*, 2010
DMT102	Diplospory-like	DNA methyltransferase	Garcia-Aguilar *et al.*, 2010
DMT103	Diplospory-like	DNA methyltransferase	Garcia-Aguilar *et al.*, 2010
MEM	Apospory-like	RNA helicase	Schmidt *et al.*, 2011
Parthenogenesis/embryogenesis			
hap	Genome elimination	nd	Hagberg and Hagberg, 1980
IG1	Genome elimination, multiple embryos	LOB domain protein	Evans, 2007
LEC1	Promotes embryogenesis	HAP3 subunit of CCAAT box-binding factor	Lotan *et al.*, 1998
BBM1	Promotes embryogenesis	AP2 domain transcription factor	Boutilier *et al.*, 2002
AtSERK1	Promotes embryogenesis	LRR receptor-like kinase	Hecht *et al.*, 2001
MSI1	Parthenogenesis	WD40 domain protein; part of PRC2 Polycomb complex	Guitton and Berger, 2005
Autonomous endosperm			
MEA	Autonomous endosperm	SET domain polycomb protein	Grossniklaus *et al.*, 1998
FIS2	Autonomous endosperm	Part of C2H2 zinc finger protein of Polycomb complex	Luo *et al.*, 1999
FIE	Autonomous endosperm	WD40 domain protein; part of Polycomb complex; EXTRA SEX COMBS (*ESC*) homolog	Ohad *et al.*, 1999
MSI1	Autonomous endosperm	See above	Guitton and Berger, 2005
Endosperm balance			
MET1	Reduction of parent of origin effect on seed size	DNA methyltransferase	Vinkenoog *et al.*, 2001
TTG2	Reduced lethality in interploidy crosses	WRKY transcription factor	Dilkes *et al.*, 2008

Apomeiosis

There are several cases of unreduced gametes arising from sexually derived mutations in plants (Bretagnolle and Thompson, 1995). Unreduced gametes can arise from a range of meiotic abnormalities; however, in most cases the unreduced female gametes do not fully retain the parental genotype and are not truly apomeiotic. It was demonstrated more recently that a mutation in a molecularly characterized gene can lead to apomeiosis for the *dyad* mutant, a hypomorphic allele of the *DYAD/SWITCH1* (*SWI1*) gene of *Arabidopsis* that encodes a truncated protein (Ravi *et al.*, 2008). This finding provided a proof of principle that alteration of a single gene of known molecular identity can lead to a functional component of apomixis including formation of viable seed. *SWI1*, similar to its maize orthologue *AMEIOTIC1* (*AM1*) (Hamant *et al.*, 2006), encodes a meiotic-specific protein required for sister chromatid cohesion and chromosome organization during meiosis. However, the efficiency of seed set in *dyad/swi1* mutants is too low to be directly applied as a synthetic model for diplospory.

An efficient model for diplosporous apomeiosis has been shown to result from the combination of mutations in three genes – *ATSPO11-1*, *ATREC8*, and *OMISSION OF SECOND DIVISION1* (*OSD1*) – that control different aspects of chromosome organization and cell division in meiosis (d'Erfurth *et al.*, 2009). *ATSPO11* encodes a type VI topoisomerase–related protein that is responsible for the formation of recombination-initiating double-strand breaks (Grelon *et al.*, 2001). Chromosomes in the *Atspo11-1* mutant fail to undergo recombination and pairing, remaining as univalents. *ATREC8*, which encodes a meiotic-specific cohesin, is required for co-orientation of sister centromeres and monopolar attachment to the spindle in meiosis I (Bai *et al.*, 1999). In the *Atspo11-1 Atrec8* double mutant, the reductional meiosis I division is replaced by an equational one equivalent to mitosis (Chelysheva *et al.*, 2005). *OSD1* is required for meiosis II, and loss of *OSD1* function leads to omission of the second division of meiosis and formation of unreduced spores. The *Atspo11-1 Atrec8 osd1* triple mutant combination has been termed Mitosis from Meiosis (MiMe) and undergoes a single division that is genetically equivalent to mitosis, resulting in the efficient production of functional apomeiotic female gametes. A similar phenotype is obtained by substituting the *osd1* mutant allele in the above combination with a *tardy asynchronous meiosis* (*tam*) mutant allele (d'Erfurth *et al.*, 2010). *TAM* encodes CYCLIN A1;2, which cooperates with OSD1 in controlling progression through meiosis I and meiosis II.

Arabidopsis ago9, *rdr6*, and *sgs3*, as discussed previously, show a loss in restriction of gametic cell fate. The maize *AGO104* gene is related to *Arabidopsis AGO9*; however, the dominant loss of function *ago104* mutant shows a different phenotype to *ago9* as altered MMC meiosis leads to the formation of functional unreduced female gametes (Singh *et al.*, 2011). The *ago104* mutant phenotype in maize resembles diplosporous apomeiosis, whereas the *Arabidopsis ago9* phenotype resembles apospory. Likewise, the *mel1* mutant of rice (Nonomura *et al.*, 2007), which is in an *Arabidopsis AGO5* related gene, shows a phenotype of meiotic arrest at leptotene. Overall, these findings point to a role for distinct branches of the small RNA pathway in control of early events in megasporogenesis and gametic cell fate specification.

A comparative expression analysis for genes encoding chromatin modifying enzymes between sexual and apomictic ovules of maize-*Tripsacum* hybrids revealed differences for six genes (Garcia-Aguilar *et al.*, 2010). Three of the genes that encode putative DNA methyltransferases (DMT102 and DMT103) and a chromomethylase (CHR106) belong to a class of genes implicated in RNA-directed DNA methylation and are strongly down-regulated in apomictic ovules. *DMT102* and *DMT103* also show down-regulation in autotetraploid maize compared with diploid suggesting that their expression is ploidy dependent, whereas *CHR106* expression was specific to apomictic ovules and independent of ploidy. A role for DNA methylation in control of meiosis and germ cell formation has been shown in the case of *dmt102* and *dmt103* mutants of maize, which show unreduced male gamete formation as well as the formation of multiple embryo sacs. The origin and ploidy of the extra embryo sac has not been determined (Garcia-Aguilar *et al.*, 2010). These findings suggest that control of DNA methylation is important for normal meiosis and gametogenic cell number in the maize ovule.

Activation of Seed Development

The early mutant screens for apomixis in *Arabidopsis* led to the identification of fertilization-independent initiation of seed development (*FIS*) class genes, which encode members of the Polycomb-related complex (see Chapter 4) (Koltunow and Grossniklaus, 2003). The *fis*-class mutants initiate endosperm development without fertilization with different degrees of expressivity. However, the frequency of embryo initiation by division of the egg cell is lower or does not occur at all for most *fis* mutants. An exception is *multicopy suppressor of ira1* (*msi1*), which gives a high frequency of embryo initiation (90%) (Guitton and Berger, 2005); however, the embryos are inviable.

An additional category of genes has been identified that when over-expressed lead to increased frequency or competence to form somatic embryos either *in planta* or in cell cultures. Several of these genes act as developmental regulators in the normal pathway of embryogenesis. Promoters of somatic embryo formation have been considered as potential candidates for engineering sporophytic apomixis as takes place in adventitious embryony. Members of this class of genes include *LEAFY COTYLEDON1* (*LEC1*) (Lotan *et al.*, 1998), *BABYBOOM1* (*BBM1*) (Boutilier *et al.*, 2002), and *Arabidopsis SOMATIC EMBRYOGENESIS RECEPTOR KINASE1* (*AtSERK1*) (Hecht *et al.*, 2001).

The introduction of apomixis in sexual plants would require addressing the problem of endosperm balance arising out of the change in ploidy of the apomeiotic female gamete. Natural apomicts have adapted themselves in several ways to address the issue of endosperm balance. The evolutionary adaptations range from a change in the fertilization process or in female gametogenesis so as to maintain the 2m:1p genome ratio that is normally required in the endosperm, or a corresponding increase in ploidy of the male gamete, or endosperm and seed development has been made insensitive to m:p ratio as in the case of autonomous apomicts (Koltunow and Grossniklaus, 2003).

Genetically equivalent to parthenogenesis is the phenomenon of uniparental genome elimination. The *INDETERMINATE GAMETOPHYTE1* (*IG1*) gene of maize encodes a lateral organ boundaries (*LOB*) domain protein required for embryo sac and leaf development, and *ig1* mutants exhibit formation of both maternal and paternal haploid embryos (Evans, 2007). Uniparental genome elimination is seen in certain interspecific as well as intraspecific hybrids and results in elimination of the entire set of chromosomes in the embryo from one parent after fertilization. The haploid inducer (*hap*) mutant of barley, which arose as an induced mutation, produces a high frequency of haploids among its progeny (Hagberg and Hagberg, 1980). The act of fertilization triggers seed development, but the genetic contribution from one parent is removed in the developing zygote/embryo after seed development has initiated. The technique of genome elimination has been used in plant breeding for making doubled haploids and for transfer of the genome from one cytoplasm to another. Genome elimination in interspecific crosses has been postulated to be caused by centromere incompatibilities (Bennett *et al.*, 1976). It has been shown more recently that alteration of the centromere-specific histone CENH3 in *Arabidopsis* causes selective elimination of the chromosome set from the mutant parent when crossed to a parent with wild-type CENH3 (Ravi and Chan, 2010). The finding provides a rational and mechanistic basis for engineering haploid inducer lines in sexual plants. Although uniparental genome elimination takes place with partial expressivity, there is potential for increasing the efficiency of elimination, and the phenomenon is of interest as part of a strategy for synthesis of apomixis.

Synthetic Clonal Seed Formation

The occurrence of apomeiosis in single or combinations of sexually derived mutants of *Arabidopsis* (*dyad* and MiMe) can be combined with *cenh3*-mediated chromosome elimination to produce fully clonal seeds (Figure 5.2). Such a combination of apomeiosis with chromosome elimination has been successfully shown to result in the formation of clonal seeds at a frequency of 35% of viable seeds (Marimuthu *et al.*, 2011). These results demonstrate a proof of principle for being able to

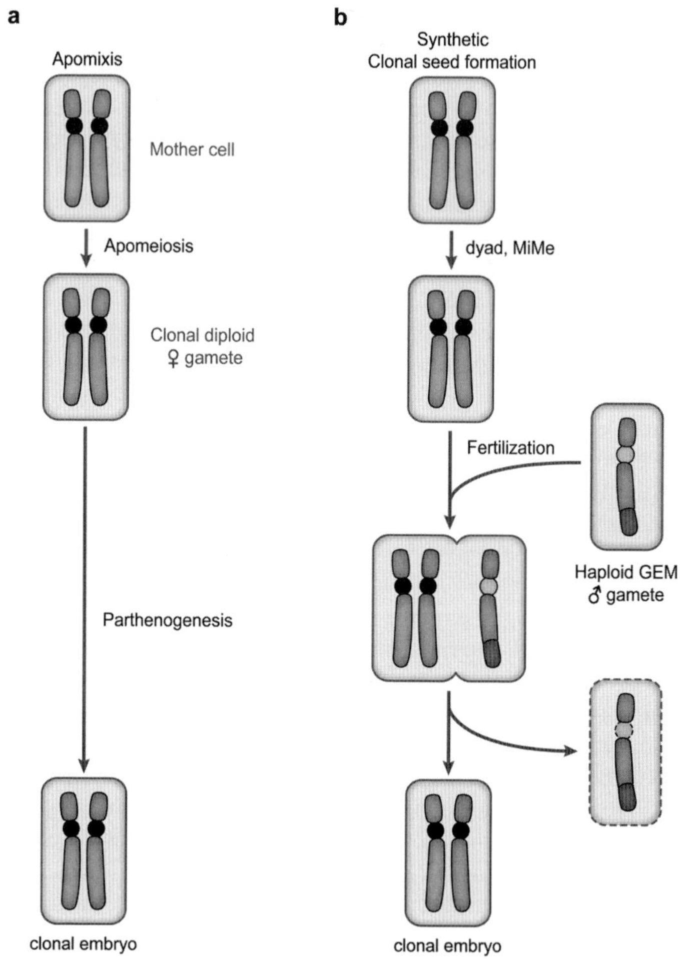

Figure 5.2 Natural and synthetic clonal reproduction through seed. *a,* In natural apomicts, clonal embryos with a maternal genotype develop without fertilization. *b,* Induction of clonal embryos in sexual plants as described by Marimuthu *et al.* (2011) involves combining mutants so that gametogenesis occurs without meiosis and subsequent fertilization with a parent whose chromosomes are modified to be eliminated after fertilization. (Modified from Marimuthu *et al.*, 2011.) (For color detail, see color plate section.)

engineer an equivalent for the outcome of apomixis by alterations in a small number of genes whose molecular identity is known.

Limitations of the system are that in some combinations the sexual pathway is no longer active in these plants as occurs in apomicts and is not easily reversible. Flexibility to switch between sexual and apomictic reproductive modes in a plant is a clear advantage for breeding apomictic crops containing multiple performance traits. The efficiency of clonal seed formation in this synthetic system is also lower than found in many apomicts, and it still relies on crossing, whereas to achieve the benefits of apomixis, the plants would need to propagate themselves without having to perform crosses. It may be possible to address some of these issues through further improvements in the methodology.

Conclusion and Future Prospects

It is apparent that the development of apomixis will require a biotechnological approach to install it in crops for which true-breeding hybrids provide a performance advantage. Analyses of apomixis control regions have shown that in several cases, loci controlling apomixis show complex structure and organization. In several species that exhibit apparently simple Mendelian inheritance of apomixis it is likely that apomixis control regions harbor more than one genetic determinant for the control of apomixis. Apomixis control in *Hieracium* and *Tripsacum* appears to be overlaid onto a background of sexual reproduction where loss of apomixis results in reversion to sexuality. These findings are consistent with the notion that apomixis results from altered control of the normal sexual program resulting in truncation or omission of events, or changes in timing of developmental events, rather than a substitution or mutational inactivation of sexual processes. The precise genetic determinants within apomixis control regions that are responsible for the trait remain to be identified. Partial sequencing of the DNA from these regions in *Pennisetum* and *Hieracium* shows the regions to be gene poor and abundant in repeat elements and transposons.

Parallel to the study of natural apomicts, analysis of sexually derived mutations has identified a number of genes whose mutation or altered expression results in phenotypes that resemble components of apomixis. Several of these genes are implicated in epigenetic control of gene expression by modification of DNA or chromatin and taken together point to the possibility that apomixis is under epigenetic control, although direct evidence in support of this remains to be obtained in the range of apomicts under study.

It has recently been possible to combine mutant alleles of molecularly characterized genes that result in apomixis related phenotypes in *Arabidopsis* to produce clonal seed, which demonstrates that it ought to be possible to synthesize apomixis in sexual plant species by manipulation of a small number of genes. Given the diversity of developmental mechanisms by which apomixis occurs, it is likely that there will be multiple ways in which apomixis may be developed as a technology for fixing heterosis and other desirable traits in crops to enable increased productivity under variable environmental conditions. Information on how to do this will come from studies on both natural apomicts and sexual model systems and strategies will also depend on agronomic features of the target crop.

References

Acuna CA, Blount AR, Quesenberry KH, Kenworthy KE, Hanna WW (2009). Bahiagrass tetraploid germplasm: reproductive and agronomic characterization of segregating progeny. Crop Sci 49: 581–588.

Acuna CA, Blount AR, Quesenberry KH, Kenworthy KE, Hanna WW (2011). Tetraploid bahiagrass hybrids: breeding technique, genetic variability and proportion of heterotic hybrids. Euphytica 179: 227–235.

Adolfsson S, Bengtsson BO (2007). The spread of apomixis and its effect on resident genetic variation. J Evol Biol 20: 1933–1940.

Akiyama Y, Conner JA, Goel S, Morishige DT, Mullet JE, Hanna WW, Ozaia-Akins P (2004). High-resolution physical mapping in *Pennisetum squamulatum* reveals extensive chromosomal heteromorphism of the genomic region associated with apomixis. Plant Physiol 134: 1733–1741.

Akiyama Y, Goel S, Conner JA, Hanna WW, Yamada-Akiyama H, Ozias-Akins P (2011). Evolution of the apomixis transmitting chromosome in *Pennisetum*. BMC Evol Biol 11: 289.

Akiyama Y, Hanna WW, Ozias-Akins P (2005). High-resolution physical mapping reveals that the apospory-specific genomic region (ASGR) in *Cenchrus ciliaris* is located on a heterochromatic and hemizygous region of a single chromosome. Theor Appl Genet 111: 1042–1051.

Albertini E, Marconi G, Reale L, Barcaccia G, Porceddu A, Ferranti F, Falcinelli M (2005). *SERK* and *APOSTART*: candidate genes for apomixis in *Poa pratensis* L. Plant Physiol 138: 2185–2199.

Albertini E, Porceddu A, Ferranti F, Reale L, Barcaccia G, Romano B, Falcinelli M (2001). Aposopory and parthenogenesis may be uncoupled in *Poa pratensis*: a cytological investigation. Sex Plant Reprod 14: 213–217.

Aliyu OM, Schranz ME, Sharbel TF (2010). Quantitative variation for apomictic reproduction in the genus *Boechera* (Brassicaceae). Am J Bot 97: 1719–1731.

Altpeter F, James VA (2005). Genetic transformation of turf-type bahiagrass (*Paspalum notatum* Flugge) by biolistic gene transfer. Int Turfgrass Soc Res J 10: 485–489.

Asker SE, Jerling L (1992). *Apomixis in Plants*. CRC Press, Boca Raton, Florida.

Bai X, Peirson, BN, Dong F, Xue C, Makaroff CA (1999). Isolation and characterization of *SYN1*, a *RAD21*-like gene essential for meiosis in *Arabidopsis*. Plant Cell 11: 417–430.

Barcaccia G, Arzenton F, Sharbel TF, Varotto S, Parrini P, Lucchin M (2006). Genetic diversity and reproductive biology in ecotypes of the facultative apomict *Hypericum perforatum L.* Heredity 96: 322–334.

Barcaccia G, Mazzucato A, Albertini E, Zethof J, Gerats A, Pezzotti M, Falcinelli M (1998). Inheritance of parthenogenesis in *Poa pratensis L.*: auxin test and AFLP linkage analyses support monogenic control. Theor Appl Genet 97: 74–82.

Barreda VD, Palazzesi L, Telleria MC, Katinas L, Crisci JV, Bremer K, Passalia MG, Corsolini R, Rodrigues Brizuela R, Bechis F (2010). Eocene Patagonia fossils of the daisy family. Science 329: 1621.

Bennett MD, Finch RA, Barclay IR (1976). The time rate and mechanism of chromosome elimination in *Hordeum* hybrids. Chromosoma 54: 175–200.

Bicknell RA (1997). Isolation of a diploid, apomictic plant of Hieracium aurantiacum. Sex Plant Reprod 10: 168–172.

Bicknell RA, Borst NK (1994). *Agrobacterium*-mediated transforamation of *Hieracium aurantiacum*. Int J Plant Sci 155: 467–470.

Bicknell RA, Borst NK, Koltunow AM (2000). Monogenic inheritance of apomixis in two *Hieracium* species with distinct developmental mechanisms. Heredity 84: 228–237.

Boutilier K, Offringa R, Sharma VK, Kieft H, Ouellet T, Zhang L, Hattori J, Liu CM, van Lammeren AA, Miki BL, Custers JB, van Lookeren Campagne MM (2002). Ectopic expression of *BABYBOOM* triggers a conversion from vegetative to embryonic growth. Plant Cell 14: 1737–1749.

Bretagnolle F, Thompson JD (1995). Gametes with the somatic chromosome number: mechanisms of their formation and role in the evolution of autopolyploid plants. New Phytol 129: 1–22.

Calderini O, Chang SB, de Jong H, Busti A, Paolocci F, Arcioni S, de Vries SC, Abma-Henkens MHC, Klein Lankhorst RM, Donnison IS, Pupilli F (2006). Molecular cytogenetics and DNA sequence analysis of an apomixis-linked BAC in *Paspalum simplex* reveal a non-pericentromere location and partial microcolinearity with rice. Theor Appl Genet 112: 1179–1191.

Campbell CS, Dickinson TA (1990). Apomixis, patterns of morphological variation, and species concepts in subfam. Maloideae (Rosaceae). Syst Bot 15: 124–135.

Canales C, Bhatt AM, Scott R, Dickinson H (2002). EXS, a putative LRR receptor kinase, regulates male germline cell number and tapetal identity and promotes seed development in *Arabidopsis*. Curr Biol 12: 1718–1727.

Carman JG (1997). Asynchronous expression of duplicate genes in angiosperms may cause apomixis, bispory, tetraspory, and polyembryony. Biol J Linn Soc 61: 51–94.

Catanach AS, Erasmuson SK, Podivinsky E, Jordan BR, Bicknell R (2006). Deletion mapping of genetic regions associated with apomixis in *Hieracium*. Proc Natl Acad Sci U S A 103: 18650–18655.

Chaudhury AM, Peacock WJ (1993). Approaches to isolating apomictic mutants in *Arabidopsis thaliana*: prospects and progress. In: Khush GS (ed). *Apomixis: Exploiting Hybrid Vigor in Rice*. International Rice Research Institute, Manilla, Philippines, pp 66–71.

Chelysheva L, Diallo S, Vezon D, Gendrot G, Vrielynck N, Belcram K, Rocques N, Márquez-Lema A, Bhatt AM, Horlow C, *et al.* (2005). AtREC8 and AtSCC3 are essential to the monopolar orientation of the kinetochores during meiosis. J Cell Sci 118: 4621–4632.

Conner JA, Goel S, Gunawan G, Cordonnier-Pratt MM, Johnson VE, Liang C, Wang H, Pratt LH, Mullett JE, Debarry J, *et al.* (2008). Sequence analysis of bacterial artificial chromosome clones from the apospory-specific genomic region of *Pennisetum* and *Cenchrus*. Plant Physiol 147: 1396–1411.

Darlington CD (1939). *The Evolution of Genetic Systems*. Cambridge University Press, Cambridge, UK.

d'Erfurth L, Jolivet S, Froger N, Catrice O, Novatchkova M, Mercier R (2009). Turning meiosis to mitosis. PLoS Biol 7: e1000124.

d'Erfurth L, Cromer L, Jolivet S, Girard C, Horlow C, Sun Y, To JP, Berchowitz LE, Copenhaver GP, Mercier R (2010). The cyclin-A CYCA1;2/TAM is required for the meiosis I to meiosis II transition and cooperates with OSD1 for the prophase to first meiotic division transition. PLoS Genet 6: e1000989.

de Wet JMJ (1968). Diploid-tetraploid-haploid cycles and the origin of variability in *Dichanthium* agamospecies. Evolution 22: 394–397.

Dilkes BP, Spielman M, Weizbauer R, Watson B, Burkart-Waco D, Scott RJ, Comai L (2008). The maternally expressed WRKY transcription factor TTG2 controls lethality in interploidy crosses of *Arabidopsis*. PLoS Biol 6: 2702–2720.

do Valle CB, Savidan YH (1996). Genetics, cytogenetics, and reproductive biology of *Brachiaria*. In: Miles JW, Maass BL, do Valle CB (eds). *Brachiaria: Biology, Agronomy and Improvement*. CIAT, Cali, Colombia, pp 147–163.

Drews GN, Koltunow AMG (2011). The female gametophyte. The *Arabidopsis* Book 9:e0155.

Dujardin M, Hanna W (1986). An apomictic polyhaploid obtained from a pearl millet × *Pennisetum squamulatum* apomictic interspecific hybrid. Theor Appl Genet 72: 33–36.

Dwivedi KK, Bhat BV, Gupta MG, Bhat SR, Bhat V (2007). Identification of a SCAR marker linked to apomixis in buffelgrass (*Cenchrus ciliaris* L.). Plant Sci 172: 788–795.

Ebina M, Nakagawa H, Yamamoto T, Araya H, Tsuruta S-I, Takahara M, Nakajima K (2005). Co-segregation of AFLP and RAPD markers to apospory in Guineagrass (*Panicum maximum* Jacq.). Grassland Sci 51: 71–78.

Ellstrand NC, Roose ML (1987). Patterns of genotypic diversity in clonal plant-species. Am J Bot 74: 123–131.

Evans MM (2007). The *INDETERMINATE GAMETOPHYTE1* gene of maize encodes a LOB domain protein required for embryo sac and leaf development. Plant Cell 19: 46–62.

Fehrer J, Krahulcovà A, Krahulec F, Hrtek J, Rosenbaumovà R, Bräutigam S (2007). Evolutionary aspects in *Hieracium* subgenus *Pilosella*. In: Hörandl E, Grossniklaus U, van Dijk PJ, Sharbel T (eds). *Apomixis: Evolution, Mechanisms and Perspectives*. Koeltz, Konigsstein, Germany, pp 359–390.

Franklin G, Oliveira M, Dias ACP (2007). Production of transgenic *Hypericum perforatum* plants via particle bombardment-mediated transformation of novel organogenic cell suspension cultures. Plant Sci 172: 1193–1203.

Gao CX, Jiang L, Folling M, Han LB, Nielsen KK (2006). Generation of large numbers of transgenic Kentucky bluegrass (*Poa pratensis* L.) plants following biolistic gene transfer. Plant Cell Rep 25: 19–25.

Garcia R, Asins MJ, Forner J, Carbonell EA (1999). Genetic analysis in *Citrus* and *Poncirus* by genetic markers. Theor Appl Genet 99: 511–518.

Garcia-Aguilar M, Michaud C, Leblanc O, Grimanelli D (2010). Inactivation of a DNA methylation pathway in maize reproductive organs results in apomixis-like phenotypes. Plant Cell 22: 3249–3267.

Goel S, Chen Z, Akiyama Y, Conner JA, Basu M, Gualtieri G, Hanna WW, Ozias-Akins P (2006). Comparative physical mapping of the apospory-specific genomic region in two apomictic grasses: *Pennisetum squamulatum* and *Cenchrus ciliaris*. Genetics 173: 389–400.

Goldman JJ, Hanna WW, Fleming GH, Ozias-Akins P (2003). Fertile transgenic pearl millet [*Pennisetum glaucum* (L.) R. Br.] plants recovered through microprojectile bombardment and phosphinothricin selection of apical meristem-, inflorescence-, and immature embryo-derived embryogenic tissues. Plant Cell Rep 21: 999–1009.

Grelon M, Vezon D, Gendrot G, Pelletier G (2001). AtSPO11-1 is necessary for efficient meiotic recombination in plants. EMBO J 20: 589–600.

Grimanelli D, Garcia M, Kazas E, Perotti E, Leblanc O (2003). Heterochronic exprression of sexual reproductive programs during apomictic development in *Tripascum*. Genetics 156: 1521–1531.

Grimanelli D, Leblanc O, Espinosa E, Perotti E, De Leon DG, Savidan Y (1998a). Non-mendelian transmission of apomixis in maize-*Tripsacum* hybrids caused by a transmission ratio distortion. Heredity 80: 40–47.

Grimanelli D, Leblanc O, Espinosa E, Perotti E, Gonzales de Leon D, Savidan Y (1998b). Mapping diplosporous apomixis in tetraploid *Tripsacum*: one gene or several genes? Heredity 80:33–39.

Grimanelli D, Leblanc O, Perotti E, Grossniklaus U (2001). Developmental genetics of gametophytic apomixis. Trends Genet 17: 597–604.

Gross-Hardt R, Lenhard M, Laux T (2002). WUSCHEL signalling functions in interregional communication during *Arabidopsis* ovule development. Genes Dev 16: 1129–1138.

Grossniklaus U, Vielle-Calzada JP, Hoeppner MA, Gagliano WB (1998). Maternal control of embryogenesis by *MEDEA*, a polycomb group gene in *Arabidopsis*. Science 280: 446–450.

Guitton E-A, Berger F (2005). Loss of function of MULTICOPY SUPPRESSOR OF IRA 1 produces nonviable parthenogentic embryos in *Arabidopsis*. Curr Biol 15: 750–754.

Gustine DL, Sherwood RT, Huff DR (1997). Apospory-linked molecular markers in buffelgrass. Crop Sci 37: 947–951.

Ha CD, Lemaux PG, Cho M-J (2001). Stable transformation of a recalcitrant Kentucky bluegrass (*Poa pratensis* L.) cultivar using mature seed-derived highly regenerative tissues. In Vitro Cell Dev Biol Plant 37: 6–11.

Hagberg A, Hagberg G (1980). High frequency of spontaneous haploids in the progeny of an induced mutation in barley. Hereditas 93: 341–343.

Hamant O, Ma H, Cande WZ (2006). Genetics of meiotic prophase I in plants. Annu Rev Plant Biol 57: 267–302.

Hanna WW (1995). Use of apomixis in cultivar development. Adv Agric 54: 33–35.

Hanna WW, Powell JB, Millot JC, Burton GW (1973). Cytology of obligate sexual plants in *Panicum maximum* Jacq. and their use in controlled hybrids. Crop Sci 13: 695–697.

Hecht V, Vielle-Calzada J-P, Hartog MV, Schmidt EDL, Boutilier K, Grossniklaus U, de Vries SC (2001). The *Arabidopsis SOMATIC EMBRYOGENESIS RECEPTOR KINASE1* gene is expressed in developing ovules and embryos and enhances embryogeneic competence in culture. Plant Physiol 127: 803–816.

Hoerandl E (1998). Species concepts in agamic complexes: applications in the *Ranunculus auricomus* complex and general perspectives. Folia Geobot 33: 335–348.

Hoerandl E (2006). The complex causality of geographical parthenogenesis. New Phytol 171: 525–538.

Hoerandl E, Paun O (2007). Patterns and sources of genetic diversity in apomictic plants: implications for evolutionary potentials. In: Horandl E, Grossniklaus U, Van Dijk PJ, Sharbel TF (eds). *Apomixis: Evolution, Mechanisms and Perspectives*. Gantner, Ruggell, Liechtenstein, pp 169–194.

Huo H, Conner JA, Ozias-Akins P (2009). Genetic mapping of the apospory-specific genomic region (ASGR) in *Pennisetum squamulatum* using retrotransposon-based molecular markers. Theor Appl Genet 119: 199–212.

Janko K, Drozd P, Flegr J, Pannell JR (2008). Clonal turnover versus clonal decay: a null model for observed patterns of asexual longevity, diversity and distribution. Evolution 62: 1264–1270.

Jessup RW, Burson BL, Burow GB, Wang YWCC, Li Z, Paterson AH, Hussey MA (2002). Disomic inheritance, suppressed recombination, and allelic interactions govern apospory in buffelgrass as revealed by genome mapping. Crop Sci 42: 1688–1694.

Johnson MTJ, Smith SD, Rausher MD (2010). Effects of plant sex on range distributions and allocation to reproduction. New Phytol 186: 769–779.

Johnson MTJ, FitzJohn RG, Smith SD, Rausher MD, Otto SP (2011). Loss of sexual recombination and segregation is associated with increased diversification in evening primroses. Evolution 65: 3230–3240.

Judd WS, Donoghue MJ, Sanders RW (1994). Angiosperm family pairs: preliminary phylogenetic analyses. Harvard Papers in Botany 5: 1–51.

Kantama L, Sharbel TF, Schranz ME, Mitchell-Olds T, de Vries S, de Jong H (2007). Diploid apomicts of the *Boechera holboellii* complex display large-scale chromosome substitutions and aberrant chromosomes. Proc Natl Acad Sci U S A 104: 14026–14031.

Kaushal P, Malaviya DR, Roy AK, Pathak S, Agrawal A, Khare A, Siddiqui SA (2008). Reproductive pathways of seed development in apomictic guinea grass (*Panicum maximum* Jacq.) reveal uncoupling of apomixis components. Euphytica 164: 81–92.

Kempe K, Gils M (2011). Pollination control technologies for hybrid breeding. Mol Breed 27: 417–437.

Kepiro JL, Roose ML (2010). AFLP markers closely linked to a major gene essential for nucellar embryony (apomixis) in *Citrus maxima* × *Poncirus trifoliata*. Tree Genet Genomes 6: 1–11.

Koltunow AM (1993). Apomixis: embryo sacs and embryos formed without meiosis and fertilization in ovules. Plant Cell 5: 1425–1437.

Koltunow AM, Bicknell RA, Chaudhury AM (1995a) Apomixis: molecular strategies for the generation of genetically identical seeds without fertilization. Plant Physiol 108: 1345–1352.

Koltunow AM, Grossniklaus U (2003). Apomixis a developmental perspective. Annu Rev Plant Biol 54: 547–574.

Koltunow AM, Johnson SD, Bicknell RA (1998). Sexual and apomictic development in *Hieracium*. Sex Plant Reprod 11: 213–230.

Koltunow AMG, Johnson SD, Bicknell RA (2000). Apomixis is not developmentally conserved in related, genetically characterized plants of varying ploidy. Sex Plant Reprod 12: 253–266.

Koltunow AMG, Johnson SD, Rodrigues JCM, Okada T, Hu Y, Tsuchiya T, Wilson S, Fletcher P, Ito K, Suzuki G, *et al.* (2011a). Sexual reproduction is the default mode in apomictic *Hieracium* subgenus *Pilosella*, in which two dominant loci function to enable apomixis. Plant J 66: 890–902.

Koltunow AMG, Johnson SD, Okada T (2011b). Apomixis in hawkweed: Mendel's experimental nemesis. J Exp Bot 62: 253–266.

Koltunow AM, Soltys K, Nito N, McClure S (1995b). Anther, ovule, seed and nucellar embryo development in *Citrus sinensis* cv. Valencia. Can J Bot 73: 1567–1582.

Labombarda P, Busti A, Caceres ME, Pupilli F, Arcioni S (2002). An AFLP marker tightly linked to apomixis reveals hemizygosity in a portion of the apomixis-controlling locus in *Paspalum simplex*. Genome 45: 513–519.

Leblanc O, Grimanelli D, Gonzalez-de-Leon D, Savidan Y (1995). Detection of the apomictic mode of reproduction in maize-*Tripsacum* hybrids using maize RFLP markers. Theor Appl Genet 90: 1198–1203.

Leblanc O, Grimanelli D, Islam-Faridi N, Berthaud J, Savidan Y (1996). Reproductive behavior in maize-*Tripsacum* polyhaploid plants: implications for the transfer of apomixis into maize. J Hered 87: 108–111.

Leblanc O, Grimanelli D, Hernandez-Rodriguez M, Galindo PA, Soriano-Martinez AM, Perotti E (2009). Seed development and inheritance studies in apomictic maize-*Tripsacum* hybrids reveal barriers for the transfer of apomixis to sexual crops. Int J Dev Biol 53: 585–596.

Li R, Qu R (2011). High throughput *Agrobacterium*-mediated switchgrass transformation. Biomass Bioenergy 35: 1046–1054.

Lieber D, Lora J, Schrempp S, Lenhard M, Laux T (2011). *Arabidopsis WIH1* and *WH2* genes act in the transition from somatic to reproductive cell fate. Curr Biol 21: 1009–1017.

Lotan T, Ohto M, Yee KM, West MA, Lo R, Kwong RW, Yamagishi K, Fischer RL, Goldberg RB, Harada JJ (1998). *Arabidopsis LEAFY COTYLEDON1* is sufficient to induce embryo development in vegetative cells. Cell 93: 1195–1205.

Luo M, Bilodeau P, Koltunow A, Dennis ES, Peacock WJ, Chaudhury AM (1999). Genes controlling fertilization-independent seed development in *Arabidopsis thaliana*. Proc Natl Acad Sci U S A 96: 296–301.

Maheshwari P (1950). *An Introduction to the Embryology of Angiosperms*. McGraw-Hill, New York, New York.

Marimuthu MPA, Jolivet S, Ravi M, Pereira L, Davda JN, Cromer L, Wang L, Nogué F, Chan SWL, Siddiqi I, Mercier R (2011). Synthetic clonal reproduction through seeds. Science 331: 876.

Martienssen R, Lippmann Z, May B, Ronemus M, Vaughn M (2004). Transposons, tandem repeats and the silencing of imprinted genes. Cold Spring Harb Sym Quant Biol 69: 371–379.

Martinez EJ, Hopp HE, Stein J, Ortiz JPA, Quarin CL (2003). Genetic characterization of apospory in tetraploid *Paspalum notatum* based on the identification of linked molecular markers. Mol Breed 12: 319–327.

Martinez EJ, Urbani MH, Quarin CL, Ortiz JPA (2001). Inheritance of apospory in bahiagrass, *Paspalum notatum*. Hereditas 135: 19–25.

Martonfi P, Brutovska R, Cellarova E, Repcak M (1996). Apomixis and hybridity in *Hypericum perforatum*. Folia Geobot Phytotax 31: 389–396.

Matzk F (1991). A novel approach to differentiated embryos in the absence of endosperm. Sex Plant Reprod 4: 88–94.

Matzk F, Meister A, Brutovska R, Schubert I (2001). Reconstruction of reproductive diversity in *Hypericum perforatum L.* opens novel strategies to manage apomixis. Plant J 26: 275–282.

Matzk F, Meister A, Schubert I (2000). An efficient screen for reproductive pathways using mature seeds of monocots and dicots. Plant J 21: 97–108.

Matzk F, Prodanovic S, Baumlein H, Schubert I (2005). The inheritance of apomixis in *Poa pratensis* confirms a five locus model with differences in gene expressivity and penetrance. Plant Cell 17: 13–24.

Nakano M, Shimizu T, Fujii H, Shimada T, Endo T, Nesumi H, Kuniga T, Omura M (2008a). Marker enrichment and construction of haplotype-specific BAC contigs for the polyembryony genomic region in *Citrus*. Breed Sci 58: 375–383.

Nakano M, Shimizu T, Kuniga T, Nesumi H, Omura M (2008b). Mapping and haplotyping of the flanking region of the polyembryony locus in *Citrus unshiu* Marcow. Journal of the Japanese Society for Horticultural Science 77: 109–114.

Nogler GA (1984a). Gametophytic apomixis. In: Johri BM (ed): *Embryology of Angiosperms*. Springer, Berlin, Germany, pp 475–518.

Nogler GA (1984b). Genetics of apospory in apomictic *Ranunculus auricomus*. V. Conclusion. Bot Helv 92: 411–423.

Nogler GA (1995). Genetics of apomixis in *Ranunculus auricomus*. VI. Epilogue. Bot Helv 105: 111–115.

Nonomura K, Miyoshi K, Suzuki T, Miyao A, Hirochika H, Kurata N (2003). The *MSP1* gene is necessary to restrict the number of cells entering into male and female sporogenesis and to initite anther wall formation in rice. Plant Cell 15: 1728–1739.

Nonomura K, Morohoshi A, Nakano M, Eiguchi M, Miyao A, Hirochika H, Kurata N (2007). A germ cell specific gene of the ARGONAUTE family is essential for the progression of premitosis, mitosis and meiosis during the male and female sporogenesis in rice. Plant Cell 19: 2583–2594.

Noyes RD (2005). Inheritance of apomeiosis (diplospory) in fleabanes (*Erigeron*, Asteraceae). Heredity 94: 193–198.

Noyes R (2006). Apomixis via recombination of genome regions for apomeiosis (diplospory) and parthenogenesis in *Erigeron* (daisy fleabane, Asteraceae). Sex Plant Reprod 19: 7–18.

Noyes RD, Baker R, Mai B (2007). Mendelian segregation for two-factor apomixis in *Erigeron annuus* (Asteraceae). Heredity 98: 92–98.

Noyes RD, Rieseberg LH (2000). Two independent loci control agamospermy (apomixis) in the triploid flowering plant *Erigeron annuus*. Genetics 155: 379–390.

Ohad N, Yadegari R, Margossian L, Hannon M, Michaeli D, Harada JJ, Goldberg RB, Fischer RL (1999). Mutations in *FIE*, a WD polycomb group gene, allow endosperm development without fertilization. Plant Cell 11: 407–416.

Okada T, Ito K, Johnson SD, Oelkers KO, Suzuki G, Houben A, Mukai Y, Koltunow AM (2011). Chromosomes carrying meiotic avoidance loci in three apomixtic eudicot *Hieracium* subgenus *Pilosella* species share structural features with two monocot apomicts. Plant Physiol 157: 1327–1341.

Olmedo-Monfil V, Duràn-Figueroa N, Arteaga-Vàzquz M, Demesa-Arévalo E, Autran D, Grimanelli D, Slotkin RK, Martienssen RA, Vielle-Calzada J-P (2010). Control of female gamete formation by a small RNA pathway in *Arabidopsis*. Nature 464: 628–632.

Ozias-Akins P (2006). Apomixis: developmental characteristics and genetics. Crit Rev Plant Sci 25: 199–214.

Ozias-Akins P, Lubbers EL, Hanna WW, McNay JW (1993). Transmission of the apomictic mode of reproduction in *Pennisetum*: co-inheritance of the trait and molecular markers. Theor Appl Genet 85: 632–638.

Ozias-Akins P, Roche D, Hanna WW (1998). Tight clustering and hemizygosity of apomixis-linked molecular markers in *Pennisetum squamulatum* implies genetic control of apospory by a divergent locus which may have no allelic form in sexual genotypes. Proc Natl Acad Sci U S A 95: 5127–5132.

Ozias-Akins P, van Dijk PJ (2007). Mendelian genetics of apomixis in plants. Annu Rev Genet 41: 509–537.

Pawlowski WP, Wang CJ, Golubovskaya IN, Szymaniak JM, Shi L, Hamant O, Zhu T, Harper L, Sheridan WF, Cande WZ (2009). Maize AMEIOTIC1 is essential for multiple early meiotic processes and likely required for the initiation of meiosis. Proc Natl Acad Sci U S A 106: 3606–3608.

Peel MD, Carman JG, Leblanc O (1997). Megasporocyte callose in apomictic buffelgrass, Kentucky bluegrass, *Pennisetum squamulatum* Fresen, *Tripsacum* L, and weeping lovegrass. Crop Sci 37: 724–732.

Pessino SC, Evans C, Ortiz JPA, Armstead I, do Valle CB, Hayward MD (1998). A genetic map of the apospory-region in *Brachiaria* hybrids: identification of two markers closely associated with the trait. Hereditas 128: 153–158.

Pessino SC, Ortiz JPA, Leblanc O, do Valle CB, Hayward MD (1997). Identification of a maize linkage group related to apomixis in *Brachiaria*. Theor Appl Genet 94: 439–444.

Pupilli F, Labombarda P, Caceres ME, Quarin CL, Arcioni S (2001). The chromosome segment related to apomixis in *Paspalum simplex* is homoeologous to the telomeric region of the long arm of rice chromosome 12. Mol Breed 8: 53–61.

Pupilli F, Martinez EJ, Busti A, Calderini O, Quarin CL, Arcioni S (2004). Comparative mapping reveals partial conservation of synteny at the apomixis locus in *Paspalum* spp. Mol Genet Genom 270: 539–548.

Ramulu KS, Sharma VK, Naumova TN, Dijkhuis P, van Lookeren Campagne MM (1999). Apomixis for crop improvement. Protoplasma 208: 196–205.

Ravi M, Chan SWL (2010). Haploid plants produced by centromere-mediated genome elimination. Nature 464: 615–618.

Ravi M, Marimuthu MPA, Siddiqi I (2008). Gamete formation without meiosis in *Arabidopsis*. Nature 451: 1121–1124.

Roche D, Chen ZB, Hanna WW, Ozias-Akins P (2001b). Non-Mendelian transmission of an apospory-specific genomic region in a reciprocal cross between sexual pearl millet (*Pennisetum glaucum*) and an apomictic F1 (*P. glaucum* × *P. squamulatum*). Sex Plant Reprod 13: 217–223.

Roche D, Cong PS, Chen ZB, Hanna WW, Gustine DL, Sherwood RT, Ozias-Akins P (1999). An apospory-specific genomic region is conserved between buffelgrass (*Cenchrus ciliaris* L.) and *Pennisetum squamulatum* Fresen. Plant J 19: 203–208.

Roche D, Hanna WW, Ozias-Akins P (2001a). Gametophytic apomixis, polyploidy, and supernumerary chromatin. Sex Plant Reprod 13: 343–349.

Rodrigues JCM, Luo M, Berger F, Koltunow AMG (2010a). Polycomb Group function in sexual and asexual seed development in angiosperms. Sex Plant Reprod 23: 123–133.

Rodrigues JCM, Okada T, Johnson SD, Koltunow AM (2010b). A *MULTICOPY SUPPRESSOR of IRA1* (*MSI1*) homologue is not associated with the switch to autonomous seed development in apomictic (asexual) *Hieracium* plants. Plant Sci 179: 590–597.

Rodrigues JCM, Tucker MR, Johnson SD, Hrmova M, Koltunow AMG (2008). Sexual and apomictic seed formation in *Hieracium* requires the plant Polycomb Group gene *FERTILIZATION INDEPENDENT ENDOSPERM*. Plant Cell 20: 2372–2386.

Rollins RC (1967). Evolutionary fate of inbreeders and nonsexuals. Am Nat 101: 343–351.

Rosenberg O (1907). Experimental and cytological studies in the Hieracia II. Cytolological studies on the apogamy in *Hieracium*. Bot Tidsskr 28: 143–170.

Savidan Y (1980). Chromosomal and embryological analyses in sexual X apomictic hybrids of *Panicum maximum* Jacq. Theor Appl Genet 57: 153–156.

Savidan Y (2000). Apomixis: genetics and breeding. Plant Breed Rev 18: 13–85.

Schallau A, Arzenton F, Johnston AJ, Hahnel U, Koszegi D, Blattner FR, Altschmied L, Haberer G, Barcaccia G, Baumlein H (2010). Identification and genetic analysis of the APOSPORY locus in *Hypericum perforatum* L. Plant J 62: 773–784.

Schiefthaler U, Balasubramanian S, Sieber P, Chevalier D, Wisman E, Schneitz K (1999). Molecular analysis of *NOZZLE* a gene involved in pattern formation and early sporogenesis during sex organ development in *Arabidopsis thaliana*. Proc Natl Acad Sci U S A 96: 11664–11669.

Schmidt A, Wuest SE, Vijverberg K, Baroux C, Kleen D, Grossniklaus U (2011). Transcriptome analysis of the *Arabidopsis* megaspore mother cell uncovers the importance of RNA helicases for plant germline development. PLoS Biol 9: e1001155.

Schranz ME, Dobes C, Koch MA, Mitchell-Olds T (2005). Sexual reproduction, hybridization, apomixis, and polyploidization in the genus *Boechera* (Brassicaceae). Am J Bot 92: 1797–1810.

Schranz ME, Kantama L, de Jong H, Mitchell-Olds T (2006). Asexual reproduction in a close relative of *Arabidopsis*: a genetic investigation of apomixis in *Boechera* (Brassicaceae). New Phytol 171: 425–438.

Schwander T, Crespi BJ (2009). Twigs on the tree of life? Neutral and selective models for integrating macroevolutionary patterns with microevolutionary processes in the analysis of asexuality. Mol Ecol 18: 28–42.

Sharbel TF, Voigt ML, Corral JM, Thiel T, Varshney A, Kumlehn J, Vogel H, Rotter B (2009). Molecular signatures of apomictic and sexual ovules in the *Boechera holboellii* complex. Plant J 58: 870–882.

Sharbel TF, Voigt ML, Corral JM, Galla G, Kumlehn J, Klukas C, Schreiber F, Vogel H, Rotter B (2010). Apomictic and sexual ovules of *Boechera* display heterochronic global gene expression patterns. Plant Cell 22: 655–671.

Sheridan WF, Golubeva EA, Abrhamova LI, Golubovskaya IN (1996). The *mac1* mutation alters the developmental fate of the hypodermal cells and their cellular progeny in the maize anther. Genetics 153: 933–941.

Sherwood RT, Berg CC, Young BA (1994). Inheritance of apospory in buffelgrass. Crop Sci 34: 1490–1494.

Silvertown J (2008). The evolutionary maintenance of sexual reproduction: evidence from the ecological distribution of asexual reproduction in clonal plants. Int J Plant Sci 169: 157–168.

Singh M, Conner JA, Zeng YJ, Hanna WW, Johnson VE, Ozias-Akins P (2010). Characterization of apomictic BC_7 and BC_8 pearl millet: meiotic chromosome behavior and construction of an ASGR-carrier chromosome-specific library. Crop Sci 50: 892–902.

Singh M, Goel S, Meeley RB, Dantec C, Parrinello H, Michaud C, Leblanc O, Grimanelli D (2011). Production of viable gametes without meiosis in maize deficient for an ARGONAUTE protein. Plant Cell 23: 443–458.

Smith RL, Grando MF, Li YY, Seib JC, Shatters RG (2002). Transformation of bahiagrass (*Paspalum notatum* Flugge). Plant Cell Rep 20: 1017–1021.

Snyder LA, Hernandez AR, Warmke HE (1955). The mechanism of apomixis in *Pennisetum ciliare* Botan Gaz 116: 209–221.

Somleva MN, Tomaszewski Z, Conger BV (2002). *Agrobacterium*-mediated genetic transformation of switchgrass. Crop Sci 42: 2080–2087.

Stein J, Pessino SC, Martinez EJ, Pia Rodriguez M, Siena LA, Quarin CL, Amelio Ortiz JP (2007). A genetic map of tetraploid *Paspalum notatum* Flugge (bahiagrass) based on single-dose molecular markers. Mol Breed 20: 153–166.

Tas ICQ, van Dijk PJ (1999). Crosses between sexual and apomictic dandelions (*Taraxacum*). I. The inheritance of apomixis. Heredity 83: 707–714.

Tester M, Langridge P (2010). Breeding technologies to increase crop production in a changing world. Science 327: 818–822.

Tucker MR, Araujo AC, Paech NA, Hecht V, Schmidt ED, Rossell JB, de Vries SC, Koltunow AM (2003). Sexual and apomictic reproduction in *Hieracium* subgenus *Pilosella* are closely interrelated developmental pathways. Plant Cell 15: 1424–1537.

van Baarlen P, van Dijk PJ, Hoekstra RF, de Jong JH (2000). Meiotic recombination in sexual diploid and apomictic triploid dandelions (*Taraxacum officinale* L.). Genome 43: 827–835.

van Dijk PJ (2003). Ecological and evolutionary opportunities of apomixis: insights from *Taraxacum* and *Chondrilla*. Phil Trans R Soc B Biol Sci 358: 1113–1121.

van Dijk PJ, Bakx-Schotman JMT (2004) Formation of unreduced megaspores (diplospory) in apomictic dandelions (*Taraxacum officinale*, s.l.) is controlled by a sex-specific dominant locus. Genetics 166: 483–492.

van Dijk P, de Jong H, Vijverberg K, Biere A (2009). An apomixis-gene's view on dandelions. In: Schon I, Martens K, van Dijk P (eds). *Lost Sex: The Evolutionary Biology of Parthenogenesis*. Springer, Dordrecht, The Netherlands, pp 475–493.

van Dijk PJ, Tas ICQ, Falque M, Bakx-Schotman T (1999). Crosses between sexual and apomictic dandelions (*Taraxacum*). II. The breakdown of apomixis. Heredity 83: 715–721.

van Dijk PJ, van Baarlen P, de Jong H (2003). The occurrence of phenotypically complementary apomixis-recombinants in crosses between sexual and apomictic dandelions (*Taraxacum officinale*). Sex Plant Reprod 16: 71–76.

Vaucheret H (2008). Plant ARGONAUTES. Trends Plant Sci 13:350–358.

Verduijn MH, van Dijk PJ, van Damme JMM (2004). The role of tetraploids in the sexual-asexual cycle in dandelions (*Taraxacum*). Heredity 93: 390–398.

Vielle-Calzada JP, Crane CF, Stelly DM (1996). Apomixis the asexual revolution. Science 274: 1322–1323.

Vijverberg K, Milanovic S, Bakx-Schotman T, van Dijk PJ (2010). Genetic fine mapping of DIPLOSPOROUS in *Taraxucum* (dandelion; Asteraceae) indicates a duplicated DIP-gene. BMC Plant Biol 10: 154.

Vijverberg K, Van der Hulst RGM, Lindhout P, van Dijk PJ (2004). A genetic linkage map of the diplosporous chromosomal region in *Taraxacum officinale* (common dandelion; Asteraceae). Theor Appl Genet 108: 725–732.

Vinkenoog R, Spielmann M, Scott RJ (2001). Autonomous endosperm development in flowering plants: how to overcome the imprinting problem? Sex Plant Reprod 14: 189–194.

Willemse MTM, van Went JL (1984). The female gametophyte. In: Johri BM (ed). *Embryology of Angiosperms*, Springer, Berlin, Germany, pp 159–196.

Yang SL, Xie LF, Mao HZ, Puah CS, Yang WC, Jiang L, Sundaresan V, Ye D (2003). *TAPETUM DETERMINANT1* is required for cell specialization in the *Arabidopsis* anther. Plant Cell 15: 2792–2804.

Yang WC, Ye D, Xu J, Sundaresan V (1999). The *SPOROCYTELESS* gene of *Arabidopsis* is required for initiation of sporogenesis and encodes a novel nuclear protein. Genes Dev 13: 2108–2117.

Zeng YJ, Conner J, Ozias-Akins P (2011). Identification of ovule transcripts from the Apospory-Specific Genomic Region (ASGR)-carrier chromosome. BMC Genomics 12: 206.

Zhao DZ, Wang GF, Speal B, Ma H (2002). The excess microsporocytes1 gene encodes a putative leucine-rich repeat receptor protein kinase that controls somatic and reproductive cell fates in the *Arabidopsis* anther. Genes Dev 16: 2021–2031.

Zhao X, de Palma J, Oane R, Gamuyao R, Luo M, Chaudhury A, Hervé P, Xue Q, Bennett J (2008). OsTDL1A binds to the LRR domain of rice receptor kinase MSP1 and is required to limit sporocyte numbers. Plant J 54: 375–387.

Zorzatto C, Chiari L, de Araujo Bitencourt G, do Valle CB, de Campos Leguizamon GO, Schuster I, Pagliarini MS (2010). Identification of a molecular marker linked to apomixis in *Brachiaria humidicola* (Poaceae). Plant Breed 129: 734–736.

6 High-Throughput Genetic Dissection of Seed Dormancy

Jose M. Barrero, Colin Cavanagh, and Frank Gubler

Introduction

Seed Dormancy – Why It Is Important?

The success of flowering plants in most ecosystems could be seen as a result of the emergence of the seed structure as a reproduction vehicle. If plants are broadly defined as immobile living organisms, in contrast to animals, the seed would be the exception to that definition. The seed, enclosing an embryonic plant and nutrient sources, represents the final stage of the plant reproduction and allows the safe dispersion of the progeny. For doing that, the seed needs to survive a challenging and changing environment and to preserve the next generation until conditions are favorable for survival.

The delay between seed formation and seed germination is one of the most important times during the entire plant cycle and has to be carefully synchronized with the environment to maximize seedling survival. This timing is principally determined by seed dormancy, which is a biological condition (physiological, morphological, and physical) that temporarily blocks germination, keeping the seed quiescent (Baskin and Baskin, 2004). Physiological dormancy is the most common form and generally includes components of embryo- and seed coat–based dormancy. Seed dormancy is an adaptive trait with high variability across species that has enormous importance in both wild and domesticated plants. Seed dormancy programs determine the ecological niche in which the seed germinates and prospers and are related to different factors such as climate, moisture, soil characteristics, light, temperature, nutrients, abiotic and biotic stress factors, and many others (Finch-Savage and Leubner-Metzger, 2006). In domesticated species, the control of seed dormancy is crucial and influences important agricultural traits, such as uniform germination and stand establishment, preharvest sprouting susceptibility (Gubler *et al.*, 2005), and seed storage requirements.

Gene Regulation of Seed Dormancy

During their formation, seeds undergo dramatic changes in their developmental program, becoming dry latent structures from highly metabolic tissues. The genetic program regulating germination in seeds is influenced by the life history of the mother plant and external environment. Given the right

Seed Genomics, First Edition. Edited by Philip W. Becraft.

environmental conditions, a nondormant seed germinates and grows, resulting in an independent autotrophic organism. During the transitions leading to germination, radical changes in hormone and protein content occur, along with changes in transcriptome profiles. Gene regulation is essential for dormancy control. There are several lines of evidence based on mutagenesis, global gene expression studies, and chemical inhibition that indicate that the messenger RNA (mRNA) that is stored late in seed development, before desiccation, will have a major role in determining the dormancy level of the seed (Finkelstein *et al.*, 2008; Holdsworth *et al.*, 2008a; Barrero *et al.*, 2010a; Weitbrecht *et al.*, 2011). Rajjou *et al.* (2004) found that seeds cannot complete germination in the presence of translational repressors, whereas *de novo* transcription was not required. Those results indicate that protein synthesis is required for germination, and consequently the stored mRNA population in the seed may have a key significance. This population of stored mRNA may be determined by the conditions in which the seed completed its development, and this influences the depth of seed dormancy. In relation to that, research in several species using mutants clearly shows that mechanisms controlling gene translation, especially those related to post-transcriptional RNA processing, maturation, and metabolism, are very important in the regulation of seed dormancy (Lu and Fedoroff, 2000; Hugouvieux *et al.*, 2001; Xiong *et al.*, 2001; Nishimura *et al.*, 2005; Grasser *et al.*, 2009; Narsai *et al.*, 2011). For example, the *hyl1* insertion mutation which knocks out miRNA processing, results in ABA hypersensitivity in seeds (Lu *et al.*, 2002).

Seed Dormancy Complexity Is a Scientific Challenge

General mechanisms or pathways associated with seed dormancy have been described, mainly in the model plant *Arabidopsis*. Microarrays have been performed to profile gene expression in dormant and nondormant seeds (from different ecotypes) along with comparison of dormant and after-ripened seeds from the same ecotype in *Arabidopsis* (Cadman *et al.*, 2006; Carrera *et al.*, 2007). However, a comprehensive model of the dormancy development program that explains how seeds acquire their dormancy, how it is maintained, and how it is broken down to allow germination has yet to be determined. The key genetic elements that keep a seed under the control of the dormancy program or the elements that are needed to initiate the germination program are unknown (Nonogaki *et al.*, 2010). Although dormancy has been intensively studied for a long time in the model plant *Arabidopsis* and in other species, we are only starting to understand what elements govern seed dormancy, and the manipulation of this trait is still a challenge. The complex web of genetic and environmental factors involved in seed dormancy and germination is wide, which could be a reason why reductionist approaches to analyzing seed dormancy have not provided a clear model explaining seed dormancy. The elements determining the dormancy level and the elements triggering germination are probably composite signals that result from interactions between genes and environment. New platforms and technologies for sequencing genomes and transcriptomes have the potential to dissect those complex interactions accurately and rapidly. Also, new computing power and visualization tools allow a more global analysis, combining systems biology approaches and coexpression studies for taking advantage of a full set of publicly available datasets. Seed dormancy needs to be studied in the context of the seed cycle because it is an intermediate step between seed maturation and germination (Gutierrez *et al.*, 2007; Angelovici *et al.*, 2010).

This chapter focuses on the genetic dissection of seed dormancy, more specifically on gene expression. We highlight how new high-throughput platforms and associated techniques can be used for revealing some of the complex mechanisms behind seed dormancy. Finally, we also review strategies for gene discovery that can be used for mapping and identifying dormancy-related genes in crop species, where the genetic dissection and manipulation of seed dormancy are even more complex.

Profiling of Transcriptomic Changes

Global gene expression analyses using microarrays have been used for dissecting several aspects of seed dormancy. Several reviews summarize the findings of those approaches (Finkelstein *et al.*, 2008; Holdsworth *et al.*, 2008a, 2008b; Barrero *et al.*, 2010a). The novelty of these genome-wide approaches was their capability of identifying gene expression profiles or fingerprints associated with the dormant and the nondormant states. Specific gene expression profiles of dormant and nondormant (after-ripened) seeds were first described for the *Arabidopsis* Cvi ecotype (Cadman *et al.*, 2006). ABA-, seed maturation– and stress-related genes were abundant in the group of genes associated with dormancy. Genes related to protein metabolism, reserve mobilization, and cell wall modification were abundant in the after-ripened set. A similar transcriptomic scenario was described in the *Ler* ecotype (Carrera *et al.*, 2007), which produces seeds with weaker dormancy than Cvi, indicating that similar changes occur during after-ripening in different ecotypes, regardless of their level of dormancy. This work introduced an interesting new tool for analyzing and visualizing the gene datasets. Because the typical *Arabidopsis* gene ontology (GO) (Gaudet *et al.*, 2009) annotation provided limited understanding regarding which class of genes was important in dormancy and germination, the authors reannotated the genes in relation to previously described roles in germination- and dormancy-related terms (Microsoft Excel TAGGIT macro) (Carrera *et al.*, 2007). This TAGGIT workflow was used for reanalyzing new and previous microarray data and has been shown to give a distinct visual gene signature for dormant and after-ripened seeds (Holdsworth *et al.*, 2008b). Also, it was used for reanalyzing microarray data generated from dry and imbibed Col-0 seeds (Nakabayashi *et al.*, 2005) providing distinct gene signatures for both conditions. In relation to the gene expression changes found during early imbibition of Col-0 seeds, it was found that genes with differential expression between imbibed and dry seeds tended to be located in chromosome clusters, suggesting that epigenetic mechanisms could be involved in the regulation of those areas (Nakabayashi *et al.*, 2005).

Microarrays also have been used for identifying gene expression changes in response to environmental factors that alter dormancy (Finch-Savage *et al.*, 2007) including after-ripening, low temperature, nitrate, and light. All those factors promote germination in *Arabidopsis,* and the transcriptomic results indicated that with the exception of after-ripening, all the other factors have similar effects on transcriptomic profiles. The after-ripening mechanism has been investigated in several studies but remains poorly understood (Iglesias-Fernandez *et al.*, 2011). Part of the problem is the difficulty of separating the dormancy and germination programs from the after-ripening per se. Studies using mutants have demonstrated that the after-ripening is a genetically controlled process and that it occurs independently of germination and dormancy. Mutant seeds that cannot complete germination, such as *comatose-1* (*cts-1*), or that do not develop seed dormancy, such as *aba deficient 1-1* (*aba1-1*) or *aba insensitive 1-1* (*abi1-1*), undergo similar after-ripening changes as wild-type seed, showing similar gene expression signatures (Carrera *et al.*, 2007, 2008).

Microarray-based transcription studies have been used for spatial profiling of transcriptomes from different seed tissues. For example, gene expression changes have been profiled in the *Arabidopsis* endosperm and embryo during germination (Penfield *et al.*, 2006), indicating a very different response in those tissues. This finding is not surprising because it is clear that different tissues of the seed have different contributions during seed dormancy and germination (Finch-Savage and Leubner-Metzger, 2006). In this sense, tissue-specific approaches can discover some genetic regulation that could be masked under the transcription of the whole seed. This tissue-specific microarray approach was performed in barley to compare dormant and after-ripened grains during early imbibition (Barrero *et al.*, 2009). Based on the results, it is proposed that the coleorhiza (a specific tissue in the embryo that surrounds the seminal roots) plays a major role in dormancy in cereals, acting as a physical barrier to root emergence in dormant grains. Similar to endosperm of dicotyledonous seeds, the coleorhiza cells undergo cell degradation in imbibed after-ripened seeds, allowing the root to emerge from the grain. Another more recent study used a similar approach in *Lepidium* for examining the role of the radicle and the micropylar endosperm cap during germination (Morris *et al.*, 2011). Both works indicate that genes involved in fundamental pathways controlling dormancy and germination, such as hormone metabolism and responsiveness or cell wall modification, are expressed in a very tight spatial and temporal pattern within the seed. These studies highlight the importance of accurate quantification of RNA populations in both time and space frames for understanding changes in seed dormancy. It is likely that translation and mRNA degradation may also be differentially regulated across different seed tissues and developmental stages.

These tissue-specific studies can now be performed with more accuracy and in a relatively high-throughput way using a new technique called INTACT (Isolation of Nuclei Tagged in Specific Cell Types) (Deal and Henikoff, 2010, 2011). This technique consists of tagging the nuclear envelop of specific cell types. This tag is a biotinylated protein, and the nuclei of interest can be isolated on extraction with the use of streptavidin-coated magnetic beads. This straightforward technique does not require expensive equipment and results in high purity of isolated nuclei. Although it has been applied only in *Arabidopsis*, it can potentially be used in any other plant species and could allow the dissection of seed dormancy at the cellular level.

Apart from microarrays, other high-throughput platforms are available to study the expression of all the transcription factors from *Arabidopsis* and rice by real-time reverse transcriptase polymerase chain reaction. This platform has been used in *Arabidopsis* for analyzing the dormant and nondormant states using C24 seeds (Barrero *et al.*, 2010b). This strategy allows the study of dormancy-related transcription factors that are often undetectable in microarray approaches because of their low expression values.

Use of New Sequencing Platforms and Associated Techniques to Study Seed Dormancy

The new high-throughput platforms for sequencing nucleic acids (known as next generation sequencing) (Lister *et al.*, 2009) are having a large impact in many fields of plant molecular biology, including seed dormancy (Weitbrecht *et al.*, 2011). Although there are not yet many studies using next generation platforms to analyze seed dormancy, they may have the potential to unmask the key elements controlling the dormancy program and dissecting poorly understood processes such as dry after-ripening or the induction of secondary dormancy.

The detailed analysis of the mRNA populations stored in the seed and how these populations change depending on the maturation conditions, during after-ripening or in response to imbibition, is now readily achievable using next generation sequencing. Of special interest is the capability of the next generation platforms for studying mRNA decay. Gene expression is a result of the balance between mRNA transcription and decay. Although the first has been extensively studied in seeds, the second has been mainly ignored, even when mutant approaches have provided strong support for the importance of mRNA stability in dormancy and germination. For example, *Arabidopsis* mutants affecting RNA capping (Hugouvieux *et al.*, 2001), mRNA splicing and degradation (Xiong *et al.*, 2001), and mRNA polyadenylated tail stability (Nishimura *et al.*, 2005) all have altered dormancy. The lack of study of mRNA decay has been partially due to difficulties in detecting it without the confounding impacts of transcription on transcript abundance. More recent studies have highlighted the importance of different mRNA decay pathways during plant growth and development, and the degree of specificity that different decay pathways show for regulating different biological processes during plant development (Belostotsky and Sieburth, 2009). General mRNA decay mechanisms include the 5′ decapping followed by a 5′-to-3′ exoribonuclease, the 3′ deadenylation followed by the exosome 3′-to-5′ digestion, and the internal cleavage followed by degradation in both directions. Prior analyses of these processes on a genome-wide scale discovered interesting details about how the life span of different mRNA species varies and how some sequences and motifs may be more susceptible to increased or reduced decay (Narsai *et al.*, 2007). Microarray studies were used to analyze the degradation of uncapped mRNA (Jiao *et al.*, 2008), but this technique had limitations for unmasking the contributions of specific decay pathways and for a reliable quantification of the mRNA population. Microarray detection and quantification is dependent on the number of probes on the microarray chip and their positions along the transcript. The new deep sequencing approaches can generate a continuous series of overlapping reads along a transcript, and the analysis of distorted distributions of those reads along transcript length could be a signature for 5′ or 3′ decay processes. Another important advance is that the absolute number of reads can be easily normalized against an external control RNA molecule, providing an opportunity to monitor relative or absolute changes in mRNA abundance with accuracy.

Post-transcriptional regulation of gene expression by microRNA-directed degradation pathways is important in many stages of plant growth and development. So far, only two microRNAs (miRNA 159 and 160) have been shown to be involved in the regulation of seed dormancy and germination (Liu *et al.*, 2007; Reyes and Chua, 2007). Current sequencing systems have the capability to reveal the degradome characteristics of mRNA populations (Meng *et al.*, 2010). Using the 5′-rapid amplification of cDNA ends (RACE), cDNA libraries of the 5′ ends of polyadenylated messengers can be readily constructed and sequenced providing a full map of the microRNA-targeted degradome. This method, called parallel analysis of RNA ends (PARE) (German *et al.*, 2008), has become popular for validating known microRNA targets and for finding new ones. This approach has been used for analyzing the microRNAs involved during seed development and germination in rice and maize (Xue *et al.*, 2009; Wang *et al.*, 2011), and it could be applied for analyzing several aspects of seed dormancy.

In combination with polysome or ribosome profiling, the new sequencing platforms allow for the identification of mRNAs that are associated with ribosomes (Ingolia *et al.*, 2009). This would generate an indirect estimation of the translation happening in a given tissue at a given time, providing a snapshot of the number of 30 nucleotide footprints occupied by ribosomes per mRNA species. mRNAs that are being actively translated have a higher number of ribosomes associated with the mRNA than those that are not being translated. The use of this technique would allow the identification of the population of mRNAs that is stored in the seed associated with ribosomes or which mRNAs are potentially translated in imbibed dormant or after-ripened seeds.

To be effective, next generation sequencing approaches need to be standardized across the seed biology community. Library construction and replication should be unified to compare results. In addition, experimental conditions used in dormancy experiments need to be standardized to allow datasets to be combined for future analyses. Protocols for different seed dormancy studies have been published more recently (Kermode, 2011).

Visualization Tools

The capability to analyze and visualize genome-wide results is limited. Microarray datasets composed of thousands of data points are difficult to visualize and summarize as a whole set, and often only partial analysis of a particular group of genes is given, leaving a large proportion of the data not fully analyzed. Next generation experiments are increasing that problem as the data points coming from the new platforms are counted in millions. Only with the use of computing power and new visualization tools will it be possible to achieve a global analysis of the datasets.

The visualization of the molecular processes associated with seed dormancy would facilitate the complex and comprehensive analysis of changes happening under different conditions. Common tools used to visualize transcriptomic changes, such as the MapMan and PageMan packages (http://MapMan.gabipd.org), have been implemented with new pathways related to seed dormancy and germination (Joosen *et al.*, 2011). These tools are expected to allow a more complete and accurate visualization and analysis of array datasets.

Of special interest is the Bio-Array Resource (BAR) developed by the University of Toronto (http://bar.utoronto.ca/welcome.htm). The BAR contains a series of Web-based tools for visualizing gene expression and protein changes. The electronic fluorescent pictographic (eFP) browser (Winter *et al.*, 2007; Bassel *et al.*, 2008) has proven to be of great value for seed scientists. The BAR website contains microarray information from dozens of seed germination experiments under different conditions and dormancy release treatments. It also allows the clustering of a gene of interest with the genes included in several TAGGIT ontology categories (Carrera *et al.*, 2007). Several other eFP browsers are also available for other species, such as poplar, *Medicago*, rice, barley, maize, and soybean.

Coexpression Studies and Systems Biology Approaches

In addition to single gene expression analyses, visualization tools have been developed to explore the interactions of cohorts of genes that share distinct expression profiles. This approach for studying gene coexpression could be very important for finding new genes involved in the control of seed dormancy and even more so in species where the gene annotation is poor. The Virtual Seed Web Resource (http://vseed.nottingham.ac.uk) contains tools for the analysis of gene networks during seed maturation, dormancy, and germination. The SeedNet network model (Bassel *et al.*, 2011) uses publicly available seed expression data from many *Arabidopsis* samples to build a genome-wide gene association clustering. This is a powerful tool that takes advantage of experiments performed by the scientific community to generate hierarchical associations between genes on the basis of their expressions in multiple seed-related analyses. SeedNet can accurately find novel seed germination and dormancy regulators and predict genetic interactions between them. In the future, the website will incorporate similar modules for the dissection of genetic interactions in a tissue-specific manner in the micropylar endosperm of *Arabidopsis* (MENet) and in the seminal root apical meristem (AxisNet). Conserved networks will also be identified by comparing gene datasets from wheat grain development and in response to abiotic stress (WheatNet). The Virtual Seed Web Resource is also implementing eFP browsers for the visualization of high-resolution gene expression changes in different tissues of germinating *Arabidopsis* and *Lepidium* seeds (Linkies *et al.*, 2009; Morris *et al.*, 2011).

Finally, a systems biology perspective can be used to formulate different scenarios related to seed dormancy and to predict the outcome of complex gene expression. This approach has been used to explore the balance and crosstalk between gibberellins and ABA, taking into account different environmental conditions (Penfield and King, 2009). It allows the formulation of different models that can be experimentally tested.

Mapping Populations for Gene Discovery

Marker-Assisted Gene Discovery

The study of seed dormancy in crop species such as cereals and the modification of this trait by breeding strategies has lagged behind the rapid progress made in model plants such as *Arabidopsis*. The use of molecular markers for improving cereal breeding has had the most success in tagging loci for large effect or Mendelian traits, such as disease resistance through the selection of markers in early-generation breeding material (e.g., F2 or early backcrosses). A far more difficult endeavor is the implementation of marker systems for polygenic traits (Baenziger *et al.*, 2006; Bernardo, 2008), such as seed dormancy. Typically, major QTLs, such as the QTL in the chromosome 4AL in wheat (Mares *et al.*, 2005), have been implemented in breeding, but often several QTLs need to be combined to be effective. The challenge for gene cloning, using traditional biparental mapping populations, either recombinant inbred lines (RILs) or doubled haploids, is that the support intervals for the QTL are typically very large. This makes it difficult to identify candidate genes or to implement them in marker assisted selection. To ensure with confidence that a given QTL is being selected, large regions of the genome are selected, and the sections of the genome are unable to recombine. Genetic background also affects the strength of a given QTL (Jannink *et al.*, 2009). There are dozens of QTL studies for seed dormancy in many species from *Arabidopsis* to wheat (Anderson *et al.*, 1993; Groos *et al.*, 2002; Alonso-Blanco *et al.*, 2003; Mares *et al.*, 2005). So far in *Arabidopsis*, only one candidate QTL gene, *Delay of Germination 1* (*DOG1*) (Bentsink *et al.*, 2006), has been cloned, and the protein has an unknown function. In rice and wheat, genes responsible for major QTLs involved in grain dormancy and testa color has been also identified (Gu *et al.*, 2011; Himi *et al.*, 2011). In wheat *Tamyb10* and in rice *Rc*, a basic helix-loop-helix transcription factor acts on the same flavonoid biosynthetic pathway responsible for the red testa color. Several other highly significant dormancy QTLs have been known for some time, and their molecular identification remains to be elucidated. Interspecies approaches for comparing syntenic regions are expected to become more common because the new sequencing platforms will provide more genomic sequences, and this will facilitate the cloning of QTLs (Somyong *et al.*, 2011).

Genome-wide Association Studies

Genome-wide association (GWAS) mapping has been used more recently as a method for detecting marker-trait associations (Mackay and Powell, 2007). Success using this approach has been mixed depending on many factors, including the availability of a sequence, size of the genome, polyploidy, and availability of high-throughput markers. In species where a genome sequence is available (e.g., *Arabidopsis* and rice), greater success has been achieved (Atwell *et al.*, 2010; Huang *et al.*, 2010).

The major advantage of GWAS is that it may be implemented within a breeding program or existing varieties, making it a very attractive option for cereal breeders. Numerous public initiatives are currently underway in wheat (http://www.triticeaegenome.eu/) and barley (http://www.agoueb.org/) (Waugh *et al.*, 2010). GWAS also has the advantage that the resolution is much higher than linkage analysis in biparental populations owing to the many generations of recombination captured in the population. This resolution generally means that an association is detected only when the marker is in tight linkage to the QTL; this increases the need for greater marker density. However, with the new sequencing capabilities available, high density marker coverage can be achieved using low pass resequencing (Huang *et al.*, 2010).

One disadvantage of GWAS is the need to control for population structure, which causes an increase in the number of false-positives (Cockram *et al.*, 2010). This problem can be partially alleviated through appropriate statistical modeling (Yu *et al.*, 2006). Association mapping can suffer from the factors that generate its precision, and for greater resolution a greater population size is required to exploit the historical recombination. Although population structure may be accounted for within a model, the risk of not identifying an important QTL that is associated with population structure will also occur.

Alternative Mapping Approaches

More recently, alternative approaches to marker-trait association identification have been proposed in experimental populations. These populations differ from the traditional approaches in that they include multiple founders and are large in size. The first is the nested association mapping (NAM) (Yu *et al.*, 2008) approach, where a series of lines are crossed to a common founder. In maize, NAM has been done using a common parent and 25 lines that were crossed generating 200 RILs per cross for a total of 5000 RILs in the population. NAM uses the power of linkage analysis (within cross) with the precision of association mapping (between families). Results are now emerging (Elshire *et al.*, 2011, Tian *et al.*, 2011), and the ability to impute marker information from a relatively sparse genotyping effort is possible in populations where a genome sequence is unavailable. Such an approach allows many founders to be used and for a trait such as dormancy allows the incorporation of a wide array of germplasm known to contain varying levels of dormancy.

The second approach being implemented is the multiparent advanced generation inter-cross (MAGIC) (Cavanagh *et al.*, 2008). This approach has been implemented in many plant species such as *Arabidopsis* (Kover *et al.*, 2009; Huang *et al.*, 2011), rice (http://irri.org/news-events/irri-news/irri-magic-team-unveils-genetic-diversity), and wheat (http://www.csiro.au/science/MAGIC), and results are beginning to emerge regarding the power of this approach for gene-trait associations. MAGIC uses multiple founders crossed in a balanced fashion to create a set of progeny lines, each having a mosaic of the founder genomes. This approach has already shown success in the identification of genomic regions controlling, among other traits, days to germination with a high degree of accuracy based on previously known genes (Kover *et al.*, 2009). MAGIC provides the ability to look at additive gene effects simultaneously along with epistatic effects. A major difference with other population designs is that each allelic effect can be assessed in a range of genetic backgrounds that occur within the population, without the need to worry about the effects of population structure. In addition, a series of near isogenic lines may be generated immediately from the RIL population after QTL identification. By identifying heterozygous individuals (for a specific QTL) within the population, the effects resulting from a range of genetic backgrounds may be tested. This is a very powerful feature of MAGIC, allowing a trackable approach to understanding the genetic factors underlying complex traits, without the need to generate further genetic resources.

Perspective

The impact of next generation sequencing is expected to speed up the identification of novel genes and alleles that control seed dormancy. Those platforms have the capability to dissect complex processes related to seed dormancy, such as after-ripening, which is going to be possible on a variety of species and conditions. Already the power of these technologies is speeding up the fine mapping of candidate QTL genes in nonsequenced crops. This will lead to widespread introgression of high dormancy germplasm, which will deliver crops that are resistant to preharvest sprouting. It is expected that cutting-edge technologies, such as cell-specific approaches, will help define the contribution of various cell types to the overall seed dormancy and help to unmask its control mechanisms. Future studies are likely to focus on how these various cell types interact to determine whether a seed germinates or remains dormant.

Acknowledgments

The authors wish to thank Dr. Maria Alonso-Peral for comments on the manuscript and for providing information for some of the sections.

References

Alonso-Blanco C, Bentsink L, Hanhart CJ, Vries HBE, Koornneef M (2003). Analysis of natural allelic variation at seed dormancy loci of Arabidopsis thaliana. Genetics 164: 711–729.

Anderson JA, Sorrells ME, Tanksley SD (1993). RFLP analysis of genomic regions associated with resistance to pre-harvest sprouting in wheat. Crop Sci 33: 453–459.

Angelovici R, Galili G, Fernie AR, Fait A (2010). Seed desiccation: a bridge between maturation and germination. Trends Plant Sci 15: 211–218.

Atwell S, Huang YS, Vilhjalmsson BJ, Willems G, Horton M, Li Y, Meng DZ, Platt A, Tarone AM, Hu TT, Jiang R, Muliyati NW, Zhang X, Amer MA, Baxter I, Brachi B, Chory J, Dean C, Debieu M, de Meaux J, Ecker JR, Faure N, Kniskern JM, Jones JDG, Michael T, Nemri A, Roux F, Salt DE, Tang CL, Todesco M, Traw MB, Weigel D, Marjoram P, Borevitz JO, Bergelson J, Nordborg M (2010). Genome-wide association study of 107 phenotypes in Arabidopsis thaliana inbred lines. Nature 465: 627–631.

Baenziger PS, Russell WK, Graef GL, Campbell BT (2006). Improving lives: 50 years of crop breeding, genetics, and cytology (C-1). Crop Sci 46: 2230–2244.

Barrero JM, Jacobsen J, Gubler F (2010a). Seed dormancy: approaches for finding new genes in cereals. In: Pua EC, Davey MR (eds). *Plant Developmental Biology—Biotechnological Perspectives*. Springer, Berlin, Germany, pp 361–381.

Barrero JM, Millar AA, Griffiths J, Czechowski T, Scheible WR, Udvardi M, Reid JB, Ross JJ, Jacobsen JV, Gubler F (2010b). Gene expression profiling identifies two regulatory genes controlling dormancy and ABA sensitivity in Arabidopsis seeds. Plant J 61: 611–622.

Barrero JM, Talbot MJ, White RG, Jacobsen JV, Gubler F (2009). Anatomical and transcriptomic studies of the coleorhiza reveal the importance of this tissue in regulating dormancy in barley. Plant Physiol 150: 1006–1021.

Baskin JM, Baskin CC (2004). A classification system for seed dormancy. Seed Sci Res 14: 1–16.

Bassel GW, Fung P, Chow TFF, Foong JA, Provart NJ, Cutler SR (2008). Elucidating the germination transcriptional program using small molecules. Plant Physiol 147: 143–155.

Bassel GW, Lan H, Glaab E, Gibbs DJ, Gerjets T, Krasnogor N, Bonner AJ, Holdsworth MJ, Provart NJ (2011). Genome-wide network model capturing seed germination reveals coordinated regulation of plant cellular phase transitions. Proc Natl Acad Sci U S A 108: 9709–9714.

Belostotsky DA, Sieburth LE (2009). Kill the messenger: mRNA decay and plant development. Curr Opin Plant Biol 12: 96–102.

Bentsink L, Jowett J, Hanhart CJ, Koornneef M (2006). Cloning of DOG1, a quantitative trait locus controlling seed dormancy in Arabidopsis. Proc Natl Acad Sci U S A 103: 17042–17047.

Bernardo R (2008). Molecular markers and selection for complex traits in plants: learning from the last 20 years. Crop Sci 48: 1649–1664.

Cadman CSC, Toorop PE, Hilhorst HWM, Finch-Savage WE (2006). Gene expression profiles of Arabidopsis Cvi seeds during dormancy cycling indicate a common underlying dormancy control mechanism (vol 46, pg 805, 2006). Plant J 47: 164.

Carrera E, Holman T, Medhurst A, Dietrich D, Footitt S, Theodoulou FL, Holdsworth MJ (2008). Seed after-ripening is a discrete developmental pathway associated with specific gene networks in Arabidopsis. Plant J 53: 214–224.

Carrera E, Holman T, Medhurst A, Peer W, Schmuths H, Footitt S, Theodoulou FL, Holdsworth MJ (2007). Gene expression profiling reveals defined functions of the ATP-binding cassette transporter COMATOSE late in phase II of germination. Plant Physiol 143: 1669–1679.

Cavanagh C, Morell M, Mackay I, Powell W (2008). From mutations to MAGIC: resources for gene discovery, validation and delivery in crop plants. Curr Opin Plant Biol 11: 215–221.

Cockram J, White J, Zuluaga DL, Smith D, Comadran J, Macaulay M, Luo Z, Kearsey MJ, Werner P, Harrap D, Tapsell C, Liu H, Hedley PE, Stein N, Schulte D, Steuernagel B, Marshall DF, Thomas WT, Ramsay L, Mackay I, Balding DJ; AGOUEB Consortium, Waugh R, O'Sullivan DM (2010). Genome-wide association mapping to candidate polymorphism resolution in the unsequenced barley genome. Proc Natl Acad Sci U S A 107: 21611–21616.

Deal RB, Henikoff S (2010). A simple method for gene expression and chromatin profiling of individual cell types within a tissue. Dev Cell 18: 1030–1040.

Deal RB, Henikoff S (2011). The INTACT method for cell type-specific gene expression and chromatin profiling in Arabidopsis thaliana. Nat Protocols 6: 56–68.

Elshire RJ, Glaubitz JC, Sun Q, Poland JA, Kawamoto K, Buckler ES, Mitchell SE (2011). A robust, simple Genotyping-by-Sequencing (GBS) approach for high diversity species. PLoS ONE 6: e19379.

Finch-Savage WE, Cadman CSC, Toorop PE, Lynn JR, Hilhorst HWM (2007). Seed dormancy release in Arabidopsis Cvi by dry after-ripening, low temperature, nitrate and light shows common quantitative patterns of gene expression directed by environmentally specific sensing. Plant J 51: 738.

Finch-Savage WE, Leubner-Metzger G (2006). Seed dormancy and the control of germination. New Phytol 171: 501–523.

Finkelstein R, Reeves W, Ariizumi T, Steber C (2008). Molecular aspects of seed dormancy. Annu Rev Plant Biol 59: 387–415.

Gaudet P, Chisholm R, Berardini T, Dimmer E, Engel SR, Fey P, Hill DP, Howe D, Hu JC, Huntley R, Khodiyar VK, Kishore R, Li D, Lovering RC, McCarthy F, Ni L, Petri V, Siegele DA, Tweedie S, Van Auken K, Wood V, Basu S, Carbon S, Dolan M, Mungall CJ, Dolinski K, Thomas P, Ashburner M, Blake JA, Cherry JM, Lewis SE, Reference Genome Group of the Gene Ontology Consortium (2009). The Gene Ontology's Reference Genome Project: a unified framework for functional annotation across species. PLoS Comput Biol 5: e1000431.

German MA, Pillay M, Jeong DH, Hetawal A, Luo SJ, Janardhanan P, Kannan V, Rymarquis LA, Nobuta K, German R, De Paoli E, Lu C, Schroth G, Meyers BC, Green PJ (2008). Global identification of microRNA-target RNA pairs by parallel analysis of RNA ends. Nat Biotechnol 26: 941–946.

Grasser M, Kane CM, Merkle T, Melzer M, Emmersen J, Grasser KD (2009). Transcript elongation factor TFIIS is involved in Arabidopsis seed dormancy. J Mol Biol 386: 598–611.

Groos C, Gay G, Perretant MR, Gervais L, Bernard M, Dedryver F, Charmet G (2002). Study of the relationship between preharvest sprouting and grain color by quantitative trait loci analysis in a white x red grain bread-wheat cross. Theor Appl Genet 104: 39–47.

Gu XY, Foley ME, Horvath DP, Anderson JV, Feng J, Zhang L, Mowry CR, Ye H, Suttle JC, Kadowaki KI, Chen Z (2011). Association between seed dormancy and pericarp color is controlled by a pleiotropic gene that regulates ABA and flavonoid synthesis in weedy red rice. Genetics 189: 1515–1524.

Gubler F, Millar AA, Jacobsen JV (2005). Dormancy release, ABA and pre-harvest sprouting. Curr Opin Plant Biol 8: 183–187.

Gutierrez L, Van Wuytswinkel O, Castelain M, Bellini C (2007). Combined networks regulating seed maturation. Trends Plant Sci 12: 294–300.

Himi E, Maekawa M, Miura H, Noda K (2011). Development of PCR markers for Tamyb10 related to R-1, red grain color gene in wheat. Theor Appl Genet 122: 1561–1576.

Holdsworth MJ, Bentsink L, Soppe WJJ (2008a). Molecular networks regulating Arabidopsis seed maturation, after-ripening, dormancy and germination. New Phytol 179: 33–54.

Holdsworth MJ, Finch-Savage WE, Grappin P, Job D (2008b). Post-genomics dissection of seed dormancy and germination. Trends Plant Sci 13: 7–13.

Huang XH, Wei XH, Sang T, Zhao QA, Feng Q, Zhao Y, Li CY, Zhu CR, Lu TT, Zhang ZW, Li M, Fan DL, Guo YL, Wang A, Wang L, Deng LW, Li WJ, Lu YQ, Weng QJ, Liu KY, Huang T, Zhou TY, Jing YF, Li W, Lin Z, Buckler ES, Qian QA, Zhang QF, Li JY, Han B (2010). Genome-wide association studies of 14 agronomic traits in rice landraces. Nat Genet 42: 961–976.

Huang XQ, Paulo MJ, Boer M, Effgen S, Keizer P, Koornneef M, van Eeuwijk FA (2011). Analysis of natural allelic variation in Arabidopsis using a multiparent recombinant inbred line population. Proc Natl Acad Sci U S A 108: 4488–4493.

Hugouvieux V, Kwak JM, Schroeder JI (2001). An mRNA cap binding protein, ABH1, modulates early abscisic acid signal transduction in Arabidopsis. Cell 106: 477–487.

Iglesias-Fernandez R, Rodriguez-Gacio MD, Matilla AJ (2011). Progress in research on dry afterripening. Seed Sci Res 21: 69–80.

Ingolia NT, Ghaemmaghami S, Newman JRS, Weissman JS (2009). Genome-wide analysis in vivo of translation with nucleotide resolution using ribosome profiling. Science 324: 218–223.

Jannink JL, Moreau L, Charmet G, Charcosset A (2009). Overview of QTL detection in plants and tests for synergistic epistatic interactions. Genetica 136: 225–236.

Jiao YL, Riechmann JL, Meyerowitz EM (2008). Transcriptome-wide analysis of uncapped mRNAs in Arabidopsis reveals regulation of mRNA degradation. Plant Cell 20: 2571–2585.

Joosen RVL, Ligterink W, Dekkers BJW, Hilhorst HWM (2011). Visualization of molecular processes associated with seed dormancy and germination using MapMan. Seed Sci Res 21: 143–152.

Kermode AR (ed) (2011). Seed dormancy. In: Methods and Protocols. Methods in Molecular Biology, Vol. 773.

Kover PX, Valdar W, Trakalo J, Scarcelli N, Ehrenreich IM, Purugganan MD, Durrant C, Mott R (2009). A multiparent advanced generation inter-cross to fine-map quantitative traits in Arabidopsis thaliana. PLoS Genet 5: e1000551.

Linkies A, Muller K, Morris K, Tureckova V, Wenk M, Cadman CSC, Corbineau F, Strnad M, Lynn JR, Finch-Savage WE, Leubner-Metzger G (2009). Ethylene interacts with abscisic acid to regulate endosperm rupture during germination: a comparative approach using Lepidium sativum and Arabidopsis thaliana. Plant Cell 21: 3803–3822.

Lister R, Gregory BD, Ecker JR (2009). Next is now: new technologies for sequencing of genomes, transcriptomes, and beyond. Curr Opin Plant Biol 12: 107–118.

Liu PP, Montgomery TA, Fahlgren N, Kasschau KD, Nonogaki H, Carrington JC (2007). Repression of AUXIN RESPONSE FACTOR10 by microRNA160 is critical for seed germination and post-germination stages. Plant J 52: 133–146.

Lu C, Fedoroff N (2000). A mutation in the arabidopsis HYL1 gene encoding a dsRNA binding protein affects responses to abscisic acid, auxin, and cytokinin. Plant Cell 12: 2351–2365.

Lu C, Han M-H, Guevara-Garcia A, Fedoroff NV (2002). Mitogen-activated protein kinase signaling in postgermination arrest of development by abscisic acid. Proc Natl Acad Sci U S A 99: 15812–15817.

Mackay I, Powell W (2007). Methods for linkage disequilibrium mapping in crops. Trends Plant Sci 12: 57–63.

Mares D, Mrva K, Cheong J, Williams K, Watson B, Storlie E, Sutherland M, Zou Y (2005). A QTL located on chromosome 4A associated with dormancy in white- and red-grained wheats of diverse origin. Theor Appl Genet 111: 1357–1364.

Meng YJ, Gou LF, Chen DJ, Wu P, Chen M (2010). High-throughput degradome sequencing can be used to gain insights into microRNA precursor metabolism. J Exp Botany 61: 3833–3837.

Morris K, Linkies A, Muller K, Oracz K, Wang XF, Lynn JR, Leubner-Metzger G, Finch-Savage WE (2011). Regulation of seed germination in the close Arabidopsis relative Lepidium sativum: a global tissue-specific transcript analysis. Plant Physiol 155: 1851–1870.

Nakabayashi K, Okamoto M, Koshiba T, Kamiya Y, Nambara E (2005). Genome-wide profiling of stored mRNA in Arabidopsis thaliana seed germination: epigenetic and genetic regulation of transcription in seed. Plant J 41: 697–709.

Narsai R, Howell KA, Millar AH, O'Toole N, Small I, Whelan J (2007). Genome-wide analysis of mRNA decay rates and their determinants in Arabidopsis thaliana. Plant Cell 19: 3418–3436.

Narsai R, Law SR, Carrie C, Xu L, Whelan J (2011). In-depth temporal transcriptome profiling reveals a crucial developmental switch with roles for RNA processing and organelle metabolism that are essential for germination in Arabidopsis. Plant Physiol 157: 1342–1362.

Nishimura N, Kitahata N, Seki M, Narusaka Y, Narusaka M, Kuromori T, Asami T, Shinozaki K, Hirayama T (2005). Analysis of ABA Hypersensitive Germination 2 revealed the pivotal functions of PARN in stress response in Arabidopsis. Plant J 44: 972–984.

Nonogaki H, Bassel GW, Bewley JD (2010). Germination—still a mystery. Plant Sci 179: 574–581.

Penfield S, King J (2009). Towards a systems biology approach to understanding seed dormancy and germination. Proc R Soc B Biol Sci 276: 3561–3569.

Penfield S, Li Y, Gilday AD, Graham S, Graham IA (2006). Arabidopsis ABA INSENSITIVE4 regulates lipid mobilization in the embryo and reveals repression of seed germination by the endosperm. Plant Cell 18: 1887–1899.

Rajjou L, Gallardo K, Debeaujon I, Vandekerckhove J, Job C, Job D (2004). The effect of alpha-amanitin on the Arabidopsis seed proteome highlights the distinct roles of stored and neosynthesized mRNAs during germination. Plant Physiol 134: 1598–1613.

Reyes JL, Chua NH (2007). ABA induction of miR159 controls transcript levels of two MYB factors during Arabidopsis seed germination. Plant J 49: 592–606.

Somyong S, Munkvold JD, Tanaka J, Benscher D, Sorrells ME (2011). Comparative genetic analysis of a wheat seed dormancy QTL with rice and Brachypodium identifies candidate genes for ABA perception and calcium signaling. Funct Integr Genomics 11: 479–490.

Tian F, Bradbury PJ, Brown PJ, Hung H, Sun Q, Flint-Garcia S, Rocheford TR, McMullen MD, Holland JB, Buckler ES (2011). Genome-wide association study of leaf architecture in the maize nested association mapping population. Nat Genet 43: 159–162.

Wang LW, Liu HH, Li DT, Chen HB (2011). Identification and characterization of maize microRNAs involved in the very early stage of seed germination. BMC Genomics 12: 154.

Waugh R, Marshall D, Thomas B, Comadran J, Russell J, Close T, Stein N, Hayes P, Muehlbauer G, Cockram J, O'Sullivan D, MacKay I, Flavell A, Ramsay L, *et al.* (2010). Whole-genome association mapping in elite inbred crop varieties. Genome 53: 967–972.

Weitbrecht K, Muller K, Leubner-Metzger G (2011). First off the mark: early seed germination. J Exp Botany 62: 3289–3309.

Winter D, Vinegar B, Nahal H, Ammar R, Wilson GV, Provart NJ (2007). An "Electronic Fluorescent Pictograph" browser for exploring and analyzing large-scale biological data sets. PLoS One 2: e718.

Xiong LM, Gong ZZ, Rock CD, Subramanian S, Guo Y, Xu WY, Galbraith D, Zhu JK (2001). Modulation of abscisic acid signal transduction and biosynthesis by an Sm-like protein in Arabidopsis. Dev Cell 1: 771–781.

Xue LJ, Zhang JJ, Xue HW (2009). Characterization and expression profiles of miRNAs in rice seeds. Nucl Acids Res 37: 916–930.

Yu JM, Holland JB, McMullen MD, Buckler ES (2008). Genetic design and statistical power of nested association mapping in maize. Genetics 178: 539–551.

Yu JM, Pressoir G, Briggs WH, Bi IV, Yamasaki M, Doebley JF, McMullen MD, Gaut BS, Nielsen DM, Holland JB, Kresovich S, Buckler ES (2006). A unified mixed-model method for association mapping that accounts for multiple levels of relatedness. Nat Genet 38: 203–208.

7 Genomic Specification of Starch Biosynthesis in Maize Endosperm

Tracie A. Hennen-Bierwagen and Alan M. Myers

Introduction

Plant reproductive function in many species relies on storage of glucosyl units in the seed in the form of starch granules. This storage is particularly relevant in cereal crops, in which endosperm tissue forms most of the seed mass. Human use of cereals for food and renewable raw materials has provided strong selection for higher yields, which are generated in large part by increased endosperm mass and elevated starch content within that tissue. As a result, endosperm metabolism in cereal crops has evolved to provide an extremely high sink strength for forming starch from photosynthate generated in the leaves and transported to the seed as sucrose. Metabolic flux into starch is stimulated by the crystallization of glucan polymers into starch granules, removing the product of the pathway from the reaction equilibrium. In addition to seed storage tissues, starch granules accumulate in photosynthetic tissues where they are formed and then degraded regularly throughout the diurnal cycle. Seed starch metabolism uses the same biosynthetic pathway and often the same enzymes as those employed in leaves; however, granule formation and growth continue over the course of development leading to the vast glucose stores in mature endosperm. Thus, an important function of plant genomes is to encode the starch metabolic pathway that operates in the seed.

Genetic research into starch biosynthesis began in essence with Mendel's identification of the pea *rugosus* (*r*) locus, which encodes a starch branching enzyme (SBE) (Bhattacharyya *et al.*, 1990). Maize genes encoding components of the starch biosynthetic pathway were identified shortly thereafter from scorable changes in seed phenotype conditioned by mutant alleles (e.g., *sugary1* [*su1*] and *waxy* [*wx1*], identified in 1901 and 1909) (Correns, 1901; Collins, 1909). Ensuing forward genetics research identified loci encoding enzymes that function in each step of the starch biosynthesis pathway. Enzyme purification coupled with molecular cloning studies revealed the amino acid sequences of numerous starch biosynthetic enzymes. Using this information to search the maize genome sequence revealed additional starch biosynthetic enzymes that had not been identified by mutations, and reverse genetics yielded loss-of-function alleles in starch biosynthetic genes that had not been found by classical genetics. From this body of research, a complete characterization of essentially all of the genetic functions that constitute the endosperm starch biosynthetic pathway is now available.

Starch is made up of deceptively complex molecules. The polymers that constitute starch are composed of only glucose subunits linked together with either $\alpha(1\rightarrow4)$ or $\alpha(1\rightarrow6)$ glycoside bonds. However, the precise arrangement of these bonds, exhibited in the linear chain length distribution

Seed Genomics, First Edition. Edited by Philip W. Becraft.

and the locations of branch points, can have dramatic effects on physical properties such as gelling temperature or digestability. These properties affect the applicability of different starches to various end uses and have economic significance. A combination of biochemical, genetic, and genomic studies is beginning to reveal the metabolic mechanisms that regulate these important properties.

This chapter briefly describes the molecular structure of starch and the enzymes that assemble glucose units into the polymers contained within starch granules; the reader is referred to in-depth reviews for a full treatment of the subjects. The chapter then presents a survey of the maize genome to describe each genetic element that encodes an enzyme involved in starch biosynthesis. The genes that are responsible for starch biosynthesis particularly in endosperm tissue are highlighted, and the likely evolutionary history leading to the genomic specification of starch biosynthesis in maize is described.

Overview of Starch Biosynthetic Pathway

Starch Structure

For reviews of starch structure, the reader is referred to Buléon *et al.* (1998), Zeeman *et al.* (2010), Tetlow (2011), and references therein. Starch accumulates in insoluble granules composed of homopolymers in which glucosyl units are linked predominantly by $\alpha(1\rightarrow4)$ glycoside bonds into chains. Such "linear" chains can be joined to each other by $\alpha(1\rightarrow6)$ glycoside bonds, introducing "branches" into the polymer structure. Two types of glucan polymer are present in starch granules. Amylose is almost entirely linear, with only a very slight frequency of branch linkages relative to the number of $\alpha(1\rightarrow4)$ bonds. Amylopectin, in contrast, is branched at a frequency of approximately 5% and displays a bimodal distribution of linear chain lengths. The relative abundance of amylose and amylopectin varies by species and tissue. In cereal endosperm starch, the granules are made up of ~75% amylopectin and 25% amylose. The structure of amylopectin imparts the semicrystalline nature of starch granules because mutants that lack amylose produce particles similar in size and shape to particles found in wild-type plants. Within amylopectin, the branch linkages are thought to be clustered near each other so that adjacent linear chains lacking branches are able to associate with each other and crystallize owing to hydrogen bonding and hydrophobic effects. Amylopectin is "semicrystalline," meaning that crystalline regions alternate with amorphous regions in which the branch linkage clusters are located. The primary crystalline units assemble into several levels of higher order structures, eventually leading to insoluble starch granules.

The enzymes that catalyze starch synthesis must be able to function with structural specificity. Branch linkages are not randomly located but rather are situated relatively close to each other, leaving unbranched regions in other areas of the polymer. The frequency of branch linkages is conserved among starches from different sources. Linear chain lengths are also specific within certain ranges, with some chains forming single crystalline units and others extending between crystalline layers to give rise to the bimodal distribution of chain lengths. Amylopectin structure can be compared with glycogen, which serves a glucose storage function in organisms other than plants. In glycogen, the chain length distribution is unimodal, the branch linkages are thought to be dispersed evenly throughout the polymer, and the branch frequency is ~10% – twice that of amylopectin. As a result of these structural differences, amylopectin adopts an insoluble semicrystalline structure, whereas glycogen is soluble within the cytosol. The fundamental structure of plant starches is the alternating regions of amorphous and crystalline lamellae, and this is conserved as a 9- to 10-nm repeat structure throughout the plant kingdom (Jenkins *et al.*, 1993). Owing to this conserved molecular architecture, starch granules can store vastly more glucose units than glycogen and provide an essential element of seed physiology and plant reproduction.

Starch Biosynthetic Enzymes

For recent reviews addressing the mechanisms of starch granule biosynthesis, the reader is referred to Zeeman *et al.* (2010), Tetlow (2011), Jeon *et al.* (2010), Keeling and Myers (2010), and Hennen-Bierwagen *et al.* (2012). The metabolic pathways for conversion of sucrose to starch are summarized in Figure 7.1. Glucose-1-P (Glc-1-P) is generated either directly from sucrose through two enzymatic steps or indirectly by conversion of fructose into triose phosphates and then back into Glc-1-P. The committed step for starch biosynthesis is catalyzed by ADP glucose pyrophosphorylase (AGPase; ATP:α-D-glucose-1-phosphate adenylyltransferase), which converts Glc-1-P plus ATP into ADP glucose (ADPGlc) plus inorganic pyrophosphate (PPi). ADPGlc serves as the glucosyl unit donor to growing linear glucan chains. Chain elongation is catalyzed by starch synthase (SS; ADPGlc:(1→4)-α-D-glucan 4-α-D-glucosyltransferase), which transfers the glucosyl unit from ADPGlc to the nonreducing end of the glucose polymer and creates a new α(1→4) linkage. Branch linkages are introduced by starch branching enzyme (SBE; α(1→4)-glucan:α(1→4)-glucan-6-glycosyltransferase). SBE catalyzes cleavage of an α(1→4)-linkage in a linear chain and

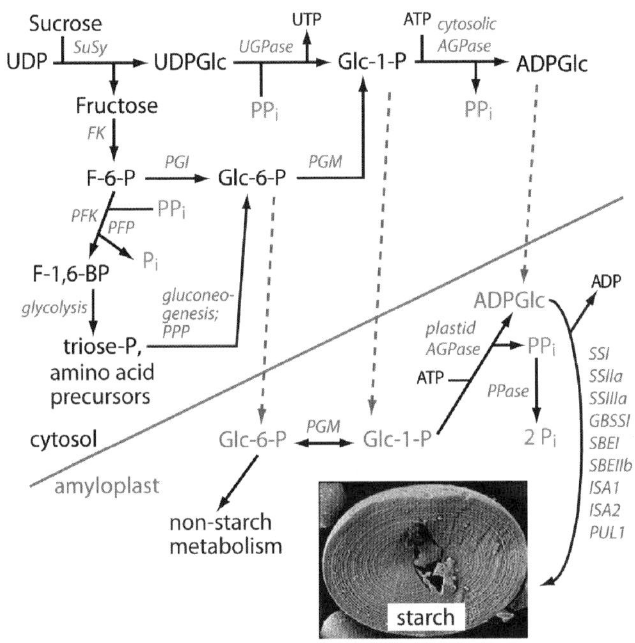

Figure 7.1 Outline of endosperm metabolism from sucrose to starch. *Gray lines* indicate transport from the cytosol into the amyloplast stroma. Glc-1-P is supplied either by UGPase in the "direct" path from sucrose or from metabolic conversion to triose phosphates and resynthesis back into hexose phosphate. Enzymes noted as being involved in starch biosynthesis have been identified by genetic analyses; this does not preclude the involvement of additional isoforms. SuSy, sucrose synthase; UGPase, UDP glucose pyrophosphorylase; FK, fructokinase; PGI, phosphoglucoisomerase; PGM, phosphoglucomutase; PFK, phosphofructokinase; PFP, pyrophosphate-dependent phosphofructokinase; PPP, pentose phosphate pathway; PPase, pyrophosphatase; F-6-P, fructose-6-phosphate; F-1,6-BP, fructose-1,6-bisphosphate. (For color detail, see color plate section.)

reattachment of the reducing end of the released fragment to a 6-hydroxyl group of either the cleaved chain or a neighboring chain, introducing an $\alpha(1\rightarrow6)$ bond. Starch debranching enzyme (DBE; $\alpha(1\rightarrow6)$-glucan hydrolase) also participates in starch biosynthesis, presumably by removal of a subset of branches introduced by SBE, forming the branch linkage arrangement thought to facilitate crystallization.

The concerted actions of SS, SBE, and DBE are responsible for generating the specific molecular architecture of amylopectin that enable crystallization and subsequent formation of the starch granule. However, the precise molecular mechanisms responsible for starch biosynthesis are unclear. A simple linear model would suggest that AGPase generates the glucose donor, SS produces linear chains, SBE creates branches, DBE removes selected branches, and finally crystallization occurs. However, such a mechanism does not seem likely to explain the specific placement of branch linkages or linear chain length distributions. Considering SS, SBE, and DBE, the products of any one reaction serve as the substrate for the other two reactions. A more realistic model suggests that SS, SBE, and DBE act simultaneously on a precursor polymer that crystallizes spontaneously and is removed from the reaction equilibrium. Crystallization and polymer growth may occur in different locations of a single precursor molecule. In support of this model, physical interactions among SS, SBE, and AGPase have been demonstrated (Tetlow *et al.*, 2004, 2008; Hennen-Bierwagen *et al.*, 2008, 2009).

Genomic Specification of Endosperm Starch Biosynthesis in Maize

The maize genome contains multiple elements encoding each of the enzymes involved in starch biosynthesis. These genes had their origin in the monophyletic event that incorporated a cyanobacterial endosymbiont to give rise to the plastid-containing organisms known as *Archaeplastida* (Deschamps *et al.*, 2008). Subsequent evolution led to divergence of the *Chloroplastida* species (i.e., green algae and land plants), distinguished by the presence of chloroplasts. At least one ortholog of each maize starch biosynthetic gene is conserved in all known *Chloroplastida* species, which indicates functional selection rather than gene duplication within specific lineages (Deschamps *et al.*, 2008). This analysis identifies conserved "classes" of each enzyme. In addition, in many instances, multiple paralogs (i.e., "isoforms") within a given class have arisen through gene duplication events. In some instances, duplicate genes (referred to as "twins") also exist within isoform groups in maize, as the result of a relatively recent whole-genome duplication specific to that evolutionary lineage. Table 7.1 specifies all known elements in the maize genome that potentially encode starch biosynthetic enzymes, each of which is discussed in the following sections.

AGPase

AGPase physiology in endosperm of grass species in the Poaceae family, including cereal crops, differs from that of other plant species and tissues owing to distinct subcellular locations of the enzyme (Geigenberger, 2011 and references therein). AGPase appears to be strictly plastidial in most plant tissues (Beckles *et al.*, 2001), whereas in cereal endosperms most AGPase activity is cytosolic and a minor isoform resides within the amyloplast (Rosti *et al.*, 2006 and references therein). Such an arrangement necessitates distinct modes of regulation and metabolic control of starch biosynthesis, including transport of ADPGlc by the protein encoded by the *brittle1 (bt1)* gene and transport of glucose phosphate (Glc-1-P and/or Glc-6-P) from the cytosol into the amyloplast (Figure 7.1).

Table 7.1 Known Elements in the Maize Genome That Potentially Encode Starch Biosynthetic Enzymes

Gene Name	Encoded Protein	Chromosome/ Coordinates[a]	Sequence References	Mutation References
sh2	AGPase LS	3R; 216,414,685–216,421,042	Bhave *et al.*, 1990; Shaw and Hannah, 1992	Mains, 1949
embL	AGPase LS	6F; 166,587,013–166,592,608	Giroux *et al.*, 1995	
leafL	AGPase LS	1R; 273,447,466–273,452,205	Yan *et al.*, 2009b; Huang *et al.*, 2010	Huang *et al.*, submitted
AGPL3	AGPase LS	7R; 23,756,240–23,767,010	Yan *et al.*, 2009b	
bt2	AGPase SS	4R; 58,954,361–58,961,481	Hannah *et al.*, 2001	Teas and Teas, 1953
embS	AGPase SS	2R; 173,380,830–173,391,388	Hannah *et al.*, 2001	Huang *et al.*, submitted
leafS	AGPase SS	1F; 221,648,603–221,653,116	Prioul *et al.*, 1994; Hannah *et al.*, 2001	Slewinski *et al.*, 2008
bt1	ADPGlc transporter	5R 112,928,080–112,930,204	Sullivan *et al.*, 1991	Mangelsdorf, 1926; Wentz, 1926
ss1	SSI	9F; 17,633,700–17,644,420	Knight *et al.*, 1998	
su2	SSIIa	6F; 113,234,637–113,238,773	Harn *et al.*, 1998	Eyster, 1934
ss2b-1	SSIIb-1	Not aligned[b]	Harn *et al.*, 1998	
ss2b-2	SSIIb-2	5R; 206,850,639–206,855,628	Yan *et al.*, 2009b	
ss2c	SSIIc	5R; 33,050,572–33,066,185	Yan *et al.*, 2009a	
du1	SSIIIa	10F; 59,506,710–59,518,288	Gao *et al.*, 1998; Lin *et al.*, 2012	Mangelsdorf, 1947
ss3b-1	SSIIIb-1	10F; 142,920,563–142,929,523	Yan *et al.*, 2009b	
ss3b-2	SSIIIb-2	2F; 8,871,415–8,890,970	Yan *et al.*, 2009b	
ss4	SSIV	8F; 125,308,442–125,317,219	Yan *et al.*, 2009b	
wx1	GBSSI	9R; 23,256,335–23,260,210	Shure *et al.*, 1983; Klösgen *et al.*, 1986	Collins, 1909
gbs2	GBSSII	7R; 34,872,733–34,879,973	Yan *et al.*, 2009b	
sbe1	SBEI	5R; 63,318,794–63,324,615	Baba *et al.*, 1991; Fisher *et al.*, 1995; Kim *et al.*, 1998	Blauth *et al.*, 2002
sbe2a	SBEIIa	2F; 58,586,007–58,596,220	Gao *et al.*, 1997	Blauth *et al.*, 2001
ae	SBEIIb	5R; 168,451,701–168,468,750	Fisher *et al.*, 1993; Stinard *et al.*, 1993	Vinyard and Bear, 1952
sbe3	SBEIII	8F; 141,990,059–141,993,024	Yan *et al.*, 2009b	
su1	ISA1	4F; 41,369,510–41,378,086	James *et al.*, 1995; Beatty *et al.*, 1997	Correns, 1901
isa2	ISA2	6F; 144,564,725–144,567,105	Dinges *et al.*, 2003	Kubo *et al.*, 2010
isa3	ISA3	7R; 129,101,533–129,113,095	Dinges *et al.*, 2003	
zpu1	PUL	2R; 108,631,513–108,681,578	Beatty *et al.*, 1999	Dinges *et al.*, 2003

[a]Chromosome coordinates refer to maize genome assembly RefGen_v2, December, 18, 2010 (http://www.plantgdb.org/ZmGDB/). The transcribed region is indicated.

[b]The cDNA encoding SSIIb-1 is specified by Genbank accession number AF019297 and is supported by multiple EST sequences in Genbank with exact sequence matches. However, the cDNA sequence does not align anywhere in the B73 maize genome sequence with near-perfect identity.

The genetic specification of AGPase in cereal species is complex and not fully understood. Plant AGPase is an $\alpha_2\beta_2$ heterotetramer of approximately 210 kDa, comprising two structurally related polypeptides referred to as the large subunit (AGPase LS) and small subunit (AGPase SS) (Morell *et al.*, 1987; Lin *et al.*, 1988; Okita *et al.*, 1990; Preiss *et al.*, 1990). The maize genome contains four genes encoding AGPase LS and three others that code for AGPase SS (Table 7.1). This set of genes has been described by multiple laboratories using various nomenclature systems. Table 7.1

defines each gene by its position in the maize genome and its nucleotide sequence, and the particular nomenclature used in this chapter is noted.

Mutations of the *shrunken2* gene (*sh2*), encoding AGPase LS, or *bt2*, encoding AGPase SS, cause reduction in endosperm starch content to ~25%–38% of normal (Laughnan, 1953; Cameron and Teas, 1954; Tsai and Nelson, 1966) and reduce total AGPase activity to ~15% of the wild-type value (Dickinson and Preiss, 1969). SH2 and BT2 are inferred to be cytosolic proteins from the findings that (1) when the BT1 ADPGlc transporter in the amyloplast membrane is defective owing to mutation, ADPGlc accumulates to high levels in endosperm cytosol; (2) cytosolic ADPGlc accumulation does not occur if *sh2* is also mutated in addition to *bt1*; and (3) neither BT2 nor SH2 possesses plastid transit peptides (Hannah, 2007).

The 15% of AGPase activity that remains in the absence of the cytosolic form could be encoded by a combination of any of the three other AGPase LS genes and two other AGPase SS genes present in maize (Table 7.1). Mutations of *embS* and *leafL* were obtained by reverse genetics and tested for effects on endosperm starch content (authors' unpublished results). In both instances, the starch content was reduced by ~6% compared with wild-type, indicating that the AGPase SS and AGPase LS encoded by these two genes contribute to AGPase activity within amyloplasts. Further analyses are necessary to test the possibility that *leafS*, *embL*, and *AGPL3* also provide plastid AGPase function in the endosperm. The dual localization of AGPase is conserved in cereal species and is presumed to provide a selective advantage. A possible explanation for such a role is modulation of the sink strength of endosperm tissue so that hexose phosphate present in the plastid can be used either for starch biosynthesis or for nonstarch metabolism (Figure 7.1).

Estimates of the relative steady-state mRNA levels determined from RNA-seq data (www.maizegdb.org) can provide evidence regarding which genes code for the AGPase activity in specific tissues. *SH2* and *BT2* transcripts are highly expressed in endosperm, with only trace amounts in embryo and essentially no transcript present in any other tissue. These data support the conclusion that cytosolic AGPase is specific to endosperm physiology. Transcripts from all five of the other genes encoding AGPase subunits accumulate to varying degrees in endosperm, although AGPL3 mRNA abundance is far lower than that for all of the other genes. These data are consistent with potential functions of all the maize AGPase genes in contributing to amyloplast AGPase activity. Multiple AGPase genes function in embryo and leaf, as shown by the fact that *embS* or *leafL* mutations reduce but do not eliminate starch content in those tissues (authors' unpublished results).

Evolution of the AGPase genes in maize has been investigated by phylogenetic comparisons (Figure 7.2*A*) (Rosti and Denyer, 2007; Georgelis *et al.*, 2008; Yan *et al.*, 2009b). AGPase LS and AGPase SS proteins from *Chloroplastida* species (i.e., green algae and land plants) are all >50% identical to each other at the amino acid level over the entire catalytic domain. These genes arose from a common ancestor present in a progenitor of the extant chloroplast-containing organisms. Duplications gave rise to at least three genes encoding AGPase LS before divergence of monocots and eudicots, as shown by orthologs present in maize, rice, *Arabidopsis*, and poplar. The maize genes *bt2* and *embS* are 90% identical, and each has an ortholog in the rice genome, which suggests that they arose as the result of a whole-genome duplication early in the evolution of the Graminacea. Subsequently, there was a second whole-genome duplication specific to maize, and this gave rise to the twin genes *bt2* and *leafS*. Accordingly, *leafS* is not present in rice. Among the four maize genes encoding AGPase LS, *sh2* and *embL* are more closely related to each other than to *leafL* or *APGL3* and are both present in rice. Thus, *sh2* and *embL* are thought to be derived from whole-genome duplication in an ancestral grass species. Orthologs of *leafL* and *AGPL3* are present in rice, maize, and eudicot species, indicating their presence in a primordial angiosperm before divergence of the monocots.

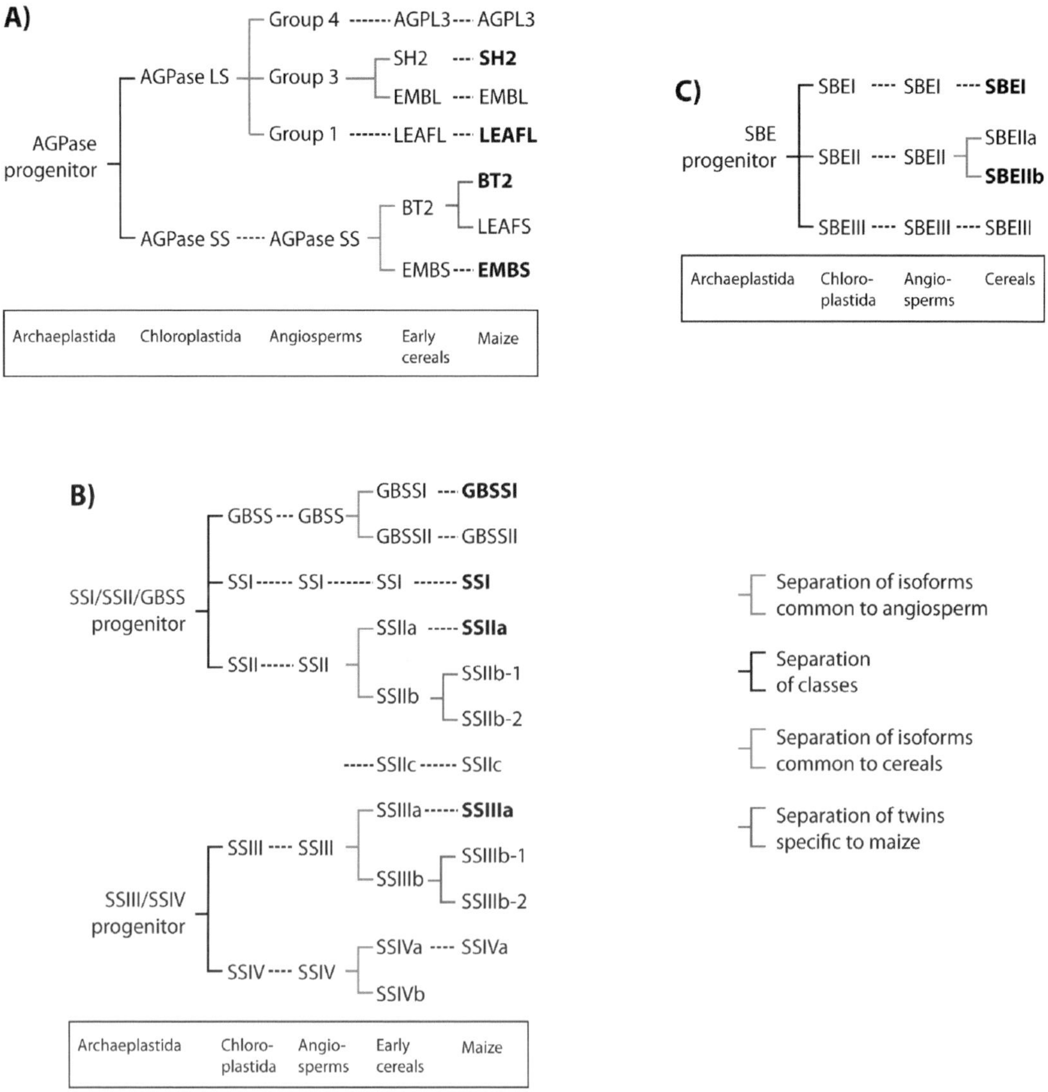

Figure 7.2 Proposed evolution of maize starch biosynthetic genes. Enzymes in bold text have been shown by genetic analysis to be involved in endosperm starch biosynthesis. Only lineages that lead to genes extant in maize are shown. *A)* AGPase. A single gene in the Archaeplastida progenitor, derived from a cyanobacterium, duplicated to form AGPase LS and AGPase SS that are conserved in all chloroplast-containing species. An angiosperm progenitor underwent further duplications to generate a minimum of three isoforms of AGPase LS and one of AGPase SS before separation of monocots and eudicots. In the Graminacea, a whole-genome duplication resulted in two isoforms of the group 3 AGPase LS and two isoforms of AGPase SS. A whole-genome duplication generated the duplicate gene encoding LEAFS that is present in maize but not in rice. *B)* SS. Two genes were present in the primoridal Archaeplastida progenitor before divergence of plastid-containing organisms. At the establishment of the Chloroplastida, these had duplicated to form five SS classes, each of which is extant in all green algae and land plants. In the Graminacea, a whole-genome duplication resulted in isoforms of GBSS, SSII, and SSIII, each of which is present in maize and rice. SSIV was also duplicated at this stage, but one copy was not retained in maize. SSIIc could have originated either in an ancestor common to both monocots and eudicots or after that division. A whole-genome duplication generated twin genes for SSIIb and SSIIIb that are present in maize but not in rice. *C)* SBE. Three classes of SBE had been formed before divergence of green algae and land plants by duplication of an ancestral gene. In general, all three classes are conserved in Chloroplastida species, although at least one exception is known. The ancestral whole-genome duplication in the Graminaceae gave rise to two isoforms of the SBE II class, both of which are present in maize and rice. (Adapted from Rosti and Denyer [2007]; Deschamps *et al.* [2008]; Georgelis *et al.* [2008]; and Yan *et al.* [2009a, 2009b].) (For color detail, see color plate section.)

Starch Synthase

Five SS classes are encoded in the maize genome, each of which is conserved in all known chloroplast-containing species (Table 7.1) (Deschamps *et al.*, 2008). Genes exist that encode two isoforms of the exclusively granule-bound enzyme, GBSS, referred to as GBSSI and GBSSII. Similarly, genes encoding three isoforms of the SSII class (SSIIa, SSIIb, and SSIIc) and two isoforms of the SSIII class (SSIIIa and SSIIIb) are present. Twin genes exist in maize for SSIIb (SSIIb-1 and SSIIb-2) and SSIIIb (SSIIIb-1 and SSIIIb-2). The SSI and SSIV classes in maize are encoded by single genes. The amino acid identity between isoforms of the same SS class ranges from 68%–81% over the catalytic domain, whereas the identity between classes ranges from 27%–49%. Classical mutations that condition scorable kernel phenotypes are known in the genes encoding GBSSI, SSIIa, and SSIIIa. From these data and further characterizations of the effects of the mutations on starch structure, it is evident that GBSSI, SSIIa, and SSIIIa all function in starch biosynthesis in endosperm. To date, mutations have not been characterized by either forward or reverse genetics in the maize genes encoding SSI, SSIIb-1, SSIIb-2, SSIIc, SSIIIb-1, SSIIIb-2, SSIV, or GBSSII. Whether or not these other SS enzymes contribute to endosperm starch biosynthesis is unknown. However, the genetic analyses indicate that none of them can fully compensate for loss of GBSSI, SSIIa, or SSIIIa.

Each conserved SS class appears to function in the generation of certain ranges of linear chain lengths within the glucan polymers in starch granules (Jeon *et al.*, 2010; Tetlow, 2011). GBSSI, encoded by *waxy1* (*wx1*), functions specifically to generate amylose. Mutations in *wx1* that prevent GBSSI expression condition complete loss of amylose, without affecting amylopectin structure or overall granule morphology. These data indicate that amylopectin is the key determinant of the semicrystalline nature of starch and of granule structure. Genetic analysis showed that SSIIa, encoded by *sugary2* (*su2*), functions in the generation of medium-length chains within amylopectin. In *su2* mutants, medium-length linear chains of degree of polymerization (DP) 16–24 are decreased in frequency and short chains of DP 8–12 are more abundant relative to wild-type. From these data, SSIIa is thought to elongate pre-existing chains in the size range of DP 8–12. This structural alteration of amylopectin is conserved in many species when SSIIa is mutated. SSIIIa, encoded by *dull1* (*du1*), appears to serve a complex role in determining amylopectin structure (Lin *et al.*, 2012). Chain length profiles show specific decreases of DP 9, DP 17, and DP 18, with increased abundance of DP 11–15 chains. Other reproducible alterations occur between DP 21 and DP 30. This pattern is highly reproducible in mutant cereal endosperm in other species including rice and barley. The effects of *du1* mutations are difficult to explain based on substrate preference for a specific chain length range. SSIII has been proposed to affect the activities of other SSs, in addition to possessing its own catalytic activity. Supporting this hypothesis, SSIIIa has been shown to be present in a high-molecular-weight complex that also contains SSIIa (Hennen-Bierwagen *et al.*, 2009).

Gene expression patterns for the SSs in maize can best be examined by surveying the publicly available RNA-seq data (www.maizegdb.org). In general, there appears to be isoform specificity that agrees with the genetic identification of enzymes required in the endosperm. For classes with multiple isoforms, at least one is expressed predominantly in endosperm, whereas the mRNAs encoding the others are most abundant in nonendosperm tissues. Endosperm-predominant expression is evident for SSIIa, SSIIIa, and GBSSI, encoded by *su2*, *du1*, and *wx1*, in agreement with genetic data indicating roles in storage starch biosynthesis. The relative levels of mRNA accumulation as estimated from RNA-seq data vary greatly for different SS classes and isoforms. The most abundant are the mRNAs encoding GBSSI and SSI, which are present at frequencies close to those of the SH2 and BT2 mRNAs. At the low end of the scale are the SSIIb-2, SSIIc, SSIIIb-1, and SSIV transcripts, which are observed ~100 times less frequently than GBSSI or SSI in maize endosperm. At this time, no RNA-seq data exist for SSIIb-1 or SSIIIb-2 in the maize genome database, and their roles remain unknown.

From genetic data in other species, particularly rice and *Arabidopsis*, SSI is thought to be responsible for generating short linear glucans within amylopectin of DP 6–10, which are elongated by SSIIa (Delvallé *et al.*, 2005; Fujita *et al.*, 2006). SSIV function has been studied in *Arabidopsis* leaf and potato tuber, where it is thought to influence granule initiation but does not affect amylopectin fine structure (Roldán *et al.*, 2007; Gamez-Arjona *et al.*, 2011). Discerning whether these same functions are provided by SSI and SSIV in cereal endosperm requires further investigation. As with the other SS classes, both SSI and SSIV are conserved in all green plants, and both are expressed at the level of mRNA accumulation in maize endosperm, so functions in seed starch biosynthesis are likely.

The *Archaeplastida* progenitor of green algae and land plants appears to have contained two separate genes encoding SS, which may have been derived from distinct prokaryotic sources (Price *et al.*, 2012). Phylogenetic analyses show clearly that genes encoding each of the five SS classes were present early in *Chloroplastida* evolution, before divergence of green algae and land plants (Figure 7.2*B*) (Deschamps *et al.*, 2008; Yan *et al.*, 2009a, 2009b). The whole-genome duplication that occurred early in the monocot lineage gave rise to the SSIIa/SSIIb, SSIIIa/SSIIIb, and GBSSI/GBSSII isoform pairs, as shown by the presence of orthologs encoding all of these proteins in both maize and rice. The late whole-genome duplication in the maize lineage generated the twin pairs SSIIb-1/SSIIb-2 and SSIIIb-1/SSIIIb-2. The evolutionary history leading to SSIIc is unclear because the duplication that gave rise to this isoform may have occurred either before the split of monocots and eudicots or specifically in the monocot lineage.

Starch Branching Enzyme

Survey of the maize genome reveals four genes that encode SBE (Table 7.1). Biochemical and genetic characterization of plant SBEs has revealed three different enzymes, termed SBEI, SBEIIa, and SBEIIb, which form two conserved enzyme classes. The SBEII class exhibits two highly homologous isoforms. A third conserved class, SBEIII was revealed by survey of the rice, *Arabidopsis*, and poplar genome sequences (Han *et al.*, 2007), and a gene encoding SBEIII also is present in maize. Genes encoding SBEII and SBEIII are present in all plant genomes analyzed, as revealed by sequence identities on the order of ≥70% when proteins are compared between species. SBEI genes are nearly universally present in plants; however, this class is not found in *Arabidopsis*. Among all of the genes encoding SS, SBE, or DBE, the lack of a SBEI-encoding gene in *Arabidopsis* is the only example known where one of the conserved enzyme classes is not encoded within a plant genome.

As seen within the SS classes, SBEs also demonstrate substrate specificities for chain length selection and transfer (Guan and Preiss, 1993; Takeda *et al.*, 1993; Guan *et al.*, 1997). *In vitro* assays showed that SBEI has a preference to transfer longer linear chains than those transferred by SBEIIb. The range of chains transferred by SBEIIb is less than DP 10, with the most abundant product branch being DP 6–7. SBEI, in contrast, transfers chains greater than DP 10. In addition, the length of the substrate chain being acted on needs to be longer for SBEI, greater than DP 16, whereas SBEIIb can cleave linear regions of DP 12 or less. SBEIIa has not been characterized by *in vitro* assays; however, the extremely high similarity to SBEIIb suggests the properties of the two isoforms will be similar.

Microarray and deep sequencing data reveal that the genes encoding SBEI, SBEIIa, SBEIIb, and SBEIII all are expressed in endosperm at the level of mRNA accumulation, at varying levels. The maize gene *amylose extender* (*ae*) encodes SBEIIb, which is expressed in endosperm at the highest level among these genes and has a major role in the determination of amylopectin structure. Mutation of *ae* results in a decrease in starch content of ~20% compared with wild-type and causes a major change in amylopectin structure such that the branch linkage frequency is decreased to a large extent and the average linear chain length is increased. SBEIIa does not compensate for loss of SBEIIb, despite the facts that the two proteins are >90% identical and that SBEIIa mRNA is present in endosperm.

Mutations of *sbe2a*, encoding SBEIIa, do not condition changes in endosperm starch content or amylopectin chain length distribution. However, SBEIIa transcript is present in endosperm at about 20% of the frequency of the SBEIIb mRNA. SBEI transcript also is present in endosperm at appreciable levels. Loss-of-function mutations in *sbe1* have a subtle effect on the structure of amylopectin, particularly with regard to the density of branches at the root of crystalline clusters (Yao *et al.*, 2004; Xia *et al.*, 2011). In rice, mutation of SBEI caused slight but reproducible increases in the frequency of short chains of less than DP 10, and slightly decreased frequency of chains of DP 12–21 and greater than DP 37 (Satoh *et al.*, 2003). These data show that SBEI affects endosperm starch structure in cereals. Transcripts encoding SBEIII are present at exceedingly low levels throughout the plant. Discerning whether or not this gene functions in starch biosynthesis in endosperm or any tissue would require loss-of-function mutations.

SBEIIb and SBEI transcripts accumulate predominantly in the endosperm, and both proteins function in seed starch biosynthesis. SBEIIa transcripts are also present in endosperm, but a function for this gene has been demonstrated only for leaf tissue. SBEIII has very low expression, and its function is unknown. Phylogenetics indicates that genes encoding SBEI, SBEII, and SBEIII all were present at the inception of the *Chloroplastida* lineage (Figure 7.2*C*) (Deschamps *et al.*, 2008; Yan *et al.*, 2009b). The SBEIIa/SBEIIb pair of isoforms is present in maize and rice, indicating they arose from the early whole-genome duplication in a cereal ancestor.

Starch Debranching Enzyme

Involvement of DBEs in starch biosynthesis was demonstrated by genetic analyses based on the "phytoglycogen-accumulation" phenotype. Loss-of-function mutations, in particular DBE classes, condition this phenotype, which entails major decreases in starch content; altered amylopectin structure; severely abnormal granule morphology; and appearance of a water-soluble glucan, termed phytoglycogen, that is similar in structure to glycogen from nonplant species (Hennen-Bierwagen *et al.*, 2012). Appreciable amounts of phytoglycogen are not detected in wild-type plants but only in plants with compromised DBE function. In addition to maize, this phenotype has been seen in rice and barley endosperm, *Arabidopsis* leaf, potato tuber, and *Chlamydomonas* cells, indicating the conserved nature of DBE function in starch biosynthesis.

Four classes of DBE are conserved in plants and green algae. One class is referred to as a "pullulanase-type DBE" (PUL), and three classes are referred to as "isoamylase-type DBEs" (ISA). Two different bacterial genes, currently extant in separate prokaryotic species, appear to be the progenitors of the four conserved plant genes. In addition to amino acid sequences, PUL and ISA differ by substrate specificity. PUL enzymes readily hydrolyze $\alpha(1\rightarrow6)$ glycosidic bonds in pullulan, a linear polymer with the structure -(Glc$\alpha(1\rightarrow4)$-Glc$\alpha(1\rightarrow4)$-Glc$\alpha(1\rightarrow6)$)-n, whereas they are inactive or have low activity toward glycogen. ISA enzymes, in contrast, are inactive toward pullulan but readily hydrolyze branch linkages within glycogen. The primordial ISA gene in plants apparently was amplified and diverged to generate ISA1, ISA2, and ISA3 very early in the evolution of chloroplast-containing organisms, before the separation of green algae and land plants. ISA1, ISA2, and ISA3 are separate classes of enzyme, each with its own conserved function, rather than isoforms that arose as the result of relatively recent gene duplications (Deschamps *et al.*, 2008).

Three genes encoding ISA and one gene encoding PUL are present in the maize genome (Table 7.1). All of the DBEs are members of a structural superfamily of proteins that includes α-amylases, and highly conserved catalytic residues of that family are generally present in the maize DBEs. The exception is ISA2, which has substitutions at multiple positions where catalytic residues are present in other members of the family. These changes render ISA2 noncatalytic for DBE activity. Despite this characteristic, however, the noncatalytic DBE class is highly conserved, indicating a function for ISA2 other than a direct enzymatic role (Hussain *et al.*, 2003).

Genetic evidence indicates function of ISA1 and ISA2 in endosperm starch biosynthesis to varying extents (Hennen-Bierwagen *et al.*, 2012). ISA1 is a key determinant of endosperm starch structure and content, as shown by the fact that mutations of the *su1* gene condition the phytoglycogen-accumulation phenotype. Loss of ISA1 also conditions phytoglycogen accumulation in rice and barley endosperm as well as *Arabidopsis* leaves and potato tubers. In contrast, loss of ISA2 has only small effects on total endosperm starch accumulation or amylopectin structure. Loss of ISA2 in maize endosperm did affect granule size during development, although by maturity the effect was relatively minor and much less severe than granules that form in *isa1-* mutants. Despite this small effect of *isa2-* single mutations, when coupled with loss of SSIIIa or with a partially functional allele of the ISA1-encoding gene, the ISA2 deficiency conditioned a phytoglycogen accumulation phenotype (Lin *et al.*, 2012). Thus, ISA2 in maize endosperm is a major determinant of starch content and structure in some, but not all, conditions.

In contrast to endosperm, ISA2 by itself is required for normal starch content in *Arabidopsis* leaves and potato tubers (Zeeman *et al.*, 1998; Bustos *et al.*, 2004). Deficiency of ISA2 in either instance causes the same phytoglycogen accumulation phenotype as mutations eliminating ISA1. This is explained by the fact that ISA1 and ISA2 function together in a heteromeric enzyme complex, and both proteins are requisite for enzyme activity in *Arabidopsis* or potato. Endosperm differs because ISA1 is enzymatically active either as a homomultimer or as a heteromultimer, and the former can sustain starch biosynthesis at normal levels. The reason for the difference between endosperm and the other tissues is unknown but could be explained either by inherent differences in ISA1 structure between the cereals and other plants or by metabolic or regulatory distinctions between the tissue types.

PUL appears to have a minor effect on starch content in endosperm. Single mutants lacking this enzyme had no discernible effect on starch content or structure. However, when coupled with a partially functional allele affecting ISA1 that does not cause a noticeable phenotype, the double mutants accumulated phytoglycogen (Dinges *et al.*, 2003). This result indicates that PUL can contribute to starch biosynthesis, although it cannot fully substitute for ISA1 as shown by the strong phytoglycogen accumulation phenotype of ISA1 loss-of-function mutations. ISA3 deficiency in endosperm has been analyzed only in rice, where the mutation had minor effects on amylopectin structure without causing reduced starch content.

Transcript encoding ISA1 is the most abundant among the DBE genes in endosperm, in excess of ISA2, PUL, or ISA3 transcripts by factors of threefold to eightfold. The ISA1 transcript is much more abundant in endosperm than in any other tissue; however, the low level seen in other tissues is close to that of ISA2, PUL, and ISA3 (www.maizegdb.org). These data are consistent with ISA1 being the predominant DBE activity involved in starch biosynthesis in endosperm. In summary, ISA1 is the major DBE active in endosperm starch accumulation, ISA2 is not normally required but does contribute a major function when other starch biosynthesis enzymes are defective, PUL provides a minor DBE biosynthetic function in this tissue, and ISA3 appears not to function in this process.

Genes encoding each DBE class were present in the progenitor of the green algae and land plants, as shown by the presence of four paralogs in any genome analyzed to date (Hennen-Bierwagen *et al.*, 2012). Isoforms within classes have not been found, so subsequent to any whole-genome duplication in the cereals one of the two copies has been lost.

Conclusion

The catalytic functions that participate in starch biosynthesis are provided by multiple conserved classes of enzymes. Each species and tissue has at least one SSI, SSII, SSIII, GBSS, and SSIV enzyme present, and similar class conservation is evident for SBE, DBE, and AGPase. Within each of the AGPase, SS, and SBE classes, multiple isoforms have arisen that appear to exhibit specific individual functions. A general conclusion is that one set of isoforms in a class functions primarily in endosperm, and a second set encoded by different genes functions in leaves and other starch-accumulating tissues. The major isoforms in maize endosperm are SSIIa, SSIIIa, GBSSI, and SBEIIb.

In contrast, some classes of starch biosynthetic enzyme exist as single entities. These include all of the DBEs, SSI, SSIV, SBEI, and SBEIII. For these enzyme classes, differences in function between the storage starch pathway in endosperm and transient starch biosynthesis in other tissues may be controlled by regulatory mechanisms specific to each tissue. One example is the DBE function provided by ISA1 and ISA2, where different assemblies of the heteromultimeric enzyme and the ISA1 homomer are evident in leaf and endosperm, even though the same genes are active in both tissues (authors' unpublished results). In the other instances, paralogous genes appear to provide tissue-specific functions.

Questions remain regarding why certain enzyme classes have not duplicated during angiosperm evolution when so many others have evolved to generate multiple isoforms, and further work is necessary to investigate this topic. A hypothesis to explain the divergence of isoforms within many enzyme classes in maize and other cereals is that endosperm-specific functions provide a strong selective advantage to seeds for reproductive fitness. The advent of large starch granules in endosperm that can persist throughout seed development and dormancy may provide such an advantage. Specialized gene functions that generate such granules, rather than the much smaller transient starch granules in leaves and other tissues, may have arisen through the gene duplication and specialization process that has occurred in the cereal lineage.

References

Baba T, Kimura K, Mizuno K, Etoh H, Ishida Y, Shida O, Arai Y (1991). Sequence conservation of the catalytic regions of amylolytic enzymes in maize branching enzyme-I. Biochem Biophys Res Commun 181: 87–94.

Beatty MK, Myers AM, James MG (1997). Genomic nucleotide sequence of a full-length wild type allele of the maize *sugary1* (*su1*) gene. Plant Physiol 115: 1731.

Beatty MK, Rahman A, Cao H, Woodman W, Lee M, Myers AM, James MG (1999). Purification and molecular genetic characterization of ZPU1, a pullulanase-type starch-debranching enzyme from maize. Plant Physiol 119: 255–266.

Beckles DM, Smith AM, ap Rees T (2001). A cytosolic ADP-glucose pyrophosphorylase is a feature of graminaceous endosperms, but not of other starch-storing organs. Plant Physiol 125: 818–827.

Bhattacharyya MK, Smith AM, Ellis TH, Hedley C, Martin C (1990). The wrinkled-seed character of pea described by Mendel is caused by a transposon-like insertion in a gene encoding starch-branching enzyme. Cell 60: 115–122.

Bhave MR, Lawrence S, Barton C, Hannah LC (1990). Identification and molecular characterization of shrunken-2 cDNA clones of maize. Plant Cell 2: 581–588.

Blauth SL, Kim KN, Klucinec J, Shannon JC, Thompson D, Guiltinan M (2002). Identification of *Mutator* insertional mutants of starch-branching enzyme 1 (sbe1) in *Zea mays* L. Plant Mol Biol 48: 287–297.

Blauth SL, Yao Y, Klucinec JD, Shannon JC, Thompson DB, Guiltinan MJ (2001). Identification of *Mutator* insertional mutants of starch-branching enzyme 2a in corn. Plant Physiol 125: 1396–1405.

Buléon A, Colonna P, Planchot V, Ball S (1998). Starch granules: structure and biosynthesis. Int J Biol Macromol 23: 85–112.

Bustos R, Fahy B, Hylton CM, Seale R, Nebane NM, Edwards A, Martin C, Smith AM (2004). Starch granule initiation is controlled by a heteromultimeric isoamylase in potato tubers. Proc Natl Acad Sci U S A 101: 2215–2220.

Cameron JW, Teas HJ (1954). Carbohydrate relationships in developing and mature endosperms of brittle and related maize genotypes. Am J Botany 41: 50–55.

Collins GN (1909). A new type of Indian corn from China. USDA Bur Plant Indust Bull 161: 1–30.

Correns C (1901). Bastarde zwischen maisrassen, mit besonder Berucksichtung der Xenien. Bibl Bot 53: 1–161.

Delvallé D, Dumez S, Wattebled F, Roldán I, Planchot V, Berbezy P, Colonna P, Vyas D, Chatterjee M, Ball S, Mérida Á, D'Hulst C (2005). Soluble starch synthase I: a major determinant for the synthesis of amylopectin in Arabidopsis thaliana leaves. Plant J 43: 398–412.

Deschamps P, Colleoni C, Nakamura Y, Suzuki E, Putaux JL, Buleon A, Haebel S, Ritte G, Steup M, Falcon LI, Moreira D, Loffelhardt W, Raj JN, Plancke C, d'Hulst C, Dauvillee D, Ball S (2008). Metabolic symbiosis and the birth of the plant kingdom. Mol Biol Evol 25: 536–548.

Deschamps P, Moreau H, Worden AZ, Dauvillee D, Ball SG (2008). Early gene duplication within chloroplastida and its correspondence with relocation of starch metabolism to chloroplasts. Genetics 178: 2373–2387.

Dickinson DB, Preiss J (1969). Presence of ADP-glucose pyrophosphorylase in *shrunken-2* and *brittle-2* mutants of maize endosperm. Plant Physiol 44: 1058–1062.

Dinges JR, Colleoni C, James MG, Myers AM (2003). Mutational analysis of the pullulanase-type debranching enzyme of maize indicates multiple functions in starch metabolism. Plant Cell 15: 666–680.

Eyster WH (1934). Genetics of *Zea mays*. Bibl Genet 11: 187–392.

Fisher DK, Boyer CD, Hannah LC (1993). Starch branching enzyme II from maize endosperm. Plant Physiol 102: 1045–1046.

Fisher DK, Kim KN, Gao M, Boyer CD, Guiltinan MJ (1995). A cDNA encoding starch branching enzyme I from maize endosperm. Plant Physiol 108: 1313–1314.

Fujita N, Yoshida M, Asakura N, Ohdan T, Miyao A, Hirochika H, Nakamura Y (2006). Function and characterization of starch synthase I using mutants in rice. Plant Physiol 140: 1070–1084.

Gamez-Arjona FM, Li J, Raynaud S, Baroja-Fernandez E, Munoz FJ, Ovecka M, Ragel P, Bahaji A, Pozueta-Romero J, Merida A (2011). Enhancing the expression of starch synthase class IV results in increased levels of both transitory and long-term storage starch. Plant Biotechnol J 9: 1049–1060.

Gao M, Fisher DK, Kim KN, Shannon JC, Guiltinan MJ (1997). Independent genetic control of maize starch-branching enzymes IIa and IIb. Isolation and characterization of a Sbe2a cDNA. Plant Physiol 114: 69–78.

Gao M, Wanat J, Stinard PS, James MG, Myers AM (1998). Characterization of *dull1*, a maize gene coding for a novel starch synthase. Plant Cell 10: 399–412.

Geigenberger P (2011). Regulation of starch biosynthesis in response to a fluctuating environment. Plant Physiol 155: 1566–1577.

Georgelis N, Braun EL, Hannah LC (2008). Duplications and functional divergence of ADP-glucose pyrophosphorylase genes in plants. BMC Evol Biol 8: 232.

Giroux M, Smith-White B, Gilmore V, Hannah LC, Preiss J (1995). The large subunit of the embryo isoform of ADP glucose pyrophosphorylase from maize. Plant Physiol 108: 1333–1334.

Guan H, Li P, Imparl-Radosevich J, Preiss J, Keeling P (1997). Comparing the properties of *Escherichia coli* branching enzymes and maize branching enzymes. Arch Biochem Biophys 342: 92–98.

Guan HP, Preiss J (1993). Differentiation of the properties of the branching isozymes from maize (*Zea mays*). Plant Physiol 102: 1269–1273.

Han Y, Sun FJ, Rosales-Mendoza S, Korban SS (2007). Three orthologs in rice, Arabidopsis, and Populus encoding starch branching enzymes (SBEs) are different from other SBE gene families in plants. Gene 401: 123–130.

Hannah LC (2007). Starch formation in the cereal endosperm. In: *Plant Cell Monographs, Endosperm*, Vol 8. Springer, Berlin, Germany, pp 179–193.

Hannah LC, Shaw JR, Giroux MJ, Reyss A, Prioul JL, Bae JM, Lee JY (2001). Maize genes encoding the small subunit of ADP-glucose pyrophosphorylase. Plant Physiol 127: 173–183.

Harn C, Knight M, Ramakrishnan A, Guan H, Keeling PL, Wasserman BP (1998). Isolation and characterization of the zSSIIa and zSSIIb starch synthase cDNA clones from maize endosperm. Plant Mol Biol 37: 639–649.

Hennen-Bierwagen TA, James MG, Myers AM (2012). Involvement of debranching enzymes in starch biosynthesis. In: Tetlow IJ (ed). *Essential Reviews in Experimental Biology: The Synthesis and Breakdown of Starch*, Vol 5. Society for Experimental Biology, London, UK.

Hennen-Bierwagen TA, Lin Q, Grimaud F, Planchot V, Keeling PL, James MG, Myers AM (2009). Proteins from multiple metabolic pathways associate with starch biosynthetic enzymes in high molecular weight complexes: a model for regulation of carbon allocation in maize amyloplasts. Plant Physiol 149: 1541–1559.

Hennen-Bierwagen TA, Liu F, Marsh RS, Kim S, Gan Q, Tetlow IJ, Emes MJ, James MG, Myers AM (2008). Starch biosynthetic enzymes from developing maize endosperm associate in multisubunit complexes. Plant Physiol 146: 1892–1908.

Huang B, Chen J, Zhang J, Liu H, Tian M, Gu Y, Hu Y, Li Y, Liu Y, Huang Y (2010). Characterization of ADP-glucose pyrophosphorylase encoding genes in source and sink organs of maize. Plant Mol Biol Reporter 29: 563–572.

Hussain H, Mant A, Seale R, Zeeman S, Hinchliffe E, Edwards A, Hylton C, Bornemann S, Smith AM, Martin C, Bustos R (2003). Three isoforms of isoamylase contribute different catalytic properties for the debranching of potato glucans. Plant Cell 15: 133–149.

James MG, Robertson DS, Myers AM (1995). Characterization of the maize gene *sugary1*, a determinant of starch composition in kernels. Plant Cell 7: 417–429.

Jenkins PJ, Cameron RE, Donald AM (1993). A universal feature in the starch granules from different botanical sources. Starch-Starke 45: 417–420.

Jeon J-S, Ryoo N, Hahn T-R, Walia H, Nakamura Y (2010). Starch biosynthesis in cereal endosperm. Plant Physiol Biochem 48: 383–392.

Keeling PL, Myers AM (2010). Biochemistry and genetics of starch synthesis. Annu Rev Food Sci Technol 1: 271–303.

Kim KN, Fisher DK, Gao M, Guiltinan MJ (1998). Genomic organization and promoter activity of the maize starch branching enzyme I gene. Gene 216: 233–243.

Klösgen RB, Gierl A, Schwarz-Sommer Z, Saedler H (1986). Molecular analysis of the *waxy* locus of *Zea mays*. Mol Gen Genet 203: 237–244.

Knight ME, Harn C, Lilley CE, Guan H, Singletary GW, MuForster C, Wasserman BP, Keeling PL (1998). Molecular cloning of starch synthase I from maize (W64) endosperm and expression in Escherichia coli. Plant J 14: 613–622.

Kubo A, Colleoni C, Dinges JR, Lin Q, Lappe RR, Rivenbark JG, Meyer AJ, Ball SG, James MG, Hennen-Bierwagen TA, Myers AM (2010). Functions of heteromeric and homomeric isoamylase-type starch debranching enzymes in developing maize endosperm. Plant Physiol 153: 956–969.

Laughnan JR (1953). The effect of the sh(2) factor on carbohydrate reserves in the mature endosperm of maize. Genetics 38: 485–499.

Lin Q, Huang B, Zhang M, Rivenbark JG, Zhang X, James MG, Myers AM, Hennen-Bierwagen TA (2012). Functional interactions between starch synthase III and isoamylase-type starch debranching enzyme in maize endosperm. Plant Physiol 158: 679–692.

Lin TP, Caspar T, Somerville CR, Preiss J (1988). A starch deficient mutant of Arabidopsis thaliana with low ADPglucose pyrophosphorylase activity lacks one of the two subunits of the enzyme. Plant Physiol 88: 1175–1181.

Mains EB (1949). Heritable characters in maize: linkage of a factor for shrunken endosperm with the *a1* factor for aleurone color. J Hered 40: 21–24.

Mangelsdorf PC (1926). The genetics and morphology of some endosperm characters in maize. Conn Agric Exp Stn Bull 279: 509–614.

Mangelsdorf PC (1947). The inheritance of amylaceous sugary endosperm and its derivatives in maize. Genetics 32: 448–458.

Morell MK, Bloom M, Knowles V, Preiss J (1987). Subunit structure of spinach leaf ADPglucose pyrophosphorylase. Plant Physiol 85: 182–187.

Okita TW, Nakata PA, Anderson JM, Sowokinos J, Morell M, Preiss J (1990). The subunit structure of potato tuber ADPglucose pyrophosphorylase. Plant Physiol 93: 785–790.

Preiss J, Danner S, Summers PS, Morell M, Barton CR, Yang L, Nieder M (1990). Molecular characterization of the *brittle-2* gene effect on maize endosperm ADPglucose pyrophosphorylase subunits. Plant Physiol 92: 881–885.

Price DC, Chan CX, Yoon HS, Yang EC, Qiu H, Weber AP, Schwacke R, Gross J, Blouin NA, Lane C, Reyes-Prieto A, Durnford DG, Neilson JA, Lang BF, Burger G, Steiner JM, Loffelhardt W, Meuser JE, Posewitz MC, Ball S, Arias MC, Henrissat B, Coutinho PM, Rensing SA, Symeonidi A, Doddapaneni H, Green BR, Rajah VD, Boore J, Bhattacharya D (2012). Cyanophora paradoxa genome elucidates origin of photosynthesis in algae and plants. Science 335: 843–847.

Prioul JL, Jeannette E, Reyss A, Gregory N, Giroux M, Hannah LC, Causse M (1994). Expression of ADP-glucose pyrophosphorylase in maize (*Zea mays* L.) grain and source leaf during grain filling. Plant Physiol 104: 179–187.

Roldán I, Wattebled F, Lucas MM, Delvallé D, Planchot V, Jiménez S, Pérez R, Ball S, D'Hulst C, Merida Á (2007). The phenotype of soluble starch synthase IV defective mutants of *Arabidopsis thaliana* suggests a novel function of elongation enzymes in the control of starch granule formation. Plant J 49: 492–504.

Rosti S, Denyer K (2007). Two paralogous genes encoding small subunits of ADP-glucose pyrophosphorylase in maize, Bt2 and L2, replace the single alternatively spliced gene found in other cereal species. J Mol Evol 65: 316–327.

Rosti S, Rudi H, Rudi K, Opsahl-Sorteberg HG, Fahy B, Denyer K (2006). The gene encoding the cytosolic small subunit of ADP-glucose pyrophosphorylase in barley endosperm also encodes the major plastidial small subunit in the leaves. J Exp Bot 57: 3619–3626.

Satoh H, Nishi A, Yamashita K, Takemoto Y, Tanaka Y, Hosaka Y, Sakurai A, Fujita N, Nakamura Y (2003). Starch-branching enzyme I-deficient mutation specifically affects the structure and properties of starch in rice endosperm. Plant Physiol 133: 1111–1121.

Shaw JR, Hannah LC (1992). Genomic nucleotide sequence of a wild-type *shrunken-2* allele of *Zea mays*. Plant Physiol 98: 1214–1216.

Shure M, Wessler S, Federoff N (1983). Molecular identification and isolation of the Waxy locus in maize. Cell 35: 225–233.

Slewinski TL, Ma Y, Baker RF, Huang M, Meeley R, Braun DM (2008). Determining the role of tie-dyed1 in starch metabolism: epistasis analysis with a maize ADP-glucose pyrophosphorylase mutant lacking leaf starch. J Hered 99: 661–666.

Stinard PS, Robertson DS, Schnable PS (1993). Genetic isolation, cloning, and analysis of a Mutator-induced, dominant antimorph of the maize amylose extender1 locus. Plant Cell 5: 1555–1566.

Sullivan TD, Strelow LI, Illingworth CA, Phillips RL, Nelson OE Jr (1991). Analysis of maize *brittle-1* alleles and a defective *Suppressor-mutator*-induced mutable allele. Plant Cell 3: 1337–1348.

Takeda Y, Guan H-P, Preiss J (1993). Branching of amylose by the branching isoenzymes of maize endosperm. Carbohydr Res 240: 253–263.

Teas HJ, Teas AN (1953). Heritable characters in maize: description and linkage of *brittle endosperm-2*. J Hered 44: 156–158.

Tetlow IJ (2011). Starch biosynthesis in developing seeds. Seed Sci Res 21: 5–32.

Tetlow IJ, Beisel KG, Cameron S, Makhmoudova A, Liu F, Bresolin NS, Wait R, Morell MK, Emes MJ (2008). Analysis of protein complexes in wheat amyloplasts reveals functional interactions among starch biosynthetic enzymes. Plant Physiol 146: 1878–1891.

Tetlow IJ, Wait R, Lu Z, Akkasaeng R, Bowsher CG, Esposito S, Kosar-Hashemi B, Morell MK, Emes MJ (2004). Protein phosphorylation in amyloplasts regulates starch branching enzyme activity and protein-protein interactions. Plant Cell 16: 694–708.

Tsai CY, Nelson OE (1966). Starch-deficient maize mutant lacking adenosine dephosphate glucose pyrophosphorylase activity. Science 151: 341–343.

Vinyard ML, Bear RP (1952). Amylose content. Maize Genet Coop News Lett 26: 5.

Wentz JB (1926). Heritable characters in maize. J Hered 17: 327–329.

Xia H, Yandeau-Nelson M, Thompson DB, Guiltinan MJ (2011). Deficiency of maize starch-branching enzyme I results in altered starch fine structure, decreased digestibility and reduced coleoptile growth during germination. BMC Plant Biol 11: 95.

Yan H, Jiang H, Pan X, Li M, Chen Y, Wu G (2009a). The gene encoding starch synthase IIc exists in maize and wheat. Plant Sci 176: 51–57.

Yan HB, Pan XX, Jiang HW, Wu GJ (2009b). Comparison of the starch synthesis genes between maize and rice: copies, chromosome location and expression divergence. Theor Appl Genet 119: 815–825.

Yao Y, Thompson DB, Guiltinan MJ (2004). Maize starch-branching enzyme isoforms and amylopectin structure. In the absence of starch-branching enzyme IIb, the further absence of starch-branching enzyme Ia leads to increased branching. Plant Physiol 136: 3515–3523.

Zeeman SC, Kossmann J, Smith AM (2010). Starch: its metabolism, evolution, and biotechnological modification in plants. Annu Rev Plant Biol 61: 209–234.

Zeeman SC, Northrop F, Smith AM, Rees T (1998). A starch-accumulating mutant of *Arabidopsis thaliana* deficient in a chloroplastic starch-hydrolysing enzyme. Plant J 15: 357–365.

8 Evolution, Structure, and Function of Prolamin Storage Proteins

David Holding and Joachim Messing

Introduction

In this chapter, we describe the origin and function of prolamin proteins, which have arisen by divergent evolution in cereal species into multiple subfunctionalized families. At this juncture, we know the positional information of prolamin genes on chromosomes of several cereals because of their sequenced genomes. However, with respect to the overlapping functions of different prolamins in their deposition mainly in the endoplasmic reticulum (ER) and, to a small extent, within storage vacuoles, most advances have been made in maize, rice, and sorghum. In particular in maize, more recent findings have taken advantage of newly developed methods for characterizing existing and new opaque endosperm mutants as well as our ability to eliminate selectively expression of whole subfamilies of prolamins with RNA interference. Upgraded tools for cell biology such as antibody production and cryofixation and freeze substitution for immunogold labeling have also given new insight into high-definition prolamin storage in the ER. Given these advances in the understanding of the genomics and cell biology of prolamins, an overarching review is timely.

Prolamin Multigene Families

Protein Classification

Seed proteins such as those of maize were mentioned in 1821 by J. Gorham but were investigated in more detail by Osborne (1924) using his novel fractionation techniques. They are classified by their solubility largely into water-soluble globulins and alcohol-soluble prolamins. Most cereal crops have a preponderance of prolamins, named for their high percentage of the amino acids proline and glutamine, because they are a major sink for assimilated nitrogen produced during photosynthesis. Cereals have a large endosperm, which persists at maturity. The endosperm arises from a second fertilization of the central cell and consists of two maternal and one paternal chromosome set. The endosperm becomes terminally differentiated, and during germination all nutrients are used to nurse the growing seedling, which arises from the embryo proper. Synthesis of prolamins in the endosperm begins around 10 days after pollination (DAP) and peaks about 18 DAP.

Prolamins are a mixture of proteins of different sizes and composition. Their composition differs among different lines of the same species as exemplified in maize (Wilson, 1989). Their initial

Seed Genomics, First Edition. Edited by Philip W. Becraft.

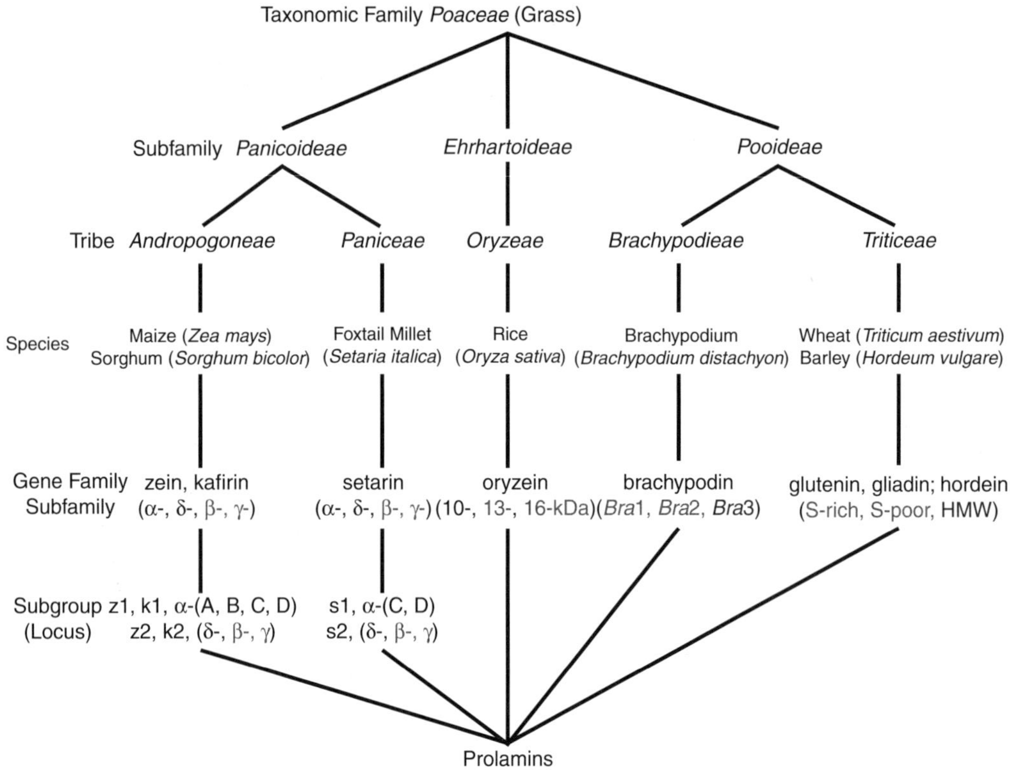

Figure 8.1 Taxonomic relationships of grass species and their prolamin genes. Classification of the subfamilies of grasses that constitute the most important cereals is shown with the different classes of prolamins. (Reproduced from Xu JH, Bennetzen JL, Messing J [2012]. Dynamic gene copy number variation in collinear regions of grass genomes. Mol Biol Evol 29: 861–871.).

classification was based on their mobility in SDS-PAGE under reducing and nonreducing conditions because many of them contain intermolecular cysteine linkages such as the high-molecular-weight (HMW) glutenins in wheat. HMW glutenins are crucial for the strength and elasticity of dough, and the smaller gliadins are crucial for the viscosity and extensibility of dough, illustrating their distinct physical properties (see Chapter 10) (Shewry and Halford, 2002). In maize, prolamins probably reached the greatest diversity of the sequenced grass genomes (Figure 8.1) and were grouped into α-zeins with 22-kDa and 19-kDa, β-zeins with 15-kDa, γ-zeins with 50-kDa, 27-kDa, and 16-kDa, and δ-zeins with 18-kDa and 10-kDa relative molecular weights. Although proteins with similar sizes could be separated by charge with isoelectric focusing, a systematic inventory in each species even at the cDNA level was difficult because of the differential expression of genes in different genetic backgrounds. Such an effort had to wait for the global analysis of entire genomes.

Evolution of Prolamins

To ensure the detection of all gene copies for each storage protein, it became important to construct redundant genomic libraries so that genomic regions could be isolated that contained tandem arrays of gene copies. Genomic regions also provided us with linked genes. They play an important role for alignment of syntenic regions from closely related species, such as those from taxonomic different subfamilies. The purpose of such alignment is to investigate insertions and deletions of gene copies. Nucleotide substitution rates of gene coding regions can be used to reconstruct the chronology of duplication events.

Such studies illustrate that across syntenic chromosomal regions of different subfamilies of the grasses such as the Pooideae (including wheat, barley, and *Brachypodium*), *Ehrhartoideae* (including rice), and Panicoideae (including Foxtail millet, sorghum, and maize), a copy of a globulin storage protein was contained in a segment of colinear genes. In contrast, the copy of the HMW glutenin storage protein was colinear only in the Pooideae and was missing in the syntenic regions of the Ehrhartoideae and Panicoideae. Phylogenetic analysis of prolamins of all the above-mentioned species indicates that this linked HMW glutenin in wheat is probably the most ancient prolamin gene copy. Given these features, it appears that water-insoluble storage proteins probably arose from water-soluble ones by unequal crossing over and were copied and inserted into other chromosomal locations. In the Pooideae, the new prolamin gene was retained in its original position adjacent to the globulin gene, whereas in the Ehrhartoideae and Panicoideae, the donor copy was deleted (Xu and Messing, 2009).

An interesting aspect of the two different modes, where the donor is retained or not, is the possible slower pace of divergence in the presence of gene conversion because HMW glutenins are present in different loci of wheat, whereas they are completely absent in the *Ehrhartoideae* and *Panicoideae*. These subfamilies instead have γ-prolamins that are closer to the gliadins in wheat, which are in yet different chromosomal locations from the glutenins. This mechanism of dispersal of gene copies is an important mechanism in the divergence of protein structure. A practical consequence is the unique property of wheat with baking quality for various food products such as noodles, unleavened breads, and most importantly leavened bread that is absent in other cereals (see Chapter 10) (Shewry and Halford, 2002). This structural differentiation of prolamins also gave rise to differential human adaptation, where 1% of Northern Europeans and an equivalent number in other populations have developed an autoimmune response against epitopes of certain wheat prolamins but not against the ones from rice or maize (Tye-Din *et al.*, 2010). All prolamins have arisen by internal duplication of a block of amino acids characterized by glutamine residues, which became already apparent when the first prolamin was sequenced (Geraghty *et al.*, 1981). It is the variability of these repeats that provides enormous numbers of epitopes, some that trigger diseases such as celiac sprue but are crucial for the physical properties of flour.

Although HMW glutenins were lost in the *Panicoideae*, they appeared to have generated a distinct new group, the α-prolamins and δ-prolamins (Figure 8.1). The latter ones are a reservoir for methionine, an essential amino acid, and are crucial for the composition of animal feed. The relatively low level of δ-zeins in feed corn requires the addition of chemically synthesized methionine, and this represents a billion-dollar supplement market to support U.S. meat production. In contrast to the δ-zeins, in which each locus comprises a single-copy gene, α-zeins are tandemly amplified in most chromosomal locations. This mode seems to be preserved within the subfamily of the *Panicoideae* as seen in the genomes of Foxtail millet, sorghum, and maize, which belong to the different tribes of the Paniceae and Angropogoneae. Intertribal syntenic alignments have also shown that dispersal occurred before and after tribal split. The oldest copy of the α-prolamins was retained in sorghum but deleted in Foxtail millet and maize. Deletion appears to have occurred through unequal crossing over of two flanking cytochrome P450 gene copies that are still present in sorghum but only as a single copy in Foxtail millet and maize with different crossover points in each species (Xu *et al.*, 2012).

Besides unequal crossing over and dispersal of gene copies into different locations, genes can also be duplicated by whole-genome duplication (WGD). Examples are γ-zeins and δ-zeins. In contrast to sorghum, maize underwent WGD about 4.8 million years ago by hybridization of two diverged progenitors that split from the progenitor of sorghum 11.9 million years ago (Swigonova *et al.*, 2004). After the split, one of the progenitors acquired an α-zein gene copy in a new chromosomal location about 7 million years ago, whereas the other one did not. When syntenic chromosomal regions from the two homeologous regions of the maize genome are aligned with sorghum, this locus is unique to one region in maize, which became maize chromosome 7S. There are also examples where dispersal occurred after WGD, which are also unique to one homeologous region on chromosome 4S. Yet another example is the oldest α-zein locus, which also had to be duplicated during WGD but is also present only in one homeologous region on maize chromosome 1 that is syntenic with sorghum chromosome 8. Because it was present before the progenitors of maize hybridized, it was concluded that it was lost after allotetraploidization of maize. Loss of homeologous gene copies has been estimated to be >50% because of selection against homeologous chromosomes during meiosis. The youngest gene copies clearly arose from tandem duplication, and in maize different inbred lines differ by copy number variation (CNV) in different loci (Xu and Messing, 2008).

Differential Expression of Gene Copies and Allelic Variation

Gene amplification contributes not only to the diversification of protein structure and function but also to the differential regulation of gene expression. The characterization of the members of the multigene family in their genomic position becomes essential. Common cDNA libraries and oligonucleotide array technologies have not been able to match individual gene copies with individual transcripts of the α-prolamins because members of recently amplified copies present in a tandem cluster are too conserved to permit the use of conventional techniques. A major insight of refined methods is that amplification also leads to gene silencing either at the transcriptional or at the post-transcriptional level (Miclaus *et al.*, 2011a). Because of the high percentage of glutamine residues in α-zeins, this is not surprising as the most common mutation is a C to T conversion. Glutamine codons are either CAG or CAA and would become stop codons TAG or TAA, which is the case. In one case, a single premature stop codon of an α-zein gene copy in W22, called csf4c5, has an intact allele with a glutamine codon in BSSS53, called azs22;8 (Llaca and Messing, 1998). Transcripts with premature stop codons have been cloned but at a very low frequency, most likely owing to decreased mRNA stability because of abortive translation (Liu and Rubenstein, 1993). Still the most common question that arises is what selective pressure is responsible for retention of silenced gene copies? We can speculate only that the silenced gene pool serves as a reserve that can be readily activated within a single generation. In such a case, we would propose gene conversion as a rapid way to reactivate gene copies (Llaca and Messing, 1998).

There are several interesting aspects concerning the regulation of the α-zein gene copies. First, CNV within maize inbred lines is the basis for a shift in gene expression. Compared with B73, the z1C1 locus in BSSS53 has two extra copies at the 3′ end of the cluster, which are strongly expressed (Song *et al.*, 2001). In hybrids of BSSS53 and B73, their expression is reduced, whereas other copies are enhanced, with a slight dosage effect because of the triploid endosperm, indicating a trans-acting compensation of gene expression, most likely at the post-transcriptional level (Song and Messing, 2003). Second, in other α-zein gene clusters, there is a major shift of expression to the most recently amplified gene copy. When endosperm is cultured, a condition that is known to reverse epigenetic silencing, the expression of older gene copies can increase, indicating their expression capacity is retained. Expression of the younger copies is reduced, similarly to the compensation effect observed in hybrid crosses described previously (Miclaus *et al.*, 2011a). Third, bisulfite sequencing of the promoter regions exhibits different methylation patterns for each locus. The dispersal of gene copies provides for a mechanism of bringing a gene copy under the influence of a new chromatin structure, which can differ between donor and the new insertion site (Miclaus *et al.*, 2011a). It is also conceivable that because of chromosomal position, different adjacent transposable elements provide fine-tuning of transcriptional regulation of gene expression as it occurred with the paralogous copy of the *P1* gene in maize (Goettel and Messing, 2010). Fourth, the two new 3′ copies at the *z1C1* locus in BSSS53 compared with B73 lost their transcriptional regulation through the O2 transcription factor as they are expressed in the homozygous *o2* mutant despite the conservation of their promoter regions (Song *et al.*, 2001). It appears that tandem amplification can provide sufficient position effect on gene expression and adaptation to allow novel transcriptional activation.

There is also CNV among γ-zeins at the 27-kDa zein locus. W64A has a single copy, whereas W22 and A188 have two copies (Das and Messing, 1987). There appear to be also allelic variations. One of the QTLs of Quality Protein Maize (QPM; also discussed with opaque mutants further on) correlates with elevated levels of the 27-kDa γ-zeins, maps close to the locus, and could be an allelic promoter version (Holding *et al.*, 2008). There is another QTL that acts in trans and could be an allele of a regulator that acts on the expression of the 27-kDa zein. Earlier studies already indicated differential transcriptional and post-transcriptional regulation of gene expression of the 27-kDa γ-zein locus (Or *et al.*, 1993).

The correlation of increased levels of the 27-kDa γ-zeins in QPM lines indicates an interesting subfunctionalization of this storage protein gene. If higher levels of the 27-kDa γ-zein can overcome the opaqueness and kernel softness in QPM, suppression of γ-zeins should revert QPM seeds to opacity. When a γ-zein RNAi event is crossed with QPM lines, they revert to opaqueness, indicating hypostasis of the QTLs and confirming that kernel hardness requires a certain level of γ-zeins (Wu *et al.*, 2010). There is an interesting twist to this role of γ-zeins; gamma zeins are the oldest prolamin genes in maize, as described previously, and there is a suggestion that they have played a role in domestication. They appear to be regulated by one of the domestication loci of Teosinte, the prolamin box-binding factor (PBF). In contrast to α-zeins-, β-zeins, and δ-zeins, γ-zeins can be activated in tissue culture (Wu and Messing, 2009). The PBF is concomitantly activated in tissue culture suggesting that PBF is insufficient to activate the younger classes of zeins. This indicates they have acquired additional cis-acting elements for other tissue-specific transcriptional activators, as expected for new versus older gene copies. PBF is suspected as one of the few domestication loci in maize versus teosinte (Jaenicke-Despres *et al.*, 2003).

Endosperm Texture and Storage of Prolamins

Early Insights into Physical Role of Zeins

Prolamins accumulate in subcellular structures, called protein bodies. Although the granular inclusions of protein bodies in the maize endosperm were first noticed in 1885 (Harz, 1885), Duvick first described their growth as cytoplasmic inclusions in 1955. Duvick (1955) described how starch grains were larger and more numerous in the central endosperm cells, and protein granules were larger and more numerous in the peripheral layers. He developed a model (summarized in

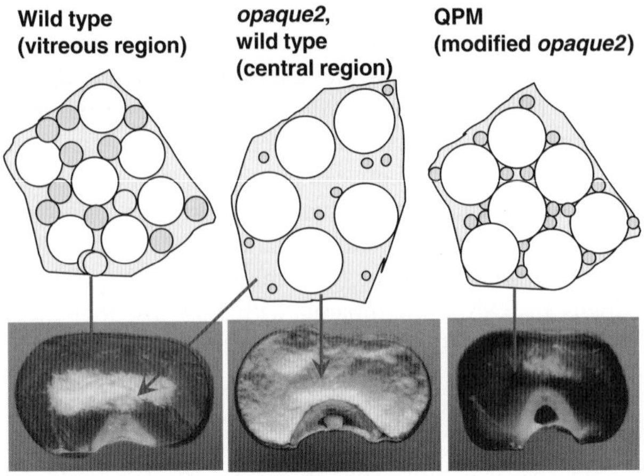

Figure 8.2 Diagram of protein body and starch grain interaction in vitreous endosperm formation. Individual cells of developing endosperm are represented with the relative size and abundance of starch grains (white spheres) and zein protein bodies (gray spheres) that are thought to result in vitreous or opaque endosperm in normal, *opaque2,* and modified *opaque2* (QPM) kernels. (For color detail, see color plate section.)

Figure 8.2) for how their size and distribution relative to that of starch grains is key to the development of the vitreous (hard) endosperm, which is such an essential agronomic trait in corn (Duvick, 1961). Duvick compared the structure of the vitreous endosperm with a box of white marbles (starch grains) densely interspersed with buckshot (protein bodies) and all bound by a transparent glue (clear viscous cytoplasm), which forms a rigid conglomerate when dry (Duvick, 1961). At the boundary between the region that will give rise to the vitreous endosperm and the central starchy (opaque) endosperm, a threshold is reached where the relatively small size of the protein bodies and the large size of the starch grains does not produce a rigid glassy matrix on dehydration but rather a friable chalky structure that is incapable of transmitting light and is opaque (Duvick, 1961). This model (Figure 8.2) and the fact that zein abundance is closely correlated with protein body size is consistent with observations that the vitreous endosperm contains much more zein than the soft central region (Hamilton *et al.*, 1951; Dombrink-Kurtzman and Bietz, 1993). Environmental conditions resulting in reduced zein synthesis, such as nitrogen depletion, result in kernels that are soft and starchy throughout (Tsai *et al.*, 1978).

Formation of ER Protein Bodies in Starchy Endosperm

The differentiation of protein structure during evolution of zein genes also determines their cellular location and contribution to subcellular and seed structures. Zein synthesis begins in starchy endosperm cells at around 8–10 days after pollination (DAP) and continues throughout endosperm development, terminating from the center outward as cells undergo programmed cell death (PCD) (Young and Gallie, 2000). The onset of expression of the different classes of zeins is not uniform (Woo *et al.*, 2001). For example, the γ-zeins are synthesized throughout the endosperm before the onset of α-zein and δ-zein expression, and this is consistent with their predicted role in priming ER-protein body formation (Woo *et al.*, 2001). All zein proteins are initially translated with signal peptides, which are subsequently cleaved after the proteins have entered the lumen of the rough endoplasmic reticulum (rER). Despite the fact that none of the zein proteins contain a canonical ER

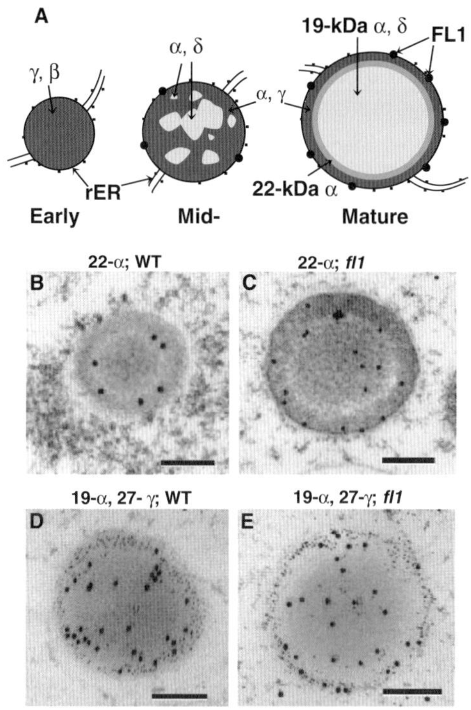

Figure 8.3 Organization of zeins in ER protein bodies. *A,* Diagram of zein distribution in early, mid-, and mature protein bodies. Small black dots in the membrane represent ribosomes, whereas large black dots represent FLOURY1 protein. *B,* Immunogold localization of 22-kDa α-zein in wild-type protein bodies (15-nm gold particles). *C,* Immunogold localization of 22-kDa α-zein in *floury1* mutant protein bodies (15-nm gold particles). *D,* Double immunogold localization of 19-kDa α-zein (15-nm) and 27-kDa α-zein (5-nm) in wild-type protein bodies. *E,* Double immunogold localization of 19-kDa α-zein (15-nm) and 27-kDa α-zein (5-nm) in *floury1* mutant protein bodies.

retention sequence, in the starchy endosperm cells zeins remain in the ER, where they are assembled into organized layered accretions referred to as protein bodies (Figure 8.3). Initially, protein bodies consist entirely of cysteine-rich β-zeins and γ-zeins cross-linked by disulfide bonds (Lending and Larkins, 1989; Lopes and Larkins, 1991). Later, the hydrophobic α-zeins and δ-zeins infiltrate the matrix of β-zeins and γ-zeins as variably sized inclusions (Figure 8.3, *middle panel*). Fully expanded mature protein bodies of 1–2 μm have an even round shape with α-zeins taking up the core surrounded by a shell of β-zeins and γ-zeins (Figure 8.3*A*, *right panel*). This model, developed from immunogold transmission electron micrographs of chemically fixed endosperm samples, did not distinguish between the 19-kDa and 22-kDa α-zeins (Lending and Larkins, 1989). However, antibodies specific to 22-kDa and 19-kDa α-zeins and cryofixation of endosperm samples revealed that these proteins have distinct patterns of accumulation (Figure 8.3, *right panel*). Although the 19-kDa α-zein is found throughout the protein body core, the 22-kDa α-zein is found only in a discrete ring at the interface between the 19-kDa α-zein–rich core and the 27-kDa γ-zein–rich peripheral region (Holding *et al.*, 2007). This is supported by a demonstrated lack of interaction between the 19-kDa and 22-kDa α-zeins (Kim *et al.*, 2002) and may suggest that a 22-kDa interface between the γ/β zein shell of the 19-kDa α-zein core is a critical structural feature. Specific RNAi suppression of the 22-kDa α-zein leads to blebbing of the protein body surface, perhaps as a result of improperly constrained 19-kDa α-zein (Wu and Messing, 2010a).

In general, investigations of the temporal and spatial patterns of zein gene expression as well as a complete analysis of all classes of zein interactions reinforced the above-described model (Woo *et al.*, 2001; Kim *et al.*, 2002). It is thought that direct interactions between zeins are essential for their retention as properly formed protein bodies (Kim *et al.*, 2002). This thought is supported by experiments that used heterologous expression of both α-zeins and γ-zeins in transgenic tobacco leaves. When α-zeins or δ-zeins were synthesized singly, both proteins appeared to be secreted and become degraded (Williamson *et al.*, 1988; Bagga *et al.*, 1995; Coleman *et al.*, 1996; Bagga *et al.*, 1997). However, if α-zeins or δ-zeins are coexpressed with β-zeins and γ-zeins, significant zein accumulation in the ER occurs (Bagga *et al.*, 1995; Coleman *et al.*, 1996; Bagga *et al.*, 1997). This suggests that β-zeins and γ-zeins are essential for the ER retention of α-zeins and δ-zeins. It was subsequently shown that the PPPVHL proline-rich repeats in the 27-kDa γ-zein N-terminal region are necessary for anchoring it within the ER (Geli *et al.*, 1994). The significance of the 27-kDa γ-zein N-terminal region for ER retention was also shown by the synthesis of a protein chimera called Zeolin, consisting of the 27-kDa γ-zein N-terminal region containing the proline-rich repeats fused to the C-terminal region of a bean phaseolin protein (a 7S globulin) (Mainieri *et al.*, 2004). 7S globulins normally form soluble homo-trimers that leave the ER and are trafficked through the Golgi and finally into storage vacuoles (Chrispeels, 1983). However, as a 27-kDa γ-zein fusion, the phaseolin portion was retained in ER-protein bodies in transgenic tobacco leaves. Corn and bean proteins when consumed together provide an adequate balance of the essential amino acids lysine and tryptophan, which are deficient in cereals, and sulfur amino acids, which are deficient in legume seeds. The zeolin study showed the potential for designer protein storage in the ER fraction (Mainieri *et al.*, 2004). However, it did not take into account that expression of significant amounts of transgenic proteins is usually difficult in seed tissues in the presence of large amounts of endogenous storage proteins (Holding and Larkins, 2008). With modern RNAi technology for reducing the levels of particular proteins, this technology has renewed potential.

The γ-zeins and β-zeins act redundantly in the cross-linked protein body shell, and severely abnormal α-zein protein bodies result from suppression of γ-zeins and β-zeins (Wu and Messing, 2010a). However, because abnormal protein bodies are still ER localized, it is likely that other nonzein factors such as FL1 that are not present in heterologous systems are also essential for ER retention. Other mechanisms besides zein interactions are involved in sequestration of prolamins within ER protein bodies. For example, in rice, differential distribution of mRNAs to different ER regions is thought to be an important prerequisite to protein body formation (Okita and Choi, 2002). Protein chaperones also play an important role. A protein called b-70, which is upregulated in the maize *fl2* mutant (described later), was identified as a homolog of the mammalian immunoglobulin binding protein, BiP (Fontes *et al.*, 1991). BiP drives the ATP-dependent folding of proteins and is itself retained in the ER lumen with a C-terminal HDEL retention sequence. Mutations that cause the aberrant accumulation of zein storage proteins (described later) cause elevation of BiP (Boston *et al.*, 1991). BiP is also demonstrated to be central to prolamin protein body formation in rice endosperm (Li *et al.*, 1993). In addition to BiP, protein disulfide isomerase (PDI) activity is essential to prolamin folding, and the abundance of PDI increases similar to BiP in response to abnormally processed zeins (Fontes *et al.*, 1991; Li and Larkins, 1996). In rice, the Cys-rich 10-kDa prolamin (crP10) is directed to the core of ER protein bodies, whereas the Cys-poor 13-kDa prolamin (cpP13) is directed to the protein body periphery (Kumamaru *et al.*, 2007), suggesting that crP10 sulfhydryl oxidations are necessary for rice protein body formation. Work in rice has shown that there are distinct PDI-like proteins (PDIL1;1 and PDIL2;3) that have nonredundant roles in protein body formation (Onda *et al.*, 2011). PDIL1;1, the ortholog of the PDI protein increased in the maize *fl2* mutant ER (Li and Larkins, 1996), is localized in the ER lumen. When PDIL1;1 is knocked out in the ESP2 mutant, large protein aggregates, consisting of incorrectly placed intermolecular disulfide bonds, form within the ER (Kawagoe *et al.*, 2009), demonstrating an important role for this protein in oxidative folding of proglutelins. It was shown that PDIL2;3 was unable to complement ESP2 and that PDIL2;3 functions specifically in placement of Cys-rich prolamins such as crP10 in the protein body core. When PDIL2;3 was knocked down, crP10 was no longer directed to the protein body core, and conversely, crp10 knockdown displaced PDIL2;3 to the lumen of the ER from its normal location in the protein body periphery supporting the specific functional interaction.

Insights from Zein Storage within Vacuoles in Aleurone Cells

The primary storage site of zeins is the starchy endosperm tissue whose cells die during maturation leaving a network of starch grains and protein bodies. The outermost endosperm layer, the aleurone, remains living and is responsible for secretion of amylases and proteases into the starchy endosperm during germination. RNA in situ hybridization experiments suggested that zein genes may also be expressed in aleurone cells (Woo *et al.*, 2001), and this was confirmed for all classes of zeins using RT-PCR on aleurone peels (Reyes *et al.*, 2011). However, in the absence of typical ER protein bodies, it was unknown if significant zein protein accumulation occurs. Immunogold transmission electron micrograph studies showed that there is negligible accumulation of 19-kDa α-zein and 27-kDa γ-zein in the aleurone, but significant amounts of 22-kDa α-zein, 15-kDa β-zein, and 18-kDa δ zein are present. These zeins exist in protein storage vacuoles (PSV) and result from the delivery of zeins not retained as ER inclusions. Zeins pass as multivesicular bodies and are incorporated into multilayer autophagic compartments before fusing into PSVs (Reyes *et al.*, 2011). Autophagy and de novo formation of prolamin storage vacuoles also occur in wheat where prolamin inclusions exit the ER and are incorporated into PSVs in the starchy endosperm cells (Levanony *et al.*, 1992). The absence of 27-kDa γ-zein and 19-kDa α-zein in maize aleurone PSVs supports previous data suggesting that zeins have defined and separable roles in ER protein body formation. Because 27-kDa zein is thought to prime ER protein body formation (Woo *et al.*, 2001), 15-kDa β-zein alone may be insufficient for this process. The 19-kDa α-zeins are the most abundant of the zeins and appear to be responsible for the bulk of protein body filling. Perhaps the rapid and extensive filling of the ER lumen with 19-kDa α-zein provides a physical blockage that facilitates the collective retention of all zeins in the ER. In the absence of 19-kDa α-zeins, it appears that the 22-kDa α-zeins alone are unable to expand protein bodies sufficiently to cause ER retention. Aleurone PSVs containing zeins contain intravacuolar membranes that are more prominent at 14 DAP and become less prominent by 22 DAP. These are likely remnants of ER membranes because ER membrane proteins TIP3-4 and FL1 were also detected in aleurone cells (Reyes *et al.*, 2011) and support the proposed specific functional association between FL1 and 22-kDa α-zein. In addition, FL1 likely plays an active role in organizing zein inclusions in aleurone PSVs. This is inferred from expression of FL1-mOrange fluorescent protein in cultured endosperm cells, in which aleurone and starchy endosperm cell types are discernible (Reyes *et al.*, 2010). In the early stages after transformation (6 days), cultured cells showed reticulate FL1 expression, whereas by 8 days, its expression was punctate and appeared to delimit the zein-rich inclusions (Reyes *et al.*, 2011).

Nonvitreous Phenotypes in Maize

Perhaps the best demonstration of the central role zein protein bodies play in vitreous endosperm formation has come from the study of opaque endosperm mutants, of which at least 18 have been described in maize, excluding the RNAi-mediated ones (Thompson and Larkins, 1993), although less than half have been described at the molecular level. These often have alterations in zein synthesis that result in protein bodies with abnormal morphology, size, or number. One opaque mutant that was described more recently results from a disruption in a gene encoding arogenate dehydrogenase, which is involved in synthesis of the amino acid tyrosine and causes pleiotropic effects on amino acid profiles and a generalized reduction in zeins (Holding *et al.*, 2010). It is clear that factors other than zeins and protein bodies are also involved in vitreous endosperm formation. The interaction of developing starch grains (amyloplasts) and protein bodies may be essential in the model described previously (Figure 8.2). In support of this, the *opaque5* mutant was recently cloned and found to encode monogalactosyldiacylglycerol synthase (MGD1) whose interruption caused a strong reduction of galactolipid abundance in the endosperm (Myers *et al.*, 2011). This resulted in changes in starch production and the appearance of compound starch granules, which could interact differently with zein protein bodies. The study underlined the importance of amyoplast membranes in proper starch granule development. As a further demonstration of the importance of zein-independent processes in vitreous endosperm formation, some of a collection of mutator transposon generated opaque endosperm mutants show no qualitative or quantitative changes in their zein profiles relative to wild-type (D. Holding, unpublished data).

Historically, maize opaque mutants have been intensely studied because of their improved grain protein quality in terms of amino acid composition, which in some cases approaches that of reference animal proteins such as milk. This is driven by reductions in the accumulation the lysine-devoid and tryptophan-devoid zein proteins, which have variable effects on protein body morphology. The *opaque2* (*o2*) and *floury2* (*fl2*) mutants were originally reported in 1935 (Emerson *et al.*, 1935) and contain double the wild-type levels of lysine and tryptophan (Mertz *et al.*, 1964; Nelson *et al.*, 1965). The *O2* gene encodes a transcriptional activator that positively regulates the expression of 22-kDa α-zein genes (Schmidt *et al.*, 1990; Ueda *et al.*, 1992) as well as many other genes (Damerval and Devienne, 1993). The reduction of zeins in *o2* does not change the overall seed protein concentration but results in increased accumulation of non-zein proteins, some of which are relatively rich in lysine and tryptophan, such as eEF1A (Lopez-Valenzuela *et al.*, 2004). At the subcellular level, the main effect of the low level of α-zeins in *o2* is a reduction in protein body number and size (Figure 8.4).

Despite its improved nutritional quality, *o2* was not commercially developed because of the undesirable agronomic properties associated with the soft kernel phenotype, such as impaired harvesting and handling characteristics, increased pathogen susceptibility, and reduced yield. However, through extensive breeding efforts in Mexico (Villegas *et al.*, 1992) and South Africa (Geevers and Lake, 1992), a new type of *o2* mutant called Quality Protein Maize (QPM) was developed in which genetic modifiers restore the vitreous kernel phenotype while maintaining the enhanced nutritional quality (Paez *et al.*, 1969; Ortega and Bates, 1983). As mentioned previously, several studies have shown that the 27-kDa γ-zein is increased twofold to threefold in QPM (Wallace *et al.*, 1990; Geetha *et al.*, 1991; Lopes and Larkins, 1991), and the degree of endosperm vitreousness closely correlates with the level of 27-kDa γ-zein protein (Lopes and Larkins, 1991). The 27-kD γ-zein gene maps to a major QTL for endosperm hardness in QPM (Holding and Larkins, 2008; Holding *et al.*, 2011). Given the evidence that 27-kDa γ-zein initiates protein body formation, it is thought that this increase results in elevation of protein body number in QPM (Lopes and Larkins, 1995; Moro *et al.*, 1995), which increases protein cross-linking and restores kernel hardness (Figure 8.2).

Another soft endosperm mutant, *opaque15* (*o15*), shows a twofold to threefold reduction in 27-kD γ-zein synthesis, resulting in a reduced protein body number but unaltered protein body size, again consistent with the 27-kD γ-zein acting in the protein body initiation (Dannenhoffer *et al.*, 1995).

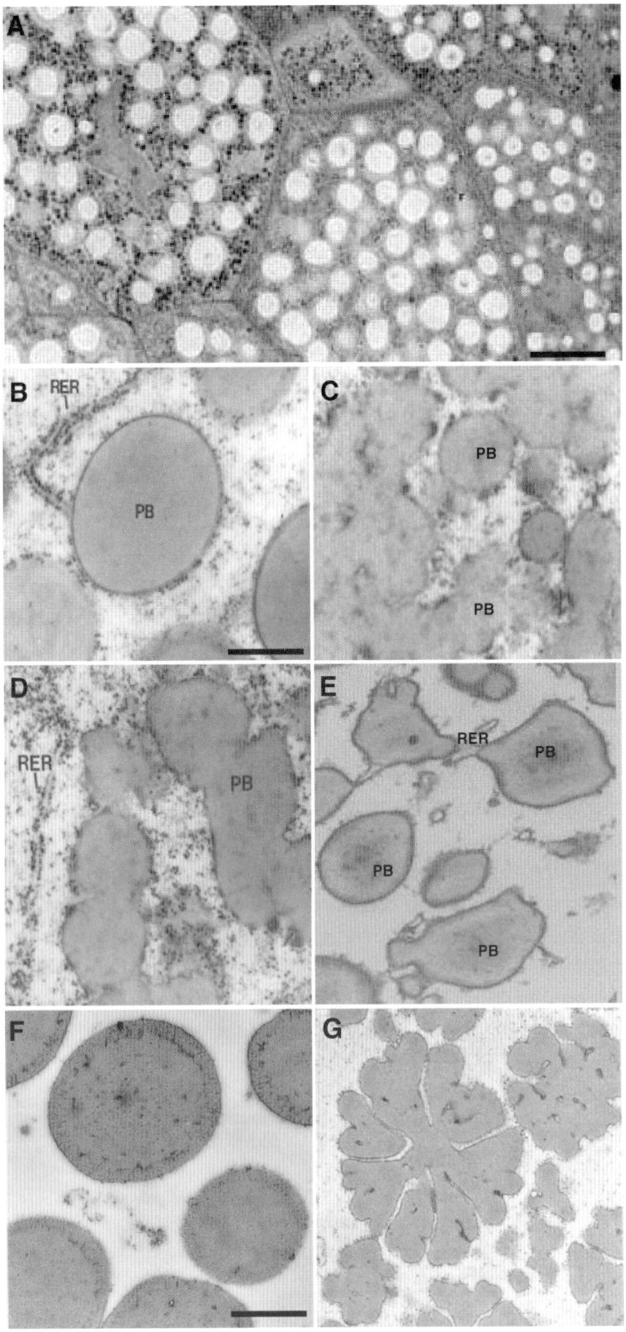

Figure 8.4 Protein body morphology in opaque endosperm mutants. All images are transmission electron micrographs. *A*, Low-magnification transmission electron micrograph section of an *o2-Spmo2o2* mutable kernel showing a boundary region between wild-type cells *(top)* with normal protein bodies and *o2/o2/o2* cells *(bottom)* that have small, less numerous, and faintly staining protein bodies. Scale bar = 50 μm. *B–E*, Protein bodies (PB) in wild-type, *fl2, De-B30,* and *Mc* developing endosperm. RER, rough endoplasmic reticulum. Scale bar in *B* is 200 nm and applies to *B–E*. *F* and *G*, Protein bodies in developing sorghum endosperm of wild-type *(F)* and HDHL mutant *(G)*. Scale bar in *F* is 500 nm and applies to *F* and *G*.

The *O15* gene maps to the same region on chromosome 7L as an *o2* modifier gene (Dannenhoffer *et al.*, 1995) suggesting that *O15* could itself be a modifier gene. In addition to γ-zein, there are several other QPM QTLs, and the extent to which γ-zein is alone sufficient for modification is unknown.

Besides *o2* and *fl2*, a third high-lysine mutant, *o7* arose spontaneously in the W22 background (McWhirter, 1971). Because of its low penetrance in other backgrounds, its molecular basis has remained unknown for a long time despite its importance. However, using transposon mutagenesis in the original background of W22, it has become possible to isolate new alleles of an acyl-CoA synthetase-like gene (ACS), located on chromosome 10L (Miclaus *et al.*, 2011b). ACS could modify the hydrophobicity of proteins so that they could locate into membranes (Black and DiRusso, 2007). The original mutation had just four amino acids in frame deletion, whereas the transposon insertion occurred in the C-terminal region, indicating that mutants might still have residual function. The truncation appeared to have a stronger effect on protein body formation than the in-frame deletion.

The recessive *o2* mutant results from a generalized reduction in α-zeins. The multigenic nature of the α-zein families explains the lack of detection of recessive mutations in the zein genes themselves. However, several dominantly acting mutations in zein genes have been described. The first of these is *fl2*, which was identified as a point mutation resulting in a single amino acid substitution in the N-terminus of a highly expressed member of the *z1C* α-zein gene family (Coleman *et al.*, 1997). This results in failed cleavage of the ER-signal peptide, and a 24-kDa α-zein species is easily identified on SDS-PAGE gels (Figure 8.5). At the cellular level, the abnormal zein is thought to accumulate at the ER membrane rather than getting recruited into the protein body core. The *Defective endosperm B30* (*De-B30*) mutant results from a similar point mutation and amino acid substitution in the ER-signal peptide region, in this case in a 19-kDa α-zein gene (Kim *et al.*, 2004). The dominant phenotype in De-B30 is caused by a mutation in a gene of low-level expression, and the mutant protein could be detected only by immunoblotting (Figure 8.5) (Kim *et al.*, 2004). *fl2* and *De-B30* have protein bodies displaying an uneven, distorted shape (Figure 8.4) (Lending and Larkins, 1992; Kim *et al.*, 2004).

The *Mucronate* (*Mc*) mutant, which also has misshapen protein bodies, results from an abnormal 16-kDa γ-zein in which a 38-bp deletion creates a frame-shift mutation and an abnormal sequence for the last 63 amino acids of the protein (Kim *et al.*, 2006). The mutant protein has a similar length to the wild-type protein but was identified using two-dimensional SDS-PAGE by virtue of its altered isoelectric characteristics. The defective 16-kD γ-zein in *Mc* shows a reduced interaction with 22-kD α-zeins (Kim *et al.*, 2006), which could disrupt the organization of zeins within the protein body. *De-B30*, *fl2*, and *Mc* each result in increased expression of genes associated with

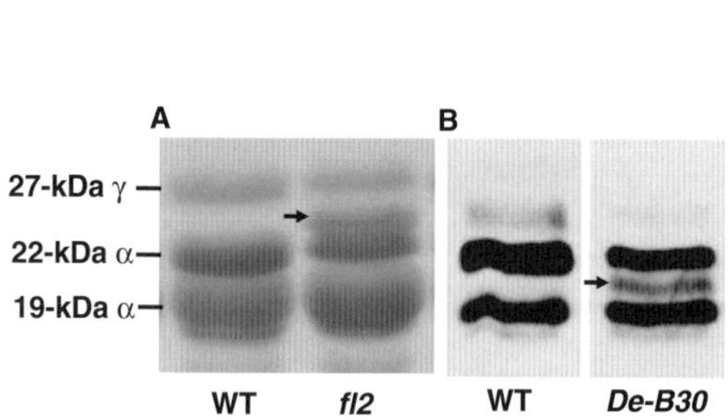

Figure 8.5 Abnormal α-zeins in *fl2* and *De-B30* mutants. *A*, Coomassie blue–stained SDS-PAGE of purified zeins showing high-level accumulation of abnormal 24-kDa α-zein in *fl2*. *B*, Protein gel blot of purified zeins labeled with general α-zein antiserum showing low-level accumulation of abnormal uncleaved 19-kDa α-zein in *De-B30*.

the unfolded protein response (UPR) (Hunter *et al.*, 2002), consistent with the expression of a malfolded, ER-localized protein.

Fl2 and *De-B30* protein bodies both exhibit lobed protein bodies (Figure 8.4), indicating that failure of a percentage of α-zeins to detach from the ER membrane prevents normal separation of discrete-sized ER accretions. However, normal amounts of γ-zeins are likely able to encase much of the α-zein, giving rise to relatively even curved surfaces on the lobed protein body structures. The *Mc* phenotype is distinct from *fl2* and *De-B30* because lobing is not the primary defect but rather more random distortions are the primary defect, often including jagged edges (Figure 8.4). This indicates that the dominant mutant 16-kDa γ-zein may interfere with the collective γ-zein and β-zein function to enclose α-zeins in the protein body core. Immunolocalization studies of each zein species in the *Mc* mutant would allow a better understanding of the cause of this phenotype.

o2 and *opaque 1* (*o1*) (Nelson *et al.*, 1965) also show UPR induction but to a lesser extent than *De-B30*, *fl2*, and *Mc* (Hunter *et al.*, 2002). *o1* has not been cloned and shows no apparent alteration in its zein profile consistent with the fact that vitreous endosperm formation involves both zein-dependent and zein-independent processes.

The *floury1* (*fl1*) mutant was first reported in 1915 (Hayes and East, 1915), but in contrast to *o2* and *fl2*, it was not widely studied because it does not have an improved lysine content. Similar *o1, fl1* does not have reduced accumulation of zeins and lacked an obvious explanation for its kernel phenotype. *FL1* encodes a novel ER protein, which is concentrated in the ER membrane surrounding zein protein bodies (Holding *et al.*, 2007). Apart from the opaque endosperm, *fl1* mutants do not share any of the phenotypes associated with previously described opaque mutants, such as general or specific reductions in zein accumulation; a constitutive UPR, as occurs in *fl2*, *DeB30*, and *Mc*; or alterations in protein body size and shape. The only defect at the cellular level is that *fl1* mutants accumulate 22-kDa α-zeins in abnormal locations within protein bodies. As mentioned earlier, the 22-kDa α-zeins occupy a discrete ring at the interface between the protein body core and the γ-zein–rich periphery (Figure 8.3). In *fl1* protein bodies, 22-kDa α-zein is randomly dispersed and is found in the center of the core as well as in the periphery, often in close association with the ER membrane (Figure 8.3). The 19-kDa α-zeins, which have a more generalized distribution throughout the protein body core, are not altered by the *fl1* mutation (Figure 8.3). Immunolocalizations appear to show increased 22-kDa α-zein accumulation, consistent with a significant increase in quantitative measurements of zein abundance in *fl1* (Holding *et al.*, 2007). These results support the proposed role for FL1 in the correct placement of the hydrophobic 22-kDa α-zeins and suggest that this location is essential for proper protein body formation and the generation of vitreous endosperm but do not rule out its involvement with zein packaging in a more general way.

Natural Variants of Sorghum

The High Digestibility High Lysine (HDHL) mutant of sorghum shares many of the same characteristics as *fl2* and *De-B30*, including an opaque kernel texture (Oria *et al.*, 2000). It has high lysine owing to reduced levels of α-kafirin proteins, which similar to the α-zeins are devoid of this essential amino acid. The molecular nature of the mutation is unknown, but its determination is a high priority because digestibility of the mutant grain is a highly important attribute. Sorghum grain is normally only ~46% digestible by humans in comparison with ~73% for maize (Maclean *et al.*, 1981), but digestibility of the mutant grain is increased to ~85% (Oria *et al.*, 2000). Although α-kafirins make up ~80% of total kafirins (Watterson *et al.*, 1993), they themselves are not especially recalcitrant to digestion (Oria *et al.*, 2000). The indigestibility is due to the highly cross-linked nature of the β-kafirins and γ-kafirins (Oria *et al.*, 1995), which normally have a peripheral location enshrouding the α-kafirins in sorghum protein bodies, similar to the arrangement of zeins in maize. Protein bodies in the HDHL mutant have a highly reticulated shape, and the γ-kafirins are found at the base of deep invaginations instead of encircling the protein bodies (Figure 8.4). This likely increases digestibility by increasing the surface area for protease action as well as unmasking the central α-kafirins from the cross linked shield of γ-kafirins (Oria *et al.*, 2000). Although the protein body phenotype is distinct from *fl2*, *De-B30*, or *Mc*, it is not unlikely that it could result from a dominant mutant kafirin, which is consistent with the elevated expression of genes that signify a UPR (B. Hamaker, personal communication). Alternatively, a mutation in an uncharacterized kafirin organizing factor analogous to Fl1 could be responsible.

Structure and Function Analysis with RNA Interference

The progress made in understanding the role of individual zein classes in ER sequestration has been limited for a number of reasons. First, there is a paucity of natural mutants with such phenotypes. One reason is the multigenic nature of the α-zein subfamilies, and the high-level expression of multiple members means that there is substantial functional redundancy (Miclaus *et al.*, 2011a), which would mask the effects of recessive mutants in individual family members. Second, the dominant mutants and *o2* all have pleiotropic effects, which complicate the evaluation of the roles of individual zeins. RNAi technology offers the ability to reduce substantially or eliminate the synthesis of entire subfamilies of zeins (Segal *et al.*, 2003).

Knockdown of 22-kDa zeins at the *z1C1* and *z1C2* loci creates a phenotype that is similar to the *o2* mutation that prevents transcription of the *z1C1* and *z1C2* loci except for CNV of the *z1C1* locus, where CNV could provide a lower penetrance of *o2*. RNAi is not only dominant but also more specific for target genes and more comprehensive. Similar to *o2* mutants, RNAi for *z1C1* and *z1C2* exhibits an opaque phenotype and a higher lysine level. In addition to opaque kernels, the reduction of 22-kDa α-zein in the z1C RNAi as described previously resulted in protein body alterations that were distinct from *o2* (Segal *et al.*, 2003; Wu and Messing, 2010a). Protein bodies were not reduced in size similar to *o2* suggesting that the 19-kDa α-zeins are responsible for the bulk of protein body filling. Small budding protuberances were observed at the protein body peripheries. This supports the suggestion that 22-kDa α-zeins are necessary for properly packaging the 19-kDa α-zeins within the γ-zein periphery.

Mutants such as *fl2*, *De-B30*, and *Mc* are semidominant. These mutations in the coding regions of specific zein genes appear to interfere with α-zein protein accumulation in *trans* leading to generalized zein shutdown as a result of an unfolded protein response. If this is the case, RNAi against a transcript of a mutated coding region should suppress the dominance of such mutations. A construct that simultaneously knocked down the related 27-kDa and 16-kDa γ-zeins prevented the expression of the *Mc* mutation, which contains a frame shift in a 16-kDa γ-zein. Because of the frame shift, the carboxy terminus of the protein has different physical properties to the native protein and is suspected to trigger the UPR. No UPR is detected when the RNAi transgene is combined with the *Mc* mutation (Wu and Messing, 2010b).

To discriminate the roles of the γ-zeins and β-zeins, which are both found in the protein body periphery, another RNAi line was made that targeted the 15-kDa β-zein (Wu and Messing, 2010a). Individually, both of these RNAi lines against the γ-zeins and β-zeins had only minor effects, slightly increasing the proportion of small, unexpanded protein bodies relative to wild-type. However, when the two transgenes were combined, plants that had substantial reductions in both γ-zeins and β-zeins exhibited irregular-shaped protein bodies, possibly indicating a threshold requirement (Wu and Messing, 2010a). This might suggest that severity of phenotype has a simple correlation with the amount of zein, which is reduced. Although γ-zeins represent the second most abundant class of zeins after α-zeins, it may be that only when both γ-zeins and β-zeins are knocked down is the zein reduction in the protein body periphery substantial enough to cause major effects. This additive effect shows that γ-zein and β-zein act with some degree of redundancy. Although not as severe, their combined affect is reminiscent of *Mc* mutants and suggests that the irregularity results from improperly constrained α-zeins. When the γ/β RNAi was combined with the z1C RNAi, severely lobed and reticulated protein bodies resulted (Wu *et al.*, 2010). This suggests that the outer γ/β layer and the intermediate 22-kDa α-zein sublayer both participate in constraining the abundant 19-kDa α-zeins. Because the γ-zeins and β-zeins are rich in cysteine, they also play a role as a sink for reduced sulfur (see later).

A double RNAi against 22-kDa and 19-kDa zeins reduces α-zein expression to lower levels and further increases lysine content, indicating the compensatory mechanism of protein accumulation in the seed (Wu and Messing, 2011, 2012). Still, even the dominant double RNAi against α-zeins is recessive to the QTLs of QPM (described earlier), consistent with their function in *o2* mutants. Breeding, however, is hampered by recessive traits. First, the recessiveness of *o2* can be overcome with RNAi, but the recessiveness of RNAi to the QTLs of QPM requires an additional linked neutral marker such as GFP to follow the inheritance of the RNAi locus. Presence of the QPM QTLs can produce green kernels containing the double RNAi against α-zein and high-lysine content, providing a new introduction strategy for QPM into diverse genetic backgrounds (Wu and Messing, 2011).

The same strategy applies to the long-term selection of high protein content, which can be 2.5 times higher than in normal lines (Moose *et al.*, 2004). Because this selection has primarily increased α-zeins, it actually has worsened protein quality. Using the double RNAi described earlier, however, α-zeins can be reduced to lower levels, whereas total protein levels remain unchanged owing to increased levels of nonzein proteins, probably secondary to post-transcriptional regulation of protein synthesis. The remaining levels of α zeins are sufficient to prevent an opaque kernel phenotype, and the increased nonzein proteins result in a balanced amino acid composition, combining three long-sought traits in cereals, high-lysine, high-protein, and hard seed (Wu and Messing, 2012).

Another zein locus that is differentially regulated at the post-transcriptional level is the δ-zein locus, which is regulated by the δ-zein regulator (*dzr1*). The regulator appears to target the UTRs of the 10-kDa δ-zein because a transgene with substituted UTRs became insensitive to *drz1* regulation (Lai and Messing, 2002). Certain inbred lines harbor a *dzr1* allele that is subject to genomic imprinting (Chaudhuri and Messing, 1994). There are also inbred lines, where δ-zeins are knocked out either by frame shift or transposon insertion, indicating that expression of δ-zein may not be essential (Wu *et al.*, 2009). Such natural knockouts have not been found for any other zein loci.

When δ-zeins are overexpressed (Lai and Messing, 2002), it occurs at the expense of γ-zeins and β-zeins (Wu *et al.*, 2012). Reciprocally, if γ-zeins and β-zeins are reduced with RNAi, δ-zeins increase. The connection here is the reduced sulfur pathway in plants. The γ-zeins are high in cysteine, which donates its sulfur moiety to methionine. So if cysteine is low because of increased γ-zein levels, less methionine is synthesized, starving δ-zein protein synthesis (Wu *et al.*, 2012).

With respect to protein body formation, the 10-kDa and 18-kDa δ-zeins colocalize with the α-zein protein core, but a specific role is unknown. When the above-described double δ-zein null mutant was crossed into the individual γ-zein and β-zein RNAi lines, the mild effects of these RNAi constructs on protein body formation were not exacerbated (Wu and Messing, 2010a). The lack of an obvious defect resulting from δ-zein elimination may reflect their naturally low abundance and, perhaps, small contribution to protein body architecture. Alternatively, it may indicate that δ-zeins have evolved simply as a store of methionine and are not essential for sequestration of other zeins. However, because the effects of the δ-zein null on the combined γ-zein and β-zein RNAi line or the triple z1C/γ-/β-RNAi line were not investigated (Wu and Messing, 2010a), a structural role for the δ-zeins cannot be ruled out.

For many years, it was believed that there is a common transcriptional activator for all zein genes, and that could be the PBF described earlier (Vicente-Carbajosa *et al.*, 1997; Wang *et al.*, 1998; Jaenicke-Despres *et al.*, 2003). Because no mutant collection has yielded a knockout mutation, it was assumed that such a mutation would produce defective kernels. RNAi against PBF could provide a useful alternative approach because it is only a reduction in gene expression and not a knockout mutation. Such an RNAi line reduces α-zein and γ-zein expression only and not the other ones (Wu and Messing, 2012). It appears that the promoters of the zein superfamily have diverged rapidly because they differ also in their response to O2 as described earlier. A combination of the PBF RNAi and the *o2* mutation gives an additive effect in the reduction of zein gene expression. The same study shows that constitutively expressed PBF and O2 can transactivate transgenic zein promoters in leaf tissue but not endogenous ones. Methylation analysis of the promoters indicates that they also differ in their epigenetic marks. Unexpectedly, one of them, the β-zein promoter, could be turned on in leaf tissue in the absence of RDR2 caused by the *mob1* mutation, which controls the siRNA pathway (Wu and Messing, 2012). These lines of experimentation with different RNAi constructs illustrate the evolution of the seed as sink for reduced nitrogen and sulfur, regulated at the post-transcriptional and transcriptional level.

Conclusion

After a hiatus of research activities in the field of storage proteins compared with the beginnings of plant biochemistry and molecular biology, progress in deeper insights of their evolution and function has gained momentum in recent years because of new methods in genomics, cell biology, and RNA interference. Unexpectedly, studies of their expression during seed development has given us plethora of mechanisms of the cell machinery in general because their accumulation uses more factors than previously believed. Therefore, these studies provide a better understanding of seed development and serve as the platform for a new green revolution.

References

Bagga S, Adams H, Kemp JD, Senguptagopalan C (1995). Accumulation of 15-kilodalton zein in novel protein bodies in transgenic tobacco. Plant Physiol 107: 13–23.

Bagga S, Adams HP, Rodriguez FD, Kemp JD, Sengupta-Gopalan C (1997). Coexpression of the maize delta-zein and beta-zein genes results in stable accumulation of delta-zein in endoplasmic reticulum-derived protein bodies formed by beta-zein. Plant Cell 9: 1683–1696.

Black PN, DiRusso CC (2007). Yeast acyl-CoA synthetases at the crossroads of fatty acid metabolism and regulation. Biochim Biophys Acta 1771: 286–298.

Boston RS, Fontes EBP, Shank BB, Wrobel RL (1991). Increased expression of the maize immunoglobulin binding-protein homolog B-70 in 3 zein regulatory mutants. Plant Cell 3: 497–505.

Chaudhuri S, Messing J (1994). Allele-specific parental imprinting of dzr1, a posttranscriptional regulator of zein accumulation. Proc Natl Acad Sci U S A 91: 4867–4871.

Chrispeels MJ (1983). The Golgi-apparatus mediates the transport of phytohemagglutinin to the protein bodies in bean cotyledons. Planta 158: 140–151.

Coleman CE, Clore AM, Ranch JP, Higgins R, Lopes MA, Larkins BA (1997). Expression of a mutant alpha-zein creates the *floury2* phenotype in transgenic maize. Proc Natl Acad Sci U S A 94: 7094–7097.

Coleman CE, Herman EM, Takasaki K, Larkins BA (1996). The maize gamma-zein sequesters alpha-zein and stabilizes its accumulation in protein bodies of transgenic tobacco endosperm. Plant Cell 8: 2335–2345.

Damerval C, Devienne D (1993). Quantification of dominance for proteins pleiotropically affected by opaque-2 in Maize. Heredity 70: 38–51.

Dannenhoffer JM, Bostwick DE, Or E, Larkins BA (1995). Opaque-15, a maize mutation with properties of a defective *opaque-2* modifier. Proc Natl Acad Sci U S A 92: 1931–1935.

Das OP, Messing JW (1987). Allelic variation and differential expression at the 27-kilodalton zein locus in maize. Mol Cell Biol 7: 4490–4497.

Dombrink-Kurtzman MA, Bietz JA (1993). Zein composition in hard and soft endosperm of maize. Cereal Chem 70: 105–108.

Duvick DN (1955). Cytoplasmic inclusions of the developing and mature maize endosperm. Am J Bot 42: 717–725.

Duvick DN (1961). Protein granules of maize endosperm cells. Cereal Chem 38: 374–385.

Emerson RA, Beadle GW, Frazer AC (1935). A summary of linkage studies in maize. In: Cornell University Agricultural Experimental Station Report 180: 3–83.

Fontes EBP, Shank BB, Wrobel RL, Moose SP, Obrian GR, Wurtzel ET, Boston RS (1991). Characterization of an immunoglobulin binding protein homolog in the maize floury-2 endosperm mutant. Plant Cell 3: 483–496.

Geetha KB, Lending CR, Lopes MA, Wallace JC, Larkins BA (1991). Opaque-2 modifiers increase gamma-zein synthesis and alter its spatial-distribution in maize endosperm. Plant Cell 3: 1207–1219.

Geevers HO, Lake JK (1992). Development of modified *opaque-2* maize in South Africa. In: Mertz ET (ed). *Quality Protein Maize*. American Society of Cereal Chemists, St. Paul, Minnesota, pp 49–78.

Geli MI, Torrent M, Ludevid D (1994). Two structural domains mediate two sequential events in gamma-zein targeting—protein endoplasmic-reticulum retention and protein body formation. Plant Cell 6: 1911–1922.

Geraghty D, Peifer MA, Rubenstein I, Messing J (1981). The primary structure of a plant storage protein: zein. Nucl Acids Res 9: 5163–5174.

Goettel W, Messing J (2010). Divergence of gene regulation through chromosomal rearrangements. BMC Genom 11: 678.

Hamilton TS, Hamilton BC, Johnson BC, Mitchell HH (1951). The dependence of the physical and chemical composition of the corn kernel on soil fertility and cropping system. Cereal Chem 28: 163–176.

Harz CO (1885). *Landwirtschaftliche Samenkunde*. Paul Perey, Berlin, Germany.

Hayes HK, East EM (1915). Further experiments on inheritance in maize. Conn Agric Exp Stn Bull 188: 1–31.

Holding DR, Hunter BG, Chung T, Gibbon BC, Ford CF, Bharti AK, Messing J, Hamaker BR, Larkins BA (2008). Genetic analysis of opaque2 modifier loci in quality protein maize. Theor Appl Genet 117: 157–170.

Holding DR, Hunter BG, Klingler JP, Wu S, Guo X, Gibbon BC, Wu R, Schulze J, Jung R, Larkins BA (2011). Characterization of opaque2 modifier QTLs and candidate genes in recombinant inbred lines derived from the K0326Y quality protein maize inbred. Theor Appl Genet 122: 783–794.

Holding DR, Larkins BA (2008). Genetic engineering of seed storage proteins. In: Bohnert HJ, Nguyen HN, Lewis NG (eds). *Advances in Plant Biochemistry and Molecular Biology*, Vol 1. Elsevier, Amsterdam, The Netherlands, pp 107–131.

Holding DR, Meeley RB, Hazebroek J, Selinger D, Gruis F, Jung R, Larkins BA (2010). Identification and characterization of the maize arogenate dehydrogenase gene family. J Exp Bot 61: 3663–3673.

Holding DR, Otegui MS, Li BL, Meeley RB, Dam T, Hunter BG, Jung R, Larkins BA (2007). The maize floury1 gene encodes a novel endoplasmic reticulum protein involved in zein protein body formation. Plant Cell 19: 2569–2582.

Hunter BG, Beatty MK, Singletary GW, Hamaker BR, Dilkes BP, Larkins BA, Jung R (2002). Maize opaque endosperm mutations create extensive changes in patterns of gene expression. Plant Cell 14: 2591–2612.

Jaenicke-Despres V, Buckler ES, Smith BD, Gilbert MT, Cooper A, Doebley J, Paabo S (2003). Early allelic selection in maize as revealed by ancient DNA. Science 302: 1206–1208.

Kawagoe Y, Onda Y, Kumamaru T (2009). ER membrane-localized oxidoreductase Ero1 is required for disulfide bond formation in the rice endosperm. Proc Natl Acad Sci U S A 106: 14156–14161.

Kim CS, Gibbon BC, Gillikin JW, Larkins BA, Boston RS, Jung R (2006). The maize Mucronate mutation is a deletion in the 16-kDa gamma-zein gene that induces the unfolded protein response. Plant J 48: 440–451.

Kim CS, Hunter BG, Kraft J, Boston RS, Yans S, Jung R, Larkins BA (2004). A defective signal peptide in a 19-kD alpha-zein protein causes the unfolded protein response and an opaque endosperm phenotype in the maize *De*-B30* mutant. Plant Physiol 134: 380–387.

Kim CS, Woo YM, Clore AM, Burnett RJ, Carneiro NP, Larkins BA (2002). Zein protein interactions, rather than the asymmetric distribution of zein mRNAs on endoplasmic reticulum membranes, influence protein body formation in maize endosperm. Plant Cell 14: 655–672.

Kumamaru T, Ogawa M, Satoh H, Okita TW (2007). *Protein Body Biogenesis in Cereal Endosperm*. Springer, Heidelberg, Germany.

Lai J, Messing J (2002). Increasing maize seed methionine by mRNA stability. Plant J 30: 395–402.

Lending CR, Larkins BA (1989). Changes in the zein composition of protein bodies during maize endosperm development. Plant Cell 1: 1011–1023.

Lending CR, Larkins BA (1992). Effect of the floury-2 locus on protein body formation during maize endosperm development. Protoplasma 171: 123–133.

Levanony H, Rubin R, Altschuler Y, Galili G (1992). Evidence for a novel route of wheat storage proteins to vacuoles. J Cell Biol 119: 1117–1128.

Li CP, Larkins BA (1996). Expression of protein disulfide isomerase is elevated in the endosperm of the maize floury-2 mutant. Plant Mol Biol 30: 873–882.

Li XX, Wu YJ, Zhang DZ, Gillikin JW, Boston RS, Franceschi VR, Okita TW (1993). Rice prolamine protein body biogenesis—a Bip-mediated process. Science 262: 1054–1056.

Liu CN, Rubenstein I (1993). Transcriptional characterization of an alpha-zein gene cluster in maize. Plant Mol Biol 22: 323–336.

Llaca V, Messing J (1998). Amplicons of maize zein genes are conserved within genic but expanded and constricted in intergenic regions. Plant J 15: 211–220.

Lopes MA, Larkins BA (1991). Gamma-zein content is related to endosperm modification in Quality Protein Maize. Crop Sci 31: 1655–1662.

Lopes MA, Larkins BA (1995). Genetic-analysis of opaque2 modifier gene activity in maize endosperm. Theor Appl Gen 91: 274–281.

Lopez-Valenzuela JA, Gibbon BC, Holding DR, Larkins BA (2004). Cytoskeletal proteins are coordinately increased in maize genotypes with high levels of eEF1A. Plant Physiol 135: 1784–1797.

Maclean WC, Deromana GL, Placko RP, Graham GG (1981). Protein-quality and digestibility of sorghum in preschool-children—balance studies and plasma-free amino-acids. J Nutr 111: 1928–1936.

Mainieri D, Rossi M, Archinti M, Bellucci M, De Marchis F, Vavassori S, Pompa A, Arcioni S, Vitale A (2004). Zeolin: a new recombinant storage protein constructed using maize gamma-zein and bean phaseolin. Plant Physiol 136: 3447–3456.

McWhirter KS (1971). A floury endosperm, high lysine locus on chromosome 10. Maize Genet Coop Newsletter 45: 184.

Mertz ET, Nelson OE, Bates LS (1964). Mutant gene that changes protein composition and increases lysine content of maize endosperm. Science 145: 279–280.

Miclaus M, Xu JH, Messing J (2011a). Differential gene expression and epiregulation of alpha zein gene copies in maize haplotypes. PLoS Genet 7: e1002131.

Miclaus M, Wu Y, Xu JH, Dooner HK, Messing J (2011b). The maize high-lysine mutant opaque7 is defective in an acyl-CoA synthetase-like protein. Genetics 189: 1271–1280.

Moose SP, Dudley JW, Rocheford TR (2004). Maize selection passes the century mark: a unique resource for 21st century genomics. Trends Plant Sci 9: 358–364.

Moro GL, Lopes MA, Habben JE, Hamaker BR, Larkins BA (1995). Phenotypic effects of opaque2 modifier genes in normal maize endosperm. Cereal Chem 72: 94–99.

Myers AM, James MG, Lin Q, Yi G, Stinard PS, Hennen-Bierwagen TA, Becraft PW (2011). Maize opaque5 encodes monogalactosyldiacylglycerol synthase and specifically affects C18:3/C18:2 galactolipids necessary for amyloplast and chloroplast function. Plant Cell 23:2331–2347.

Nelson OE, Mertz ET, Bates LS (1965). Second mutant gene affecting the amino acid pattern of maize endosperm proteins. Science 150: 1469–1470.

Okita TW, Choi SB (2002). mRNA localization in plants: targeting to the cell's cortical region and beyond. Curr Opin Plant Biol 5: 553–559.

Onda Y, Nagamine A, Sakurai M, Kumamaru T, Ogawa M, Kawagoe Y (2011). Distinct roles of protein disulfide isomerase and P5 sulfhydryl oxidoreductases in multiple pathways for oxidation of structurally diverse storage proteins in rice. Plant Cell Online 23: 210–223.

Or E, Boyer SK, Larkins BA (1993). opaque2 modifiers act post-transcriptionally and in a polar manner on gamma-zein gene expression in maize endosperm. Plant Cell 5: 1599–1609.

Oria MP, Hamaker BR, Axtell JD, Huang CP (2000). A highly digestible sorghum mutant cultivar exhibits a unique folded structure of endosperm protein bodies. Proc Natl Acad Sci U S A 97: 5065–5070.

Oria MP, Hamaker BR, Schull JM (1995). In-vitro protein digestibility of developing and mature sorghum grain in relation to alpha-kafirin, beta-kafirin, and gamma-kafirin disulfide cross-linking. J Cereal Sci 22: 85–93.

Ortega EI, Bates LS (1983). Biochemical and agronomic studies of two modified hard-endosperm opaque-2 maize (Zea-mays-L) populations. Cereal Chem 60: 107–111.

Osborne T (1924). *The Vegetable Proteins*.

Paez AV, Helm JL, Zuber MS (1969). Lysine content of *opaque-2* maize kernels having different phenotypes. Crop Sci 9: 251–253.

Reyes FC, Chung T, Holding DR, Jung R, Vierstra R, Otegui MS (2011). Delivery of prolamins to the protein storage vacuole in maize aleurone cells. Plant Cell 23: 769–784.

Reyes FC, Sun BM, Guo HN, Gruis D, Otegui MS (2010). Agrobacterium tumefaciens-mediated transformation of maize endosperm as a tool to study endosperm cell biology. Plant Physiol 153: 624–631.

Schmidt RJ, Burr FA, Aukerman MJ, Burr B (1990). Maize regulatory gene opaque-2 encodes a protein with a leucine-zipper motif that binds to zein DNA. Proc Natl Acad Sci U S A 87: 46–50.

Segal G, Song RT, Messing J (2003). A new opaque variant of maize by a single dominant RNA-interference-inducing transgene. Genetics 165: 387–397.

Shewry PR, Halford NG (2002). Cereal seed storage proteins: structures, properties and role in grain utilization. J Exp Botany 53: 947–958.

Song R, Llaca V, Linton E, Messing J (2001). Sequence, regulation, and evolution of the maize 22-kD alpha zein gene family. Genome Res 11: 1817–1825.

Song R, Messing J (2003). Gene expression of a gene family in maize based on noncollinear haplotypes. Proc Natl Acad Sci U S A 100: 9055–9060.

Swigonova Z, Lai J, Ma J, Ramakrishna W, Llaca V, Bennetzen JL, Messing J (2004). Close split of sorghum and maize genome progenitors. Genome Res 14: 1916–1923.

Thompson GA, Larkins BA (1993). Characterization of the zein genes and their regulation in maize endosperm. In: Walbot V, Freeling M (eds). *The Maize Handbook*. Springer, Heidelberg, Germany, pp 639–647.

Tsai CY, Huber DM, Warren HL (1978). Relationship of kernel sink for N to maize productivity. Crop Sci 18: 399–404.

Tye-Din JA, Stewart JA, Dromey JA, Beissbarth T, van Heel DA, Tatham A, Henderson K, Mannering SI, Gianfrani C, Jewell DP, Hill AV, McCluskey J, Rossjohn J, Anderson RP (2010). Comprehensive, quantitative mapping of T cell epitopes in gluten in celiac disease. Sci Transl Med 2: 41ra51.

Ueda T, Waverczak W, Ward K, Sher N, Ketudat M, Schmidt RJ, Messing J (1992). Mutations of the 22- and 27-kD zein promoters affect transactivation by the Opaque-2 protein. Plant Cell 4: 701–709.

Vicente-Carbajosa J, Moose SP, Parsons RL, Schmidt RJ (1997). A maize zinc-finger protein binds the prolamin box in zein gene promoters and interacts with the basic leucine zipper transcriptional activator Opaque2. Proc Natl Acad Sci U S A 94: 7685–7690.

Villegas E, Vasil SK, Bjarmarson M (1992). Quality protein maize—what is it and how was it developed? In: Mertz ET (ed). *Quality Protein Maize*. American Society of Cereal Chemists, St. Paul, Minnesota, pp 27–48.

Wallace JC, Lopes MA, Paiva E, Larkins BA (1990). New methods for extraction and quantitation of zeins reveal a high content of gamma-zein in modified ppaque-2 maize. Plant Physiol 92: 191–196.

Wang Z, Ueda T, Messing J (1998). Characterization of the maize prolamin box-binding factor-1 (PBF-1) and its role in the developmental regulation of the zein multigene family. Gene 223: 321–332.

Watterson JJ, Shull JM, Kirleis AW (1993). Quantitation of alpha-karafins, beta-karafins, and gamma-kafirins in vitreous and opaque endosperm of sorghum-bicolor. Cereal Chem 70: 452–457.

Williamson JD, Galili G, Larkins BA, Gelvin SB (1988). The synthesis of a 19-kilodalton zein protein in transgenic petunia plants. Plant Physiol 88: 1002–1007.

Wilson C. (1989). Linkages among zein genes determined by isoelectric focusing. Theor Appl Genet 77: 217–226.

Woo YM, Hu DWN, Larkins BA, Jung R (2001). Genomics analysis of genes expressed in maize endosperm identifies novel seed proteins and clarifies patterns of zein gene expression. Plant Cell 13: 2297–2317.

Wu Y, Goettel W, Messing J (2009). Non-Mendelian regulation and allelic variation of methionine-rich delta-zein genes in maize. Theor Appl Genet 119: 721–731.

Wu Y, Holding DR, Messing J (2010). Gamma-zein is essential for endosperm modification in quality protein maize. Proc Natl Acad Sci U S A 107: 12810–12815.

Wu Y, Messing J (2009). Tissue-specificity of storage protein genes has evolved with younger gene copies. Maydica 54: 409–415.

Wu Y, Messing J (2010a). Rescue of a dominant mutant with RNA interference. Genetics 186: 1493–1496.

Wu Y, Messing J (2010b). RNA interference-mediated change in protein body morphology and seed opacity through loss of different zein proteins. Plant Physiol 153: 33–347.

Wu Y, Messing J (2011). Novel genetic selection system for quantitative trait loci of quality protein maize. Genetics 188: 1019–1022.

Wu Y, Messing J (2012). RNA interference can rebalance the nitrogen sink of maize seeds without losing hard endosperm. PLoS One 7: e32850.

Wu Y, Wang W, Messing J (2012). Balancing of sulfur storage in maize seed. BMC Plant Biol 12: 77.

Xu JH, Bennetzen JL, Messing J (2012). Dynamic gene copy number variation in collinear regions of grass genomes. Mol Biol Evol 29: 861–871.

Xu JH, Messing J (2008). Organization of the prolamin gene family provides insight into the evolution of the maize genome and gene duplications in grass species. Proc Natl Acad Sci U S A 105: 14330–14335.

Xu JH, Messing J (2009). Amplification of prolamin storage protein genes in different subfamilies of the Poaceae. Theor Appl Genet 119: 1397–1412.

Young TE, Gallie DR (2000). Programmed cell death during endosperm development. Plant Mol Biol 44: 283–301.

9 Improving Grain Quality: Wheat

Peter R. Shewry

Introduction

The annual world production of cereals during the period 2007–2009 was almost 2500 million tonnes, of which about 27% was wheat (a mean of >660 million tonnes) (http://faostat.fao.org/). In terms of total production, wheat is the third most important cereal crop in the world, after maize and rice (812 million and 677 million tonnes pa during 2007–2009). However, wheat is grown over a much wider geographical and climatic range than rice and maize, from 67° N in Scandinavia and Russia to 45° S in Argentina, including higher elevations in the tropics (Feldman *et al.*, 1995). Wheat is the staple food in much of the temperate world, including North America and Europe, as well as the Middle East and North Africa. Demand for wheat products is also increasing in regions of Asia and sub-Saharan Africa, which are climatically unsuited for wheat production.

Wheat has three main end uses – human food, livestock feed, and industrial uses (notably brewing, distilling, and biofuel production) – which vary in relative importance in different countries and differ in their quality requirements. However, before considering these, it is necessary first to discuss the structure and composition of the grain.

Grain Structure and Composition

The mature wheat grain comprises three parts: endosperm, embryo, and outer layers (Figure 9.1). The endosperm accounts for ~90% of the dry weight of the mature grain, with the outer aleurone layer (consisting of a single layer of cells in wheat) accounting for ~6.5% and the starchy endosperm accounting for ~83% of the grain dry weight (Table 9.1) (Barron *et al.*, 2007). The aleurone cells have thick walls, which account for about half of their dry weight and high contents of minerals bound to phytate (~15% of the aleurone dry weight) (Hemery *et al.*, 2009), protein (~23% dry weight) (Jensen and Martens, 1983), and lipids (4.6%–8.9% dry weight) (Chung *et al.*, 2009) but little or no starch. By contrast, the starchy endosperm accounts for ~83% of the grain dry weight and is rich in starch (~78% of the tissue dry weight) (Hemery *et al.*, 2009) and protein (which varies in amount with nutrition but is generally about 10% dry weight). However, the starchy endosperm is not a homogeneous tissue but comprises several cell types with gradients in the content and composition of the major components; this is illustrated for protein, starch, and cell wall polysaccharides in Figure 9.2. In particular, the outer two or three layers of starchy endosperm cells (called

Seed Genomics, First Edition. Edited by Philip W. Becraft.

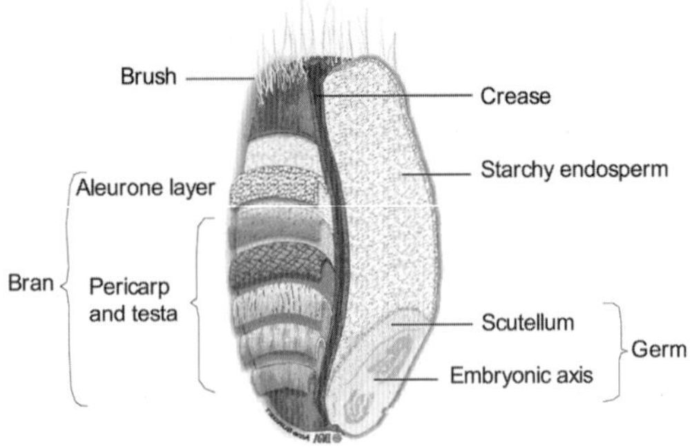

Figure 9.1 Component tissues of wheat grain. (From Surget and Barron [2005], with permission.) (For color detail, see color plate section.)

sub-aleurone cells) are rich in protein, which can account for >50% of the dry weight of flours derived from these cells (Kent, 1966), and contain little starch.

The embryo accounts for ~3% of the grain dry weight (Table 9.1) and resembles the aleurone in being rich in lipid and protein, whereas the outer layers (consisting of the outer and inner pericarp, testa, and nucellar epidermis [also called hyaline layer]) comprise 7%–8% of the grain dry weight (Table 9.1) and are rich in cell wall polysaccharides: up to 70% of the nucellar epidermis, 50% of the outer pericarp, and 44% of the testa + inner pericarp + nucellar epidermis (Barron *et al.*, 2007). The embryo, aleurone, and, to a lesser extent, the outer layers are enriched in many components that are known or considered to have health benefits for humans, including dietary fiber, minerals, vitamins, and phytochemicals. This has implications for grain processing and for the development of wheats with enhanced health benefits. The literature on these "bioactive" components is vast; Piironen *et al.* (2009) have provided an excellent account of the distributions of micronutrients and phytochemicals in the wheat grain, and Saulnier *et al.* (2007) and Stone and Morell (2009) have reviewed the distribution and properties of cell wall polysaccharides, which are the major components of wheat dietary fiber.

Most of the wheat used for human consumption is dry milled, using a complex series of rolling and sieving processes that are designed to separate the white flour derived from the starchy endosperm from the bran that contains the other botanical components. The germ (embryo) is usually present in

Table 9.1 Histological Composition of Wheat Mature Grains from Two Wheat Cultivars (Caphorn and Crousty) (% Dry Weight)

	Caphorn	Crousty
Embryo	3.0	3.2
Embryonic axis	1.5	1.7
Scutellum	1.5	1.5
Endosperm	89.2	90.1
Starchy endosperm	82.7	83.7
Aleurone layer	6.5	6.4
Outer layers	7.8	6.7
Intermediate layer[a]	3.8	3.2
Outer pericarp	4.0	3.5

[a]Composed of nucellar epidermis (hyaline layer), testa, and inner pericarp.
From Barron *et al.* (2007), with permission.

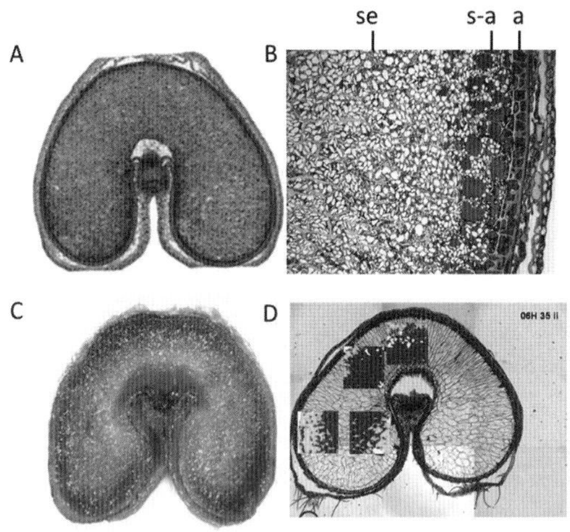

Figure 9.2 Transverse sections of developing caryopses of wheat, showing gradients in composition in the endosperm. *A* and *B*, Low-magnification and high-magnification images of a transverse section of the whole caryopsis at about 28 days after anthesis, stained with toluidine blue to show protein. Aleurone *(a)* and subaleurone *(s-a)* cells are rich in protein, and central starchy endosperm *(se)* cells are rich in starch granules (unstained). *C*, Expression of low-molecular-weight glutein subunit-promoter-GUS fusion construct in a developing endosperm, showing high expression in the subaleurone and outer starchy endosperm cells. *D*, Overlaying of Fourier-transform infra-red microspectroscopy images onto a section of cell walls only at 35 days after anthesis. The images are colored to show the distributions of high-substitution (shown in blue) and low-substitution (shown in green) forms of arabinoxylan in the cell wall. (*A* and *B*, Courtesy Cristina Sanchez-Gritsch and Paola Tosi, Rothamsted Research; *C*, courtesy Caroline Sparks and Huw Jones, Rothamsted Research; *D*, from Toole *et al.* [2010] with permission.) (For color detail, see color plate section.)

the bran fraction but can be recovered separately using specialized milling procedures. The recovery of white flour (the "extraction rate") is usually ~75% of the grain weight. Increasing this rate results in increasing proportions of the bran fraction in the flour, with "brown" flours having extraction rates of ~85% and wholemeal flours having extraction rates of 100%. In practice, these high extraction flours are usually produced by blending flour and bran fractions produced by milling, although the outer layers can also be removed before milling by friction (peeling) or abrasion (pearling), facilitating the recovery of clean aleurone-enriched fractions.

End Use Quality

The grain quality requirements are determined by the end use, which in this chapter are considered under four headings: food processing, human health, livestock feed, and fermentation for fuel and beverages. Table 9.2 provides a summary of the requirements for different end uses.

Food Processing

Wheat is consumed by humans only after extensive preparation. The most widely consumed products are bread and noodles, which are produced from hexaploid bread wheat (*Triticum aestivum*), and pasta, which is produced from tetraploid durum wheat (*T. turgidum* var *durum*). For these products, the grain is milled and the flour is mixed with water to form dough, which is either baked into bread, with or without prior fermentation, or formed into pasta or noodles. The major quality criterion for

Table 9.2 Wheat Quality Requirements

	Usage (Million Tons per Year)	Starch	Protein		Cell Wall Fiber	Texture (Milling)	Other
			Total	Gluten			
Breadmaking	5.6[a]	–	High	Strong	Low	Hard	
Biscuits and cakes		–	Medium	Weak	Low	Soft	
Human health		High amylose	–	–	High	–	High micronutrients
Distilling	0.7	High	Low	Weak	Low	Soft	
Biofuel	1.5–2.0	High	Low	Weak	Low	Soft	
Livestock feed	6	High	Low	Weak	Low	Soft	Protein quality Low phytate

[a]Breadmaking and biscuits and cakes combined.

these end uses is dough strength, which is determined largely by the amount and composition of the gluten proteins.

The gluten proteins form a network in dough that provides cohesive properties and, in the case of leavened bread, allows the entrapment of carbon dioxide released during fermentation (proofing). This entrapment expands the dough to give a porous structure, which is fixed during baking to give an acceptable crumb structure to the finished loaf. Although durum wheat grains are inevitably hard in texture, bread wheat lines differ in texture, and hard wheats are preferred for breadmaking because the strong adhesion between the starch granules and gluten proteins results in starch damage during milling, allowing water absorption during breadmaking. Lower protein contents, weaker dough strength, and soft texture are preferred for making cakes and biscuits.

Although cell wall arabinoxylan (AX) accounts only for ~1.3%–2.7% of the dry weight of white flour (Gebruers *et al.*, 2008), it is known to have both positive and negative effects on breadmaking performance (Courtin and Delcour, 2002). Bakers often use xylanases during breadmaking to optimize the structure, solubility, and viscosity of the flour AX.

Human Health

Both short-term dietary interventions and longer term epidemiological studies have shown that whole-grain foods are able to provide protection against the risk of many chronic diseases, particularly the combination of conditions known as the metabolic syndrome (which include type 2 diabetes, obesity, and cardiovascular disease (Krauss *et al.*, 2000; Nugent, 2004; de Munter *et al.*, 2007; Mellen *et al.*, 2008)). Although this protection may result from a combination of beneficial components, including phytochemicals, the evidence for the effects of dietary fiber and for the importance of both soluble and insoluble forms is particularly compelling (Slavin, 2004; Lunn and Buttriss, 2007; Topping, 2007; Wood, 2007; Buttress and Stokes, 2008; Anderson *et al.*, 2009; Fardet, 2010). It has also been suggested that the slow release in the colon of ferulic acid bound to dietary fiber has health benefits (Vitaglione *et al.*, 2008).

Because wheat products are major sources of dietary fiber in many diets (e.g., cereals and cereal products account for >40% of the daily intake in adults in the United Kingdom (Buttriss and Stokes, 2008) and bread alone for 20% (Steer *et al.*, 2008)), the improvement of the dietary fiber content and composition of wheat flour is an attractive approach to improve the health of large populations of consumers with modest cost. Reducing the glycemic index of foods is important in relation to decreasing the risk of type 2 diabetes, and high soluble dietary fiber may contribute to this by increasing the intestinal viscosity and decreasing the rate of absorption of sugars. An alternative, and complementary, approach to reducing the glycemic index is to develop wheats with resistant starch. Resistant starch has been classified into four types, with high amylose starch contributing to types RS2 (resistant to digestion) and RS3 (retrograded starch).

Livestock Feed

Wheat is widely used for livestock feed, particularly in Europe, where it is combined with protein-rich feeds from legumes or oilseeds. In many countries, feed wheat is considered primarily as a source of starch (calories), and the emphasis in breeding feed varieties is on high yield (i.e., high starch) and low protein.

The quality of the grain for livestock feed is also affected by the fiber content, with high levels of dietary fiber components, and particularly of soluble fiber, being detrimental to feed quality owing to high viscosity. This is a particular problem for chickens and other poultry in which it leads to sticky feces, and feed processors may use endoxylanase and β-glucanase enzymes to reduce the viscosity of the feed (Pettersson and Åman, 1989). However, this is an expensive option compared with developing low-fiber feed grain. Distillers dried grains and solubles (DDGS) from distilling and biofuel production (see subsequently) are also enriched in fiber (>30% dry weight), which may result in high viscosity and require enzyme treatment when used for livestock feed.

Distilling and Biofuel Production

Substantial volumes of wheat are used for alcohol production, either for distilling for beverages or for bioethanol production. A low content of dietary fiber is desirable for these processes, not only to increase alcohol yield (because low dietary fiber grain contains a higher proportion of starch) but also to reduce the viscosity, which results in technical problems during processing. A high fiber content also reduces the quality of the DDGS for feed, as discussed previously.

Redesigning the Grain

It is impossible to develop a single type of wheat suitable for all end uses. Consequently, most wheat breeders in the United Kingdom and other European countries maintain two or more parallel breeding programs: for high-yielding, low-protein types for livestock feed and for distilling and bioethanol production; for high-protein types with strong gluten and hard texture for breadmaking; and, in some cases, for soft wheats with weak gluten for biscuits and cakes. Although most research has focused on single components and end uses, it is now possible to take a more integrated approach, ascertaining the processes and mechanisms that determine the development and structure and composition of the mature grain. These should ultimately allow a more rational approach to be taken, in which the grain can be specifically designed for different end uses.

Manipulation of Grain Protein Content and Quality

Grain Protein Content

Selection for protein content can result in massive differences between cereal lines, the classic illustration being the long-term Illinois experiment in which 70 generations of selection resulted in maize lines with protein contents ranging from 4.4%–26.6% (see Chapter 13) (Dudley *et al.*, 1974). More modest differences result when selection is carried out as part of a conventional breeding program; breadmaking and feed wheats in the United Kingdom differ in protein content by ~2% dry weight when grown under similar conditions (Snape *et al.*, 1993). Breeders have also exploited genetic sources of high protein in wild species related to wheat, including an *Aegilops* species, which donated a high-protein gene to the Kansan variety Plainsman V (Finney, 1978), and, in particular, wild lines of emmer (*Triticum turgidum* var *dicoccoides*), a tetraploid species related to cultivated emmer and durum wheats (Avivi, 1978). The latter is the source of the *Gpc-B1* gene, which has been transferred into commercial wheats (Khan *et al.*, 1989, 2000; Mesfin *et al.*, 2000). The *Gpc-B1* gene encodes a transcription factor that accelerates senescence of the vegetative tissue, increasing the mobilization and transfer of nitrogen and minerals (iron and zinc) into the grain (Distelfeld *et al.*, 2006; Uauy *et al.*, 2006). However, the impact of *Gpc-B1* on grain yield is still not established, with negative (Chee *et al.*, 2001) and neutral (Mesfin *et al.*, 2000) effects being reported. Although *Gpc-B1* accounts for 70% of the variation in grain protein content in crosses (Chee *et al.*, 2001; Joppa *et al.*, 1997), no other major loci have been discovered with QTL being reported on numerous chromosomes. Worland and Snape (2001) suggested that <30% of the variation in grain protein content in winter wheats in the United Kingdom could be explained by known genes.

Grain Protein Quality

The content of essential amino acids is relevant to the use of wheat for livestock and for humans where wheat forms a major part of the diet, with wheat resembling other cereals in being particularly deficient in lysine (see Chapter 8) (Shewry, 2007). However, the main target for improving wheat protein quality is for breadmaking, which is largely determined by the gluten proteins.

Gluten Proteins

Wheat gluten can be defined as the proteinaceous mass that remains when dough is washed to remove soluble components, starch, and other cellular material. It comprises about 80% protein, 10% starch (which is probably entrapped within the protein), and 10% other material (e.g., lipids, minerals, polysaccharides) on a dry weight basis, with most of the protein components being prolamin storage proteins from the starchy endosperm cells. These proteins come together when wheat flour is mixed with water, forming a continuous network in the dough. This network provides the unique viscoelastic properties that enable the dough to be processed in bread, other baked products, pasta, and noodles. The quality of wheat for breadmaking is often limited by low elasticity (strength), particularly wheat grown in cooler temperate climates, resulting in poor gas retention during proofing, low loaf volume, and dense crumb structure.

The gluten proteins are classified into two broad groups, called gliadins and glutenins, which are present in about equal amounts. These fractions were initially defined on the basis of their solubility (gliadins) or insolubility (glutenins) in aqueous alcohols (e.g., 70% ethanol), which result from their presence as monomers and polymers. The polymeric glutenin proteins have been studied in the most detail because dough strength is positively related to the amount and size of the high-molecular-weight (HMW) glutenin polymers (Wieser *et al.*, 2006).

Glutenins and Dough Strength

The glutenin polymers are stabilized by interchain disulfide bonds, which can be reduced to release groups of HMW and low-molecular-weight subunits. Although the HMW subunits are the least abundant group of gluten proteins, accounting for ~10% of total gluten (see subsequently) and ~20% of glutenin, they have been studied in the greatest detail because of their role in determining gluten and dough elasticity. This role is demonstrated by two lines of evidence. First, the HMW subunits are concentrated in HMW glutenin polymers, which are known to contribute to dough strength (Huebner and Wall, 1976; Field *et al.*, 1983). Second, genetic studies have shown that allelic variation in HMW subunit composition is correlated with differences in dough strength and breadmaking quality (Burnouf and Bouriquet, 1980; Moonen *et al.*, 1982; Cressey *et al.*, 1987; Lawrence *et al.*, 1987; Payne, 1987).

HMW Subunits of Glutenin

Cultivars of bread wheat contain three, four, or five HMW subunits of glutenin. These are encoded by the *Glu-1* loci on the long arms of the group 1 chromosomes, with each locus comprising two genes, which encode an x-type and a y-type subunit. The x-type and y-type proteins were initially defined by their mobility on SDS-PAGE (y-type being faster) (Payne *et al.*, 1981) but are now known also to differ in their amino acid sequences, including numbers of cysteine residues available for polymer formation (Shewry *et al.*, 2003). The genes encoding x-type and y-type subunits are tightly linked, and recombination occurs only rarely.

The variation in HMW subunit number in bread wheat results from gene silencing, with 1Ay subunits never being expressed and 1Ax and 1By subunits being expressed in some lines only. However, 1Ay subunits may be present in wild-type diploid and tetraploid wheats (Waines and Payne, 1987; Levy *et al.*, 1988; Margiotta *et al.*, 1998), and the number of HMW subunits expressed in bread wheat could be increased by introgression of 1Ay subunits from these species.

As mentioned earlier, genetic variation in HMW subunit number and composition is correlated with differences in dough strength and breadmaking performance, and this probably results from both quantitative and qualitative effects. First, the expression of a 1Ax subunit is associated with good quality compared with the null (silent) form, as is the expression of subunit 1Bx7 in combination with subunit 1By8 or subunit 1By9, rather than subunit 1Bx7 alone (Payne *et al.*, 1987). It has been shown that the individual subunit proteins each account, on average, for ~2% of the total grain protein (Seilmeier *et al.*, 1991; Halford *et al.*, 1992), so these effects of subunit number may have a quantitative basis, with a higher number of expressed genes resulting in more HMW subunit protein and greater dough strength.

However, other effects of HMW subunits on dough strength appear to be qualitative, resulting from allelic differences in HMW subunit structure. This has been studied in most detail for the HMW subunit pairs encoded by the *Glu-D1* locus. Two "allelic pairs" of HMW subunits encoded by chromosome 1D occur widely in commercially grown bread wheat varieties, called 1Dx5 + 1Dy10 and 1Dx2 + 1Dy12. The 1Dx 5 + 1Dy10 allelic pair is associated with superior quality, and this effect appears to result from the presence of a higher proportion of HMW glutenin polymers, rather than more total glutenin polymers or total protein (Gupta and MacRitchie, 1994). The effect on HMW polymers may relate to the fact that subunit 1Dx5 contains an additional cysteine residue compared with subunit 1Dx2 (Anderson *et al.*, 1989), which could lead to an increase in the number of interchain disulfide bonds that are formed and the formation of larger or more highly cross-linked glutenin polymers.

Increasing Dough Strength by Transgenesis

The availability of genes encoding HMW subunits and the demonstration of a relationship between the number of expressed genes and dough strength (discussed earlier) have resulted in numerous reports of the expression of HMW transgenes in wheat, including some of the very first reports of wheat transformation (Altpeter *et al.*, 1996; Blechl and Anderson, 1996; Barro *et al.*, 1997; Alvarez *et al.*, 2000). This work has been reviewed extensively, including two detailed more recent accounts (Blechl and Jones, 2009; Jones *et al.*, 2009), and is only briefly summarized here.

All studies have used endogenous HMW subunit promoters, and expression levels tend to be similar to or greater than those of the endogenous HMW subunit genes. However, this may be misleading because workers tend to discard lines showing low levels of transgene expression.

The expression of a transgene encoding subunit 1Ax1 can result in increased dough strength (Barro *et al.*, 1997; Vasil *et al.*, 2001), although the effect is more limited in commercial cultivars with high intrinsic dough strength (Alvarez *et al.*, 2001; Field *et al.*, 2008; Rakszegi *et al.*, 2008). High-level expression of the transgene may also lead to an overstrong phenotype (Rakszegi *et al.*, 2008).

Subunits 1Dx5 and 1Dy10 are always present as a pair, and the expression of a subunit 1Dx5 transgene on its own usually results in overstrong properties (Rooke *et al.*, 1999; Alvarez *et al.*, 2001; Rakszegi *et al.*, 2005; Blechl *et al.*, 2007), which appear to relate to an increase in glutenin cross-linking (Popineau *et al.*, 2001). This overstrong effect can be ameliorated by the incorporation of purified subunit 1Dy10 into the dough (Butow *et al.*, 2003) or by coexpression of the 1Dx5 and 1Dy10 transgenes (Blechl *et al.*, 2007). Expression of the 1Dy10 transgene on its own may also result in increased dough strength, similar to the effect observed with the subunit 1Ax1 transgene (Blechl *et al.*, 2007; León *et al.*, 2009). By contrast, Graybosch *et al.* (2011) showed that field grown lines expressing the subunit 1Dy10 transgene tended to show overstrong characteristics, whereas lines expressing 1Dx5 and 1Dy10 tended to lack well-defined peaks in Mixograph analysis.

León *et al.* (2010) compared lines expressing various combinations of the 1DAx1, 1Dx5, and 1Dy10 transgenes. They showed that the combination of the 1Dx5 and 1Dy10 transgenes gave better dough properties than the combinations of the 1Dx5 and 1Ax1 or 1Ax1 and 1Dy10 transgenes. The combination of the subunits 1Dx5 + 1Dy10 transgenes also gave greater strength than the expression of all three transgenes.

These studies demonstrate that transgenic expression of HMW subunit genes can be used to increase the dough strength of wheat, although the precise effect varies with the selection of subunits, the expression level, and the genetic background (including the intrinsic dough strength and composition of endogenous HMW subunits).

The expression levels of the HMW subunit transgenes appear to be stable over multiple generations (Barro *et al.*, 2002; Rakszegi *et al.*, 2005; Shewry *et al.*, 2006; Rakszegi *et al.*, 2008), whereas the limited number of field trials that have been carried out show that they differ little from control lines in agronomic performance (Barro *et al.*, 2002; Bregitzer *et al.*, 2006; Shewry *et al.*, 2006; Graybosch *et al.*, 2011). Finally, analyses of metabolite profiles (Baker *et al.*, 2006) and gene expression patterns (Baudo *et al.*, 2006) show that they are "substantially equivalent" to nontransgenic wheat and should be acceptable to regulatory authorities.

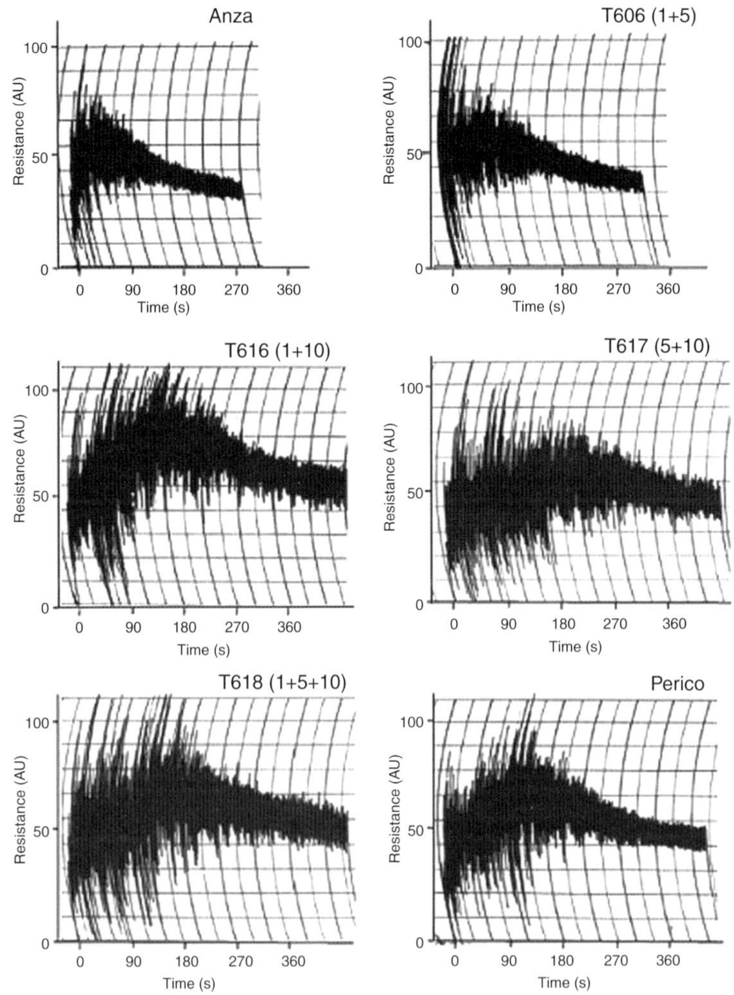

Figure 9.3 Comparison of the Mixograph curves of transgenic lines expressing additional HMW subunits (indicated in parentheses) with the control line Anza and a commercial wheat cultivar Perico. The Mixograph records the increase in stress as dough is mixed to its maximum resistance and the subsequent decrease in stress on overmixing. Numerous measurements are taken that relate to dough strength with mixing time (measured in seconds) being positively correlated with dough strength. In the experiment shown, the mixing time was 35 seconds for Anza; 128 seconds for Perico; and 63 seconds, 133 seconds, 185 seconds, and 156 seconds for transgenic lines T606 (1Ax1 + 1Dx5 transgenes), T616 (1Ax1 + 1Dy10 transgenes), T167 (1Dx5 + 1Dy10 transgenes), and T618 (1Ax1 + 1Dx5 + 1Dy10 transgenes), demonstrating that the transgenic lines compare favorably with the commercial control, with T617 having the greatest strength. (From León *et al.* [2010] with permission.)

Manipulation of Grain Texture

Kernel texture has been described as "the most important single characteristic that affects the functionality of a common wheat" (Pomeranz and Williams, 1990), determining how the grain behaves during milling and the suitability of the flour for various end uses. Between ~60% and

80% of the variation in hardness in bread wheat in controlled by allelic variation at the *Hardness* (*Ha*) locus, which is located on chromosome 5D of bread wheat, with minor hardness loci having been mapped onto numerous other chromosomes (Sourdille *et al.*, 1996; Galande *et al.*, 2001; Turner *et al.*, 2004; Weightman *et al.*, 2008). The *Ha* locus contains two genes encoding small (~15,000 kDa) sulfur-rich proteins, which are characterized by the presence of short tryptophan-rich motifs and have been named puroindoline (Pin) a and b (Douliez *et al.*, 2000). These proteins are present in greater amounts on the surface of starch granules prepared from mature grain of soft wheats than from hard wheats, and it has been suggested that they determine softness by preventing adhesion between the starch granule surface and the gluten protein matrix (Greenwell and Schofield, 1986). Consequently, the starch granules are more easily released from soft wheat during milling, requiring less energy input and with little damage. The equivalent regions of chromosomes 5A and 5B of both hexaploid bread and tetraploid wheats have independent deletions of the genes encoding puroindolines (Chantret *et al.*, 2005) and cultivated durum wheat, and other tetraploid wheats are always hard.

Most hard genotypes of bread wheat have either a deletion of the Pin a gene (*Pina-D1*) or one of six point mutations in the Pin b gene (*PinB-D1*) leading to an amino acid substitution in the encoded protein (Morris, 2002), with other types of mutations occurring more rarely (Bhave and Morris, 2008; Morris and Bhave, 2008). The point mutations are proposed to affect the binding of the puroindoline to the starch, but despite a massive volume of research, the mechanism of binding of Pins to starch granules and the impact of mutations on this remain poorly understood.

Direct effects of Pin a and b on grain softness have been demonstrated by transgenic expression in rice, a species that lacks Pins and has hard grain (Krishnamurthy and Giroux, 2001), and in hard bread wheat (Beecher *et al.*, 2002; Hogg *et al.*, 2004; Martin *et al.*, 2008). Studies have shown the existence of at least four minor variant forms of Pins encoded by loci on the long arms of chromosomes 7A, 7B, and 7D, with the 7A component mapping close to a QTL for hardness (Wilkinson *et al.*, 2008; Chen *et al.*, 2010). Other minor hardness QTL could relate to aspects of grain structure such as cell wall composition and mechanical properties.

Development of Wheat with Resistant Starch

As described in detail in Chapter 7, starch is a mixture of two glucose polymers: amylose, which comprises single unbranched (1→4) α-linked chains of up to several thousand glucose units, and amylopectin, which is highly branched (with (1→6) α-linkages and (1→4) α-linkages) and may comprise >100,000 glucose unit residues. In most species, including wheat, amylose and amylopectin occur in a ratio of 1:3 amylose to amylopectin, with only limited variation in their proportions between genotypes or in relation to the environment. High amylose starches are attractive for developing healthy foods because they are more slowly digested and become resistant on cooking, both of which lead to reduced glycemic index in the human gastrointestinal tract.

Amylose and amylopectin are both synthesized from ADP glucose, but the pathways differ. This allows the adoption of two strategies for producing high amylose starch, either increasing amylose synthesis by transgenic overexpression or reducing amylopectin synthesis by transgenesis or mutagenesis.

The synthesis of amylose in the wheat starchy endosperm requires only a single enzyme, granule bound starch synthase form Ia (BGSSIa) (Stone and Morell, 2009). In theory, it should be possible to increase amylose content substantially by a single transgenic event. However, this approach has not been reported, and work so far has focused on reducing amylopectin synthesis.

The *Starch granule protein-1* (*Sgp-1*) genes located on chromosomes 7AS, 7BS, and 7DS of bread wheat (Chao *et al.*, 1989; Yamamori and Endo, 1996) encode forms of the starch synthase II (SSII) enzyme, which are involved in the synthesis of medium-length chains in amylopectin (Stone and Morell, 1999). The encoded proteins (SGP-A1, SGP-B1 and SGP-D1) are readily identified by SDS-PAGE of proteins extracted from purified starch granules, facilitating the identification of null mutant forms and the stacking of mutations at the three loci to result in a high amylose phenotype in hexaploid bread wheat. A modest but significant increase in the proportion of amylose, from ~29% in the Chinese Spring control lines to ~37%, resulted (Yamamori *et al.*, 2000). However, the total starch content was also reduced, from 60.7% in the control Chinese Spring wheat to 49.4% in the triple mutant. The structure of the starch was also affected, with an increase in the proportion of short chains (degree of polymerization [DP] 6–10) and a decrease in longer chains (DP 11–25).

Hung *et al.* (2005) demonstrated that the incorporation of 10%, 30%, and 50% high amylose (35%) flour into bread resulted in increases in the content of resistant starch, from 0.9% dry weight in the control bread to 1.65%, 2.6%, and 3.0%. The availability of genomic sequences of the *Sgp-1* genes has also allowed the use of TILLING (targeting induced mutations in genomes (Slade *et al.*, 2005)) to identify further allelic mutations in the *Sgp-1* genes in a mutagenized wheat population (Sestili *et al.*, 2010).

A much greater impact on the amylose content of wheat starch was achieved by transgenesis, using RNAi suppression to reduce simultaneously the activities of starch branching enzymes (SBE) II (SBEIIa and SBEIIb) that are responsible for synthesizing (1→6) linkages in amylopectin (Stone and Morell, 2009). The resulting line had 74.4% amylose compared with 25.5% in the untransformed control with slightly reduced total starch content (43.4% compared with 52%) (Regina *et al.*, 2006). By contrast to the *Gsp-1* lines discussed previously, the debranched starch had a decreased proportion of short chains (DP 4–12) and an increased proportion of longer chains of DP >12. Regina *et al.* (2006) also reported that several indexes of colon health (digesta wet weight, short-chain fatty acid pools, and fecal short-chain fatty acid excretion) were increased by twofold in rats fed the wholemeal transgenic wheat compared with the control.

The practical application of high amylose wheats in food production for human consumption has been limited by the requirement to incorporate the phenotype into lines that have commercially acceptable yields and are acceptable to consumers. At the present time, neither the *Gsp-1* mutants nor the SBEII transgenics fulfill these requirements.

Improving Content and Composition of Dietary Fiber

Resistant starch and fructans may be classified as dietary fiber because "they are resistant to digestion and absorption in the human small intestine with partial or complete fermentation in the large intestine" (AACC, 2001). However, other definitions restrict the term to the nonstarch polysaccharides of plant cell walls, and this narrower definition is used here.

Wheat Grain Cell Walls

The tissues of the wheat grain differ in their content and composition of cell wall polysaccharides, as summarized in Table 9.3. The total content of nonstarch polysaccharides, including fructans, in wheat flour (derived from the starchy endosperm) is ~3.5%–4% dry weight (Knudsen, 1997; Haskå *et al.*, 2008), with the major components being cell wall polysaccharides. These comprise ~20%

Table 9.3 Compositions of Cell Wall Types in Wheat Grain Tissues (% Dry Weight)

	Cell Walls (% Dry Weight)	Cell Wall Components (% Total Polysaccharide)						Reference
		Cellulose	Lignin	Pectin	AX[a]	β-glucan[b]	Glucomannan	
Starchy endosperm	2–3	2	0	–	70	20	7	Mares and Stone, 1973; Loosveld *et al.*, 1998
Total bran (outer layers and aleurone)		29	8	–	64	6	–	Selvendran *et al.*, 1980
Aleurone	40	2–4	0	–	62–65	29–34	–	Bacic and Stone, 1981; Rhodes and Stone, 2002; Antoine *et al.*, 2003
Outer pericarp (beeswing)		30	12	–	60	–	–	Du Pont and Selvendran, 1987
Straw	85	37–40	14–17	0.5	39	–	–	Lawther *et al.*, 1995; Sun *et al.*, 2005

[a]Arabinoxylan varies in structure with highly cross-linked and substituted forms (GAX) being present in outer layers and straw.
[b](1→3,1→4)-β-D-glucan. Traces of (1→3)-β-D-glucan (callose) may also be present in grain fractions.
Modified from Shewry *et al.* (2010).

(1→3,1→4)-β-D-glucans (β-glucan), 70% arabinoxylan (AX), 2%–4% cellulose, and 2%–7% glucomannan (Stone and Morell, 2009). Andersson *et al.* (1992) showed that the glucose content of the nonstarch polysaccharides (which was presumably derived from β-glucan) ranged from 0.25%–0.63% dry weight (mean 0.44% dry weight). More detailed analyses of AX were reported by Gebruers *et al.* (2008), who determined total AX (TOT-AX) and water-extractable AX (WE-AX) in flours of 150 wheats grown on the same site in Hungary. TOT-AX ranged from 1.35%–2.75% dry weight (mean 1.93% dry weight) and WE-AX from 0.30%–1.40% dry weight (mean 0.51% dry weight). No lignin is present in white flour, and little is known about the extent of variation in the contents of cellulose and glucomannan.

Barron *et al.* (2007) showed that the sugars released on hydrolysis of the nonstarch polysaccharides present in the isolated aleurone layer accounted for 34.0% and 42.4% of the dry weight in two cultivars, with the major sugars being arabinose, xylose, and glucose. These analyses agree with the data in Table 9.3 (which are taken from Stone and Morell, 2009) showing that the composition of the aleurone cell walls is similar to that of the starchy endosperm cells except that the proportion of AX is slightly lower and that of β-glucan slightly higher.

The outer layers of the wheat grain, comprising the nucellar epidermis, testa, and inner and outer pericarp, have been studied in less detail. However, available data indicate that the pericarp (which is the major component of the outer layers) has a similar composition to vegetative tissues, comprising mainly cellulose, lignin, and glucuronoarabinoxylans with little or no β-glucan (Table 9.3) (Barron *et al.*, 2007; Stone and Morell, 2009).

Three approaches can be taken to increase the dietary intake of fiber in wheat products. The first is to increase the consumption of whole-grain foods, or foods made from flours that contain a proportion of the aleurone fraction. Most food products are made from white flour, which corresponds to ~70%–75% of the whole grain. Increasing the flour extraction rate to ≥80% results in a browner flour with an increasing proportion of the aleurone and outer layers. The second approach is to remove and fractionate the bran and to recombine defined aleurone fractions with white flour. However, both of these approaches may adversely affect the functional properties of the flour and the appearance and

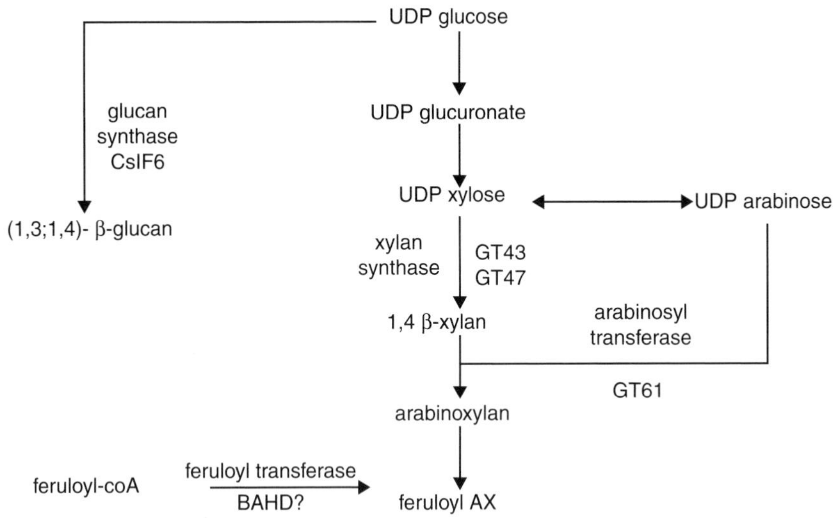

Figure 9.4 Proposed pathway of arabinoxylan synthesis. The gene families that have been proposed to encode xylan synthase, arabinosyl transferase, and feruloyl transferase are indicated.

consumer acceptability of the product and may have cost implications. It is more attractive in the long run to improve the fiber content and composition of white flour (i.e., the starchy endosperm). We have adopted the latter approach, by identifying candidate genes for key steps in β-glucan and AX synthesis (Figure 9.4) and characterizing their functions by RNAi suppression in the starchy endosperm of transgenic wheat.

Manipulating β-glucan

(1→3,1→4)-β-D-glucan (β-glucan) comprises glucose residues joined by (1→3) and (1→4) linkages (Figure 9.5*A*). Single (1→3) linkages are usually separated by two or three (1→4) linkages, but longer stretches of (1→4) linked glucan of up to 14 units have been reported for wheat bran β-glucan (Liu *et al.*, 2006). Such regions are sometimes referred to as "cellulose-like" because cellulose is (1→4)-β-D-glucan without any (1→3) linkages. The distribution of (1→3) and (1→4) linkages may determine the solubility of β-glucan and the viscosity of aqueous extracts (Lazaridou and Biliaderis, 2007).

Burton *et al.* (2006) showed that transformation of *Arabidopsis* (a species that does not normally synthesize β-glucan) with the *CSLF6* gene of rice resulted in the accumulation of β-glucan in leaves, and we have since shown that RNAi suppression of the endogenous *CSLF6* gene in developing wheat endosperm resulted in a mean decrease in total β-glucan of 42% in five transgenic lines (Nemeth *et al.*, 2010). Similarly, Burton *et al.* (2011) showed that overexpression of *CSLF6* in transgenic barley resulted in increases of >80% in the content of β-glucan in seeds.

Although these studies demonstrate that *CSLF6* encodes a β-glucan synthase, it is unclear whether this enzyme catalyzes the synthesis of both (1→3) and (1→4) linkages. However, the RNAi suppression of *CSLF6* in transgenic wheat resulted in a reduction in the molecular weight profile of the β-glucan extracted by hot water, whereas it had little effect on either the proportion of the total fraction that was extracted or the ratio of glucan fragments of DP 3 and DP 4 released by

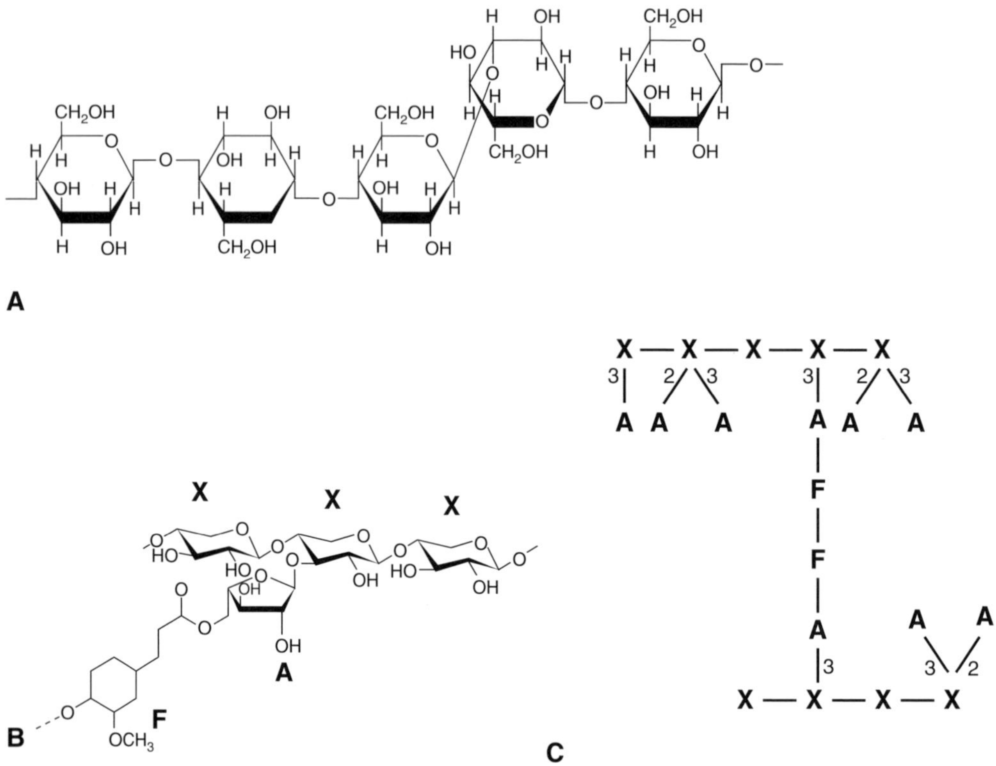

Figure 9.5 Structures of cell wall polysaccharides of wheat starchy endosperm. *A,* (1-3,1-4)-β-D-glucan. *B,* Arabinoxylan structure. *C,* Arabinoxylan schematic. A, arabinose substitution; F, ferulic acid substitution; X, (1→4) linked β-D-xylopyranosyl units.

digestion of the total β-glucan with lichenase (an *endo*-(1-3)(1-4)-β-D-glucan-4-glucanohydrolase, which was increased only slightly) (Nemeth *et al.*, 2010).

Manipulating Arabinoxylan

Arabinoxylan has a more complex structure than β-glucan, comprising a backbone of β-(1-4)-linked D-xylopyranosyl residues, which may be substituted with arabinose at the 0-3 position or at the 0-2 and 0-3 positions (Figure 9.5*B* and *C*). The 0-3-linked arabinose may also be esterified with ferulic acid or, more rarely, *p*-coumaric acid, linked to the 0-5 position. Adjacent ferulic acid groups present on AX may form diferulate (or, more rarely, triferulate) cross-links, via an oxidative process that generates a stable ether bond.

This complex structure provides a basis for considerable diversity in AX structure, which may result in differences in solubility, viscosity, and health benefits. In particular, highly arabinosylated AX may be more soluble in water owing to steric hindrance of hydrogen bonding between the xylan backbone, whereas cross-linking results in the formation of highly hydrated but insoluble gels.

Mitchell *et al.* (2007) used a bioinformatics approach to identify candidate genes for the biosynthesis and ferulylation of AX, including analysis of expression profiles in developing whole grain using

Affymetrix arrays (Toole *et al.*, 2010) and in developing starchy endosperm using high-throughput sequencing (Pellny *et al.*, 2012). This approach identified genes in the glucosyl transferase (GT) 43 and GT 47 families as encoding xylan synthases and in the GT 61 family as encoding arabinosyl transferase (Figure 9.4), and these identifications have been confirmed by RNAi suppression in transgenic wheat (Anders *et al.*, 2012; unpublished results of Wilkinson M, Lovegrove A, Freeman J, Pellny TK, Weimar T, Baker J, Shewry PR and Mitchell RAC). Similarly, genes in the acyl-CoA transferase superfamily (PF02458) were predicted to encode arabinoxylan feruloyl transferases (Figure 9.4). Piston *et al.* (2010) showed that the simultaneous downregulation of four genes in this family resulted in a modest but significant decrease in the ferulylation of cell wall AX in leaves of rice plants.

Our knowledge of dietary fiber synthesis in wheat is still incomplete, with important questions relating to the determination of the relative contents of β-glucan and AX, the linkage structure and solubility of β-glucan, and the fine structure (particularly the specificity of arabinosylation) and ferulylation of AX. Nevertheless, we should be in a position within the next few years to develop wheats with specific AX contents, compositions, and properties, exploiting a combination of natural variation, mutagenesis, and transgenesis.

Conclusion

This chapter has focused on aspects of grain quality that are determined by starch, proteins, and cell wall polysaccharides, which are the three major groups of components in the grain and together account for ≥90% of the grain dry weight. All are now well understood and amenable to manipulation by classical genetics or modern molecular approaches. However, this does not mean that they are the only quality determinants or even the most important for some applications.

Wheat is also an important source of minerals (particularly iron, zinc, and selenium), vitamin E (tocols), and B vitamins (including folates) for the human diet; a range of phytochemicals (including phenolic acids, lignans, flavonoids, and alkylresorcinols) may also have health benefits. By contrast, the accumulation of free asparagine may contribute to the formation of acrylamide (a carcinogen) during processing. Unraveling the control of grain development and composition would also facilitate the manipulation of this wider and more challenging range of targets.

Acknowledgments

This research is funded by the Biotechnology and Biological Sciences Research Council (BBSRC).

References

AACC (2001). The definition of dietary fibre. Cereal Foods World 46: 112.

Altpeter F, Vasil V, Srivastava V, Vasil IK (1996). Integration and expression of the high-molecular-weight glutenin subunit 1Ax1 gene into wheat. Nat Biotechnol 14: 1155–1159.

Alvarez ML, Gomez M, Carrillo JM, Vallejos RH (2001). Analysis of dough functionality of flours from transgenic wheat. Mol Breeding 8: 103–108.

Alvarez ML, Guelman S, Halford NG, Lustig S, Reggiardo MI, Ryabushkina N, Shewry PR, Stein J, Vallejos RH (2000). Silencing of HMW glutenins in transgenic wheat expressing extra HMW subunits. Theor Appl Genet 100: 319–327.

Anders N, Wilkinson M, Lovegrove A, Freeman J, Tryfona T, Pellny TK, Weimar T, Baker J, Shewry PR, Dupree P, Mitchell RAC (2012). Glycosyl transferases in family 61 mediate arabinofuranosyl transfer onto xylan in grasses. Proc Natl Acad Sci U S A 109: 989–993.

Anderson JW, Baird P, Davis RH Jr, Ferreri S, Knudtson M, Koraym A, Waters V, Williams CL (2009). Health benefits of dietary fiber. Nutr Rev 67: 188–205.

Anderson OD, Greene FC, Yip RE, Halford NG, Shewry PR, Malpica-Romero JM (1989). Nucleotide sequences of the two high-molecular-weight glutenin genes from the D-genome of a hexaploid bread wheat, *Triticum aestivum* L. cv Cheyenne. Nucleic Acids Res 17: 461–462.

Andersson R, Westerlund E, Tilly A-C, Åman P (1992). Natural variations in the chemical composition of white flour. J Cereal Sci 17: 183–199.

Antoine C, Peyron S, Mabille F, Lapierre C, Bouchet B, Abecassis J, Rouau X (2003). Individual contribution of grain outer layers and their cell wall structure of the mechanical properties of wheat bran. J Agric Food Chem 51: 2026–2033.

Avivi L (1978). High grain protein content in wild tetraploid wheat *Triticum dicoccoides* Korn. In: Fifth International Wheat Genetics Symposium, New Delhi, India. 23–28 February 1978, pp 372–380.

Bacic A, Stone BA (1981). Chemistry and organization of aleurone cell wall components from wheat and barley. Aust J Plant Physiol 8: 475–495.

Baker JM, Hawkins ND, Ward JL, Lovegrove A, Napier JA, Shewry PR, Beale MH (2006). A metabolomic study of substantial equivalence of field-grown genetically modified wheat. Plant Biotech J 4: 381–392.

Barron C, Surget A, Rouau X (2007). Relative amounts of tissues in mature wheat (*Triticum aestivum* L.) grain and their carbohydrate and phenolic acid composition. J Cereal Sci 45: 88–96.

Barro F, Barcelo P, Lazzeri P, Shewry PR, Martin A, Ballesteros J (2002). Field evaluation and agronomic performance of transgenic wheat. Theor Appl Genet 105: 980–984.

Barro F, Woke L, Beâkeâ SF, Gras P, Tatham AS, Fido RJ, Lazzeri P, Shewry PR, Barcelo P (1997). Transformation of wheat with HMW subunit genes results in improved functional properties. Nat Biotechnol 15: 1295–1299.

Baudo MM, Lyons R, Powers S, Pastori GM, Edwards KJ, Holdsworth MJ, Shewry PR (2006). Transgenesis has less impact on the transcriptome of wheat grain than conventional breeding. Plant Biotechnol J 4: 369–380.

Beecher B, Bettge A, Smidansky E, Giroux MJ (2002). Expression of wild-type *PinB* sequence in transgenic wheat complements a hard phenotype. Theor Appl Genet 105: 870–877.

Bhave M, Morris CF (2008). Molecular genetics of puroindolines and related genes: allelic diversity in wheat and other grasses. Plant Mol Biol 66: 205–219.

Blechl A, Anderson OD (1996). Expression of a novel high-molecular-weight glutenin subunit gene in transgenic wheat. Nat Biotech 14: 875–879.

Blechl A, Jones HD (2009). Transgenic applications in wheat improvement. In: Carver BF (ed). *Wheat Science and Trade.* Wiley-Blackwell, Ames, Iowa, pp 397–435.

Blechl A, Lin J, Nguyen S, Chan R, Anderson OD, Dupont FM (2007). Transgenic wheats with elevated levels of Dx5 and/or Dy10 high-molecular-weight glutenin subunits yield doughs with increased mixing strength and tolerance. J Cereal Sci 45: 172–183.

Bregitzer P, Blechl AE, Fiedler D, Lin J, Sebesta P, Fernandez de Soto J, Chicaiza O, Dubcovsky J (2006). Changes in high molecular weight glutenin subunit composition can be genetically engineered without affecting wheat agronomic performance. Crop Sci 46: 1553–1563.

Burnouf T, Bouriquet R (1980). Glutenin subunits of genetically related European hexaploid wheat cultivars: their relation to breadmaking quality. Theor Appl Genet 58: 107–111.

Burton RA, Wilson SM, Hrmova M, Harvey AJ, Shirley NJ, Medhurst A, Stone BA, Newbigin EJ, Bacic A, Fincher GB (2006). Cellulose synthase-like *CSLF* genes mediate the synthesis of cell wall (1,3)(1,4)-β-D-glucans. Science 311: 1940–1942.

Burton RA, Collins HM, Kibble NAJ, Smith JA, Shirley NJ, Jobling SA, Henderson M, Singh RR, Pettolino F, Wilson SM, Bird AR, Topping DL, Bacic A, Fincher GB (2011). Over-expression of specific HvCslF cellulose synthase-like genes in transgenic barley increases the levels of cell wall (1,3;1,4)-β-D-glucans and alters their fine structure. Plant Biotech J 9: 117–135.

Butow B, Tatham AS, Savage AWJ, Gilvert SM, Shewry PR, Solomon RG, Békés F (2003). Creating a balance—the incorporation of a HMW glutenin subunit into transgenic wheat lines. J Cereal Sci 38: 181–187.

Buttress JL, Stokes CS (2008). Dietary fibre and health: an overview. Nutr Bull 33, 186–200.

Chantret N, Salse J, Sabot F, Rahman S, Bellec A, Laubin B, Dubois I, Dossat C, Sourdille P, Joudrier P, Gautier M-F, Cattolico L, Beckert M, Aubourg S, Weissenbach J, Caboche M, Bernard M, Leroy P, Chalhoub B (2005). Molecular basis of evolutionary events that shaped the *Hardness* locus in diploid and polyploidy wheat species (Triticum and Aegilops). Plant Cell 17: 1033–1045.

Chao S, Sharp PJ, Worland AJ, Warham EJ, Koebner RMD, Gale MD (1989). RFLP-based genetic maps of wheat homeologous group 7 chromosomes. Theor Appl Genet 78: 495–504.

Chee PW, Elias EM, Anderson JA, Kianian SF (2001). Evaluation of high grain protein concentration QTL from Triticum turgidum L. var. dicoccoides in an adapted durum wheat background. Crop Sci 41: 295–301.

Chen F, Beecher BS, Morris CF (2010). Physical mapping and a new variant of *Puroindoline b-2* genes in wheat. Theor Appl Genet 120: 745–751.

Chung OK, Ohm JB, Ram MS, Park S-H, Howitt CA (2009). Wheat lipids. In: Shewry PR, Khan K (eds). *Wheat: Chemistry and Technology*. AACC, St Paul, Minnesota, pp 363–399.

Courtin CM, Delcour JA (2002). Arabinoxylans and endoxylases in wheat flour bread-making. J Cereal Sci 35: 225–243.

Cressey PJ, Campbell WP, Wrigley CW, Griffin WB (1987). Statistical correlations between quality attributes and grain-protein composition for 60 advanced lines of crossbred wheat. Cereal Chem 64: 299–301.

de Munter JSL, Hu FB, Spielgelman D, Frantz M, van Dam RM (2007). Whole grain, bran and germ intake and risk of type 2 diabetes: a prospective cohort study and systematic review. PLoS Med 4: 1389–1395.

Distelfeld A, Uauy C, Fahima T, Dubcovsky J (2006). Physical map of the wheat high-grain content gene Gpc-B1 and development of a high-throughput molecular marker. New Phytol 169: 753–763.

Douliez J-P, Michon T, Elmorjani K, Marion D (2000). Structure, biological and technological functions of lipid transfer proteins and indolines, the major lipid binding proteins from cereal kernels. J Cereal Sci 32: 1–20.

Dudley JW, Lambert RJ, Alexander DE (1974). Seventy generations of selection for oil and protein concentration in the maize kernel. In: Dudley JW (ed). *Seventy Generations of Selection for Oil and Protein in Maize*. Crop Science Society of America, Madison, Wisconsin, pp 181–212.

Du Pont MS, Selvendran RR (1987). Hemicellulosic polymers from the cell walls of beeswing wheat bran. Part I, polymers solubilised by alkali at 2°. Carbohydr Res 163: 99–113.

Fardet A (2010). New hypotheses for the health-protective mechanisms of whole-grain cereals: what is beyond fibre? Nutr Res Rev 23: 65–134.

Feldman M, Lupton FGH, Miller TE (1995). Wheats. In: Smartt J, Simmonds N (eds). *Evolution of Crop Plants*, ed 2, Longman Scientific & Technical, Harlow, UK, pp 184–192.

Field JM, Bhandari D, Bonet A, Underwood C, Darlington H, Shewry PR (2008). Introgression of transgenes into a commercial cultivar confirms differential effects of HMW subunits 1Ax1 and 1Dx5 on gluten proteins. J Cereal Sci 48: 457–463.

Field JM, Shewry PR, Miflin BJ (1983). Solubilization and characterization of wheat gluten proteins: correlations between the amount of aggregated proteins and baking quality. J Sci Food Agric 34: 370–377.

Finney KF (1978). Genetically high protein hard winter wheat. The Bakers Digest June, 32–35.

Galande AA, Tiwari R, Ammiraju JSS, Santra DK, Lagu MD, Rao VS, Gupta VS, Misra BK, Nagarajan S, Ranjekar PK (2001). Genetic analysis of kernel hardness in bread wheat using PCR-based markers. Theor Appl Genet 103: 601–606.

Gebruers K, Dornez E, Boros D, Fraś A, Dynkowska W, Bedö Z, Rakszegi M, Delcour JA, Courtin CM (2008). Variation in the content of dietary fibre and components thereof in wheats in the HEALTHGRAIN diversity screen. J Agric Food Chem 56: 9740–9749.

Graybosch RA, Seabourn B, Chen YR, Blechl AE (2011). Quality and agronomic effects of three high-molecular-weight glutenin subunit transgenic events in winter wheat. Cereal Chem 88: 95–102.

Greenwell P, Schofield JD (1986). A starch granule protein associated with endosperm softness in wheat. Cereal Chem 63: 379–380.

Gupta RB, MacRitchie F (1994). Allelic variations of glutenin subunit and gliadin loci, *Glu-1, Glu-3* and *Gli-1* of common wheats. II. Biochemical basis of the allelic effects on dough properties. J Cereal Sci 19: 19–29.

Halford NG, Field JM, Blair H, Urwin P, Moore K, Robert L, Thompson R, Flavell RB, Tatham AS, Shewry PR (1992). Analysis of HMW glutenin subunits encoded by chromosome 1A of bread wheat (*Triticum aestivum* L.) indicates quantitative effects on grain quality. Theor Appl Genet 83: 373–378.

Haskå L, Nyman M, Andersson R (2008). Distribution and characterisation of fructan in wheat milling fractions. J Cereal Sci 48: 768–774.

Hemery Y, Lullien-Pellerin V, Rouau X, Abecassis J, Samson M-F, Åman P, von Reding W, Spoerndli C, Barron C (2009). Biochemical markers: efficient tools for the assessment of wheat grain tissue proportions in milling fractions. J Cereal Sci 49: 55–64.

Hogg AC, Sripo T, Beecher B, Martin JM, Giroux MJ (2004). Wheat puroindolines interact to form friabilin and control wheat grain hardness. Theor Appl Genet 108: 1089–1097.

Huebner FR, Wall JS (1976). Fractionation and quantitative differences of glutenin from wheat varieties varying in baking quality. Cereal Chem 53: 258–269.

Hung PV, Yamamori M, Morita N (2005). Formation of enzyme-resistant starch in bread as affected by high-amylose wheat flour substitutions. Cereal Chem 82: 690–694.

Jensen SA, Martens H (1983). The botanical constituents of wheat and wheat milling fractions. II. Quantification by amino acids. Cereal Chem 60: 170–177.

Jones HD, Sparks CA, Shewry PR (2009). Transgenic manipulation of wheat quality. In: Khan K, Shewry PR (eds). *Wheat: Chemistry and Technology*, ed 3, AACC, St Paul, Minnesota, pp 437–451.

Joppa LR, Du C, Hart GE, Hareland GA (1997). Mapping gene(s) for grain protein in tetraploid wheat (*Triticum turgidum* L.) using a population of recombinant inbred chromosome lines. Crop Sci 37: 1586–1589.

Kent NL (1966). Subaleurone endosperm cells of high protein content. Cereal Chem 43: 585–601.

Khan IA, Procunier JD, Humphreys DG, Tranquilli G, Schlatter AR, Marcucci-Poltri S, Frohberg R, Dubcovsky J (2000). Development of PCR-based markers for a high grain protein content gene from *Triticum turgidum* spp. *dicoccoides* transferred to bread wheat. Crop Sci 40: 518–524.

Khan K, Frohberg R, Olsen T, Huckle L (1989). Inheritance of gluten protein components of high protein hard red spring wheat lines derived from *Triticum turgidum* var. *dicoccoides*. Cereal Chem 166: 397–401.

Knudsen KEB (1997). Carbohydrate and lignin contents of plant materials used in animal feeding. Anim Feed Sci Tech 67: 319–338.

Krauss RM, Eckel RH, Howard B, Appel LJ, Daniels SR, Deckelbaum RJ, Erdman JW Jr, Kris-Etherton P, Goldberg IJ, Kotchen TA, Lichtenstein AH, Mitch WE, Mullis R, Robinson K, Wylie-Rosett J, St Jeor S, Suttie J, Tribble DL, Bazzarre TL (2000). AHA Dietary Guidelines: Revision 2000: A statement for Healthcare professionals from the Nutrition Committee of the American Heart Association. Stroke 31: 2751–2766.

Krishnamurthy K, Giroux MJ (2001). Expression of wheat puroindoline genes in transgenic rice enhances grain softness. Nat Biotech 19: 162–166.

Lawrence GJ, Moss HJ, Shepherd KW, Wrigley CW (1987). Dough quality of biotypes of eleven Australian wheat cultivars that differ in high-molecular-weight glutenin subunit composition. J Cereal Sci 6: 99–101.

Lawther JM, Sun R, Banks WB (1995). Extraction, fractionation, and characterization of structural polysaccharides from wheat straw. J Agric Food Chem 43: 667–675.

Lazaridou A, Biliaderis CG (2007). Molecular aspects of cereal β-glucan functionality: physical properties, technological applications and physiological effects. J Cereal Sci 46: 101–118.

León E, Aouni R, Piston F, Rodriguez-Quijano M, Shewry PR, Martin A, Barro F (2010). Stacking HMW-GS transgenes in bread wheat: combining subunit 1Dy10 gives improved mixing properties and dough functionality. J Cereal Sci 51: 13–20.

León E, Marin S, Gimenez MJ, Piston F, Rodriguez-Quijano M, Shewry PR, Barro F (2009). Mixing properties and dough functionality of transgenic lines of a commercial wheat cultivar expressing the 1Ax1, 1Dx5 and 1Dy10 HMW glutenin subunit genes. J Cereal Sci 49: 148–156.

Levy AA, Galili G, Feldman M (1988). Polymorphism and genetic control of high molecular weight glutenin subunits in wild tetraploid wheats *Triticum turgidum* var. dicoccoides. Heredity 61: 63–72.

Liu W, Cui SW, Kakuda Y (2006). Extraction, fractionation, structural and physical characterization of wheat β-D-glucans. Carbohyd Polym 63: 408–416.

Loosveld A, Maes C, van Casteren WHM, Schols HA, Grobet PJ, Delcour JA (1998). Structural variation and levels of water-extractable arabinogalactan-peptide in European wheat flours. Cereal Chem 75: 815–819.

Lunn J, Buttriss JL (2007). Carbohydrates and dietary fibre. Nutr Bull 32: 21–64.

Mares DJ, Stone BA (1973). Studies on wheat endosperm. I. Chemical composition and ultrastructure of the cell walls. Aust J Biol Sci 26: 793–812.

Margiotta AB, Urbano M, Colaprico G, Turchetta T, Lafiandra D (1998). Variation of high molecular weight glutenin subunits in tetraploid wheats of genomic formula. In: Slinkard AE (ed). *Proceedings 9th International Wheat Genetic Symposium,* Vol 4. AAGG, Saskatoon, Canada.

Martin J, Meyer F, Smidansky E, Wanjugi H, Blechl A, Giroux MJ (2008). Complementation of the *Pina* (null) allele with the wild type *Pina* sequence restores a soft phenotype in transgenic wheat. Theor Appl Genet 113: 1563–1570.

Mellen BP, Walsh TF, Herrington DM (2008). Whole grain intake and cardiovascular disease: a meta analysis. Nutr Metab Cardiovasc 18: 283–290.

Mesfin A, Frohberg RC, Khan K, Olson TC (2000). Increased grain protein content and its association with agronomic and end-use quality in two hard red spring wheat populations derived from *Triticum turgidum* L. var. *dicoccoides*. Euphytica 116: 237–242.

Mitchell RAC, Dupree P, Shewry PR (2007). A novel bioinformatics approach identifies candidate genes for the synthesis and feruloylation of arabinoxylan. Bioinformatics 144: 43–53.

Moonen JHE, Scheepstra A, Graveland A (1982). Use of SDS-sedimentation test and SDS-polyacrylamide gel electrophoresis for screening breeders' samples of wheat for breadmaking quality. Euphytica 31: 677–690.

Morris CF (2002). Puroindolines: the molecular genetic basis of wheat grain hardness. Plant Mol Biol 48: 633–647.

Morris CF, Bhave M (2008). Reconciliation of D-genome puroindoline allele designations with current DNA sequence data. J Cereal Sci 48: 277–287.

Nemeth C, Freeman J, Jones HD, Sparks C, Pellny TK, Wilkinson MD, Dunwell J, Andersson AAM, Åman P, Guillon F, Saulnier L, Mitchell RAC, Shewry PR (2010). Down-regulation of the *CSLF6* gene results in decreased (1,3;1,4)-β-D glucan in endosperm of wheat. Plant Physiol 152: 1209–1218.

Nugent AP (2004). The metabolic syndrome. Nutr Bull 29: 36–43.

Payne PI (1987). Genetics of wheat storage proteins and the effect of allelic variation on bread-making quality. Ann Rev Plant Physiol 38: 141–153.

Payne PI, Holt LM, Worland AJ, Law CN (1981). Structural and genetical studies on the high-molecular-weight subunits of wheat glutenin. Part III. Telocentric mapping of the subunit genes on the long arms of the homologous group 1 chromosomes. Theor Appl Genet 63: 129–138.

Payne PI, Nightingale MA, Krattiger AF, Holt LM (1987). The relationship between HMW glutenin subunit composition and the breadmaking quality of British grown wheat varieties. J Sci Food Agric 40: 51–65.

Pellny TK, Lovegrove A, Freeman J, Tosi P, Love CG, Knox JP, Shewry PR, Mitchell RAC (2012). Cell walls of developing wheat (*Triticum aestivum* L.) endosperm: comparison of composition and RNA Seq transcriptome. Plant Physiol 158: 612–627.

Pettersson D, Åman P (1989). Enzyme supplementation of a poultry diet containing rye and wheat. Br J Nutr 62: 139–149.

Piironen V, Lampi A-M, Ekholm P, Salmenkallio-Marttila M, Liukkonen K-H (2009). Micronutrients and phytochemicals in wheat grain. In: Shewry PR, Khan K (eds). *Wheat: Chemistry and Technology*. AACC, St Paul, Minnesota, pp 179–222.

Piston F, Uauy C, Fu L, Langston J, Labavitch J, Dubcovsky J (2010). Down-regulation of four putative arabinoxylan fruloyl transferase genes from family PF02458 reduces ester-linked ferulate content in rice cell walls. Planta 231: 677–691.

Pomeranz Y, Williams PC (1990). Wheat hardness: its genetic, structural and biochemical background, measurement and significance. In: Pomeranz Y (ed). *Advances in Cereal Science and Technology*, Vol 10. AACC, St Paul, MN, pp 471–544.

Popineau Y, Deshayes G, Lefebvre J, Fido R, Tatham AS, Shewry PR (2001). Prolamin aggregation, gluten viscoelasticity, and mixing properties of transgenic wheat lines expressing 1Ax and 1Dx high molecular weight glutenin subunit transgenes. J Agric Food Chem 49: 395–401.

Rakszegi M, Békés F, Lang L, Tamas L, Shewry PR, Bedo Z (2005). Technological quality of transgenic wheat expressing an increased amount of a HMW glutenin subunit. J Cereal Sci 42: 15–23.

Rakszegi M, Pastori G, Jones HD, Békés F, Butow B, Lang L, Bedo Z, Shewry PR (2008). Technological quality of field grown transgenic lines of commercial wheat cultivars expressing the 1Ax1 HMW glutenin subunit gene. J Cereal Sci 47: 310–321.

Regina A, Bird A, Topping D, Bowden S, Freeman J, Barsby T, Kosar-Hashemi B, Li Z, Rahman S, Morell M (2006). High-amylose wheat generated by RNA interference improves indices of large-bowel health in rats. Proc Natl Acad Sci U S A 103: 3546–3551.

Rhodes DI, Stone BA (2002). Proteins in walls of wheat aleurone cells. J Cereal Sci 36: 83–101.

Rooke L, Békés F, Fido R, Barro F, Gras P, Tatham AS, Barcelo P, Lazzeri P, Shewry PR (1999). Overexpression of a gluten protein in transgenic wheat results in greatly increased dough strength. J Cereal Sci 30: 115–120.

Saulnier L, Sado P-E, Branlard G, Charmet G, Guillon F (2007). Wheat arabinoxylans: exploiting variation in amount and composition to develop enhanced varieties. J Cereal Sci 46: 261–281.

Seilmeier W, Belitz H-D, Wieser H (1991). Separation and quantitative determination of high-molecular-weight subunits of glutenin from different wheat varieties and genetic variants of the variety Sicco. Zeitschrift FuÈr Lebensmittel-Untersuchung Und-Forschung 192: 124–129.

Selvendran RR, Ring SG, O'Neill MA, Du Pont MS (1980). Composition of cell wall material from wheat bran used in clinical feeding trails. Chem Ind (London) 22: 885–888.

Sestili F, Botticella E, Bedo Z, Phillips A, Lafiandra D (2010). Production of novel allelic variation for genes involved in starch biosynthesis through mutagenensis. Mol Breeding 25: 145–154.

Shewry PR (2007). Improving the protein content and composition of cereal grain. J Cereal Sci 46: 239–250.

Shewry PR, Freeman J, Wilkinson M, Pellny T, Mitchell RAC (2010). Challenges and opportunities for using wheat for biofuel production. In: Halford NG, Karp A (eds). *Energy Crops*. RSC Publishing, Cambridge, UK, pp 13–26.

Shewry PR, Halford NG, Tatham AS, Popineau Y, Lafiandra D, Belton P (2003). The high molecular weight subunits of wheat glutenin and their role in determining wheat processing properties. Adv Food Nutr Res 45: 221–302.

Shewry PR, Powers S, Field JM, Fido RJ, Jones HD, Arnold GM, West J, Lazzeri PA, Barcelo P, Barro F, Tatham AS, Békés, F, Butow B, Darlington H (2006). Comparative field performance over three years and two sites of transgenic wheat lines expressing HMW subunit transgenes. Theor Appl Genet 113: 128–136.

Slade AJ, Fuerstenberg SI, Loeffler D, Steine MN, Facciotti D (2005). A reverse genetic, nontransgenic approach to wheat crop improvement by TILLING. Nat Biotechnol 23: 75–81.

Slavin J (2004). Dietary fiber and the prevention of cancer. In: Remacle C, Reusens B (eds). *Functional Foods, Ageing and Degenerative Disease*. Woodhead Publishing Ltd, Cambridge, UK, pp 628–644.

Snape JW, Hyne V, Aitken K (1993). Targeting genes in wheat using marker mediated approaches. In: Li ZS, Xin ZY (eds). Proceedings of Eighth International Wheat Genetics Symposium, Beijing, China, pp 749–759.

Sourdille P, Perretant MR, Charmet G, Leroy P, Gautier MF, Joudrier P, Nelson JC, Sorrells ME, Bernard M (1996). Linkage between RFLP markers and genes affecting kernel hardness in wheat. Theor Appl Genet 93: 580–586.

Steer T, Thane C, Stephen A, Jebb S (2008). Bread in the diet: consumption and contribution to nutrient intakes of British adults. Proc Nutr Soc 67: E363.

Stone B, Morell MK (2009). Carbohydrates. In: Shewry PR, Khan K (eds). *Wheat: Chemistry and Technology*. AACC, St Paul, MN, pp 299–362.

Sun X-F, Sun RC, Flower P, Baird MS (2005). Extraction and characterization of original lignin and hemicelluloses from wheat straw. J Agric Food Chem 53: 860–870.

Surget A, Barron C (2005). Histologie du grain de blé. Indust Céréales 145: 3–7.

Toole GA, Le Gall G, Colquhoun IJ, Nemeth C, Saulnier L, Lovegrove A, Pellny T, Freeman J, Mitchell RAC, Mills ENC, Shewry PR (2010). Compositional and spatial changes in cell wall composition in developing grains of wheat cv. Hereward. Planta 232: 677–689.

Topping D (2007). Cereal complex carbohydrates and their contribution to human health. J Cereal Sci 46: 220–229.

Turner AS, Bradburne RP, Fish L, Snape JW (2004). New quantitative trait loci influencing grain texture and protein content in bread wheat. J Cereal Sci 40: 51–60.

Uauy C, Distelfeld A, Fahima T, Blechl A, Dubcovsky J (2006). A NAC gene regulating senescence improves grain protein, zinc and iron content in wheat. Science 314: 1298–1301.

Vasil IK, Bean S, Zhao J, McCluskey P, Lookhart G, Zhao H-P, Altpeter F, Vasil V (2001). Evaluation of baking properties and gluten protein composition of field grown transgenic wheat lines expressing high molecular weight glutenin gene 1Ax1. J Plant Physiol 158: 521–528.

Vitaglione P, Napolitano A, Fogliano V (2008). Cereal dietary fibre: a natural functional ingredient to deliver phenolic compounds into the gut. Trends Food Sci Tech 19: 451–463.

Waines JG, Payne PI (1987). Electrophoretic analysis of the high-molecular-weight glutenin subunits of *Triticum monococcum, T. urartu* and the A genome of bread wheat (*T. aestivum*). Theor Appl Genet 74: 71–76.

Weightman RM, Millar S, Alava J, Foulkes MJ, Fish L, Snape JW (2008). Effects of drought and the presence of the 1BL/1RS translocation on grain vitreosity, hardening and protein content in winter wheat. J Cereal Sci 47: 457–468.

Wieser H, Bushuk W, MacRitchie F (2006). The polymeric glutenins. In: Wrigley C, Bekes F, Bushuk W (eds). *Gliadin and Glutenin. The Unique Balance of Wheat Quality*. AACC International, St Paul, Minnesota, pp 213–240.

Wilkinson M, Wan Y, Tosi P, Leverington M, Snape J, Mitchell RAC, Shewry PR (2008). Identification and genetic mapping of variant forms of puroindoline b expressed in developing wheat grain. J Cereal Sci 48: 722–728.

Wood J (2007). Cereal β-glucans in diet and health. J Cereal Sci 46: 230–238.

Worland T, Snape JW (2001). Genetic basis of worldwide wheat varietal improvement. In: Bonjean AP, Angus WJ (eds). *The World Wheat Book—A History of Wheat Breeding*. Groupe Limagrain, Paris, France, pp 59–100.

Yamamori M, Endo TR (1996). Variation of starch granule proteins and chromosome mapping of their coding genes in common wheat. Theor Appl Genet 93: 275–281.

Yamamori M, Fujita S, Hayakawa K, Matsuki J, Yasui T (2000). Genetic elimination of a starch granule protein, SGP-1, of wheat generates an altered starch with apparent high amylose. Theor Appl Genet 101: 21–29.

10 Legume Seed Genomics: How to Respond to the Challenges and Potential of a Key Plant Family?

Mélanie Noguero, Karine Gallardo, Jérôme Verdier, Christine Le Signor, Judith Burstin, and Richard Thompson

Introduction

What Is Special about Legume Seeds?

The Fabales represent the third largest flowering plant family, with members ranging from tiny forage species to impressive rainforest trees. In addition to their biodiversity, legumes are of particular importance because of their ability to fix atmospheric nitrogen by a symbiotic relationship with rhizobacteria. Many legumes also benefit from mycorrhizal symbioses that supply phosphate and other mineral ions and help regulate water homeostasis. Legume seeds have always been an important source of protein in the human diet, and current agricultural practice also relies on legumes as a source of protein for animal feed. For example, peas are used for human consumption and for animal feed, particularly for poultry and to a lesser extent pigs. Compared with soybean, peas contain less protein (22%–25% as opposed to 35%). In addition to their widespread use as a source of fixed nitrogen in crop rotations and pastures and as a source of protein, cultivated legumes have a wide range of other uses, the most important of which are as renewable fuel and biomaterial sources and for combating desertification.

The use of legume crops as a source of nitrogen is crucial for sustainable agriculture, synthetic N fertilizer production being an energy-intensive process. Despite their potential, the area sown to legumes is small in many countries and the breeding effort modest compared with cereals or other major crops. Many legume seed crops contain antinutritional components, which can compromise their utility in feed and food, and removal of these components constitutes an important breeding objective (Jezierny *et al.*, 2010). The efforts made in improving seed protein content, bioavailability, and nutritional quality of legume seeds have been summarized by Burstin *et al.* (2011). Genomics can contribute to the breeding effort by providing tools for the identification of genes encoding key traits such as seed protein content and quality and by generating markers applicable for selection in many legume crops.

Oilseed and Starch Pulses

Legumes employ different and species-specific strategies for carbon storage, some storing principally lipids, such as soybean, and others storing mainly starch, such as pea, chickpea, and faba

Seed Genomics, First Edition. Edited by Philip W. Becraft.

bean; a third group stores carbon skeletons mainly as protein (e.g., *Medicago truncatula*). Grains of noncrop legumes generally also contain appreciable quantities of carbon in the form of cell wall carbohydrates. Legume seed feedstocks are important energy as well as protein sources, and legume breeding contrives to maximize seed calorific content in addition to its protein content. Because peas accumulate starch and not lipids, they have a lower calorific value than soybeans. In their favor, they do not require irrigation in Europe, and they have a measure of cold tolerance compared with soybean. Breeders have long known that in soybean the two major reserve substances, protein and triglyceride, are metabolically linked, and their level can be selected as a trait. A shortfall of accumulation of a major reserve substance may limit the availability of critical nutrients for the postgermination seedling. Suppression of seed storage protein production results in compensating nonstorage protein accumulation, shown by mutation-induced suppression or genetic modification of storage protein synthesis in soybean (Kinney *et al.*, 2001; Takahashi *et al.*, 2003), all resulting in rebalancing protein content by increased accumulation of other seed proteins. Seed protein composition is also strongly influenced by nutrient availability, as shown by the ability of legume seeds to compensate for low levels in the relatively poor 11S globulins under sulfur limiting conditions by increasing 7S globulins as a way to maintain seed protein content (Sexton *et al.*, 1998). The accumulation of different seed storage components (protein, lipid, and starch) is inversely linked in that increased protein content is accompanied by reduced lipid deposition and vice versa (Schmidt and Herman, 2008). Lipid and protein reserves are kept within narrow limits, presumably corresponding to the quantities required for germination.

Development of Genomics Tools

Genome Sequences and Phylogeny

Legume seed genomics is anchored in the pioneering studies on two model species, *M. truncatula* and *Lotus japonicus*. These two forage legumes were chosen primarily for their small genome sizes (*M. truncatula* 375 Mb (Young *et al.*, 2011), *L. japonicus* 470 Mb (Sato *et al.*, 2008)) and for their ease of genetic transformation (Barker *et al.*, 1990; Handberg and Stougaard, 1992). In the first instance, they were chosen to study the rhizobial symbioses they undergo and subsequently generally adopted as legume models.

A consortium was launched in 2003 to sequence the *M. truncatula* genome (http://www.medicagohapmap.org/?genome), and the Lotus genome sequencing project was started in parallel (http://www.kazusa.or.jp/lotus/index.html). Sato *et al.* (2008) reported 315.1Mb of sequences corresponding to 67% of the genome and covering ~90% of the gene space, and an update on this sequencing effort was presented subsequently by Sato *et al.* (2010). The Medicago genome sequence (Young *et al.*, 2011) reveals the presence of ~60,000 genes for both species at a density of 12.6/100 kb for *M. truncatula* compared with 17.4/100 kb for *L. japonicus*, the difference being due to the higher proportion of mainly retroposon-type repeated sequences in the *M. truncatula* gene-rich regions. The *M. truncatula* genome also contains a higher proportion of local gene duplication than the other sequenced legumes. The Medicago and Lotus genomic sequencing efforts both have followed the classical approach of assembling a BAC contig and sequencing BAC-by-BAC, with, for Medicago, alignment via optical mapping. More recently, massively parallel sequencing has been adopted to finish these sequences and to sequence other legumes including crop species. The first published genomic legume sequence was that of soybean (Schmutz *et al.*, 2010), and the *M. truncatula* sequence followed (Young *et al.*, 2011), with numerous other species genomes in the pipeline (Table 10.1), the first to appear being that of the pigeonpea (Singh *et al.*, 2012). These sequences will fuel the generation of a new series of genomic resources, replacing those currently used based on unassembled genomic sequences and EST sequencing.

Sequence comparisons between *M. truncatula* and *L. japonicus* show the presence of extensive synteny, frequently extending over an entire chromosome arm. In contrast, a few regions show very little syntenic relationship, owing to extensive fine-scale rearrangements. The comparison also revealed the existence of an ancient whole-genome duplication (WGD) ~58 Myr. ago, which predated the Medicago-Lotus divergence (or predated speciation within the legume clade) and which is reflected in a corresponding second, lower level of synteny arising from matches between the paralogous segments between the two species (Cannon *et al.*, 2006). Both *M. truncatula* and *L. japonicus* belong to the Papilionid clade of the Fabaceae; the Medicago genus is in the Galegoid branch, including pea, clover, chickpea, pigeonpea, lentil, and broad bean, whereas Lotus is more closely related to the tropical legumes soybean and Phaseolus. The extensive synteny observed within these two groups (Hougaard *et al.*, 2008; Young and Udvardi, 2009; Young *et al.*, 2011) enables the transfer of knowledge from model species to crops. This process has been facilitated by the development of sets of gene-based anchor markers (Aubert *et al.*, 2006; Hougaard *et al.*, 2008) that are used to define syntenic "mileposts" when comparing related genetic maps. The phylogenetic similarity between *M. truncatula* and pea is reflected in close sequence similarity at the DNA level and in a high degree of synteny, which facilitates transfer of markers from the model to the crop species. In soybean, physical maps for *Glycine max* and its undomesticated ancestor, *Glycine soja* were built that will serve as a framework for ordering sequence fragments, comparative genomics, cloning genes, and evolutionary analyses of legume genomes (Ha *et al.*, 2012).

Legume Seed sRNA Databases

Although many miRNA sequencing projects are in progress, there are currently only two reports in the literature specifically on miRNAs expressed in legume seeds, on soybean (Song *et al.*, 2011) and *Phaseolus vulgaris* (Pelaez *et al.*, 2012). Because this situation is likely to change rapidly, we also provide references to plant miRNA prediction algorithms and databases that will be updated and that include seed-expressed miRNAs (Mhuantong *et al.*, 2009; Zhang *et al.*, 2010; Meng *et al.*, 2011; Bielewicz *et al.*, 2012). Legume-specific miRNA families have been identified in *M. truncatula* and *Phaseolus vulgaris* (Arenas-Huertero *et al.*, 2009; Jagadeeswaran *et al.*, 2009) but are not yet functionally characterized.

Reverse Genetics Resources Available for Model and Nonmodel Legumes

Numerous reverse genetics resources have been developed for legumes (Tadege *et al.*, 2009; Young and Udvardi, 2009) and the availability of genome sequences will accelerate the development of these tools in the near future. Two main complementary approaches were used . One approach produced point mutations throughout the genome that can be screened by Targeting Induced Local Lesions in Genomes (TILLING) in various protein domains of candidate genes. TILLING offers the opportunity to study *in vivo* protein function, identify enzyme active sites, and design novel alleles for quality improvement. The second approach created insertion or deletion mutations, which proved useful to understand the function of candidate genes when nonlethal phenotypes are obtained. TILLING resources have been created for many model and crop legumes, including pea (Dalmais

Table 10.1 Overview of Genomics Resources Available for Legume Species as of June 2012

Name	Synonym	Ploidy	Chromosome Number	Genome Size (Mbp)	ESTs (result of NCBI search May 2012)	Markers Resources	References for Marker Resources	Transcriptome Resources	Proteome Resources	Genome Resources
Acacia sp.		diploid	n = 13		11,569	Acacia mangium Willd: RFPL (219) SRR (33) Acacia senegal (L.) Willd: RAPD (10) ISSRs (5) SSRs (11)	Assoumane *et al.*, 2009 Butcher *et al.*, 2000 Josiah *et al.*, 2008			http://www.ukm.my/acacia/index.html
Arachis hypogaea (AB genome)	peanut, groundnut, earth nut, goober pea, mani, monkey nut, runner peanut, Spanish peanut, Valencia peanut, Virginia peanut	tetraploid	n = 10	2,813	178,490	SSRs (7482)	Pandey *et al.*, 2012	Zhang *et al.* (2012)	Guo *et al.* (2008), Kottapalli *et al.* (2008), Schmidt *et al.* (2009)	
Arachis duranensis (A-genome)						SSRs (1329) AFLP (148) SNPs (1144) RFLPs (121) AFLPs (148)	Pandey *et al.*, 2012			
Arachis ipaënsis (B-genome)					SSRs (598)	Pandey *et al.*, 2012				
Cajanus cajan	pigeon pea, kadios	diploid	n = 11	858	25,314	SSRs (46)	Odeny *et al.*, 2007 Saxena *et al.*, 2010	Dubey *et al.* (2011)		Singh *et al.* (2012)
Cicer arietinum	chickpea, Bengal gram, calvance pea, chestnut bean, chich, chich-pea, dwarf pea, garavance, garbanza, garbanzo bean, garbanzos, gram, gram pea, homes, hamaz, nohub, lablabi, shimbra, yellow gram	diploid	n = 8	740	44,157	RIL: C. arietinum (ICC 4958) × C. reticulatum (PI 489777): SSRs (311) SNPs (220) EST-SSRs (51) CISRs (87) CAPS (143)	Nayak *et al.*, 2010 Gujaria *et al.*, 2011	Garg *et al.* (2011)	Pandey *et al.* (2006, 2008), Bhushan *et al.* (2006)	
Cyamopsis tetragonoloba	guar bean, cluster bean, gawaar, gwaar ki phalli									
Dolichos lablab	hyacinth bean, bonavist, bataw, lablab									

[illegible]	[illegible] soyabean, edamame	[illegible]	[illegible]	[illegible]	[illegible]	[illegible] Williams 82: SNPs (4991) SSRs (874)Glycine max (L.) Merr. : SNPs (2928) SSRs (1014) RFLP (709)	[illegible] 2010 Choi *et al.*, 2007	[illegible] (2004), Choi *et al.* (2007), Libault *et al.* (2010), Mochida *et al.* (2010), Severin *et al.* (2010), Jones *et al.* (2010) Asakura *et al.* (2012), Sha *et al.* (2012)	[illegible] (2005), Agrawal *et al.* (2008), Hajduch *et al.* (2011)	[illegible] .net/soybean http://rsoy.psc.riken.jp/ http://soybase.org/ http://harvest.ucr.edu/
Lens culinaris	lentil, black lentil, brown lentil, green lentil, green mungbean, large-seeded lentil, red mungbean, small-seeded lentil, wild lentil, yellow lentil, adas, mercimek, messer, masser, heramame	diploid	n = 7	4,063	9,513	SSRs (46)	Hamwiech *et al.*, 2009	Kaur *et al.* (2011)	Scippa *et al.* (2010)	
Lotus japonicus		diploid	n = 6	472	242,432	SRRs (788) dCAPS (80)	Sato *et al.*, 2008		Dam *et al.* (2009), Nautrup-Pedersen *et al.* (2010)	http://www.kazusa.or.jp/lotus/ http://www.shigen.nig.ac.jp/bean/lotusjaponicus/
Lupinus angustifolius	Blue lupin	diploid	n = 20	740		SNP (8207)		Foley *et al.* (2011)		
Lupinus albus	sweet lupin	diploid	n = 25	455					Magni *et al.* (2007)	
Medicago sativa	alfalfa	tetraploid	n = 8		12,371					
Medicago truncatula		diploid	n = 8	~ 500	269,238	SNPs (3000000) SSRs (317)	Branca *et al.*, 2011 Mun *et al.*, 2006 Young *et al.*, 2009	Benedito *et al.* (2008), Kakar *et al.* (2008), Verdier *et al.* (2008) He *et al.* (2009)	Gallardo *et al.* (2007), Colditz and Braun (2010)	http://www.medicago.org/ http://www.medicagohapmap.org/index.php
Phaseolus acutifolius	Tepary bean, tepari bean									
Phaseolus lunatus	Lima bean, butter bean, patani									
Phaseolus vulgaris	commun bean, common field bean, kidney bean, navy, habichuela, snap bean	diploid	n = 11		123,988	EST-SSRs (302)	Garcia *et al.*, 2011	Yin *et al.* (2011)	De la Fuente *et al.* (2011)	

(continued)

Table 10.1 (*Continued*)

Name	Synonym	Ploidy	Chromosome Number	Genome Size (Mbp)	ESTs (result of NCBI search May 2012)	Markers Resources	References for Marker Resources	Transcriptome Resources	Proteome Resources	Genome Resources
Pisum sativum	pea, dry pea, Chinese pea, Chinese pea pod, Chinese snow pea, edible-podded pea, edible pod pea, podded pea, snap pea, snow pea, sugar snap pea, batani, chicharo, erbese, ater, pois, takamany borso, pisello, holoan, mange-tout, papdi	diploid	n = 7	4,778	18,576	SSRs (434) EST-SSRs (171)	Loridon *et al.*, 2005 Mishra *et al.*, 2012	Franssen *et al.* (2011), Kaur *et al.* (2012)	Bourgeois *et al.* (2011)	Bordat *et al.* (2011)
Trifolium pratense	red clover	diploid	n = 7	440	38,109	SSRs (1414) AFLP (181) RFP (204)	Isobe *et al.*, 2009			
Trifolium repens	white clover	allopoliploid	n = 8	956	46	SSRs (415) AFLP (283)	Khanlou *et al.*, 2011 Zhang *et al.*, 2007			
Vicia faba	broad bean, horse bean, fava bean, bell bean, field bean, tic bean	diploid	n = 6	~ 13000	5,415	EST-SSRs (11) ITAP (135) RADP (238) SSRs (191) AFLP (36)	Diaz-Ruiz *et al.*, 2010 Ellwood *et al.*, 2008 Gong *et al.* 2011	Kaur *et al.* (2012)		
Vicia sativa	vetch, commun vetch									
Vigna angularis	*Phaseolus angularis*, Adzuki bean, azuki bean, Adanka bean, danka bean	diploid	n = 11	540	25	SSRs (191 AFLP (36)	Kaga *et al.*, 2008			
Vigna unguiculata	*Vigna sinensis*, cowpea, asparagus bean, black eyed pea, black eyed bean, crowder pea, field pea, southern pea, frijole, lobhia, kibal, nieve, paayap	diploid	n = 11	620	187,487	SNPs (298) SSRs (102)	Gupta *et al.*, 2010 Timko *et al.*, 2008	Diouf *et al.* (2011)	Nogueira *et al.* (2007)	
Vigna radiata	*Phaseolus aureus*, mung bean, black dahl, black gram, black mung, golden gram, gram bean, green gram, mungo, red mung bean, urd, chop suey bean http://www.cropsreview .com/grain-legumes .html	diploid	n = 11	580	829	RIL: Berken × ACC41 SSRs (97) RFLPs (76) RAPDs (4) STSs (2)	Zhao *et al.*, 2010	Moe *et al.* (2011)		

et al., 2008), soybean (Cooper *et al.*, 2008), *M. truncatula* (Le Signor *et al.*, 2009), *L. japonicus* (Perry *et al.*, 2003), Cicer (Tadege *et al.*, 2009; Tapan *et al.*, 2012), Phaseolus (Porch *et al.*, 2009), peanut (Knoll *et al.*, 2011), and pigeonpea (Varshney *et al.*, 2010).

De-TILLING (Rogers *et al.*, 2009) is an adaptation of the TILLING procedure that was used to screen deletions in *M. truncatula* created by fast neutron mutagenesis. A population of 156,000 *M. truncatula* plants was structured as 13 towers each representing 12,000 M2 plants. Three-dimensional pooling allows efficient location of mutants from within the towers. Mutants were detected from this population at a rate of 29% using five targets per gene.

Insertion mutant resources have been created for *M. truncatula* by incorporating the retrotransposon Tnt1 from tobacco (d'Erfurth *et al.*, 2003). These resources have been developed by the Noble foundation, which provides a reverse genetic platform permitting screening of 12,000 lines, a total of 300,000 insertions, either by an in silico screen of sequenced Tnt1 insertion borders or by PCR on DNA bulks from the collection (Tadege *et al.*, 2008; http://bioinfo4.noble.org/mutant/). Endogenous retrotransposons have been mobilized in *M. truncatula* (Rakocevic *et al.*, 2009) and in *L. japonicus* (Fukai *et al.*, 2012) and form the basis of further insertion mutant collections. In contrast, this approach, to our knowledge, has not been successfully applied to major crop legumes for which no insertion mutant collections exist.

Virus-induced gene silencing (VIGS) is another strategy that has been developed for functional genomics in model and crop legumes by at least three laboratories (Constantin *et al.*, 2004; Igarashi *et al.*, 2009; Varallyay *et al.*, 2010). The most promising approach uses agro-infection to furnish virus during inoculation. These methods involve transient transformation, generally of an RNAi construct, and are particularly suited to the rapid detection of qualitative changes in plant morphology. Their use for specifically studying legume seed phenotypes has not been reported so far. Nishizawa *et al.* (2010) took an interesting alternative route by transforming somatic embryos with a vector containing an antisense construction for beta-conglycinin and observed reduced conglycinin expression in transformed embryos.

Applications of Genomics Tools to Legume Seed Biology

Characterizing the Proteome and Transcriptome of Developing Seeds

Large-scale proteomics and transcriptomics analyses of seed gene expression have been carried out on several legume species, including both models and crops. Lei *et al.* (2011) created a useful legume-specific protein database (LegProt, http://bioinfo.noble.org/manuscript-support/legumedb/) that regroups much of the proteomics information obtained, whereas Li *et al.* (2012) set up Legume IP, a platform for transcriptomics data from the model legume species. In this chapter, we have selected studies from some of the best-researched legumes.

In soybean, a high-throughput proteomic approach using two-dimensional gel electrophoresis (2DGE) and matrix assisted laser desorption ionization (MALDI-TOF) was employed to determine the expression profile and identity of 679 proteins during seed filling (Hajduch *et al.*, 2005). There were 14 major functional categories. Proteins involved in metabolism, protein destination and storage, metabolite transport, and disease and defense were the most abundant. For each functional category, a composite expression profile was obtained. An overall decrease in metabolism-related proteins versus an increase in proteins associated with destination and storage was observed during seed filling. A user-intuitive database (http://oilseedproteomics.missouri.edu) was developed to access these data for soybean and other oilseeds (currently oilseed rape, *Arabidopsis,* castor bean). More recently, this group combined 2DGE and semicontinuous multidimensional protein identification technology (Sec-MudPIT) coupled with liquid chromatography–mass spectrometry, which decreased the redundancy of the sequence data obtained (Agrawal *et al.*, 2008). Comparison of the dataset with a parallel study of rapeseed (*Brassica napus*) (Hajduch *et al.*, 2011) uncovered an increased expression of glycolytic and fatty acid biosynthetic proteins in rapeseed compared with soybean, which the authors suggest could be responsible for higher oil in rapeseed owing to an increased commitment of hexoses to glycolysis and eventually to *de novo* fatty acid synthesis pathways.

More recently, a 27K soybean microarray resource developed by Vodkin *et al.* (2004), which included immature cotyledon ESTs, was used for monitoring seed development by Jones *et al.* (2010). These authors first noted the high abundance of transcription factor transcripts in the mature seed, which are thought to be stored in the dry seed in preparation for germination. An Affymetrix genechip soybean genome array (http://www.affymetrix.com) bearing 37,500 transcript sequences, plus 15,800 transcripts for *Phytophthora sojae* and 7500 for *Heterodera glycines* (cyst nematode pathogen) transcripts was also made available and was used by Asakura *et al.* (2012) to profile soybean transcription throughout seed development. High-throughput sequencing was used to analyze normalized cDNA libraries from a series of stages of soybean seed development (Sha *et al.*, 2012). Soybean transcriptome data are accessible at SoyXpress (Cheng and Stromvik, 2008) and at LegumeIP, a site that integrates information from several model legumes (Li *et al.*, 2012). Detailed seed transcriptome information for soybean and *Phaseolus coccineus* is also available at http://seedgenenetwork.net/ site. This project has exploited laser capture microdissection (LCM) to obtain RNA samples from many component tissues of the seed that have been converted to cDNA and subjected to high-throughput sequencing (Le *et al.*, 2007). LCM allows a much finer assignment of gene expression profile to cell type than is possible with hand-dissection. Le *et al.* (2007) identified many region-specific transcripts and could show that coregulation was higher for genes expressed from the same region. They also undertook the first detailed study of regulation of suspensor gene expression.

Dam *et al.* (2009) characterized the development of seeds in the model legume *Lotus japonicus*. Protein, oil, starch, phytic acid, and ash contents were determined, which indicated that the composition of mature Lotus seed is more similar to soybean than to pea, consistent with its phylogenetic proximity to the Phaseoleae tribe rather than the Galegoid tribe. Two-dimensional polyacrylamide gel electrophoresis and gel-based liquid chromatography–mass spectrometry were used to identify proteins. Five legumins, LLP1 to LLP5, and two convicilins, LCP1 and LCP2, were identified by MALDI-TOF mass spectrometry. For two distinct developmental phases, seed filling and desiccation, 665 and 181 unique proteins corresponding to gene accession numbers were identified for the two phases. All of the proteome data are available at http://www.cbs.dtu.dk/cgi-bin/lotus/db.cgi (Lotus Experiment Database). A second Web interface (http://bioinfoserver.rsbs.anu.edu.au/utils/PathExpress4legumes/) was set up to collect and collate all protein identifications for Lotus, Medicago, and soybean seed proteomes.

The same group extended their analysis and included a comparative analysis of the pod proteome, identifying 604 pod proteins and 965 seed proteins, including 263 proteins distinguishing the pod (Nautrup-Pedersen *et al.*, 2010). The complete dataset is publicly available at the Lotus Experiment Database website, where spots in a reference map are linked to experimental data, such as matched peptides, quantification values, and gene accessions. Identified pod proteins represented enzymes from 85 different metabolic pathways, including storage globulins and a late embryogenesis abundant protein. In contrast to seed maturation, pod maturation was associated with decreasing total protein content, especially proteins involved in protein biosynthesis and photosynthesis. Proteins detected only in pods included three enzymes participating in the urea cycle and four in nitrogen and amino group metabolism, highlighting the importance of nitrogen metabolism during pod development. Additionally, five legume seed proteins previously unassigned in the glutamate metabolism pathway were identified, consistent with the important role of the pod in nutrient, and particularly N, relocation, and in contrast to the N-deposition occurring in developing seeds. A Lotus gene expression atlas (LjGEA) has been set up, similar to that available for Medicago and including 79 tissues and five stages of seed development (J. Verdier, personal communication).

To investigate processes leading to desiccation tolerance in seeds, 16k-microarrays were used to monitor changes in the transcriptome of desiccation-sensitive radicles of *Medicago truncatula* seeds during incubation in a polyethylene glycol solution (Buitink *et al.*, 2006). During the incubation, >1300 genes were differentially expressed. A large number of genes involved in carbon metabolism are expressed during the re-establishment of desiccation tolerance. Quantification of carbon reserves confirms that lipids, starch, and oligosaccharides were mobilized, coinciding with the production of sucrose during the early osmotic adjustment. Several regulatory genes typically expressed during abiotic/drought stresses were also upregulated during maturation, arguing for the partial overlap of abscisic acid–dependent and abscisic acid–independent regulatory pathways involved in both drought and desiccation tolerance.

The availability of *M. truncatula* genomic sequence has permitted the development of an Affymetrix chip representative of the transcriptome (Benedito *et al.*, 2008). The data obtained with this array, including samples taken at six time points through seed development, have been assembled and are publicly available on the Medicago Gene Atlas site (He *et al.*, 2009). The relative insensitivity of chip hybridizations has motivated the development of a qRT-PCR platform for *M. truncatula* transcription factor (TF) genes (Kakar *et al.*, 2008). This platform was used by Verdier *et al.* (2008) to analyze the expression of >700 *M. truncatula* genes encoding putative TFs throughout seven stages of seed development. By applying a k-means clustering to the expression data, six classes of TF profiles could be identified, corresponding to the developmental stages. Four different TF classes correlated with the start of storage protein mRNA synthesis, vicilin-type protein synthesis, legumin A synthesis, and legumin K synthesis, these events being sequential in legume seed development. In complement, by purifying nuclei from *M. truncatula* developing seeds at the beginning of the seed filling stage, Repetto *et al.* (2008) obtained sequence identifications of >100 nuclear-localized proteins. Most were ribosomal proteins or other proteins involved in rRNA synthesis in the nucleolus, but numerous other chromatin-related proteins and transcription factors were also represented.

First attempts have been made to compare proteomics with transcriptomics data from developing *M. truncatula* seeds by Gallardo *et al.* (2007), who classified genes according to kinetic profiles of expression. Although for many genes the protein encoded had a temporal expression profile similar to that of the corresponding transcript, there were some exceptions. In cases where transcript accumulation is not accompanied by the corresponding protein, several explanations are possible. The most likely is that there is a translational control on protein accumulation operating. This may be analogous to that observed for stored mRNAs accumulated in maturing seeds for use during germination. miRNAs found in developing seeds (see later) may be implicated in translational control of some mRNAs. The protein may also be rapidly degraded or processed and not be detected in its intact form. Where the protein appears to accumulate at stages where there is little of the corresponding mRNA, it is assumed that the protein is stable, but the mRNA is unstable and rapidly turned over. A double-labeling allowing measuring of the *de novo* synthesis of mRNA and of proteins would provide additional information useful to understand the regulation of seed protein accumulation.

Seed Metabolomics

Currently, publications on legume seed metabolomics are essentially limited to applications of metabolomics to understand the effect of single-gene mutations on seed composition. An example of this type of publication is Vigeolas *et al.* (2008), who examined the effect of reduced PA2 albumin accumulation on pea seed composition. In an analysis by chromatography–mass spectrometry of soluble metabolites in the line lacking PA2 albumin, there was a shift in the relative proportions of carbon metabolites from sugars to amino acids, whereas later in seed development, sugar levels were similar to the wild-type line, but there were increases in several amino acids, particularly arginine, a precursor of spermidine, which was accompanied by decrease in the latter. In addition, there was an overall increase in protein content in mature seeds. Consistent with this result, arginine decarboxylase (ADC) activity was reduced in PA2 minus lines. However, the ADC locus was unlinked to that of PA2, ruling out a direct effect of the PA2 locus mutation on ADC.

Frank *et al.* (2009) used a chromatography–based approach to compare the metabolite profiles of two low phytic acid (lpa) soybean mutants and their wild-types. Both the lpa mutants had major phytic acid reductions. For Gm-lpa-TW75-1, decreased accumulation of lower inositol phosphates was also observed, consistent with a mutation in *myo*-inositol phosphate synthase. Consideration of the metabolic changes observed for Gm-lpa-ZC-2 (accumulation of lower inositol phosphates and increased *myo*-inositol contents) indicated a mutation event affecting the latter biosynthetic steps leading to phytic acid, possibly in the inositol phosphate kinase gene. The study demonstrated the applicability of metabolite profiling for the detection of changes in the metabolite phenotype induced by mutation breeding.

Modifying Seed Characteristics by Manipulating Gene Expression

Protein Composition

Several groups have obtained soybean mutants defective in storage protein accumulation in the developing seed (Takahashi *et al.*, 1994; Hayashi *et al.*, 1998; Krishnan *et al.*, 2001; Kim *et al.*, 2008; Hayashi *et al.*, 2009; Lee *et al.*, 2011). These mutants include trans-acting dominant and recessive loci controlling 7S Globulin (beta-conglycinin) accumulation, and cis-acting alleles at these loci (Teraishi *et al.*, 2001; Jegadeesan *et al.*, 2012).

To analyze the rate-limiting role of amino acids for seed protein synthesis, a *Vicia faba* amino acid permease, VfAAP1, has been ectopically expressed in pea (*Pisum sativum*) and *Vicia narbonensis* seeds under the control of the legumin B4 promoter (Rolletschek *et al.*, 2005). In mature seeds, starch is unchanged, but total nitrogen is 10%–25% higher, which affects mainly globulin, vicilin, and legumin, rather than albumin synthesis. Pea seeds overexpressing a sucrose transporter also show stimulated storage protein biosynthesis at the level of transcripts, proteins, and protein bodies. It was concluded that sucrose functions as both a signal and a fuel to stimulate storage protein accumulation (Rosche *et al.*, 2005). By RNA interference, Schmidt *et al.* (2011) suppressed the synthesis of soybean (Glycine max) glycinin and conglycinin storage proteins. The storage protein knockdown (SP−) seeds resemble the wild-type and maintain wild-type levels of protein and storage triglycerides. The SP− soybeans were evaluated with systems biology techniques of proteomics, metabolomics, and transcriptomics using both microarray and RNA-Seq. Proteomic analysis shows that rebalancing of protein content largely results from the selective increase in the accumulation of only a few proteins: Kunitz trypsin inhibitor, soybean lectin, and the immunodominant soybean allergen P34 or Gly m Bd 30k. A similar set of proteins greatly increases in accumulation in a cross between two soybean lines that carry conglycinin and glycinin null alleles (Takahashi *et al.*, 2003). The rebalancing of protein composition occurs with small alterations to the seed's transcriptome and metabolome.

Major legume seed proteins are very low in cysteine and tryptophan, the latter being an essential amino acid. Efforts have been made to manipulate amino acid composition, to increase the accumulation of sulfur-rich proteins or to introduce a sulfur-rich foreign protein (Krishnan, 2005; Kita *et al.*, 2010). Insertion by transgenesis of a protein "sink" rich in cysteine residues results in a decrease in other sulfur compounds, such as free sulfur amino acids and glutathione (Tabe and Droux, 2002). Generally, only modest increases in seed cysteine content were obtained, suggesting amino acid supply may be limiting. To address this possibility, Tan *et al.* (2010) increased phloem transport of *S*-methylmethionine, by expressing a yeast *S*-methylmethionine transporter from the AAP1 promoter in transgenic pea. The transgenics had increased plant vigor and increased phloem transport of both sulfur and nitrogen. Seed number was also increased, but the seed sulfur content was unchanged. Tabe *et al.* (2010) demonstrated that cysteine biosynthesis in legumes can be modified by manipulating sulfur assimilatory pathways. More recently, Kim *et al.* (2012) overexpressed an enzyme in the cysteine biosynthetic pathway, O-acetylserine sulfhydrylase, in transgenic soybean. The transformed seeds had significantly increased free and protein-bound cysteine contents. Elevated levels of methionine in seeds of narbon bean (*Vicia narbonensis* L.) was also obtained by generating double transformants, which express a methionine-rich storage protein and a feedback-insensitive bacterial aspartate kinase. Aspartate kinase is an enzyme in the methionine biosythesis pathway, which in plants is feedback-inhibited by several intermediate precursors (Demidov *et al.*, 2003).

Studies on the regulation of storage protein gene expression in the common bean (*Phaseolus vulgaris*) date from the early 1980s, making it one of the best-characterized systems in terms of promoter-TF interactions despite the absence for most of this period of a homologous stable genetic transformation protocol. Of particular interest is the existence of bean lines defective in accumulation of major storage protein classes (Blair *et al.*, 2007), which have been used as the basis for improving seed amino acid composition (Taylor *et al.*, 2008). Marsolais *et al.* (2010) and Yin *et al.* (2011) have carried out a study of common bean seed composition for mutants defective in storage protein accumulation. Yin *et al.* (2011) generated an EST collection from RNA of developing common bean seeds and used this database to investigate the basis for elevated sulfur content in lines lacking major sulfur-poor proteins. They observed increases in sulfur-rich proteins and in starch and raffinose metabolic enzymes and a downregulation of proteins of the secretory pathway. The reduced levels of seed storage proteins were accompanied by an increase in sulfur amino acid content in genetically related lines. Reduced phaseolin, phytohemagglutinin, and arcelin were mainly compensated by increases in legumin, the mannose lectin FRIL, and α-amylase inhibitors. The increased S-amino acid content was due principally to legumin, albumin-1 and albumin-2 and defensin, whereas gamma-glutamyl-S-methyl-cysteine was reduced. Traffic through the protein secretory pathway was disturbed, as shown by changes in various proteins implicated in its functioning.

Oil Composition

In addition to modification of protein content and composition, there is interest in increasing soybean oil content and modifying its composition (see Chapter 11). Wang *et al.* (2007) screened soybean Dof type transcription factors that were expressed during seed development and found two of these that increased seed fatty acid accumulation when overexpressed in *Arabidopsis*. Several groups have identified mutations in the omega-6 fatty acid desaturase gene, which converts linoleic acid to oleic acid. This is of interest for providing high oleic acid oils. Dierking and Bilyeu (2009) screened mutations in the omega-6 fatty acid desaturase gene, FAD 2-1A, from a Soybean TILLING collection and obtained a mutant allele affected in a conserved region of the enzyme. Seeds of this mutant had an altered fatty acid composition, with increased proportion of oleic acid and decreased linoleic acid, as predicted.

Manipulation of Starch Deposition

The effect of repressing ADP-glucose pyrophosphorylase (AGP) synthesis by RNAi in pea seeds was analyzed using transcript and metabolite profiling to monitor the effects that reduced carbon flow into starch has on carbon-nitrogen metabolism and related pathways (Weigelt *et al.*, 2009). Repressing AGP has several consequences. It appears to enhance provision of precursors such as acetyl-coenzyme A and organic acids, which promote amino acid and storage protein biosynthesis as well as pathways fed by cytosolic acetyl-coenzyme A, such as cysteine biosynthesis and fatty acid elongation/metabolism. The resulting higher nitrogen (N) demand depletes transient N storage pools, specifically asparagine and arginine, and leads to N limitation. Seed size is reduced, although protein content is increased. Increased sugar accumulation appears to stimulate cell proliferation, and deregulation of starch biosynthesis resulted in increased mitochondrial metabolism and osmotic stress. Although multiple stress responses are elicited, these occur in a controlled fashion, and premature senescence or apoptosis is not seen.

Vriet *et al.* (2010) have obtained mutations in genes involved in starch biosynthesis from a *Lotus japonicus* TILLING collection. Although this article is primarily focused on starch metabolism in vegetative tissues, it does report on the seed phenotype of a mutant in *GWD* (glucan water dikinase), an enzyme involved in starch degradation. The homozygous mutant plant showed a strong maternal effect with a high proportion of aborted seed and pods, showing the importance of glucan water dikinases in starch degradation and the importance of this process in seed development, presumably in remobilizing carbon in maternal tissues, there being no starch accumulation in the mature seed of *L. japonicus*.

Antinutritionals

Among the main breeding objectives for pulse legumes is the elimination by genetic selection or genetic manipulation of certain seed components that reduce digestibility. In most cases, progress has been made by forward genetic screens. Null alleles of the major trypsin inhibitor *Tri* genes of pea have been routinely selected for Pea breeding in recent years, based on a PCR screen (Page *et al.*, 2002; Chinoy *et al.*, 2011). A naturally occurring mutant allele in pea of the poorly digested and possibly allergenic PA2 albumin is also available (Vigeolas *et al.*, 2008). However, the removal of these proteins has to be balanced against their proposed contributions to seed insect and fungal resistance (Srinivasan *et al.*, 2005; Da Silva *et al.*, 2010). Members of the raffinose family oligosaccharides (RFOs) accumulate in maturing seeds of many legumes, where they appear to contribute to desiccation tolerance. RFOs are poorly digested by the intestinal flora of monogastric animals and cause flatulence – hence efforts to eliminate their production. Dierking and Bilyeu (2009) screened a soybean TILLING population for EMS mutations in the Raffinose Synthase gene *rs2*. They obtained a mutant that had 24% stachyose and 37% raffinose levels compared with wild-type, accompanied by an increase in seed sucrose concentration. The residual RFOs are probably due to the presence of a second RS gene. Polowick *et al.* (2009) took a different, transgenic, approach by overexpressing an alpha-galactosidase from Coffea in pea seeds. Using this strategy, they obtained similar reductions in RFOs to those observed in the soybean *rs2* TILLING mutant.

Understanding Regulation of Gene Expression in Seeds

Functional analyses of legume seed promoters began in the 1980s with work on the phaseolin (common bean storage protein) gene from Hall's laboratory (Slightom *et al.*, 1983). The phaseolin promoter was functional in the heterologous tobacco system, where it conferred embryo-specific expression (Senguptagopalan *et al.*, 1985). There followed detailed promoter dissection that identified several cis-acting elements in the phaseolin promoter and in that of an unidentified seed protein gene from *Vicia faba* USP (Fiedler *et al.*, 1993). Hall's group showed that activation of the phaseolin promoter required at least two steps: the binding of an activator transcription factor (TF), PvALF, that modified chromatin conformation on the promoter and an ABA-dependent step of gene activation (Li *et al.*, 1999). Subsequently, these two steps were shown to correspond to histone lysine acetylation and methylation events in the associated nucleosomes on the Phas promoter that potentiate gene activation (Ng *et al.*, 2006). More comprehensive analyses of the TFs interacting with seed-expressed promoters in legumes have been made possible by the availability of transcriptome data.

In contrast to studies on maize or *Arabidopsis*, most legume TFs involved in seed development and seed filling have not been identified via screening for defective seed mutant phenotypes but initially by sequence homologies with previously characterized TFs. To restrict the number of TFs being analyzed, generally only legume TFs coexpressed (in time and in tissue) with putative target promoters have been studied. More recent global transcriptomics analyses have confirmed the overall similarity of the seed TF complement of *M. truncatula* with that of *Arabidopsis* seeds, consistent with their similar ontogenesis and morphologies, although very few functional analyses have yet been reported (Udvardi *et al.*, 2007). Verdier *et al.* (2008) analyzed the expression by q-RT-PCR of >700 *M. truncatula* genes encoding putative TFs throughout seven stages of seed development. Of 169 TFs expressed during seed filling, the tissue specificity within the seed was examined for the 41 most highly expressed sequences. To identify possible target genes for these TFs, the data were combined with a microarray-derived entire transcriptome dataset. This study identified 17 TFs preferentially expressed in individual seed tissues and 135 corresponding coexpressed genes, including possible targets. Some of the TFs, which coexpressed with storage protein mRNAs, were homologous to those already known to regulate seed storage protein synthesis in *Arabidopsis*. Putative orthologues of the "master regulators" of seed development LEC1, LEC1-LIKE, FUS3, and ABI3 could be identified in *M. truncatula*, showing an overall conservation of gene-specific transcription factor sequences in the seeds of the two genera. However, an obvious candidate for a LEC2 orthologue was not found in *M. truncatula*. *M. truncatula*, similar to other legumes, has a delayed expression of the legumin-class storage proteins with respect to the vicilin class of storage proteins, and a class of TFs could be specifically assigned to this phase of expression.

An alternative approach to narrowing down the gene candidates for encoding a given trait is to look for genetic linkage. Gillman *et al.* (2011) used genetic mapping and candidate gene association in a RIL population and a panel of soybean lines with defined coloration (seed coat and hilum, pubescence, and flower) to determine the R gene controlling black or brown seed coat coloration secondary to anthocyanin derivatives in soybean to be an R2R3 MYB gene (Glyma09g36990). Similarly, Verdier *et al.* (2012) have identified an R2R3Myb, MtrPAR1, which controls proanthocyanidin production in *Medicago truncatula* seed coats. By overexpressing this TF in transgenic hairy roots, they were able to accumulate proanthocyanidin ectopically in the roots. These studies may offer potential for eliciting proanthocyanidin accumulation in vegetative tissues of forage crops, such as Lucerne/alfalfa, to combat pasture bloat in ruminants.

In contrast to *Arabidopsis*, many crop legumes (pea, faba bean, common bean, lupin) accumulate starch in the mature seed. It will be of interest to see whether dedicated TFs control this process.

Identifying Loci Controlling Seed Traits
The *M. truncatula* model has been exploited to identify genes important for seed maturation (Vandecasteele *et al.*, 2011). The maturation of the seed comprises several distinguishable stages: acquisition of desiccation tolerance, acquisition of seed dormancy, determination of longevity, and determination of germinative seed vigor. In contrast to *Arabidopsis*, maturing legume seeds accumulate large quantities of raffinose family oligosaccharides (RFOs), which are thought to play a role in these processes and provide a useful complement to studies on other plant models. Vandecasteele *et al.* (2011) used two recombinant inbred line (RIL) populations of *M. truncatula* to map QTL for physiological traits related to germination. Several of the QTLs were found colocated with QTL controlling the Sucrose/RFO ratio and suggested that a decreased sucrose/RFO ratio might be a positive determinant of seed vigor. One of these populations, named LR4, has been used to reveal a large number of QTL controlling seed weight and composition (K. Gallardo, personal communication) and to map QTL for seed mineral concentration (Sankaran *et al.*, 2009). The concentrations and content of seed Ca, Cu, Fe, K, Mg, Mn, P, and Zn was determined in the LR4 population, leading to the identification of 46 QTLs for seed mineral concentration and of 26 QTLs for seed mineral content. The availability of genomic sequences in this species will facilitate the positional cloning of genes controlling these traits. The Medicago HapMap project (http://www.medicagohapmap.org/) is sequencing 384 inbred lines using Solexa technology to detect SNPs at high density and provide a platform for association mapping.

Several laboratories have used molecular markers to carry out QTL analyses of soybean seed size and composition, including Li *et al.* (2007), who identified loci controlling protein and oil content, and Teng *et al.* (2009), who identified loci controlling seed weight. The availability of EST sequences and high-throughput sequencing of the *G. max* genome have permitted the development of high-density linkage maps and a panel for QTL mapping using SNPs (Choi *et al.*, 2007; Hyten *et al.*, 2010). A PQL (protein quantity loci) analysis of mature pea seeds was performed that identified QTLs for several hundreds of protein spots detected on 2D gels from 157 recombinant inbred lines (Bourgeois *et al.*, 2011). Most protein quantity loci mapped in clusters, suggesting that the accumulation of the major storage protein families was under the control of a limited number of loci. Some loci controlled both seed protein composition and protein content, and a locus on LGIIa appears to be a major regulator of protein composition and of *in vitro* protein digestibility. With the objective of identifying genes controlling such traits in grain legumes, the power of the translational genomics approach from highly characterized model plants to poorly characterized crop plants was underlined by Bordat *et al.* (2011). This approach, which integrates genetics and genomics data, provides a valuable source of markers to saturate a zone of interest. A broader view of pea genome evolution was obtained by revealing syntenic relationships between pea and sequenced genomes. Blocks of synteny were identified that gave new insights into the evolution of chromosome structure in Papillionoids and Eudicots.

Future Challenges

Network Construction

The seed represents the sum of the metabolic activities of all the plant organs contributing to its production. This also applies to the embryo and endosperm, which depend on the other seed tissues for their metabolite supply and development. Although it is simple to identify loss-of-function mutations affecting seed development, the relationship between the genes encoded is not always evident, partly because of the interactions between the component tissues.

To establish connections between interdependent gene products, many groups have developed methods for identifying networks of interaction. This approach is likely to be crucial for seed biology in general and important for legume seed biology where the loci affecting nutrient balance in the whole plant are particularly important. The method of Bassel *et al.* (2011) exploited 138 previously obtained transcriptomics datasets on gene expression during *Arabidopsis* germination. A coexpression analysis using significance analysis of microarrays (SAM) (Tusher *et al.*, 2001) revealed two distinct networks of coexpression, for genes expressed in germinating and nongerminating seeds. By analyzing the highest degree nodes (hubs) of the network, these authors were able to identify a series of putative regulators of seed germination. The same methodology is being used for *Medicago truncatula* seed expression datasets (J. Buitink, personal communication). Once nonconditional sources of information are used, it becomes necessary to distinguish the mode of interaction of network elements. One possible approach was taken by Junker *et al.* (2010), who adopted engineering terminology that allows the representation of a series of operationally different relationships.

Integrating Physiological Modeling

To incorporate physiological and metabolic data into functional seed models, it is necessary first to acquire data in three dimensions within the developing seed and second to conceive programs capable of handling and representing the data (Gubatz *et al.*, 2007). Data in three dimensions can be approximated by using optical sections, for example, based on serial optical (confocal microscopy) or derived from mechanical sections. The datasets can be obtained for macromolecular seed constituents such as specific proteins by immunocytochemistry or for transcripts by in situ hybridization, and for insoluble carbohydrates or lipids as well as for cellular structures using staining methods. Advances in nuclear magnetic resonance imaging have extended this noninvasive 3D technique to allow for the first time the identification of metabolite distributions within the legume seed (Melkus *et al.*, 2009) (see Chapter 13). The adoption of this type of method, along with high-throughput enzyme assays, is vital for confirming the occurrence and extent of metabolic processes that are often only inferred from gene expression data.

References

Agrawal GK, Hajduch M, Graham K, Thelen JJ (2008). In-depth investigation of the soybean seed-filling proteome and comparison with a parallel study of rapeseed. Plant Physiology 148: 504–518.

Arenas-Huertero C, Perez B, Rabanal F, Blanco-Melo D, De la Rosa C, Estrada-Navarrete G, Sanchez F, Covarrubias AA, Reyes JL (2009). Conserved and novel miRNAs in the legume Phaseolus vulgaris in response to stress. Plant Molecular Biology 70: 385–401.

Asakura T, Tamura T, Terauchi K, Narikawa T, Yagasaki K, Ishimaru Y, Abe K (2012). Global gene expression profiles in developing soybean seeds. Plant Physiology and Biochemistry: PPB/Societe Francaise de Physiologie Vegetale 52: 147–153.

Assoumane AA, Vaillant A, Mayaki AZ, Verhaegen D (2009). Isolation and characterization of microsatellite markers for Acacia senegal (L.) Willd., a multipurpose arid and semi-arid tree. Molecular Ecology Resources 9: 1380–1383.

Aubert G, Morin J, Jacquin F, Loridon K, Quillet MC, Petit A, Rameau C, Lejeune-Henaut I, Huguet T, Burstin J (2006). Functional mapping in pea, as an aid to the candidate gene selection and for investigating synteny with the model legume Medicago truncatula. Theoretical and Applied Genetics 112: 1024–1041.

Bassel GW, Lan H, Glaab E, Gibbs DJ, Gerjets T, Krasnogor N, Bonner AJ, Holdsworth MJ, Provart NJ (2011). Genome-wide network model capturing seed germination reveals coordinated regulation of plant cellular phase transitions. Proceedings of the National Academy of Sciences of the United States of America 108: 9709–9714.

Benedito VA, Torres-Jerez I, Murray JD, Andriankaja A, Allen S, Kakar K, Wandrey M, Verdier J, Zuber H, Ott T, Moreau S, Niebel A, Frickey T, Weiller G, He J, Dai XB, Zhao PX, Tang YH, Udvardi MK (2008). A gene expression atlas of the model legume Medicago truncatula. Plant Journal 55: 504–513.

Bielewicz D, Dolata J, Zielezinski A, Alaba S, Szarzynska B, Szczesniak MW, Jarmolowski A, Szweykowska-Kulinska Z, Karlowski WM (2012). mirEX: a platform for comparative exploration of plant pri-miRNA expression data. Nucleic Acids Research 40: D191–D197.

Blair MW, Porch T, Cichy K, Galeano CH, Lariguet P, Pankhurst C, Broughton W (2007). Induced mutants in common bean (Phaseolus vulgaris), and their potential use in nutrition quality breeding and gene discovery. Israel Journal of Plant Sciences 55: 191–200.

Bordat A, Savois V, Nicolas M, Salse J, Chauveau A, Bourgeois M, Potier J, Houtin H, Rond C, Murat F, Marget P, Aubert G, Burstin J (2011). Translational genomics in legumes allowed placing in silico 5460 unigenes on the pea functional map and identified candidate genes in Pisum sativum L. G3 (Bethesda, Md.) 1: 93–103.

Bourgeois M, Jacquin F, Cassecuelle F, Savois V, Belghazi M, Aubert G, Quillien L, Huart M, Marget P, Burstin J (2011). A PQL (protein quantity loci) analysis of mature pea seed proteins identifies loci determining seed protein composition. Proteomics 11: 1581–1594.

Branca A, Paape TD, Zhou P, Briskine R, Farmer AD, Mudge J, Bharti AK, Woodward JE, May GD, Gentzbittel L, Ben C, Denny R, Sadowsky MJ, Ronfort J, Bataillon T, Young ND, Tiffin P (2011). Whole-genome nucleotide diversity, recombination, and linkage disequilibrium in the model legume Medicago truncatula. Proceedings of the National Academy of Sciences of the United States of America 108: E864–E870.

Buitink J, Leger JJ, Guisle I, Vu BL, Wuilleme S, Lamirault G, Le Bars A, Le Meur N, Becker A, Kuester H, Leprince O (2006). Transcriptome profiling uncovers metabolic and regulatory processes occurring during the transition from desiccation-sensitive to desiccation-tolerant stages in Medicago truncatula seeds. Plant Journal 47: 735–750.

Burstin J, Gallardo K, Mir RR, Varshney RK, Duc G (2011). Improving protein content and nutrition quality. In: Pratap A, Kumar J (eds). *Biology and Breeding of Food Legumes*. CAB International, Wallingford, United Kingdom, pp 314–328.

Cannon SB, Sterck L, Rombauts S, Sato S, Cheung F, Gouzy J, Wang X, Mudge J, Vasdewani J, Schiex T, Spannagl M, Monaghan E, Nicholson C, Humphray SJ, Schoof H, Mayer KFX, Rogers J, Quétier F, Oldroyd GE, Debellé F, Cook DR, Retzel EF, Roe BA, Town CD, Tabata S, Van de Peer Y, Young ND (2006). Legume genome evolution viewed through the Medicago truncatula and Lotus japonicus genomes. Proceedings of the National Academy of Sciences of the United States of America 103: 14959–14964.

Cheng KCC, Stromvik MV (2008). SoyXpress: A database for exploring the soybean transcriptome. Bmc Genomics 9.

Chinoy C, Welham T, Turner L, Moreau C, Domoney C (2011). The genetic control of seed quality traits: effects of allelic variation at the Tri and Vc-2 genetic loci in Pisum sativum L. Euphytica 180: 107–122.

Choi I-Y, Hyten DL, Matukumalli LK, Song Q, Chaky JM, Quigley CV, Chase K, Lark KG, Reiter RS, Yoon M-S, Hwang E-Y, Yi S-I, Young ND, Shoemaker RC, van Tassell CP, Specht JE, Cregan PB (2007). A soybean transcript map: gene distribution, haplotype and single-nucleotide polymorphism analysis. Genetics 176: 685–696.

Clarke BR, Denyer K, Jenner CF, Smith AM (1999). The relationship between the rate of starch synthesis, the adenosine 5′-diphosphoglucose concentration and the amylose content of starch in developing pea embryos. Planta 209: 324–329.

Colditz F, Braun H-P (2010). Medicago truncatula proteomics. Journal of Proteomics 73: 1974–1985.

Constantin GD, Krath BN, MacFarlane SA, Nicolaisen M, Johansen IE, Lund OS (2004). Virus-induced gene silencing as a tool for functional genomics in a legume species. Plant Journal 40: 622–631.

Cooper JL, Till BJ, Laport RG, Darlow MC, Kleffner JM, Jamai A, El-Mellouki T, Liu S, Ritchie R, Nielsen N, Bilyeu KD, Meksem K, Comai L, Henikoff S (2008). TILLING to detect induced mutations in soybean. Bmc Plant Biology 8.

d'Erfurth I, Cosson V, Eschstruth A, Lucas H, Kondorosi A, Ratet P (2003). Efficient transposition of the Tnt1 tobacco retrotransposon in the model legume Medicago truncatula. Plant Journal 34: 95–106.

Da Silva P, Rahioui I, Laugier C, Jouvensal L, Meudal H, Chouabe C, Delmas AF, Gressent F (2010). Molecular requirements for the insecticidal activity of the plant peptide pea albumin 1 subunit b (PA1b). Journal of Biological Chemistry 285: 32689–32694.

Dalmais M, Schmidt J, Le Signor C, Moussy F, Burstin J, Savois V, Aubert G, Brunaud V, de Oliveira Y, Guichard C, Thompson R, Bendahmane A (2008). UTILLdb, a Pisum sativum in silico forward and reverse genetics tool. Genome Biology 9.

Dam S, Laursen BS, Ornfelt JH, Jochimsen B, Staerfeldt HH, Friis C, Nielsen K, Goffard N, Besenbacher S, Krusell L, Sato S, Tabata S, Thogersen IB, Enghild JJ, Stougaard J (2009). The proteome of seed development in the model legume Lotus japonicus. Plant Physiology 149: 1325–1340.

De La Fuente M, Borrajo A, Bermudez J, Lores M, Alonso J, Lopez M, Santalla M, De Ron AM, Zapata C, Alvarez G (2011). 2-DE-based proteomic analysis of common bean (Phaseolus vulgaris L.) seeds. Journal of Proteomics 74: 262–267.

Demidov D, Horstmann C, Meixner M, Pickardt T, Saalbach I, Galili G, Muntz K (2003). Additive effects of the feed-back insensitive bacterial aspartate kinase and the Brazil nut 2S albumin on the methionine content of transgenic narbon bean (Vicia narbonensis L.). Molecular Breeding 11: 187–201.

Diaz-Ruiz R, Torres AM, Satovic Z, Gutierrez MV, Cubero JI, Roman B (2010). Validation of QTLs for Orobanche crenata resistance in faba bean (Vicia faba L.) across environments and generations. Theoretical and Applied Genetics 120: 909–919.

Dierking EC, Bilyeu KD (2009). New sources of soybean seed meal and oil composition traits identified through TILLING. Bmc Plant Biology 9.

Diouf D (2011). Recent advances in cowpea Vigna unguiculata (L.) Walp. "omics" research for genetic improvement. African Journal of Biotechnology 10: 2803–2810.

Dubey A, Farmer A, Schlueter J, Cannon SB, Abernathy B, Tuteja R, Woodward J, Shah T, Mulasmanovic B, Kudapa H, Raju NL, Gothalwal R, Pande S, Xiao YL, Town CD, Singh NK, May GD, Jackson S, Varshney RK (2011). Defining the transcriptome assembly and its use for genome dynamics and transcriptome profiling studies in pigeonpea (Cajanus cajan L.). DNA Research 18: 153–164.

Fiedler U, Filistein R, Wobus U, Baumlein H (1993). A complex ensemble of cis-regulatory elements controls the expression of a vicia-faba non-storage protein gene. Plant Molecular Biology 22: 669–679.

Foley RCFRC, Gao LL, Spriggs A, Soo LYC, Goggin DE, Smith PMC, Atkins CA, Singh KB (2011). Identification and characterisation of seed storage protein transcripts from Lupinus angustifolius. Bmc Plant Biology 11.

Frank T, Noerenberg S, Engel K-H (2009). Metabolite profiling of two novel low phytic acid (lpa) soybean mutants. Journal of Agricultural and Food Chemistry 57: 6408–6416.

Franssen SU, Shrestha RP, Braeutigam A, Bornberg-Bauer E, Weber APM (2011). Comprehensive transcriptome analysis of the highly complex Pisum sativum genome using next generation sequencing. Bmc Genomics 12.

Fukai E, Soyano T, Umehara Y, Nakayama S, Hirakawa H, Tabata S, Sato S, Hayashi M (2012). Establishment of a Lotus japonicus gene tagging population using the exon-targeting endogenous retrotransposon LORE1. Plant Journal 69: 720–730.

Gallardo K, Firnhaber C, Zuber H, Hericher D, Belghazi M, Henry C, Kuster H, Thompson R (2007). A combined proteome and transcriptome analysis of developing Medicago truncatula seeds. Molecular and Cellular Proteomics 6: 2165–2179.

Garcia RAV, Rangel PN, Brondani C, Martins WS, Melo LC, Carneiro MS, Borba TCO, Brondani RPV (2011). The characterization of a new set of EST-derived simple sequence repeat (SSR) markers as a resource for the genetic analysis of Phaseolus vulgaris. Bmc Genetics 12.

Garg R, Patel RK, Jhanwar S, Priya P, Bhattacharjee A, Yadav G, Bhatia S, Chattopadhyay D, Tyagi AK, Jain M (2011). Gene discovery and tissue-specific transcriptome analysis in chickpea with massively parallel pyrosequencing and web resource development. Plant Physiology 156: 1661–1678.

Gillman JD, Tetlow A, Lee J-D, Shannon JG, Bilyeu K (2011). Loss-of-function mutations affecting a specific Glycine max R2R3 MYB transcription factor result in brown hilum and brown seed coats. Bmc Plant Biology 11.

Gubatz S, Dercksen VJ, Bruess C, Weschke W, Wobus U (2007). Analysis of barley (Hordeum vulgare) grain development using three-dimensional digital models. Plant Journal 52: 779–790.

Gujaria N, Kumar A, Dauthal P, Dubey A, Hiremath P, Prakash AB, Farmer A, Bhide M, Shah T, Gaur PM, Upadhyaya HD, Bhatia S, Cook DR, May GD, Varshney RK (2011). Development and use of genic molecular markers (GMMs) for construction of a transcript map of chickpea (Cicer arietinum L.). Theoretical and Applied Genetics 122: 1577–1589.

Guo BZ, Liang XQ, Chung SY, Holbrook CC, Maleki SJ (2008). Proteomic analysis of peanut seed storage proteins and genetic variation in a potential peanut allergen. Protein and Peptide Letters 15: 567–577.

Gupta SK, Gopalakrishna T (2010). Development of unigene-derived SSR markers in cowpea (Vigna unguiculata) and their transferability to other Vigna species. Genome 53: 508–523.

Ha J, Abernathy B, Nelson W, Grant D, Wu X, Nguyen HT, Stacey G, Yu Y, Wing RA, Shoemaker RC, Jackson SA (2012). Integration of the draft sequence and physical map as a framework for genomic research in soybean (Glycine max (L.) Merr.) and wild soybean (Glycine soja Sieb. and Zucc.). G3 (Bethesda, Md.) 2: 321–329.

Hajduch M, Ganapathy A, Stein JW, Thelen JJ (2005). A systematic proteomic study of seed filling in soybean: establishment of high-resolution two-dimensional reference maps, expression profiles, and an interactive proteome database. Plant Physiology 137: 1397–1419.

Hajduch M, Matusova R, Houston NL, Thelen JJ (2011). Comparative proteomics of seed maturation in oilseeds reveals differences in intermediary metabolism. Proteomics 11: 1619–1629.

Hamwieh A, Udupa SM, Sarker A, Jung C, Baum M (2009). Development of new microsatellite markers and their application in the analysis of genetic diversity in lentils. Breeding Science 59: 77–86.

Handberg K, Stougaard J (1992). Lotus-Japonicus, an autogamous, diploid legume species for classical and molecular-genetics. Plant Journal 2: 487–496.

Hayashi M, Harada K, Fujiwara T, Kitamura K (1998). Characterization of a 7S globulin-deficient mutant of soybean (Glycine max (L.) Merrill). Molecular and General Genetics 258: 208–214.

Hayashi M, Kitamura K, Harada K (2009). Genetic mapping of Cgdef gene controlling accumulation of 7s globulin (beta-conglycinin) subunits in soybean seeds. Journal of Heredity 100: 802–806.

He J, Benedito VA, Wang MY, Murray JD, Zhao PX, Tang YH, Udvardi MK (2009). The Medicago truncatula gene expression atlas web server. Bmc Bioinformatics 10.

Hofer J, Turner L, Moreau C, Ambrose M, Isaac P, Butcher S, Weller J, Dupin A, Dalmais M, Le Signor C, Bendahmane A, Ellis N (2009). Tendril-less regulates tendril formation in pea leaves. Plant Cell 21: 420–428.

Hougaard BK, Madsen LH, Sandal N, Moretzsohn MD, Fredslund J, Schauser L, Nielsen AM, Rohde T, Sato S, Tabata S, Bertioli DJ, Stougaard J (2008). Legume anchor markers link syntenic regions between Phaseolus vulgaris, Lotus japonicus, Medicago truncatula and Arachis. Genetics 179: 2299–2312.

Hyten DL, Choi I-Y, Song Q, Specht JE, Carter TE Jr, Shoemaker RC, Hwang E-Y, Matukumalli LK, Cregan PB (2010). A high density integrated genetic linkage map of soybean and the development of a 1536 universal soy linkage panel for quantitative trait locus mapping. Crop Science 50: 960–968.

Igarashi A, Yamagata K, Sugai T, Takahashi Y, Sugawara E, Tamura A, Yaegashi H, Yamagishi N, Takahashi T, Isogai M, Takahashi H, Yoshikawa N (2009). Apple latent spherical virus vectors for reliable and effective virus-induced gene silencing among a broad range of plants including tobacco, tomato, Arabidopsis thaliana, cucurbits, and legumes. Virology 386: 407–416.

Isobe S, Koelliker R, Hisano H, Sasamoto S, Wada T, Klimenko I, Okumura K, Tabata S (2009). Construction of a consensus linkage map for red clover (Trifolium pratense L.). Bmc Plant Biology 9.

Jagadeeswaran G, Zheng Y, Li YF, Shukla LI, Matts J, Hoyt P, Macmil SL, Wiley GB, Roe BA, Zhang WX, Sunkar R (2009). Cloning and characterization of small RNAs from Medicago truncatula reveals four novel legume-specific microRNA families. New Phytologist 184: 85–98.

Jegadeesan S, Yu K, Woodrow L, Wang Y, Shi C, Poysa V (2012). Molecular analysis of glycinin genes in soybean mutants for development of gene-specific markers. Theoretical and Applied Genetics 124: 365–372.

Jezierny D, Mosenthin R, Bauer E (2010). The use of grain legumes as a protein source in pig nutrition: A review. Animal Feed Science and Technology 157: 111–128.

Jones SI, Gonzalez DO, Vodkin LO (2010). Flux of transcript patterns during soybean seed development. Bmc Genomics 11.

Josiah CC, George DO, Eleazar OM, Nyamu WF (2008). Genetic diversity in Kenyan populations of Acacia senegal (L.) wild revealed by combined RAPD and ISSR markers. African Journal of Biotechnology 7: 2333–2340.

Kaga A, Isemura T, Tomooka N, Vaughan DA (2008). The genetics of domestication of the azuki bean (Vigna angulatis). Genetics 178: 1013–1036.

Kakar K, Wandrey M, Czechowski T, Gaertner T, Scheible WR, Stitt M, Torres-Jerez I, Xiao YL, Redman JC, Wu HC, Cheung F, Town CD, Udvardi MK (2008). A community resource for high-throughput quantitative RT-PCR analysis of transcription factor gene expression in Medicago truncatula. Plant Methods 4.

Kaur S, Cogan NOI, Pembleton LW, Shinozuka M, Savin KW, Materne M, Forster JW (2011). Transcriptome sequencing of lentil based on second-generation technology permits large-scale unigene assembly and SSR marker discovery. Bmc Genomics 12.

Kaur S, Pembleton LW, Cogan NOI, Savin KW, Leonforte T, Paull J, Materne M, Forster JW (2012). Transcriptome sequencing of field pea and faba bean for discovery and validation of SSR genetic markers. Bmc Genomics 13.

Khanlou KM, Vandepitte K, Asl LK, Van Bockstaele E (2011). Towards an optimal sampling strategy for assessing genetic variation within and among white clover (Trifolium repens L.) cultivars using AFLP. Genetics and Molecular Biology 34: 252–258.

Kim MY, Lee S, Van K, Kim TH, Jeong SC, Choi IY, Kim DS, Lee YS, Park D, Ma J, Kim WY, Kim W-S, Chronis D, Juergens M, Schroeder AC, Hyun SW, Jez JM, Krishnan HB (2012). Transgenic soybean plants overexpressing O-acetylserine sulfhydrylase accumulate enhanced levels of cysteine and Bowman-Birk protease inhibitor in seeds. Planta 235: 13–23.

Kim W-S, Ho HJ, Nelson RL, Krishnan HB (2008). Identification of several gy4 nulls from the USDA Soybean Germplasm Collection provides new genetic resources for the development of high-quality tofu cultivars. Journal of Agricultural and Food Chemistry 56: 11320–11326.

Kinney AJ, Jung R, Herman EM (2001). Cosuppression of the alpha subunits of beta-conglycinin in transgenic soybean seeds induces the formation of endoplasmic reticulum-derived protein bodies. Plant Cell 13: 1165–1178.

Kita Y, Nakamoto Y, Takahashi M, Kitamura K, Wakasa K, Ishimoto M (2010). Manipulation of amino acid composition in soybean seeds by the combination of deregulated tryptophan biosynthesis and storage protein deficiency. Plant Cell Reports 29: 87–95.

Knoll JE, Ramos ML, Zeng YJ, Holbrook CC, Chow M, Chen SX, Maleki S, Bhattacharya A, Ozias-Akins P (2011). TILLING for allergen reduction and improvement of quality traits in peanut (Arachis hypogaea L.). Bmc Plant Biology 11.

Kottapalli KR, Payton P, Rakwal R, Agrawal GK, Shibato J, Burow M, Puppala N (2008). Proteomics analysis of mature seed of four peanut cultivars using two-dimensional gel electrophoresis reveals distinct differential expression of storage, anti-nutritional, and allergenic proteins. Plant Science 175: 321–329.

Krishnan HB (2001). Characterization of a soybean Glycine max (L.) Merr. mutant with reduced levels of Kunitz trypsin inhibitor. Plant Science 160: 979–986.

Krishnan HB (2005). Engineering soybean for enhanced sulfur amino acid content. Crop Science 45: 454–461.

Le BH, Wagmaister JA, Kawashima T, Bui AQ, Harada JJ, Goldberg RB (2007). Using genomics to study legume seed development. Plant Physiology 144: 562–574.

Le Signor C, Savois V, Aubert G, Verdier J, Nicolas M, Pagny G, Moussy F, Sanchez M, Baker D, Clarke J, Thompson R (2009). Optimizing TILLING populations for reverse genetics in Medicago truncatula. Plant Biotechnol J 7: 430–441.

Lee KJ, Kim J-B, Kim SH, Ha B-K, Lee B-M, Kang S-Y, Kim DS (2011). Alteration of seed storage protein composition in soybean glycine max (L.) Merrill mutant lines induced by gamma-irradiation mutagenesis. Journal of Agricultural and Food Chemistry 59: 12405–12410.

Lei ZT, Dai XB, Watson BS, Zhao PX, Sumner LW (2011). A legume specific protein database (LegProt) improves the number of identified peptides, confidence scores and overall protein identification success rates for legume proteomics. Phytochemistry 72: 1020–1027.

Li GF, Bishop KJ, Chandrasekharan MB, Hall TC (1999). beta-Phaseolin gene activation is a two-step process: PvALF-facilitated chromatin modification followed by abscisic acid-mediated gene activation. Proceedings of the National Academy of Sciences of the United States of America 96: 7104–7109.

Li J, Dai X, Liu T, Zhao PX (2012). LegumeIP: an integrative database for comparative genomics and transcriptomics of model legumes. Nucleic Acids Research 40: D1221–D1229.

Li W, Sun D, Du Y, Chen Q, Zhang Z, Qiu L, Sun G (2007). Quantitative trait loci underlying the development of seed composition in soybean (Glycine max L. Merr.). Genome 50: 1067–1077.

Libault M, Farmer A, Joshi T, Takahashi K, Langley RJ, Franklin LD, He J, Xu D, May G, Stacey G (2010). An integrated transcriptome atlas of the crop model Glycine max, and its use in comparative analyses in plants. Plant Journal 63: 86–99.

Loridon K, McPhee K, Morin J, Dubreuil P, Pilet-Nayel ML, Aubert G, Rameau C, Baranger A, Coyne C, Lejeune-Henaut I, Burstin J (2005). Microsatellite marker polymorphism and mapping in pea (Pisum sativum L.). Theoretical and Applied Genetics 111: 1022–1031.

Magni C, Scarafoni A, Herndl A, Sessa F, Prinsi B, Espen L, Duranti M (2007). Combined 2D electrophoretic approaches for the study of white lupin mature seed storage proteome. Phytochemistry 68: 997–1007.

Marsolais F, Pajak A, Yin FQ, Taylor M, Gabriel M, Merino DM, Ma V, Kameka A, Vijayan P, Pham H, Huang SZ, Rivoal J, Bett K, Hernandez-Sebastia C, Liu QA, Bertrand A, Chapman R (2010). Proteomic analysis of common bean seed with storage protein deficiency reveals up-regulation of sulfur-rich proteins and starch and raffinose metabolic enzymes, and down-regulation of the secretory pathway. Journal of Proteomics 73: 1587–1600.

Melkus G, Rolletschek H, Radchuk R, Fuchs J, Rutten T, Wobus U, Altmann T, Jakob P, Borisjuk L (2009). The metabolic role of the legume endosperm: a noninvasive imaging study. Plant Physiology 151: 1139–1154.

Meng YJ, Gou LF, Chen DJ, Mao CZ, Jin YF, Wu P, Chen M (2011). PmiRKB: a plant microRNA knowledge base. Nucleic Acids Research 39: D181–D187.

Mhuantong W, Wichadakul D (2009). MicroPC (mu PC): a comprehensive resource for predicting and comparing plant microRNAs. Bmc Genomics 10.

Mishra RK, Gangadhar BH, Nookaraju A, Kumar S, Park SW (2012). Development of EST-derived SSR markers in pea (Pisum sativum) and their potential utility for genetic mapping and transferability. Plant Breeding 131: 118–124.

Mochida K, Yoshida T, Sakurai T, Yamaguchi-Shinozaki K, Shinozaki K, Tran LSP (2010). LegumeTFDB: an integrative database of Glycine max, Lotus japonicus and Medicago truncatula transcription factors. Bioinformatics 26: 290–291.

Moe KT, Chung J-W, Cho Y-I, Moon J-K, Ku J-H, Jung J-K, Lee J, Park Y-J (2011). Sequence information on simple sequence repeats and single nucleotide polymorphisms through transcriptome analysis of mungbean. Journal of Integrative Plant Biology 53: 63–73.

Mun JH, Kim DJ, Choi HK, Gish J, Debelle F, Mudge J, Denny R, Endre G, Saurat O, Dudez AM, Kiss GB, Roe B, Young ND, Cook DR (2006). Distribution of microsatellites in the genome of Medicago truncatula: a resource of genetic markers that integrate genetic and physical maps. Genetics 172: 2541–2555.

Nautrup-Pedersen G, Dam S, Laursen BS, Siegumfeldt AL, Nielsen K, Goffard N, Staerfeldt HH, Friis C, Sato S, Tabata S, Lorentzen A, Roepstorff P, Stougaard J (2010). Proteome analysis of pod and seed development in the model legume Lotus japonicus. Journal of Proteome Research 9: 5715–5726.

Nelson MN, Moolhuijzen PM, Boersma JG, Chudy M, Lesniewska K, Bellgard M, Oliver RP, Swiecicki W, Wolko B, Cowling WA, Ellwood SR (2010). Aligning a new reference genetic map of Lupinus angustifolius with the genome sequence of the model legume, Lotus japonicus. DNA Research 17: 73–83.

Ng DWK, Chandrasekharan MB, Hall TC (2006). Ordered histone modifications are associated with transcriptional poising and activation of the phaseolin promoter. Plant Cell 18: 119–132.

Nishizawa K, Takagi K, Teraishi M, Kita A, Ishimoto M (2010). Application of somatic embryos to rapid and reliable analysis of soybean seed components by RNA interference-mediated gene silencing. Plant Biotechnology 27: 409–420.

Nogueira FCS, Goncalves EF, Jereissati ES, Santos M, Costa JH, Oliveira-Neto OB, Soares AA, Domont GB, Campos FAP (2007). Proteome analysis of embryogenic cell suspensions of cowpea (Vigna unguiculata). Plant Cell Reports 26: 1333–1343.

Odeny DA, Jayashree B, Ferguson M, Hoisington D, Crouch J, Gebhardt C (2007). Development, characterization and utilization of microsatellite markers in pigeonpea. Plant Breeding 126: 130–136.

Page D, Aubert G, Duc G, Welham T, Domoney C (2002). Combinatorial variation in coding and promoter sequences of genes at the Tri locus in Pisum sativum accounts for variation in trypsin inhibitor activity in seeds. Molecular Genetics and Genomics 267: 359–369.

Pandey A, Choudhary MK, Bhushan D, Chattopadhyay A, Chakraborty S, Datta A, Chakraborty N (2006). The nuclear proteome of chickpea (Cicer arietinum L.) reveals predicted and unexpected proteins. Journal of Proteome Research 5: 3301–3311.

Pelaez P, Trejo MS, Iniguez LP, Estrada-Navarrete G, Covarrubias AA, Reyes JL, Sanchez F (2012). Identification and characterization of microRNAs in Phaseolus vulgaris by high-throughput sequencing. Bmc Genomics 13.

Perry JA, Wang TL, Welham TJ, Gardner S, Pike JM, Yoshida S, Parniske M (2003). A TILLING reverse genetics tool and a web-accessible collection of mutants of the legume Lotus japonicus. Plant Physiology 131: 866–871.

Polowick PL, Baliski DS, Bock C, Ray H, Georges F (2009). Over-expression of alpha-galactosidase in pea seeds to reduce raffinose oligosaccharide content. Botany-Botanique 87: 526–532.

Porch TG, Blair MW, Lariguet P, Galeano C, Pankhurst CE, Broughton WJ (2009). Generation of a mutant population for TILLING common bean genotype BAT 93. Journal of the American Society for Horticultural Science 134: 348–355.

Rakocevic A, Mondy S, Tirichine L, Cosson V, Brocard L, Iantcheva A, Cayrel A, Devier B, Abu El-Heba GA, Ratet P (2009). MERE1, a low-copy-number copia-type retroelement in Medicago truncatula active during tissue culture. Plant Physiology 151: 1250–1263.

Repetto O, Rogniaux H, Firnhaber C, Zuber H, Kuster H, Larre C, Thompson R, Gallardo K (2008). Exploring the nuclear proteome of Medicago truncatula at the switch towards seed filling. Plant Journal 56: 398–410.

Rogers C, Wen JQ, Chen RJ, Oldroyd G (2009). Deletion-based reverse genetics in Medicago truncatula. Plant Physiology 151: 1077–1086.

Rolletschek H, Hosein F, Miranda M, Heim U, Gotz KP, Schlereth A, Borisjuk L, Saalbach I, Wobus U, Weber H (2005). Ectopic expression of an amino acid transporter (VfAAP1) in seeds of Vicia narbonensis and pea increases storage proteins. Plant Physiology 137: 1236–1249.

Rosche EG, Blackmore D, Offler CE, Patrick JW (2005). Increased capacity for sucrose uptake leads to earlier onset of protein accumulation in developing pea seeds. Functional Plant Biology 32: 997–1007.

Sankaran RP, Huguet T, Grusak MA (2009). Identification of QTL affecting seed mineral concentrations and content in the model legume Medicago truncatula. Theoretical and Applied Genetics 119: 241–253.

Sato S, Isobe S, Tabata S (2010). Structural analyses of the genomes in legumes. Current Opinion in Plant Biology 13: 146–152.

Sato S, Nakamura Y, Kaneko T, Asamizu E, Kato T, Nakao M, Sasamoto S, Watanabe A, Ono A, Kawashima K, Fujishiro T, Katoh M, Kohara M, Kishida Y, Minami C, Nakayama S, Nakazaki N, Shimizu Y, Shinpo S, Takahashi C, Wada T, Yamada M, Ohmido N, Hayashi M, Fukui K, Baba T, Nakamichi T, Mori H, Tabata S (2008). Genome structure of the legume, Lotus japonicus. DNA Research 15: 227–239.

Saxena KB, Sultana R, Mallikarjuna N, Saxena RK, Kumar RV, Sawargaonkar SL, Varshney RK (2010). Male-sterility systems in pigeonpea and their role in enhancing yield. Plant Breeding 129: 125–134.

Schmidt H, Gelhaus C, Latendorf T, Nebendahl M, Petersen A, Krause S, Leippe M, Becker WM, Janssen O (2009). 2-D DIGE analysis of the proteome of extracts from peanut variants reveals striking differences in major allergen contents. Proteomics 9: 3507–3521.

Schmidt MA, Barbazuk WB, Sandford M, May G, Song ZH, Zhou WX, Nikolau BJ, Herman EM (2011). Silencing of soybean seed storage proteins results in a rebalanced protein composition preserving seed protein content without major collateral changes in the metabolome and transcriptome. Plant Physiology 156: 330–345.

Schmidt MA, Herman EM (2008). Proteome rebalancing in soybean seeds can be exploited to enhance foreign protein accumulation. Plant Biotechnology Journal 6: 832–842.

Schmutz J, Cannon SB, Schlueter J, Ma J, Mitros T, Nelson W, Hyten DL, Song Q, Thelen JJ, Cheng J, Xu D, Hellsten U, May GD, Yu Y, Sakurai T, Umezawa T, Bhattacharyya MK, Sandhu D, Valliyodan B, Lindquist E, Peto M, Grant D, Shu S, Goodstein D, Barry K, Futrell-Griggs M, Abernathy B, Du J, Tian Z, Zhu L, Gill N, Joshi T, Libault M, Sethuraman A, Zhang X-C, Shinozaki K, Nguyen HT, Wing RA, Cregan P, Specht J, Grimwood J, Rokhsar D, Stacey G, Shoemaker RC, Jackson SA (2010). Genome sequence of the palaeopolyploid soybean. Nature 463: 178–183.

Scippa GS, Rocco M, Ialicicco M, Trupiano D, Viscosi V, Di Michele M, Arena S, Chiatante D, Scaloni A (2010). The proteome of lentil (Lens culinaris Medik.) seeds: discriminating between landraces. Electrophoresis 31: 497–506.

Senguptagopalan C, Reichert NA, Barker RF, Hall TC, Kemp JD (1985). Developmentally regulated expression of the bean beta-phaseolin gene in tobacco seed. Proceedings of the National Academy of Sciences of the United States of America 82: 3320–3324.

Severin AJ, Woody JL, Bolon YT, Joseph B, Diers BW, Farmer AD, Muehlbauer GJ, Nelson RT, Grant D, Specht JE, Graham MA, Cannon SB, May GD, Vance CP, Shoemaker RC (2010). RNA-Seq atlas of Glycine max: a guide to the soybean transcriptome. Bmc Plant Biology 10.

Sexton PJ, Paek NC, Shibles RM (1998). Effects of nitrogen source and timing of sulfur deficiency on seed yield and expression of 11S and 7S seed storage proteins of soybean. Field Crops Research 59: 1–8.

Sha A-H, Li C, Yan X-H, Shan Z-H, Zhou X-A, Jiang M-L, Mao H, Chen B, Wan X, Wei W-H (2012). Large-scale sequencing of normalized full-length cDNA library of soybean seed at different developmental stages and analysis of the gene expression profiles based on ESTs. Molecular Biology Reports 39: 2867–2874.

Singh NK, Gupta DK, Jayaswal PK, Mahato AK, Dutta S, Singh S, Bhutani S, Dogra V, Singh BP, Kumawat G, Pal JK, Pandit A, Singh A, Rawal H, Kumar A, Prashat GR, Khare A, Yadav R, Raje RS, Singh MN, Datta S, Fakrudin B, Wanjari KB, Kansal R, Dash PK, Jain PK, Bhattacharya R, Gaikwad K, Mohapatra T, Srinivasan R, Sharma TR (2012). The first draft of the pigeonpea genome sequence. Journal of Plant Biochemistry and Biotechnology 21: 98–112.

Slightom JL, Sun SM, Hall TC (1983). Complete nucleotide-sequence of a French bean storage protein gene—phaseolin. Proceedings of the National Academy of Sciences of the United States of America–Biological Sciences 80: 1897–1901.

Song QX, Liu YF, Hu XY, Zhang WK, Ma BA, Chen SY, Zhang JS (2011). Identification of miRNAs and their target genes in developing soybean seeds by deep sequencing. Bmc Plant Biology 11.

Sreenivasulu N, Borisjuk L, Junker BH, Mock H-P, Rolletschek H, Seiffert U, Weschke W, Wobus U (2010). Barley grain development: toward an integrative view. In: Jeon KW (ed). International Review of Cell and Molecular Biology, Vol 281, pp 49–89.

Srinivasan A, Giri AP, Harsulkar AM, Gatehouse JA, Gupta VS (2005). A Kunitz trypsin inhibitor from chickpea (Cicer arietinum L.) that exerts anti-metabolic effect on podborer (Helicoverpa armigera) larvae. Plant Molecular Biology 57: 359–374.

Tabe L, Wirtz M, Molvig L, Droux M, Hell R (2010). Overexpression of serine acetlytransferase produced large increases in O-acetylserine and free cysteine in developing seeds of a grain legume. Journal of Experimental Botany 61: 721–733.

Tabe LM, Droux M (2002). Limits to sulfur accumulation in transgenic lupin seeds expressing a foreign sulfur-rich protein. Plant Physiology 128: 1137–1148.

Tadege M, Wang TL, Wen JQ, Ratet P, Mysore KS (2009). Mutagenesis and beyond! Tools for understanding legume biology. Plant Physiology 151: 978–984.

Tadege M, Wen JQ, He J, Tu HD, Kwak Y, Eschstruth A, Cayrel A, Endre G, Zhao PX, Chabaud M, Ratet P, Mysore KS (2008). Large-scale insertional mutagenesis using the Tnt1 retrotransposon in the model legume Medicago truncatula. Plant Journal 54: 335–347.

Takahashi K, Banba H, Kikuchi A, Ito M, Nakamura S (1994). An induced mutant line lacking the alpha-subunit of beta-conglycinin in soybean glycine-max (L) merrill. Breeding Science 44: 65–66.

Takahashi M, Uematsu Y, Kashiwaba K, Yagasaki K, Hajika M, Matsunaga R, Komatsu K, Ishimoto M (2003). Accumulation of high levels of free amino acids in soybean seeds through integration of mutations conferring seed protein deficiency. Planta 217: 577–586.

Tan QM, Zhang LZ, Grant J, Cooper P, Tegeder M (2010) Increased phloem transport of S-methylmethionine positively affects sulfur and nitrogen metabolism and seed development in pea plants. Plant Physiology 154: 1886–1896.

Tapan K, Bharadwaj C, Jain PK, Srinivasan R, Varshney R, Kumar J, Shailesh T, Tara Satyavathi C (2012). Targeting induced local lesions in genomics (TILLING): application of traditional mutagenesis for chickpea drought genomics. 2012. In: Summaries of papers presented in National Seminar on Indian Agriculture: Preparedness for Climate Change, March 24–25, 2012, NASC Complex, New Delhi, India. https://www.integratedbreeding.net/keywords/chickpea.

Taylor M, Chapman R, Beyaert R, Hernandez-Sebastia C, Marsolais F (2008). Seed storage protein deficiency improves sulfur amino acid content in common bean (Phaseolus vulgaris L.): redirection of sulfur from gamma-glutamyl-S-methyl-cysteine. Journal of Agricultural and Food Chemistry 56: 5647–5654.

Teng W, Han Y, Du Y, Sun D, Zhang Z, Qiu L, Sun G, Li W (2009). QTL analyses of seed weight during the development of soybean (Glycine max L. Merr.). Heredity 102: 372–380.

Teraishi M, Takahashi M, Hajika M, Matsunaga R, Uematsu Y, Ishimoto M (2001). Suppression of soybean beta-conglycinin genes by a dominant gene, Scg-1. Theoretical and Applied Genetics 103: 1266–1272.

Timko MP, Rushton PJ, Laudeman TW, Bokowiec MT, Chipumuro E, Cheung F, Town CD, Chen XF (2008). Sequencing and analysis of the gene-rich space of cowpea. Bmc Genomics 9.

Tusher VG, Tibshirani R, Chu G (2001). Significance analysis of microarrays applied to the ionizing radiation response (vol 98, pg 5116, 2001). Proceedings of the National Academy of Sciences of the United States of America 98: 10515–10515.

Udvardi MK, Kakar K, Wandrey M, Montanari O, Murray J, Andriankaja A, Zhang J-Y, Benedito V, Hofer JMI, Chueng F, Town CD (2007). Legume transcription factors: global regulators of plant development and response to the environment. Plant Physiology 144: 538–549.

Vandecasteele C, Teulat-Merah B, Morere-Le Paven MC, Leprince O, Vu BL, Viau L, Ledroit L, Pelletier S, Payet N, Satour P, Lebras C, Gallardo K, Huguet T, Limami AM, Prosperi JM, Buitink J (2011). Quantitative trait loci analysis reveals a correlation between the ratio of sucrose/raffinose family oligosaccharides and seed vigour in Medicago truncatula. Plant Cell and Environment 34: 1473–1487.

Varallyay E, Lichner Z, Safrany J, Havelda Z, Salamon P, Bisztray G, Burgyan J (2010). Development of a virus induced gene silencing vector from a legumes infecting tobamovirus. Acta Biologica Hungarica 61: 457–469.

Varshney RK, Penmetsa RV, Dutta S, Kulwal PL, Saxena RK, Datta S, Sharma TR, Rosen B, Carrasquilla-Garcia N, Farmer AD, Dubey A, Saxena KB, Gao J, Fakrudin B, Singh MN, Singh BP, Wanjari KB, Yuan M, Srivastava RK, Kilian A, Upadhyaya HD, Mallikarjuna N, Town CD, Bruening GE, He G, May GD, McCombie R, Jackson SA, Singh NK, Cook DR (2010). Pigeonpea genomics initiative (PGI): an international effort to improve crop productivity of pigeonpea (Cajanus cajan L.). Molecular Breeding 26: 393–408.

Verdier J, Kakar K, Gallardo K, Le Signor C, Aubert G, Schlereth A, Town CD, Udvardi MK, Thompson RD (2008). Gene expression profiling of M-truncatula transcription factors identifies putative regulators of grain legume seed filling. Plant Molecular Biology 67: 567–580.

Verdier J, Zhao J, Torres-Jerez I, Ge S, Liu C, He X, Mysore KS, Dixon RA, Udvardi MK (2012). MtPAR MYB transcription factor acts as an on switch for proanthocyanidin biosynthesis in Medicago truncatula. Proceedings of the National Academy of Sciences of the United States of America 109: 1766–1771.

Vigeolas H, Chinoy C, Zuther E, Blessington B, Geigenberger P, Domoney C (2008). Combined metabolomic and genetic approaches reveal a link between the polyamine pathway and albumin 2 in developing pea seeds. Plant Physiology 146: 74–82.

Vodkin LO, Khanna A, Shealy R, Clough SJ, Gonzalez DO, Philip R, Zabala G, Thibaud-Nissen F, Sidarous M, Stromvik MV, Shoop E, Schmidt C, Retzel E, Erpelding J, Shoemaker RC, Rodriguez-Huete AM, Polacco JC, Coryell V, Keim P, Gong G, Liu L, Pardinas J, Schweitzer P (2004). Microarrays for global expression constructed with a low redundancy set of 27,500 sequenced cDNAs representing an array of developmental stages and physiological conditions of the soybean plant. Bmc Genomics 5.

Vriet C, Welham T, Brachmann A, Pike M, Pike J, Perry J, Parniske M, Sato S, Tabata S, Smith AM, Wang TL (2010). A suite of Lotus japonicus starch mutants reveals both conserved and novel features of starch metabolism. Plant Physiology 154: 643–655.

Wang H-W, Zhang B, Hao Y-J, Huang J, Tian A-G, Liao Y, Zhang J-S, Chen S-Y (2007). The soybean Dof-type transcription factor genes, GmDof4 and GmDof11, enhance lipid content in the seeds of transgenic Arabidopsis plants. Plant Journal 52: 716–729.

Weigelt K, Kuester H, Rutten T, Fait A, Fernie AR, Miersch O, Wasternack C, Emery RJN, Desel C, Hosein F, Mueller M, Saalbach I, Weber H (2009). ADP-glucose pyrophosphorylase-deficient pea embryos reveal specific transcriptional and metabolic changes of carbon-nitrogen metabolism and stress responses. Plant Physiology 149: 395–411.

Yin F, Pajak A, Chapman R, Sharpe A, Huang S, Marsolais F (2011). Analysis of common bean expressed sequence tags identifies sulfur metabolic pathways active in seed and sulfur-rich proteins highly expressed in the absence of phaseolin and major lectins. Bmc Genomics 12: 268.

Yin FQ, Pajak A, Chapman R, Sharpe A, Huang S, Marsolais F (2011). Analysis of common bean expressed sequence tags identifies sulfur metabolic pathways active in seed and sulfur-rich proteins highly expressed in the absence of phaseolin and major lectins. Bmc Genomics 12.

Young ND, Debelle F, Oldroyd GED, Geurts R, Cannon SB, Udvardi MK, Benedito VA, Mayer KFX, Gouzy J, Schoof H, Van de Peer Y, Proost S, Cook DR, Meyers BC, Spannagl M, Cheung F, De Mita S, Krishnakumar V, Gundlach H, Zhou SG, Mudge J, Bharti AK, Murray JD, Naoumkina MA, Rosen B, Silverstein KAT, Tang HB, Rombauts S, Zhao PX, Zhou P, Barbe V, Bardou P, Bechner M, Bellec A, Berger A, Berges H, Bidwell S, Bisseling T, Choisne N, Couloux A, Denny R, Deshpande S, Dai XB, Doyle JJ, Dudez AM, Farmer AD, Fouteau S, Franken C, Gibelin C, Gish J, Goldstein S, Gonzalez AJ, Green PJ, Hallab A, Hartog M, Hua A, Humphray SJ, Jeong DH, Jing Y, Jocker A, Kenton SM, Kim DJ, Klee K, Lai HS, Lang CT, Lin SP, Macmil SL, Magdelenat G, Matthews L, McCorrison J, Monaghan EL, Mun JH, Najar FZ, Nicholson C, Noirot C, O'Bleness M, Paule CR, Poulain J, Prion F, Qin BF, Qu CM, Retzel EF, Riddle C, Sallet E, Samain S, Samson N, Sanders I, Saurat O, Scarpelli C, Schiex T, Segurens B, Severin AJ, Sherrier DJ, Shi RH, Sims S, Singer SR, Sinharoy S, Sterck L, Viollet A, Wang BB, Wang KQ, Wang MY, Wang XH, Warfsmann J, Weissenbach J, White DD, White JD, Wiley GB, Wincker P, Xing YB, Yang LM, Yao ZY, Ying F, Zhai JX, Zhou LP, Zuber A, Denarie J, Dixon RA, May GD, Schwartz DC, Rogers J, Quetier F, Town CD, Roe BA (2011). The Medicago genome provides insight into the evolution of rhizobial symbioses. Nature 480: 520–524.

Young ND, Udvardi M (2009). Translating Medicago truncatula genomics to crop legumes. Current Opinion in Plant Biology 12: 193–201.

Zhang JA, Liang S, Duan JL, Wang J, Chen SL, Cheng ZS, Zhang Q, Liang XQ, Li YR (2012). De novo assembly and characterisation of the transcriptome during seed development, and generation of genic-SSR markers in Peanut (Arachis hypogaea L.). Bmc Genomics 13.

Zhang Y, Sledge MK, Bouton JH (2007). Genome mapping of white clover (Trifolium repens L.) and comparative analysis within the Trifolieae using cross-species SSR markers. Theoretical and Applied Genetics 114: 1367–1378.

Zhang Z, Yu J, Li D, Zhang Z, Liu F, Zhou X, Wang T, Ling Y, Su Z (2010). PMRD: plant microRNA database. Nucleic Acids Research 38: D806–D813.

Zhao D, Cheng X, Wang L, Wang S, Ma Y (2010). Integration of mungbean (Vigna radiata) genetic linkage map. Acta Agronomica Sinica 36: 932–939.

11 Cotton Fiber Genomics

Xueying Guan and Z. Jeffrey Chen

Introduction

Cotton is the largest source of renewable textile fiber and an important oil crop. Cotton fibers make comfortable apparel, beautiful home furnishings, durable paper money such as U.S. dollars (75% cotton and 25% linen), and innovative products such as plastics and digital screens. World consumption of cotton fiber is ~115 million bales or ~27 million metric tons per year (National Cotton Council, Cotton Council International, http://www.cottonusa.org/index.cfm). Cotton is also a major oilseed crop. Because of its flavor stability, cottonseed oil is used for making mayonnaise, salad oil, and salad dressing. Nonhydrogenated cottonseed oil is also used as a deep frying medium to reduce trans fatty acid content in French fries (Daniel *et al.*, 2005). However, raw cottonseed oil has to be extensively treated to reduce the level of gossypol, the consumption of which may produce toxic effects in humans, monogastric animals, and insects (Sunilkumar *et al.*, 2006; Mao *et al.*, 2007). In addition, cottonseed meal (after removal of oil) is the second most abundant plant protein feed available throughout the United States, after soybean meal (National Cottonseed Products Association, www.cottonseed.com). It is usually used for animal feed and in organic fertilizers. Modifying cottonseed for food and feed could profoundly enhance nutrition and livelihoods of millions of people in food-challenged economies.

Cotton fiber is seed hair or trichome and an ideal model for the study of plant cell elongation and cell wall and cellulose biosynthesis (Delmer, 1999; Kim and Triplett, 2001). Each seed produces ~25,000 cotton fibers, which represent 25%–30% of epidermal cells on a cotton ovule (Basra and Malik, 1984; Wilkins and Jernstedt, 1999). Each cotton fiber is a singular and elongated cell from the epidermal layer of the ovule and can reach 1–3 cm in length, which is probably the longest plant cell type (Kim and Triplett, 2001). The fiber is composed of nearly pure cellulose, the largest component of plant biomass. The completion of cotton fiber development and maturation takes 50–60 days, from fiber cell initiation, cell elongation, primary cell wall and secondary cell wall formation, to dehydration and maturation.

Cotton is a good model for understanding the functional and agronomic significance of polyploidy and genome size variation within the *Gossypium* genus. Fiber trait is a good example of natural and artificial selection during cotton genome evolution. Greater than 95% of the annual cotton crop worldwide is upland or American cotton (*Gossypium hirsutum* L.), and the extra-long staple or Pima cotton (*Gossypium barbadense* L.) accounts for <2% (National Cotton Council, 2010). Genetic and genomic improvement of cotton fiber production and processing will ensure that this

Seed Genomics, First Edition. Edited by Philip W. Becraft.

natural-renewable product will be competitive with petroleum based synthetic fibers. This chapter reviews molecular and genomic studies on cotton fiber development with an emphasis on cotton fiber initiation.

Cotton Fiber Development

Cotton fibers are unicellular seed trichomes differentiated from the ovule epidermal layer. Fiber development is a highly regulated process that includes four overlapping phases termed fiber initiation, fiber elongation, secondary wall biosynthesis, and maturation (Figure 11.1A) (Basra and Malik, 1984; Tiwari and Wilkins, 1995; Wilkins and Jernstedt, 1999; Kim and Triplett, 2001). In *G. hirsutum*, lint fibers develop before or on the day of anthesis (defined as 0 days post anthesis [DPA]) (green-colored cells, Figure 11.1D), and the process is quasisynchronized in each developing ovule and among ovules within each ovary (boll) (Figure 11.1B). Fuzz fiber development usually occurs in a later stage but varies among genotypes (Figure 11.1E); fuzz fibers cease developing shortly after (brown-colored cells, Figure 11.1E) and remain attached to seeds after ginning (Figure 11.1F, lower middle). During ginning, lint fibers are spun into yarn, and fuzz fibers become trash. In fiberless mutants such as naked seed (*N1N1*), fiber initiation is delayed (Figure 11.1B), and no fuzz fiber is formed (Figure 11.1F, lower right corner) (Lee *et al.*, 2006). The fiber initials for lint fiber continue to grow without cell division for ~16–25 days (fiber elongation), concurrently with the biosynthesis of cellulose, which is the primary composition of cell walls in fibers (Figure 11.1C) (John and Crow, 1992; Kim and Triplett, 2001). The elongated fiber cells can range in length from 1–3 cm (Figure 11.1F) (Kim and Triplett, 2001). Fibers in *G. barbadense* may reach lengths of 6 cm. Fiber cells from *G. hirsutum* range in diameter from 11–22 μm, and the fiber length is 1000–3000 times longer than their width. Finally, fiber cells start to mature at ~50–60 DPA when cotton bolls open (Figure 11.1C). Long mature fibers (lint fibers) can be detached from the seeds and used for textile and other applications.

Among four distinctive stages of fiber development, fiber initiation and fiber elongation are extensively investigated. Despite recent advances, the molecular basis of fiber initiation remains poorly understood. Fiber initiation is a quasisynchronous process that occurs rapidly right after anthesis. Fiber cell differentiation from protodermal cells may start even before anthesis (about 2–3 days before anthesis) (Graves and Stewart, 1988a, 1988b). After application with auxin and gibberellic acid (GA), ovules harvested at 3 DPA and 2 DPA can produce fiber on culture media (Graves and Stewart, 1988b). Cell fate determination (fiber cells or nonfiber cells) precedes the formation of morphologically visible fiber cell initials (Figure 11.1D and E), and ~15%–25% of the epidermal cells become spinnable fibers (Kim and Triplett, 2001; Lee *et al.*, 2006). Fiber initiation and elongation are orchestrated in each fiber cell through changes in gene expression and intercellular signaling pathways (Lee *et al.*, 2006). Fiber elongation requires an increased level of protein synthesis for rapid cell growth and development (Basra and Malik, 1984).

Roles for Transcription Factors in Development of *Arabidopsis* Leaf Trichomes, Seed Hairs, and Cotton Fibers

Together with *Arabidopsis* leaf trichome, root hair, and pollen tube (Larkin *et al.*, 2003; Hülskamp, 2004), cotton fiber cells are used as a model to study plant cell differentiation (Kim and Triplett, 2001; Lee *et al.*, 2007). After an ovular protodermal cell has initiated to become a fiber cell, the

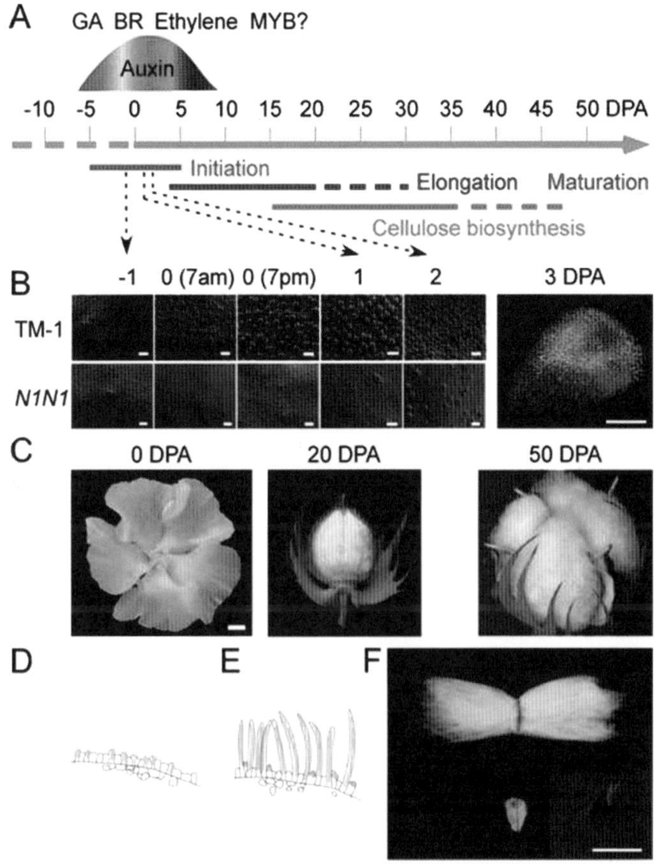

Figure 11.1 Cotton fiber development. *A,* Continuous and overlapping stages of cotton fiber development, including fiber cell initiation, elongation, cellulose (secondary cell wall) biosynthesis, and maturation. During elongation and maturation stages, cellulose biosynthesis is the dominant cellular metabolic process. (Secondary cell wall accumulation starts by depositing the cellulose under the primary cell wall after 15 DPA.) Regulation of phytohormones, including auxin, brassinosteroid (BR), and ethylene and transcription regulators, including MYB proteins, is implicated in cotton fiber cell initiation. *B,* Scanning electron microscope images of cotton fiber initiation processes. TM-1, *G. hirsutum* L. TM-1; N1N1, naked seed mutant. The ovules were collected at −1, 0 (7 a.m. and 7 p.m.), and 1 and 2 days after anthesis (0 = on the day of anthesis). Bar = 20 μm; bar = 100 μm for TM-1 ovule at 3 DPA. *C,* TM-1 cotton flower on 0 DPA, boll on 20 DPA and 50 DPA. *D,* Diagram of cotton fiber lint cell initiation on 0 DPA (green). *E,* Diagram of fuzz fiber cell initiation on 5–7 DPA (orange) along with lint fiber (green). *F,* Mature cotton lint and fuzz fiber on TM-1 seed. *Inset,* Mature seed of the fiberless mutant *XuFL.* (For color detail, see color plate section.)

cell nucleus starts a process of endoreduplication, the cell elongates, and primary and secondary cell walls are formed through biosynthesis of cellulose that can account for >95% of the dry fiber weight after dehydration (Meinert and Delmer, 1977; Kim and Triplett, 2001). An *Arabidopsis* leaf trichome cell undergoes a similar process of nuclear endoreduplication, cell extension, and cell wall biosynthesis (Hülskamp, 2004). The main differences are that leaf trichome development is exposed to light, and the trichome forms 0–3 branches; whereas cotton fiber cell development proceeds in dark conditions under the cover and compact of carpel walls, and the cotton fiber is unbranched.

At the molecular level, cotton seed hair development shares many similarities with *Arabidopsis* leaf trichome development (Lee *et al.*, 2007), which is mediated by a "trichome activation complex"

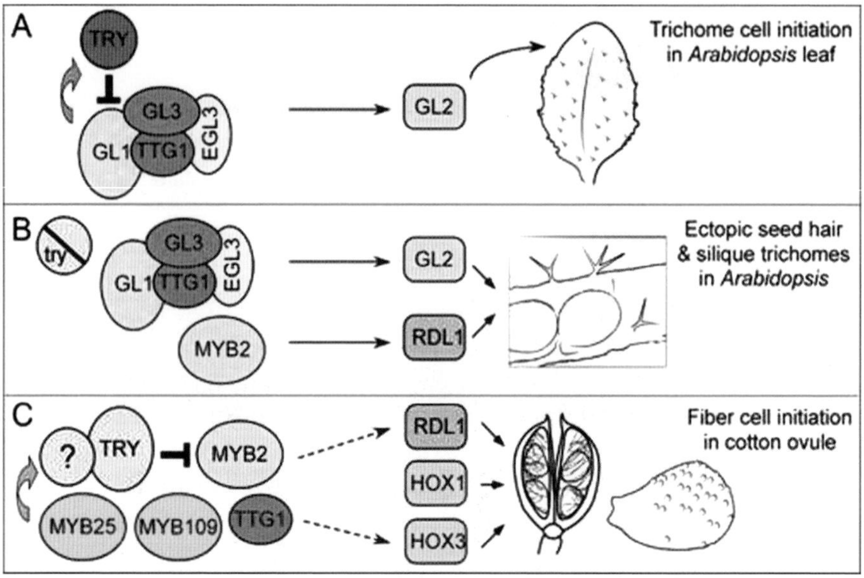

Figure 11.2 Models for leaf trichome and seed hair development. *A,* The trichome initiation complex consisting of GL1, GL3, EGL3, TTG1, and a negative regulator TRY promotes GL2 expression, leading to initiation of leaf trichomes in *Arabidopsis thaliana*. *B,* Cotton fiber genes, including *MYB2* and *RDL1,* stimulate the development of ectopic seed hairs and silique trichomes in *A. thaliana*. *C,* Multiple transcription factors identified in cotton fibers are shown to play roles in the development of seed hair and fiber in cotton. (From Guan *et al.* [2011].) (For color detail, see color plate section.)

(Figure 11.2A) (Szymanski *et al.*, 2000; Esch *et al.*, 2004; Hülskamp, 2004; Ishida *et al.*, 2008; Pesch and Hülskamp, 2009). *Arabidopsis* trichome cell fate is determined by positive transcription regulators, including a R2R3 MYB protein GLABROUS1 (GL1) (Oppenheimer *et al.*, 1991; Larkin *et al.*, 1993; Larkin *et al.*, 1994), a WD repeated protein TTG1 (Walker *et al.*, 1999), and a bHLH protein GL3 (Payne *et al.*, 2000) and its homolog EGL3. These proteins form a complex (Payne *et al.*, 2000; Zhang *et al.*, 2003; Zimmermann *et al.*, 2004) to drive expression of the downstream homeodomain transcription factor gene *GL2* (Szymanski and Marks, 1998). The transcription factor complex also activates another small single repeat R3 MYB transcription factor TRY (Hülskamp *et al.*, 1994; Esch *et al.*, 2004). TRY can competitively disrupt the binding between GL1 and GL3, to function as a feedback negative regulator that suppresses trichome formation (Schellmann *et al.*, 2002; Esch *et al.*, 2003). *Arabidopsis* seeds normally do not develop hair, so little is known about additional genes and regulatory pathways responsible for seed hair formation.

In the cotton genome, all homologs of *Arabidopsis* trichome positive regulators have been identified (Figure 11.2B). *G. hirsutum* MYB109 (GhMYB109) and GhMYB2 are putative MYB transcription factors present in cotton fibers (Suo *et al.*, 2003). Overexpression of *G. arboreum MYB2* (*GaMYB2*) complements a *gl1* mutation (trichomeless) and restores the trichome phenotype in *Arabidopsis thaliana* (Wang *et al.*, 2004). Cotton fiber–related genes *MYB2* and *RDL1* serve as positive regulators for both leaf trichome and seed hair development (Figure 11.2B). In cotton fiber cells, multiple regulators may determine and promote the cell fate for seed hair. The functional homologs of *TTG1*, *GL3*, and *GL2* from the cotton genome are also cloned and studied. Genome-wide analysis of gene expression showed that *G. hirsutum MYB2*, *MYB25*, *MYB109*, *HOX1*, *HOX2*, and *TTG1* are associated with seed hair development (Figure 11.2C). Individual and interactive roles for these factors in seed hair development are largely unknown. It is likely that multiple factors are required for the initiation of trichomes on reproductive organs such as siliques and hairs on seeds. One gene in *A. thaliana* has multiple homologs in cotton because of its polyploid origin. For example, at least four homologs of *TTG1* (Humphries *et al.*, 2005) and three homologs of *GL2* (Guan *et al.*, 2008) are present in the cotton genome. Searches for homologs of negative regulators have not identified good candidates. These data suggest that *Arabidopsis* and cotton use similar transcription factors for regulating the development of leaf trichomes and seed hairs.

A more recent study has demonstrated that cotton fiber–related genes promote development of seed hair and silique trichomes (Guan *et al.*, 2011). Overexpression of *GhMYB2* and its downstream gene, *GhRDL1*, activated trichome and seed hair pathways in the *try* mutant *A. thaliana*. TRIPTYCHON (TRY) is a R3MYB protein that inhibits trichome formation, and *try* mutants form trichomes in clusters instead of as evenly spaced individuals (Hülskamp *et al.*, 1994; Esch *et al.*, 2004). The transgenic plants developed ectopic trichomes on siliques (Figure 11.3A and B) and

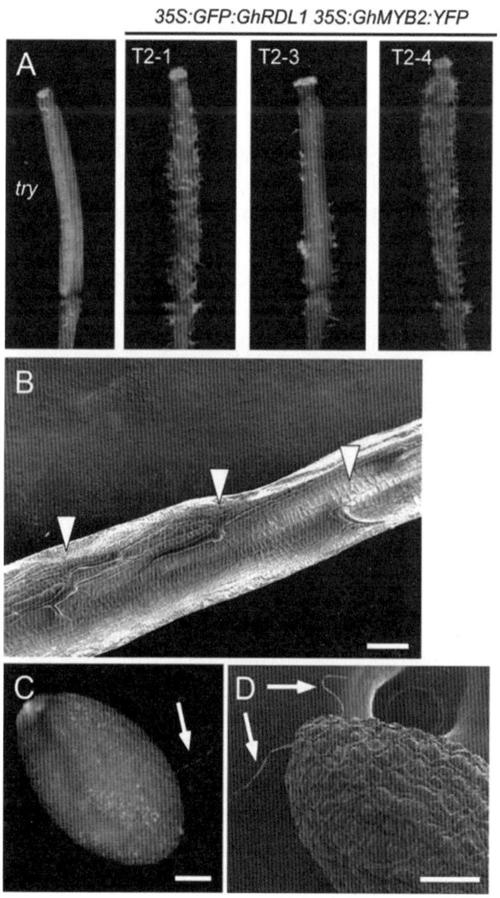

Figure 11.3 Development of ectopic trichomes inside and outside of siliques in transgenic *Arabidopsis thaliana* plants overexpressing *GhMYB2* and *GhRDL1*. *A,* Ectopic trichomes outside of siliques in transgenic *A. thaliana* plants. *B,* Ectopic trichomes inside of siliques *(arrows)* in transgenic *A. thaliana* plants. *C* and *D,* Seed hair observed in transgenic *A. thaliana* seeds under light microscope *(C)* and Scanning electron microscope *(D)*. (From Guan *et al.*, [2011].) (For color detail, see color plate section.)

seed hair in seeds (Figure 11.3C and D), providing strong evidence for cotton fiber–related genes in trichome and seed hair development. Overexpressing *GhMYB2* or *GhRDL1* alone in the *A. thaliana try* mutant that produces multibranched trichomes leads to ~5% of transgenic seeds producing seed hair. Overexpressing both *GhMYB2* and *GhRDL1* results in ectopic trichomes on outer and inner sides of siliques and seed hair development in 10% of the transgenic seeds. GFP:GhRDL1 and GhMYB2:YFP fusion proteins are colocalized in nuclei of ectopic trichomes.

In addition to results in homologs of transcription factor genes in trichome pathways, many cotton genes are found to be highly expressed during fiber development (Wu *et al.*, 2006; An *et al.*, 2007; Chen and Guan, 2011). These genes include *GhMYB25,* which is predominately expressed in ovules and in fiber cell initials (Wu *et al.*, 2006). GhMYB25 is a homolog of AmMIXTA/AmMYBML1 that controls conical cell and trichome differentiation in *Antirrhinum majus* petals (Noda *et al.*, 1994; Perez-Rodriguez *et al.*, 2005). Suppressing *GhMYB25*-like gene expression by RNA interference in cotton impairs fiber development (Walford *et al.*, 2011). *G. barbadense* ML1 (GbML1), a L1 box binding protein present only in cotton, binds GbMYB25 in *G. barbadense*. The data suggest additional and possibly novel cotton genes are required for fiber cell initiation.

Fiber Cell Expansion through Cell Wall Biosynthesis

Cotton fiber cells elongate rapidly shortly after fertilization up to ~20 DPA. During this process, a fiber cell can expand >2 cm in length through primary cell wall expansion and accumulate cellulose for biosynthesis of the secondary cell wall. This elongation process is rapid and associated with elevated expression levels of genes in the functional categories of cell skeleton, carbohydrate metabolism, and expansin (Ruan *et al.*, 2001; Shi *et al.*, 2006; An *et al.*, 2007; Gou *et al.*, 2007; Haigler *et al.*, 2009).

It is well known that functional activities of expansin, xyloglucan, endo-(1,4)-β-D-glucanase, and reactive oxygen species (ROS) are important to the enlargement of primary cell walls (Cosgrove, 2005). Because the primary cell wall expansion determines the cotton fiber cell length, the primary cell wall is critical for the cotton fiber quality. Six cotton *expansin A* genes are expressed differentially in the rapid elongation stage of fiber development (An *et al.*, 2007). High expression levels of cotton *expansin* genes during early stages of cotton fiber development coincide with the active biosynthesis of primary cell wall. Three genes encoding xyloglucan endotransglycosylases/hydrolases (XTH) in cotton are expressed in elongating cotton fiber cells (Lee *et al.*, 2007). *XTH* transcript levels are increased in transgenic cotton plants overexpressing *GhXTH1*, and these plants produce longer fibers (Lee *et al.*, 2010). Two groups of genes – *GhAPXs* encoding ascorbate peroxidase and *GhPOXs* encoding class III peroxidase – are actively expressed during cotton fiber elongation and related to the accumulation of ROS, such as hydroxyl radical (—OH) (Qin *et al.*, 2008; Mei *et al.*, 2009). Genome-wide gene expression analysis found an evolutionary role of ROS in the fiber trait (Hovav *et al.*, 2008). Three genes encoding peroxidases were consistently expressed in domesticated and wild cotton species with long fibers, but expression was not detected in wild species with short fibers. *GhAPX1* and *GhPOX1* are highly induced in cotton fiber cells specifically in the rapid elongation stage (Qin *et al.*, 2008; Mei *et al.*, 2009). ROS functions as wall-loosening agents in living cells. These genes may affect fiber length by participating in primary cell wall extension.

As substrates for biosynthesis of polysaccharides (primary cell wall) and cellulose (secondary cell wall), glucose, fructose, galactose, and sucrose accumulate in rapidly growing cotton fiber cells. At 9 DPA, >90% metabolites in rapidly elongating fiber cells are glucose, fructose, galactose, and sucrose (Gou *et al.*, 2007). This trend of composition is relatively stable in cotton fiber cells from 6–18 DPA. Correspondingly, the genes encoding sucrose and other carbohydrate synthases and transporters are actively expressed during the fiber elongation stage. The genes that have been investigated encode *G. hirsutum* sucrose transporter1 (GhSUT1) (Ruan *et al.*, 2001), sucrose synthase (GhSUS) (Ruan *et al.*, 2003), vacuolar invertase1 (GhVIN1) (Wang *et al.*, 2010), and β-1,3-glucanase (GhGluc) (Ruan *et al.*, 2004). Sucrose and potassium are major osmotic solutes that are imported into cotton fiber from cells underneath the seed coat. The transport is either through plasmodesmata (PD) or through transporters across the cell wall and plasma membrane, which is facilitated by tonoplast H^+-ATPases (Ruan *et al.*, 2001). *GhKT* encodes a cotton potassium transporter. GhKT and GhSUT1 together may facilitate transport of sucrose and potassium through PD into cotton fiber cells to maintain the turgor pressure that drives the cotton fiber elongation (Ruan *et al.*, 2001, 2004). Sucrose is also one of the substrates for cell wall biosynthesis. In fiber cells, sucrose is degraded into UDP-glucose and fructose by sucrose synthase in the cytoplasm or hydrolyzed by acid invertase into glucose and fructose in the vacuole (Delmer, 1999; Salnikov *et al.*, 2001; Wang *et al.*, 2010). During cellulose biosynthesis, actin and tubulin are uniformly distributed along with sucrose synthases (Salnikov *et al.*, 2003). Transgenic cotton expressing antisense *GhSUS* produces short cotton fibers (Ruan *et al.*, 2003). Overexpressing *GhVIN1* and repressing *GhVIN1* by *ghvin1*-RNAi in transgenic cotton shows positive correlation of *GhVIN1* expression levels with cotton fiber length (Wang *et al.*, 2010). These data suggest that carbohydrate metabolism and transport play roles in fiber cell elongation.

Sucrose is the initial carbon source for cellulose biosynthesis. The cellulose synthase (CESA) uses UDP-glucose as a substrate to produce cellulose (Delmer, 1999). Secondary cell wall deposition begins at 15–18 DPA. Mature cotton fibers consist of >95% of pure cellulose.

Regulation of Phytohormones during Cotton Fiber Development

Cotton fiber development is effected by most plant hormones (Lee *et al.*, 2007). Auxin and GA are known to affect fiber cell development (Beasley, 1973). Under *in vitro* culture conditions, a combination of 500 nM auxin [indole-3-acetic acid (IAA)] with 5.0 mM GA is found to be optimal for fiber production (Beasley and Ting, 1974). Ethylene and brassinosteroid (BA) can also positively affect fiber growth in cultured ovules at certain concentrations (Sun *et al.*, 2005; Shi *et al.*, 2006; Clark *et al.*, 2010). Microarray analysis of gene expression in elongating cotton fibers identified an enrichment of genes in ethylene biosynthesis and signaling pathways (Shi *et al.*, 2006). Further analysis found a role for ethylene in fiber cell elongation. In cotton, ethylene affects auxin biosynthesis and transport (Beyer and Morgan, 1969), and ethylene function is affected by GA activity (Morgan and Durham, 1975). It is likely there are interactive roles for different growth hormones in cotton fiber cell initiation and development.

The mechanisms for how phytohormones stimulate cotton fiber development are poorly understood. Genome-wide analysis of expressed sequence tags (ESTs) and transcriptomes suggests that some genes involved in phytohormone biosynthetic and signaling pathways are highly expressed at different stages of cotton fiber development. In cotton fiber EST collections, the enriched genes include homologs of GA biosynthetic genes—*GA20ox, GA3ox,* and *KO*—and genes involved in GA signaling pathways, such as *GAI, RGL2, RGL1, DDF1, PHOR1, RSG, POTH1*, *PKL, GL1, GAMYB, AGAMOUS*, and *LUE1* (Yang *et al.*, 2006). DELLA protein, an important GA signaling transduction signal component (Richards *et al.*, 2001), also accumulates during early stages of cotton fiber development (Guan *et al.*, 2011).

A list of phytohormone-related genes identified from two microarray experiments is shown in Table 11.1. These genes were expressed at high levels in ovules (−2 DPA) and fibers (3–20 DPA). Although the function of individual genes in cotton fiber development has yet to be elucidated, this

Table 11.1 Alternative Expressed Phytohormone-Related Genes in Cotton Fiber

Phytohormone	Gene	Description	Reference
GA	*GA20ox1*	Gibberellin 20-oxidase 1	1
	GA3ox1	Gibberellin 3-hydroxylase 1 mRNA	1
	DELLA	Negative regulators of gibberellin acid (GA) *signaling*	2
Auxin	*ARF2*	Auxin response factor 2	2
	ARF6	Auxin response factor 6	2
	ARF11	Auxin response factor 11	2
Ethylene	*RTE1*	Reversion-to-ethylene sensitivity1	2
	ERF3	Ethylene responsive element binding factor 3	2
	ERF4	Ethylene responsive element binding factor 4	2
	EIN2	Ethylene insensitive 2	2
	ACO1	The 1-aminocyclopropane-1-carboxylic acid oxidase 1	1
	ACO2	The 1-aminocyclopropane-1-carboxylic acid oxidase 2	1
	ACO3	The 1-aminocyclopropane-1-carboxylic acid oxidase 3	1
BR	*BRI1*	Brassinosteroid insensitive 1	2
	SMT1	24-sterol C-methyltransferase (SMT1)	1
	DET2	Steroid 5-alpha-reductase	1
ABA	*ABI5*	ABA insensitive5	2

[1]2 Shi *et al.*, 2006.
[2]Guan *et al.*, 2011.

group of phytohormone genes whose expression is enriched in fiber tissues suggests complex and interactive roles for these major phytohormones in cotton fiber cell development.

Auxin response factor (ARF) family gene transcripts are enriched in cotton EST collections and microarray analysis (Yang *et al.*, 2006; Lee *et al.*, 2007; Guan *et al.*, 2011). *ARF2*, *ARF6*, and *ARF11* are actively expressed in the cotton fiber cells in the early development stage (Guan *et al.*, 2011). Synthetic NAA suppresses secondary cell cellulose synthesis but increases cell elongation (Singh *et al.*, 2009). A study employed different promoters to drive expression of the IAA biosynthetic gene *iaaM* and control endogenous auxin in cotton ovules (Zhang *et al.*, 2011). The promoters that are able to increase IAA from −2 to 0 DPA in the ovule epidermal layers result in cotton plants that produce more and longer fiber cells and finer fibers. This suggests both timing and location of auxin production are important for cotton fiber production (Figure 11.1A).

Cotton Fiber Genes in Diploid and Tetraploid Cotton

The genus *Gossypium* includes ~45 diploid (2n = 2x = 26) and five tetraploid (2n = 4x = 52) species (Percival *et al.*, 1999). Diploid species fall into eight genomic groups (A–G and K). The African clade (A, B, E, and F genomes) originated in Africa and Asia, whereas the D-genome clade is indigenous to the Americas. A third diploid clade (C, G, and K genomes) is found in Australia. The most widely cultivated cotton is tetraploid. Polyploidization is estimated to have occurred 1–2 million years ago (Mya) (Wendel and Cronn, 2003), giving rise to five extant allotetraploid species. Two cultivated species, *G. hirsutum* L. and *G. barbadense* L., are classic natural allotetraploids that arose in the New World from interspecific hybridization between an A-genome–like ancestral African species and a D-genome–like American species. The closest extant relatives of the original tetraploid progenitors are the A-genome species *G. herbaceum* L. (A1) and *G. arboreum* L. (A2) and the D-genome species *G. raimondii* (D5) Ulbrich (Brubaker *et al.*, 1999). The A-genome species produce spinnable fiber and are cultivated on limited scale, whereas the D-genome species do not (Applequist *et al.*, 2001). Both the A-subgenome and the D-subgenome in the allotetraploids contribute to superior fiber traits (Jiang *et al.*, 1998; Saha *et al.*, 2006; Yang *et al.*, 2006). The fiber in allotetraploids is much longer and stronger compared with that in diploids, suggesting activation or silencing of homoeologous fiber-related genes via genetic and epigenetic mechanisms (Wendel and Cronn, 2003; Adams and Wendel, 2005; Chen, 2007; Lee *et al.*, 2007).

The genomic interactions in allotetraploid cotton are predicted to affect gene expression related to superior fiber quality. The existing data are debatable. Based on the distribution of fiber-related quantitative trait loci (QTLs), some studies indicate that contributions of the D subgenome are more significant to lint fiber development than the contributions of the A subgenome in allotetraploid cotton (Jiang *et al.*, 1998; Rong *et al.*, 2007), although the D genome ancestor does not produce fiber. However, other studies suggest that the A subgenome is more important than D subgenome for the fiber trait based on analysis of QTLs, EST collections, and gene expression (Mei *et al.*, 2004; Ulloa *et al.*, 2005; Yang *et al.*, 2006). Additional results argue that the two subgenomes are equally important for fiber development (Han *et al.*, 2006; Xu *et al.*, 2010). These discrepancies are probably caused by the lack of cotton genome sequence and partially informed knowledge about fiber gene expression. Distribution of QTLs in the cotton genome is uneven, and many QTLs are clustered in some regions of chromosomes (Mei *et al.*, 2004; Ulloa *et al.*, 2005; Rong *et al.*, 2007; Wang *et al.*, 2007). QTL clusters for fiber yield and quality may suggest multiple *cis*-acting and *trans*-acting factors that affect cotton fiber cell growth and development.

Further genetic improvement of cotton fiber production greatly depends on genome sequences. The strategy for sequencing cotton genomes has been outlined by the cotton community (Chen *et al.*, 2007). Toward that goal, physical maps of two diploid progenitors, *G. arboreum* and *G. raimondii*, and the most widely cultivated allotetraploid cotton, *G. hirsutum*, have been developed. *G. raimondii* has been prioritized for complete genome sequencing in collaboration with the Department of Energy (DOE) Joint Genome Institute (Chen *et al.*, 2007; Lin *et al.*, 2010). This physical map consists of 13,662 BAC-end sequences and 1585 contigs that are anchored to a cotton consensus linkage map with 2828 probes. The Gossypium raimondii genome was just seqeunced (Wang *et al.* 2012), and another version is expected to be made available soon. The construction of several BAC libraries and physical maps of the A-genome species *G. arboreum* and the allotetraploid species *G. hirsutum* are also underway (Ruan *et al.*, 2004). These physical maps will provide frameworks for sequence assembly and ultimate sequencing of allotetraploid cotton genomes.

Roles for Small RNAs in Cotton Fiber Development

Small RNAs, including small interfering RNAs (siRNAs), microRNAs (miRNAs), and trans-acting siRNAs (ta-siRNAs), affect gene expression and epigenetic regulation in plants and animals (Baulcombe, 2004; Chapman and Carrington, 2007; Chen, 2009). Many transcription factor and phytohormone regulator genes such as *ARFs* are regulated by miRNAs (Mallory *et al.*, 2005; Gutierrez *et al.*, 2009), and these target genes are actively expressed in cotton fibers (Guan *et al.*, 2011), suggesting a role for small RNAs in cotton fiber development.

Small RNAs have been investigated in cotton ovules at 0–10 DPA, and 583 unique sequences were identified (Abdurakhmonov *et al.*, 2008). Most of these small RNA sequences (62.1%) are 24-nt long. Using *Gossypium* EST databases, further analysis found 22 conserved miRNA precursors, including 15 from *G. hirsutum* (AADD), 5 from *G. raimondii* (DD), and 1 from *G. herbaceum* (AA) and *G. arboreum* (AA) (Khan Barozai *et al.*, 2008). A comparative analysis of small RNA expression changes between the wild-type and fiberless (*fl*) mutant in ovules at 0–10 DPA revealed 22 conserved candidate miRNA families including 111 members (Devor *et al.*, 2009). Some miRNA families display differential expression in ovules between the wild-type and *fl* mutant (Kwak *et al.*, 2009).

High-throughput sequencing analysis of small RNAs revealed differential expression of siRNAs and miRNA in fiber-bearing tissues (cotton ovules at −3 through 3 DPA cotton ovule) and nonfiber tissues (leaves) (Pang *et al.*, 2009). During early stages of cotton fiber development, 24-nt siRNAs are highly enriched, and 21-nt miRNAs are severely repressed. The enrichment of 24-nt siRNAs (78%–84% of total small RNA reads in fiber-bearing ovules) and relative to leaves suggests a role for these siRNAs that are yet to be uncovered in fiber cell development. Most of these 24-nt siRNAs are derived from transposable elements (TEs) and TE-associated genes. TEs are associated with gene silencing and RNA-directed DNA methylation (RdDM) (Mette *et al.*, 2000; Law and Jacobsen, 2010). Although the origin of these 24-nt small RNAs is unknown owing to the lack of the cotton genome sequence, the enrichment of 24-nt small RNAs in cotton ovules and fibers is reminiscent of maternal inheritance of a major class of 24-nt siRNAs whose biogenesis is dependent on RNA polymerase IV (PolIV or p4) (or p4-siRNAs) in developing seeds (Mosher *et al.*, 2009). Rapid and dynamic changes in siRNA and miRNA expression in cotton fibers are reminiscent of changes observed in *Arabidopsis* allotetraploids and their progenitors (Ha *et al.*, 2009). Cotton fibers are derived from maternal tissues in the epidermal layer of seeds. Enrichment of 24-nt siRNA in cotton ovules may suggest roles for these siRNAs in RdDM and gene regulation during cotton ovule development, fiber differentiation, and rapid growth and regulation of homoeologous genes in the allotetraploid genome.

Repression of miRNAs in cotton ovules and fibers is associated with upregulation of many genes encoding transcription factors and phytohormone regulators (Pang *et al.*, 2009). Target genes have been identified for >30 families of miRNAs, and transcript cleavage events have been documented for some miRNAs. These miRNAs cleave target gene transcripts that encode transcription factors, such as homeodomain (HD) proteins, APETAL2 protein, and phytohormone response factors such as ARFs. These results suggest that miRNAs play a role for regulating expression of these transcription factor genes in cotton fiber cell development. One of the new identified cotton miRNAs, Gh-miR2948, targets a gene encoding a sucrose synthase-like protein, suggesting a novel role for this miRNA in carbohydrate metabolism in cotton fibers. Although several new miRNAs and miRNA targets have also been identified, understanding the genome-wide landscape of regulatory pathways that are programmed by siRNAs and miRNAs during fiber cell development requires the cotton genome sequence, which is predicted to be available in the near future (Chen *et al.*, 2007; Lin *et al.*, 2010).

Conclusion

Economically, cotton is the largest source of renewable fibers for the textile and home furnishing industries, and cottonseed oil provides unique features of flavor and stability in cooking and food preparation. Biologically, cotton fiber is a model system to study plant cell differentiation, expansion, and wall biosynthesis. Analysis of cotton fibers is leading to a significant translational genomic interface with leaf trichomes, seed fibers, xylem elements, and other cells with thick walls. The cotton genus *Gossypium* is also a model for study of plant genome evolution, including polyploidization. With anticipated completion of genome sequences in diploid and allotetraploid cotton, genomic research in cotton will reveal new knowledge and information with respect to cell fate determination, cell elongation, cellulose biosynthesis, and polyploidization. Roles for phytohormone regulation and small RNAs in cotton cell differentiation and elongation are expected to generate broader impacts in the basic understanding of cell biology and small RNA biology as well as in the improvement of cotton quality and yield through application of genome-enabled biotechnological tools.

References

Abdurakhmonov IY, Devor EJ, Buriev ZT, Huang L, Makamov A, Shermatov SE, Bozorov T, Kushanov FN, Mavlonov GT, Abdukarimov A (2008). Small RNA regulation of ovule development in the cotton plant, G. hirsutum L. BMC Plant Biol 8: 93.

Adams KL, Wendel JF (2005). Polyploidy and genome evolution in plants. Curr Opin Plant Biol 8: 135–141.

An C, Saha S, Jenkins JN, Scheffler BE, Wilkins TA, Stelly DM (2007). Transcriptome profiling, sequence characterization, and SNP-based chromosomal assignment of the EXPANSIN genes in cotton. Mol Genet Genomics 278: 539–553.

Applequist WL, Cronn R, Wendel JF (2001). Comparative development of fiber in wild and cultivated cotton. Evol Dev 3: 3–17.

Basra A, Malik CP (1984). Development of the cotton fiber. Int Rev Cytol 89: 65–113.

Baulcombe D (2004). RNA silencing in plants. Nature 431: 356–363.

Beasley CA (1973). Hormonal regulation of growth in unfertilized cotton ovules. Science 179: 1003–1005.

Beasley CA, Ting IP (1974). The effects of plant growth substances on *in vitro* fiber development from unfertilized cotton ovules. Am J Bot 61: 188–194.

Beyer EM, Morgan PW (1969). Ethylene modification of an auxin pulse in cotton stem sections. Plant Physiol 44: 1690–1694.

Brubaker CL, Bourland FM, Wendel JF (1999). The origin and domestication of cotton. In: Smith CW, Cothren JT (eds). *Cotton: Origin, History, Technology, and Production*. John Wiley & Sons, New York, New York, pp 3–32.

Chapman EJ, Carrington JC (2007). Specialization and evolution of endogenous small RNA pathways. Nat Rev Genet 8: 884–896.

Chen X (2009). Small RNAs and their roles in plant development. Annu Rev Cell Dev Biol 25: 21–44.

Chen ZJ (2007). Genetic and epigenetic mechanisms for gene expression and phenotypic variation in plant polyploids. Annu Rev Plant Biol 58: 377–406.

Chen ZJ, Guan X (2011). Auxin boost for cotton. Nat Biotechnol 29: 407–409.

Chen ZJ, Scheffler BE, Dennis E, Triplett BA, Zhang T, Guo W, Chen X, Stelly DM, Rabinowicz PD, Town CD, Arioli T, Brubaker C, Cantrell RG, Lacape JM, Ulloa M, Chee P, Gingle AR, Haigler CH, Percy R, Saha S, Wilkins T, Wright RJ, Van Deynze A, Zhu Y, Yu S, Abdurakhmonov I, Katageri I, Kumar PA, Mehboob Ur R, Zafar Y, Yu JZ, Kohel RJ, Wendel JF, Paterson AH (2007). Toward sequencing cotton (Gossypium) genomes. Plant Physiol 145: 1303–1310.

Clark G, Torres J, Finlayson S, Guan X, Handley C, Lee J, Kays JE, Chen ZJ, Roux SJ (2010). Apyrase (nucleoside triphosphate-diphosphohydrolase) and extracellular nucleotides regulate cotton fiber elongation in cultured ovules. Plant Physiol 152: 1073–1083.

Cosgrove DJ (2005). Growth of the plant cell wall. Nat Rev Mol Cell Biol 6: 850–861.

Daniel DR, Thompson LD, Shriver BJ, Wu CK, Hoover LC (2005). Nonhydrogenated cottonseed oil can be used as a deep fat frying medium to reduce trans-fatty acid content in French fries. J Am Diet Assoc 105: 1927–1932.

Delmer DP (1999). Cellulose biosynthesis: exciting times for a difficult field of study. Annu Rev Plant Physiol Plant Mol Biol 50: 245–276.

Devor EJ, Huang L, Abdukarimov A, Abdurakhmonov IY (2009). Methodologies for in vitro cloning of small RNAs and application for plant genome(s). Int J Plant Genomics 2009: 915061.

Esch JJ, Chen M, Sanders M, Hillestad M, Ndkium S, Idelkope B, Neizer J, Marks MD (2003). A contradictory GLABRA3 allele helps define gene interactions controlling trichome development in Arabidopsis. Development 130: 5885–5894.

Esch JJ, Chen MA, Hillestad M, Marks MD (2004). Comparison of TRY and the closely related At1g01380 gene in controlling Arabidopsis trichome patterning. Plant J 40: 860–869.

Gou JY, Wang LJ, Chen SP, Hu WL, Chen XY (2007). Gene expression and metabolite profiles of cotton fiber during cell elongation and secondary cell wall synthesis. Cell Res 17: 422–434.

Graves DA, Stewart JM (1988a). Analysis of the protein constituency of developing cotton fibers. J Exp Bot 39: 59–69.

Graves DA, Stewart JM (1988b). Chronology of the differentiation of cotton (*Gossypium hirsutum* L.) fiber cells. Planta 175: 254–258.

Guan X, Lee JJ, Pang M, Shi X, Stelly DM, Chen ZJ (2011). Activation of Arabidopsis seed hair development by cotton fiber-related genes. PLoS One 6: e21301.

Guan XY, Li QJ, Shan CM, Wang S, Mao YB, Wang LJ, Chen XY (2008). The HD-Zip IV gene GaHOX1 from cotton is a functional homologue of the Arabidopsis GLABRA2. Physiol Plant 134: 174–182.

Gutierrez L, Bussell JD, Pacurar DI, Schwambach J, Pacurar M, Bellini C (2009). Phenotypic plasticity of adventitious rooting in Arabidopsis is controlled by complex regulation of AUXIN RESPONSE FACTOR transcripts and microRNA abundance. Plant Cell 21: 3119–3132.

Ha M, Lu J, Tian L, Ramachandran V, Kasschau KD, Chapman EJ, Carrington JC, Chen X, Wang XJ, Chen ZJ (2009). Small RNAs serve as a genetic buffer against genomic shock in *Arabidopsis* interspecific hybrids and allopolyploids. Proc Natl Acad Sci U S A 106: 17835–17840.

Haigler CH, Singh B, Wang G, Zhang D (2009). Genomics of cotton fiber secondary wall deposition and cellulose biogenesis. In: Paterson AH (ed): *Genetics and Genomics of Cotton, Plant Genetics and Genomics: Crops and Models 3*. Springer Science + Business Media, New York, New York, pp 385–417.

Han Z, Wang C, Song X, Guo W, Gou J, Li C, Chen X, Zhang T (2006). Characteristics, development and mapping of Gossypium hirsutum derived EST-SSRs in allotetraploid cotton. Theor Appl Genet 112: 430–439.

Hovav R, Udall JA, Chaudhary B, Hovav E, Flagel L, Hu G, Wendel JF (2008). The evolution of spinnable cotton fiber entailed prolonged development and a novel metabolism. PLoS Genet 4: e25.

Hülskamp M (2004). Plant trichomes: a model for cell differentiation. Nat Rev Mol Cell Biol 5: 471–480.

Hülskamp M, Misra S, Jurgens G (1994). Genetic dissection of trichome cell development in Arabidopsis. Cell 76: 555–566.

Humphries JA, Walker AR, Timmis JN, Orford SJ (2005). Two WD-repeat genes from cotton are functional homologues of the Arabidopsis thaliana TRANSPARENT TESTA GLABRA1 (TTG1) gene. Plant Mol Biol 57: 67–81.

Ishida T, Kurata T, Okada K, Wada T (2008). A genetic regulatory network in the development of trichomes and root hairs. Annu Rev Plant Biol 59: 365–386.

Jiang C, Wright RJ, El-Zik KM, Paterson AH (1998). Polyploid formation created unique avenues for response to selection in *Gossypium*. Proc Natl Acad Sci U S A 95: 4419–4424.

John ME, Crow LJ (1992). Gene expression in cotton (Gossypium hirsutum L.) fiber: cloning of the mRNAs. Proc Natl Acad Sci U S A 89: 5769–5773.

Khan Barozai MY, Irfan M, Yousaf R, Ali I, Qaisar U, Maqbool A, Zahoor M, Rashid B, Hussnain T, Riazuddin S (2008). Identification of micro-RNAs in cotton. Plant Physiol Biochem 46: 739–751.

Kim HJ, Triplett BA (2001). Cotton fiber growth in planta and in vitro: Models for plant cell elongation and cell wall biogenesis. Plant Physiol 127: 1361–1366.

Kwak PB, Wang QQ, Chen XS, Qiu CX, Yang ZM (2009). Enrichment of a set of microRNAs during the cotton fiber development. BMC Genomics 10: 457.

Larkin JC, Brown ML, Schiefelbein J (2003). How do cells know what they want to be when they grow up? Lessons from epidermal patterning in Arabidopsis. Ann Rev Plant Biol 54: 403–430.

Larkin JC, Oppenheimer DG, Lloyd AM, Paparozzi ET, Marks MD (1994). Roles of the GLABROUS1 and TRANSPARENT TESTA GLABRA aenes in Arabidopsis trichome development. Plant Cell 6: 1065–1076.

Larkin JC, Oppenheimer DG, Pollock S, Marks MD (1993). Arabidopsis GLABROUS1 gene requires downstream sequences for function. Plant Cell 5: 1739–1748.

Law JA, Jacobsen SE (2010). Establishing, maintaining and modifying DNA methylation patterns in plants and animals. Nat Rev Genet 11: 204–220.

Lee JJ, Hassan OS, Gao W, Wei NE, Kohel RJ, Chen XY, Payton P, Sze SH, Stelly DM, Chen ZJ (2006). Developmental and gene expression analyses of a cotton naked seed mutant. Planta 223: 418–432.

Lee JJ, Woodward AW, Chen ZJ (2007b). Gene expression changes and early events in cotton fibre development. Ann Bot (Lond) 100: 1391–1401.

Lee J, Burns TH, Light G, Sun Y, Fokar M, Kasukabe Y, Fujisawa K, Maekawa Y, Allen RD (2010) Xyloglucan endotransglycosylase/hydrolase genes in cotton and their role in fiber elongation. Planta, 232(5): 1191–1205.

Lin L, Pierce GJ, Bowers JE, Estill JC, Compton RO, Rainville LK, Kim C, Lemke C, Rong J, Tang H, Wang X, Braidotti M, Chen AH, Chicola K, Collura K, Epps E, Golser W, Grover C, Ingles J, Karunakaran S, Kudrna D, Olive J, Tabassum N, Um E, Wissotski M, Yu Y, Zuccolo A, ur Rahman M, Peterson DG, Wing RA, Wendel JF, Paterson AH (2010). A draft physical map of a D-genome cotton species (Gossypium raimondii). BMC Genomics 11: 395.

Mallory AC, Bartel DP, Bartel B (2005). MicroRNA-directed regulation of Arabidopsis AUXIN RESPONSE FACTOR17 is essential for proper development and modulates expression of early auxin response genes. Plant Cell 17: 1360–1375.

Mao YB, Cai WJ, Wang JW, Hong GJ, Tao XY, Wang LJ, Huang YP, Chen XY (2007). Silencing a cotton bollworm P450 monooxygenase gene by plant-mediated RNAi impairs larval tolerance of gossypol. Nat Biotechnol 25: 1307–1313.

Mei M, Syed NH, Gao W, Thaxton PM, Smith CW, Stelly DM, Chen ZJ (2004). Genetic mapping and QTL analysis of fiber-related traits in cotton (Gossypium). TAG 108: 280–291.

Mei W, Qin Y, Song W, Li J, Zhu Y (2009). Cotton GhPOX1 encoding plant class III peroxidase may be responsible for the high level of reactive oxygen species production that is related to cotton fiber elongation. J Genet Genomics 36: 141–150.

Meinert MC, Delmer DP (1977). Changes in biochemical composition of the cell wall of the cotton fiber during development. Plant Physiol 59: 1088–1097.

Mette MF, Aufsatz W, van der Winden J, Matzke MA, Matzke AJ (2000). Transcriptional silencing and promoter methylation triggered by double-stranded RNA. Embo J 19: 5194–5201.

Morgan PW, Durham JI (1975). Ethylene-induced leaf abscission is promoted by gibberellic acid. Plant Physiol 55: 308–311.

Mosher RA, Melnyk CW, Kelly KA, Dunn RM, Studholme DJ, Baulcombe DC (2009). Uniparental expression of PolIV-dependent siRNAs in developing endosperm of Arabidopsis. Nature 460: 283–286.

Noda K, Glover BJ, Linstead P, Martin C (1994). Flower colour intensity depends on specialized cell shape controlled by a Myb-related transcription factor. Nature 369: 661–664.

Oppenheimer DG, Herman PL, Sivakumaran S, Esch J, Marks MD (1991). A myb gene required for leaf trichome differentiation in Arabidopsis is expressed in stipules. Cell 67: 483–493.

Pang M, Woodward AW, Agarwal V, Guan X, Ha M, Ramachandran V, Chen X, Triplett BA, Stelly DM, Chen ZJ (2009). Genome-wide analysis reveals rapid and dynamic changes in miRNA and siRNA sequence and expression during ovule and fiber development in allotetraploid cotton (Gossypium hirsutum L). Genome Biol 10: R122.

Payne CT, Zhang F, Lloyd AM (2000). GL3 encodes a bHLH protein that regulates trichome development in arabidopsis through interaction with GL1 and TTG1. Genetics 156: 1349–1362.

Percival AE, Wendel JF, Stewart JM (1999). Taxonomy and germplasm resources. In: Smith CW, Cothren JT (eds). *Cotton: Origin, History, Technology, and Production*. John Wiley & Sons, New York, New York, pp 33–63.

Perez-Rodriguez M, Jaffe FW, Butelli E, Glover BJ, Martin C (2005). Development of three different cell types is associated with the activity of a specific MYB transcription factor in the ventral petal of Antirrhinum majus flowers. Development 132: 359–370.

Pesch M, Hülskamp M (2009). One, two, three... models for trichome patterning in Arabidopsis? Curr Opin Plant Biol 12: 587–592.

Qin YM, Hu CY, Zhu YX (2008). The ascorbate peroxidase regulated by H(2)O(2) and ethylene is involved in cotton fiber cell elongation by modulating ROS homeostasis. Plant Signal Behav 3: 194–196.

Richards DE, King KE, Ait-Ali T, Harberd NP (2001). How gibberellin regulates plant growth and development: a molecular genetic analysis of gibberellin signaling. Annu Rev Plant Physiol Plant Mol Biol 52: 67–88.

Rong J, Feltus FA, Waghmare VN, Pierce GJ, Chee PW, Draye X, Saranga Y, Wright RJ, Wilkins TA, May OL, Smith CW, Gannaway JR, Wendel JF, Paterson AH (2007). Meta-analysis of polyploid cotton QTL shows unequal contributions of subgenomes to a complex network of genes and gene clusters implicated in lint fiber development. Genetics 176: 2577–2588.

Ruan Y, Llewellyn D, Furbank R (2001). The control of single-celled cotton fiber elongation by developmentally reversible gating of plasmodesmata and coordinated expression of sucrose and k(+) transporters and expansin. Plant Cell 13: 47–60.

Ruan YL, Llewellyn DJ, Furbank RT (2003). Suppression of sucrose synthase gene expression represses cotton fiber cell initiation, elongation, and seed development. Plant Cell 15: 952–964.

Ruan YL, Xu SM, White R, Furbank RT (2004). Genotypic and developmental evidence for the role of plasmodesmatal regulation in cotton fiber elongation mediated by callose turnover. Plant Physiol 136: 4104–4113.

Saha S, Raska DA, Stelly DM (2006). Upland cotton (*Gossypium hirsutum* L.) x Hawaiian cotton (*G. tomentosum* Nutt. ex. Seem) F_1 hybrid hypoaneuploid chromosome substitution series. J Cotton Sci 10: 146–154.

Salnikov VV, Grimson MJ, Delmer DP, Haigler CH (2001). Sucrose synthase localizes to cellulose synthesis sites in tracheary elements. Phytochemistry 57: 823–833.

Salnikov VV, Grimson MJ, Seagull RW, Haigler CH (2003). Localization of sucrose synthase and callose in freeze-substituted secondary-wall-stage cotton fibers. Protoplasma 221: 175–184.

Schellmann S, Schnittger A, Kirik V, Wada T, Okada K, Beermann A, Thumfahrt J, Jurgens G, Hülskamp M (2002). TRIPTYCHON and CAPRICE mediate lateral inhibition during trichome and root hair patterning in Arabidopsis. Embo J 21: 5036–5046.

Shi YH, Zhu SW, Mao XZ, Feng JX, Qin YM, Zhang L, Cheng J, Wei LP, Wang ZY, Zhu YX (2006). Transcriptome profiling, molecular biological, and physiological studies reveal a major role for ethylene in cotton fiber cell elongation. Plant Cell 18: 651–664.

Singh B, Cheek HD, Haigler CH (2009). A synthetic auxin (NAA) suppresses secondary wall cellulose synthesis and enhances elongation in cultured cotton fiber. Plant Cell Rep 28: 1023–1032.

Sun Y, Veerabomma S, Abdel-Mageed HA, Fokar M, Asami T, Yoshida S, Allen RD (2005). Brassinosteroid regulates fiber development on cultured cotton ovules. Plant Cell Physiol 46: 1384–1391.

Sunilkumar G, Campbell LM, Puckhaber L, Stipanovic RD, Rathore KS (2006). Engineering cottonseed for use in human nutrition by tissue-specific reduction of toxic gossypol. Proc Natl Acad Sci U S A 103: 18054–18059.

Suo J, Liang X, Pu L, Zhang Y, Xue Y (2003). Identification of GhMYB109 encoding a R2R3 MYB transcription factor that expressed specifically in fiber initials and elongating fibers of cotton (Gossypium hirsutum L.). Biochim Biophys Acta 1630: 25–34.

Szymanski DB, Lloyd AM, Marks MD (2000). Progress in the molecular genetic analysis of trichome initiation and morphogenesis in *Arabidopsis*. Trends Plant Sci 5: 214–219.

Szymanski DB, Marks MD (1998). GLABROUS1 overexpression and TRIPTYCHON alter the cell cycle and trichome cell fate in Arabidopsis. Plant Cell 10: 2047–2062.

Tiwari SC, Wilkins TA (1995). Cotton (*Gossypium hirsutum*) seed trichomes expand via diffuse growing mechanism. Can J Bot 73: 746–757.

Ulloa M, Saha S, Jenkins JN, Meredith WR Jr, McCarty JC Jr, Stelly DM (2005). Chromosomal assignment of RFLP linkage groups harboring important QTLs on an intraspecific cotton (Gossypium hirsutum L.) Joinmap. J Hered 96: 132–144.

Walford SA, Wu Y, Llewellyn DJ, Dennis ES (2011). GhMYB25-like: a key factor in early cotton fibre development. Plant J 65: 785–797.

Walker AR, Davison PA, Bolognesi-Winfield AC, James CM, Srinivasan N, Blundell TL, Esch JJ, Marks MD, Gray JC (1999). The TRANSPARENT TESTA GLABRA1 locus, which regulates trichome differentiation and anthocyanin biosynthesis in Arabidopsis, encodes a WD40 repeat protein. Plant Cell 11: 1337–1350.

Wang B, Guo W, Zhu X, Wu Y, Huang N, Zhang T (2007). QTL mapping of yield and yield components for elite hybrid derived-RILs in upland cotton. J Genet Genomics 34: 35–45.

Wang L, Li XR, Lian H, Ni DA, He YK, Chen XY, Ruan YL (2010). Evidence that high activity of vacuolar invertase is required for cotton fiber and Arabidopsis root elongation through osmotic dependent and independent pathways, respectively. Plant Physiol 154: 744–756.

Wang S, Wang JW, Yu N, Li CH, Luo B, Gou JY, Wang LJ, Chen XY (2004). Control of plant trichome development by a cotton fiber MYB gene. Plant Cell 16: 2323–2334.

Wang, K. *et al.* (2012) The draft genome of a diploid cotton Gossypium raimondii. Nature Genetics 44, 1098–1103.

Wendel JF, Cronn RC (2003). Polyploidy and the evolutionary history of cotton. Adv Agronomy 78: 139–186.

Wilkins TA, Jernstedt JA (1999). 9. Molecular genetics of developing cotton fibers. In: Basra AM (ed). *Cotton Fibers*. Hawthorne Press, New York, New York, pp 231–267.

Wu Y, Machado AC, White RG, Llewellyn DJ, Dennis ES (2006). Expression profiling identifies genes expressed early during lint fibre initiation in cotton. Plant Cell Physiol 47: 107–127.

Xu Z, Yu JZ, Cho J, Yu J, Kohel RJ, Percy RG (2010). Polyploidization altered gene functions in cotton (Gossypium spp.). PLoS ONE 5: e14351.

Yang SS, Cheung F, Lee JJ, Ha M, Wei NE, Sze SH, Stelly DM, Thaxton P, Triplett B, Town CD, Chen ZJ (2006). Accumulation of genome-specific transcripts, transcription factors and phytohormonal regulators during early stages of fiber cell development in allotetraploid cotton. Plant J 47: 761–775.

Zhang F, Gonzalez A, Zhao M, Payne CT, Lloyd A (2003). A network of redundant bHLH proteins functions in all TTG1-dependent pathways of Arabidopsis. Development 130: 4859–4869.

Zhang M, Zheng X, Song S, Zeng Q, Hou L, Li D, Zhao J, Wei Y, Li X, Luo M, Xiao Y, Luo X, Zhang J, Xiang C, Pei Y (2011). Spatiotemporal manipulation of auxin biosynthesis in cotton ovule epidermal cells enhances fiber yield and quality. Nat Biotechnol 29: 453–458.

Zimmermann IM, Heim MA, Weisshaar B, Uhrig JF (2004). Comprehensive identification of Arabidopsis thaliana MYB transcription factors interacting with R/B-like BHLH proteins. Plant J Cell Mol Biol 40: 22–34.

12 Genomic Changes in Response to 110 Cycles of Selection for Seed Protein and Oil Concentration in Maize

Christine J. Lucas, Han Zhao, Martha Schneerman, and Stephen P. Moose

Introduction

Identification of the genetic mechanisms underlying phenotypic variation between and within populations is essential to understanding natural selection and evolution. Changes in gene expression are an important source of phenotypic variation and may arise as a result of a combination of *cis*-regulatory and *trans*-regulatory variation within complex genetic networks (Wittkopp, 2005). This complexity has hindered identification of genomic targets of both natural and artificial selection. Long-term phenotypic selection experiments in plants and animals provide an invaluable resource for studying evolutionary change and regulatory variation, with implications for enhancing the efficiency of genetic gain in plant and animal breeding (Dudley and Lambert, 2004). Plants have the additional unique advantage of regeneration, where replicate selection experiments may be conducted in parallel. The Illinois Long-Term Selection Experiment for grain composition is a classic in plant genetics; >800 cycles of selection have been conducted from the same source population during the past century (Figure 12.1).

Previous quantitative trait loci (QTL) mapping studies suggest that changes in both protein and oil concentration in the Illinois experiment are governed by many genes (Laurie *et al.*, 2004; Dudley *et al.*, 2007) with small phenotypic effect. However, there is also evidence that variation at relatively few loci can be associated with significant phenotypic differences. Advances in high-density genotyping coupled with functional genomics approaches enable a more thorough understanding of the genetic basis for the dramatic phenotypic responses observed. In this chapter, we provide an update on more recent phenotypic responses within the Illinois Long-Term Selection Experiment and related populations, a review of past efforts to characterize the genetic architecture of selection response, and a preliminary view of gene expression changes resulting from divergent selection for grain protein, and we introduce the use of promoter-reporter genes as tools to understand how selection has altered the regulation of seed storage protein genes. Collectively, these approaches illustrate that long-term selection for kernel composition has targeted variation in both trans-acting regulatory factors and specific genes known to influence oil or protein concentrations.

Background on the Illinois Long-Term Selection Experiment

The Illinois Long-Term Selection Experiment was initiated in 1896 at the University of Illinois by Cyril G. Hopkins and was the first planned attempt to improve cereal grain quality through

Seed Genomics, First Edition. Edited by Philip W. Becraft.

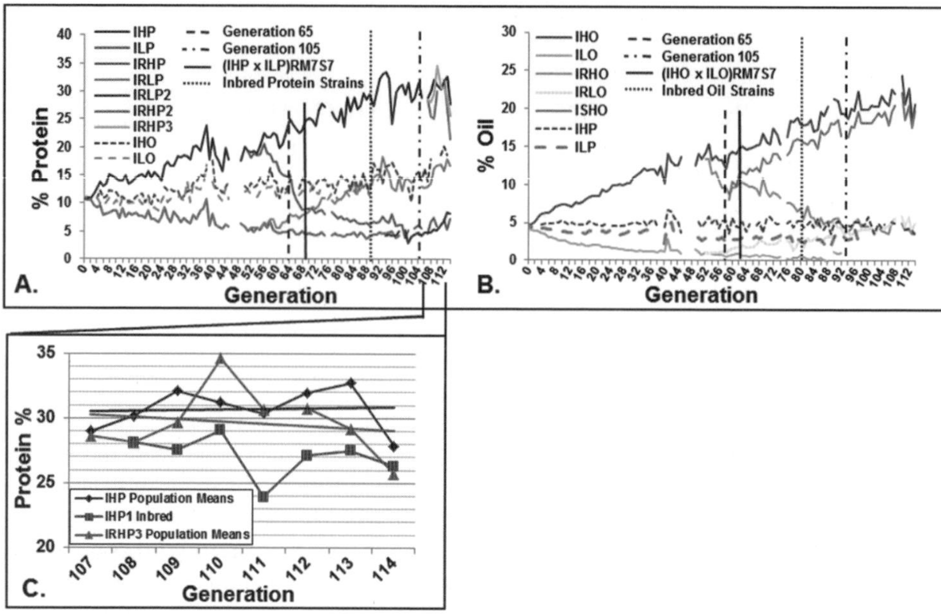

Figure 12.1 *A* and *B,* Selection responses in the Illinois Protein Strains *(A)* and Illinois Oil Strains *(B)*. Protein concentrations for IHO and ILO are plotted as *dashed lines* in *(A)*, and oil concentrations for IHP and ILP are plotted as *dashed lines* in *(B)*. The vertical lines highlight important generations of the experiment: generation 70, from which QTL mapping populations have been derived from crosses of selected strains; generation 90, as a source of inbred lines derived from either the Illinois Protein or Oil Strains; generation 65, the oldest for which seed is still available; and generation 105, a recent generation used for comparison to generation 65. *C,* Expanded view of the last eight cycles of selection response in IHP and Illinois Reverse High Protein 3 compared with IHP1 inbred line. (For color detail, see color plate section.)

breeding (Moose *et al.*, 2004). It took <10 generations to do so (Smith, 1908). The objective was later modified to determine the limits of selection for grain protein and oil concentrations in maize, and the experiment has since undergone 114 generations of recurrent selection, making it the longest running continuous genetics experiment in higher plants. Briefly, the experiment began by measuring grain protein and oil concentrations from 163 ears of the open-pollinated maize variety, "Burr's White." The 24 ears with the highest protein concentrations were selected as the parents for Illinois High Protein (IHP), the 24 with the highest oil concentrations for Illinois High Oil (IHO), and the 12 ears lowest for these traits became parents of Illinois Low Protein (ILP) and Illinois Low Oil (ILO) (Hopkins, 1899). Although the breeding scheme has varied throughout the experiment, the methods employed were chosen to minimize inbreeding. During the past 40 cycles, selected ears from each strain were divided into two groups, with controlled pollinations performed only between groups and using each individual plant as a parent only once to generate at least 60 ears from each strain for analysis. Similar cycles of selection have been conducted annually to the present except for during World War II. The resulting four populations span the known extremes for grain protein and oil compositions; compared with an initial population mean of 11% grain protein and a range of 8%–14%, IHP kernels contain >32% protein and ILP kernels only 4% protein (Figure 12.1*A*). Similarly, compared with an initial population mean of 5% grain oil and a range of 3%–5%, IHO kernels contain >20% oil, and ILO kernels contain 1% oil (Figure 12.1*B*). Additional details about selection procedures throughout the course of the experiment can be found in Dudley and Lambert

(2004), and the historical impacts of the experiment to the fields of plant breeding and genetics can be found in Moose *et al.* (2004).

To determine the extent of genetic variability and phenotypic response remaining after 48 generations of forward selection, the direction of selection in the IHP, ILP, IHO, and ILO populations was reversed using the same method of selection as the forward strains (Woodworth *et al.*, 1952). Low protein ears were selected from IHP to create Illinois Reverse High Protein (IRHP), and high protein ears were selected from ILP to create Illinois Reverse Low Protein (IRLP) (Figure 12.1*A*). Illinois Reverse High Oil (IRHO) and Illinois Reverse Low Oil (IRLO) were created in the same way from IHO and ILO (Figure 12.1*B*). Illinois Switchback High Oil (ISHO) was initiated from cycle 55 of IRLO with the same overall goal of determining genetic variation. The response to selection began to plateau in ILP by cycle 60, but selection was continued until cycle 93, when the strain was discontinued because of poor germination and lack of phenotypic response. To determine the level of genetic variation remaining in ILP, Illinois Reverse Low Protein 2 (IRLP2) selection was initiated from cycle 90. The ILO selections were also discontinued because of a lack of method for detecting such small differences in oil (<1%). Two new reverse selections are introduced here, Illinois Reverse High Protein 2 and 3 (IRHP2 and IRHP3), which were initiated from cycle 103 of IHP by selecting for low protein individuals. They were created for three reasons: (1) to estimate the extent of genetic variability remaining in IHP after >100 cycles of forward selection, (2) to assess the impact of soil N fertility on phenotypic means and selection response, and (3) to initiate new replicated populations where DNA and seeds of the founding individuals are preserved for future analysis of genetic responses to selection.

In addition to being analyzed for their respective traits, grain protein concentrations were measured from the bulks of IHO and ILO ears (Figure 12.1*A*). Similarly, grain oil concentrations were measured from the bulks of IHP and ILP ears (Figure 12.1*B*). The relatively limited change in these phenotypes throughout the experiment confirms that oil and protein concentrations are not significantly correlated, and IHP and ILP serve as unselected controls for the oil strains. Likewise, IHO and ILO serve as unselected controls for the protein strains. The unselected controls provide an estimate of the impacts of environmental variability and the extent of genetic drift in these populations. The mean protein concentrations for IHO and ILO show a slight positive trend since 1896 and range from 8%–20% protein throughout the experiment. The oil means of IHP and ILP have varied only by 2%–7% oil. The plots also reveal that throughout the selection experiment, protein is more sensitive to environmental fluctuations than oil. For example, mean protein concentrations fluctuated widely during generations 90–98, where 1993 conditions appeared to be a highly favorable for protein deposition but 1995 conditions were very poor.

Phenotypic Responses to Selection

The phenotypic responses to artificial selection of the Illinois Protein and Oil Strains can be described as steady and prolonged, with the exception of IRHP. This observation suggests a complex genetic architecture and high levels of genetic variation. The sustained progress and continuous phenotypic response observed in these strains is characteristic of quantitative traits controlled by many genes and consistent with the results of genetic mapping studies using progenies created by the cross of IHP x ILP and IHO x ILO. Estimates of genetic variability and genetic gain based on quantitative genetics theory have been reported at various times throughout the experiment and are reviewed by Dudley *et al.* (2007). IHP, IHO, and SHO still exhibited significant realized heritability, and IHO and ISHO also exhibited significant change per generation. An attempt was made to assess genetic variability of ILP via genetic response in IRLP2, which was initiated three cycles before the last cycle of selection in ILP. However, these results were inconclusive.

Accounting for environmental effects becomes more important as phenotypic responses to selection begin to plateau, and the magnitude of phenotypic differences owing to environmental factors becomes relatively large compared with genetic response. A new approach to assess genetic gain is to normalize changes in population means to values measured in homozygous inbred lines derived from the selected populations. For the Illinois Protein Strains, inbred lines derived from generation 90 of the Illinois Protein Strains (named IHP1, ILP1, IRHP1, and IRLP1) and described in Uribelarrea *et al.* (2004) can be used. Each of these inbred lines has been grown on a yearly basis since 2004 (generation 104) in the same field plots, and kernel composition has been estimated by near infrared reflectance (NIR). Because they are genetically fixed inbreds, observed phenotypic differences can be attributed to environmental or possibly epigenetic variation. Figure 12.1*C* illustrates differences in protein concentration between the IHP1 inbred line and the IHP and IRHP3 populations since generation 104. Except for generation 111, the protein concentrations for the IHP1 inbred are consistently between 28% and 30%. The IHP population mean continues to exhibit a slight positive trend and is always higher than IHP1. Conversely, IRHP3 appears to exhibit slightly lower concentrations than IHP in the last three cycles. Inbred lines from the most recent cycles of the Illinois Oil Strains are also being generated, as indicated by the vertical dotted lines in Figure 12.1.

Phenotypic selection will continue in the Illinois Protein and Oil Strains until a limit to selection is definitively observed. As genomics tools become increasingly available and QTL identified for these traits, it would be interesting to determine if genetic gain can be enhanced in these populations or perhaps revived in ILP using marker assisted selection.

Additional Traits Affected by Selection

Numerous additional traits have been affected by selection in the Illinois experiment. It was quickly observed that selection for grain protein concentration operated indirectly on other traits (Hopkins *et al.*, 1903; Smith, 1908). Grain starch concentration and kernel size and hence grain yield were each shown to be inversely related to grain protein concentration in all of the Illinois Protein Strains (Below *et al.*, 2004). Selection for high protein has also resulted in greater lodging and shorter plant height and increases in successful germination and tillering (Woodworth *et al.*, 1952). These traits are oppositely affected by selection for low protein, where the poor germination of ILP was one of the reasons it was discontinued.

Whole-plant nitrogen metabolism may also be altered by selection for grain protein concentration. The nitrogen assimilated by the plant during vegetative development accumulates and is stored as kernel protein until needed as a source of nitrogen by the developing seedling. Physiological changes affecting nitrogen metabolism in response to selection for grain protein concentration were initially noted by Hoener and DeTurk (1938) and have been reviewed by Below *et al.* (2004). The results of these studies demonstrate elevated nitrogen uptake, nitrogen assimilation by seedling leaves, and nitrogen remobilization from source to seed sink tissues of IHP compared with ILP. Uribellarrea *et al.* (2004, 2007) also found that hybrids derived from the Illinois Protein Strains exhibited differences in agronomic nitrogen use efficiency. IHP was also more efficient at remobilizing nitrogen than IRLP. One underlying cause of these physiological differences is changes in the activities of the enzymes involved in these pathways (Dembinski *et al.*, 1991; Below *et al.*, 2004). Because of the inverse relationship between protein and starch, activities of key carbon and metabolism and starch biosynthesis enzymes have also been examined for response to selection, including ADP-glucose pyrophosphorylase, that exhibits enhanced activity in ILP (Reggiani *et al.*, 1985; Below *et al.*, 2004).

Finally, increased soil nitrogen availability, owing to use of fertilizers beginning with generation 53, has been proposed as one environmental factor that has contributed to continued increases in grain protein concentrations within IHP (Dudley and Lambert, 2004). Cereal seed protein concentrations are known to increase after nitrogen fertilization, and this has been observed in hybrids derived from the Illinois Protein Strains (Uribelarrea *et al.*, 2004). One reason for creating IRHP2 and IRHP3 was to test directly the effect of nitrogen fertilizer application on protein concentration. These strains are derived from IHP, which should be most sensitive to soil nitrogen levels. IRHP2 plants were grown for 3 consecutive years (2004–2006) without supplemental nitrogen in a nitrogen-responsive field, and IRHP3 plants were grown in adjacent plots supplemented with 200 kg N per hectare. The protein concentrations of the IRHP2 population were 2.6%–4.5% lower than IRHP3 over the 3 years. Therefore, although N fertilizer may account for some of the recent increase in grain protein concentration, it does not fully explain the continued phenotypic response of IHP throughout the experiment.

Unlimited Genetic Variation?

The continuous phenotypic response to selection in the Illinois Selection Strains may result from existing genetic variation or arise from novel mutations and recombination. It can occur at the DNA, RNA, protein, or epigenetic levels. Here we review estimates of genetic variation in the Protein and Oil Strains and discuss possible reasons why genetic variation has been maintained despite intense phenotypic selection. One inherent source of genetic variation in quantitative traits arises owing to an extensive number of genes and alleles regulating the trait, where selection has not yet fixed all alleles contributing favorably to the phenotype. This is a plausible scenario in the Illinois experiment because of the following measures taken to minimize inbreeding and retain variation: first, a large number (163) of individuals were used to create the original population; second, the strains are derived from an open-pollinated variety, Burr's White; third, a mating design that minimizes inbreeding has been employed throughout the experiment. Furthermore, the maize genome exhibits a high level of both nucleotide and structural genetic diversity (Chia *et al.*, 2012). Maize also exhibits a low level of linkage disequilibrium (LD), where estimated average decay distances ranging from 1–10 kb have been reported among maize populations.

It is also likely that selection has targeted different genes throughout the course of the experiment. Although fixation of all favorable alleles is unlikely for the above-described reasons, the genes contributing the largest proportion of phenotypic variance or with high allele frequencies are more likely to become fixed, exhausting at least some initial variation. For sustained gains, additional targets with smaller effects or low allele frequencies may become increasingly important. Selection may also act on novel genetic variation arising from spontaneous mutation and recombination. However, as discussed by Laurie *et al.* (2004), responses due to new mutation are typically accompanied by drastic changes in phenotype, a pattern inconsistent with the observations of steady gain in the Illinois Protein and Oil Strains. The rapid decline in IRHP may be an exception to this observation and merits further discussion (see later). Structural variation is yet another possibility, where unequal crossing over during meiosis could generate novel alleles. Intraspecific structural variation has been observed in maize using cytogenetic and flow cytometric approaches, but sequence-based approaches allow for higher resolution studies. Whole-genome, array-based, comparative genomic hybridization (CGH) is one approach for studying structural variation at the genome scale and has become possible with the sequencing of the B73 genome. A high rate of copy number variation (CNV) and presence/absence variation (PAV) was observed between maize inbred lines B73 and Mo17 (Springer *et al.*, 2009). CGH could be used to investigate genome-wide structural variation between the Illinois Selection Strains.

Selection can also act on epigenetic regulatory variation. Three major mechanisms for initiation and maintenance of epigenetic silencing known to occur in plants are DNA methylation, histone modification, and RNA-associated silencing (Egger *et al.*, 2004). DNA methylation is one underlying mechanism of genomic imprinting, a type of epigenetic modification causing differential expression of a gene depending on the sex of the parent that transmits it. Two types of genomic imprinting exist in plants: allele imprinting, where only alleles from a certain genetic background are affected by parent-of-origin-specific gene expression, and locus imprinting, where all known alleles from different backgrounds are under parent-of-origin control (Gehring *et al.*, 2004) (see Chapter 4). The α-zein genes in maize are known to be regulated by maternal imprinting (Chaudhuri and Messing, 1994), which in plants occurs primarily in the endosperm tissue (Gehring *et al.*, 2004). It has been hypothesized that the evolutionary role of maternal imprinting in the endosperm might be to maintain control of gene expression affecting kernel growth and development because of the dependency of kernel development on the maternal organ, the cob (Alleman and Doctor, 2000). Maternal imprinting of the α-zein genes occurs via allele imprinting, where the zeins exhibit differential methylation patterns depending on their parental heritage. It is proposed that methylation of paternal alleles inhibits zein gene expression by altering the chromatin structure of certain zein genes, whereas maternal alleles are demethylated and derepressed (Lund *et al.*, 1995; Lauria *et al.*, 2004); for review, see Bird and Wolffe, 1999. Parental imprinting is also known to regulate expression of an allele of the *dzr1* locus, a posttranscriptional regulator of the 10-kDa δ-zeins (Chaudhuri and Messing, 1994). Additional studies are required to determine if epigenetic variation has contributed to the phenotypic variation observed in the Illinois Selection Strains, and testing for differential methylation of the zein genes is a good starting point. RNA-associated silencing is also being investigated via RNASeq as a potential mechanism of epigenetic regulatory variation of the Selection Strains.

Genetic Response to Selection: QTL Mapping in the Crosses of IHP x ILP and IHO x ILO

It is apparent from previous research that long-term selection for grain protein concentration has altered numerous related traits, and from this it may be hypothesized that many genes have also been affected (Moose *et al.*, 2004). However, the number and identity of genes remain unknown. Genomics approaches will be increasingly important for understanding the genetic and molecular changes that have altered the genomes of the Illinois Selection Strains as a result of long-term artificial selection. QTL mapping is one approach for studying the genetic architecture of quantitative traits. Several QTL mapping studies have been conducted using populations created by crossing individuals from cycle 70 of IHP and ILP or IHO and ILO (Dudley, 1977) and are reviewed by Dudley *et al.* (2007). QTL influencing protein (Goldman *et al.*, 1993, 1994; Dijkhuizen *et al.*, 1998; Sene *et al.*, 2001; Dudley *et al.*, 2004) and oil (Laurie *et al.*, 2004) have been identified. The results of these studies suggest the presence of many QTLs having small phenotypic additive effects, which is consistent with theoretical estimates for the number of effective genetic factors based on genetic variances, 102–178 genetic factors for protein and 14–69 factors for oil (Moose *et al.*, 2004).

One QTL mapping study using a cross of IHO (cycle 90) and a derivative of ILO that had also been selected for early maturity identified a major-effect QTL on chromosome 6 associated with the ratio of oleic to linoleic fatty acid content QTL that was linked to a region containing the *linoleic acid 1* (*ln11*) locus (Alrefai *et al.*, 1995). Another study also identified a major-effect QTL on chromosome 6 that accounted for 11% and 9.5% of the variation associated with maize seed oil content and embryo oil concentration (Zheng *et al.*, 2008). The mapping population used in this latter study was derived from the cross of a high-oil inbred line, ASKC28IB1, where IHO was 1 of 56 varieties included in the base population of ASK (Lambert *et al.*, 2004), and a normal-oil inbred line, PH09B. Fine-mapping revealed the presence of a gene encoding an acyl-CoA:diacylglycerol acyltransferase (DGAT1-2), an enzyme that catalyzes the final step in oil synthesis. The ASKC28IB1 allele of DGAT1-2 was shown to contain several polymorphisms, including a phenylalanine insertion at position 469, which was responsible for the increased oil content and embryo oil concentration. Because higher oleic acid content was also associated with near-isogenic lines homozygous for the ASKC28IB1 allele, the DGAT1-2 region was sequenced in IHO. The results indicate that the *ln1* locus in the previous study and the ASKC28IB1 allele of DGAT1-2 are one and the same. It was also shown that although ancestral varieties of maize, including 46 accessions of teosinte lines, contain the ASKC28IB1 allele, most modern inbred varieties of maize, including B73 and Mo17, contain the PH09B allele. Introgression of this DGAT1-2 allele increased kernel oil concentrations in the inbred parents of a current elite Chinese hybrid (Chia *et al.*, 2012).

Identification of factors regulating protein has been more difficult, but strategies for increasing the precision of marker-QTL associations have been attempted. Strong selection in artificial selection experiments is known to increase linkage disequilibrium (LD), which can reduce the precision of associations between markers and QTL and between QTL. Random mating within a mapping population is a strategy for reducing LD through recombination and the breaking up of linkages (Dudley, 1994). Owing to the intensity of selection in the Illinois Selection Strains, random mating of the progeny of the cross of IHP x ILP and IHO x ILO was a strategy employed to reduce LD and break up coupling-phase linkages among genes controlling these traits. Studies by Dudley *et al.* (2004) for IHP x ILP and Willmot *et al.* (2006) for IHO x ILO showed large reductions in the percentage of markers declared significant between the Syn0 (one generation of random mating the F1) and Syn4 (four generations of random mating the F1) populations. These studies revealed the need for a larger number of markers with which to associate the increased number of linkage blocks incurred as a result of recombination.

Subsequent mapping studies employed ~500 SNP markers genotyped on populations following either 7 (IHP x ILP) (Dudley *et al.*, 2007) or 10 generations of random mating (IHO x ILO) (Laurie *et al.*, 2004; Clark *et al.*, 2006). However, given the high frequency of recombination events and reductions in LD within these random-mated populations, 500 markers was likely insufficient to detect with high probability most marker-trait associations. In addition, these more recent genetic mapping studies are complicated by population structure owing to the use of multiple parents from the respective selected populations and the phenotyping of segregating families rather than inbred lines. Despite these past efforts to map genomic regions regulating protein and oil, the number and identity of genes influencing these traits remain largely unknown.

New Mapping Population: Illinois Protein Strain Recombinant Inbreds

To overcome the limitations of the previously described mapping populations, a new recombinant inbred mapping population was created: the Illinois Protein Strain Recombinant Inbred (IPSRI) population. The IPSRI population consists of 500 recombinant inbred lines derived from the cross of cycle 70 IHP and ILP plants, followed by seven generations of random mating (Dudley *et al.*, 2007) and six generations of inbreeding (Moose laboratory). This population exhibits a normal phenotypic distribution, ranging from 5%–24% protein, and captures most phenotypic variation in the IHP and ILP populations at cycle 70. It also consists of a large number of individuals, which should provide adequate power to detect small-effect loci. Finally, six generations of inbreeding greatly reduced the phenotypic noise owing to segregation within each family.

The IPSRI population will permit replication of defined genotypes and be useful in approaches combining linkage and association mapping with the application of high density markers. Results from a genome-wide scan using 384 simple sequence repeat markers permit preliminary estimates for the degree of divergence among the inbred lines derived from the Illinois protein strains. Analysis of allelic diversity in the inbred protein strains revealed that ~50% of the 384 SSR markers are polymorphic for IHP and ILP, and only 4% showed allele frequency differences among the strains that are consistent with the changes in protein concentration. Using the data for the 500 SNP markers previously genotyped on the progenitor families of the IPSRIs, it was possible to assess population structure and to reduce the original population of 500 to a subset of 138 individuals that effectively represents the phenotypic and allelic variation in the original population (Figure 12.2). This subset of 138 IPSRIs is being used in ongoing mapping studies, which are discussed in greater detail later.

A definitive requirement of subsequent mapping studies that use an advanced random mated population, such as the IPSRIs, is a larger number of molecular markers. The development of low-cost, high-throughput sequencing technologies provides numerous options for IPSRI marker development. A genotyping array for 50,000 maize SNPs was developed for diversity analysis and high density linkage mapping in maize (Ganal *et al.*, 2011). However, marker information may be limited because of the inevitable ascertainment bias that arises with selection of lines for marker development within the array platforms. Because the IPSRIs are derived from nineteenth-century maize germplasm, the lines chosen for SNP arrays may not capture the allelic diversity present

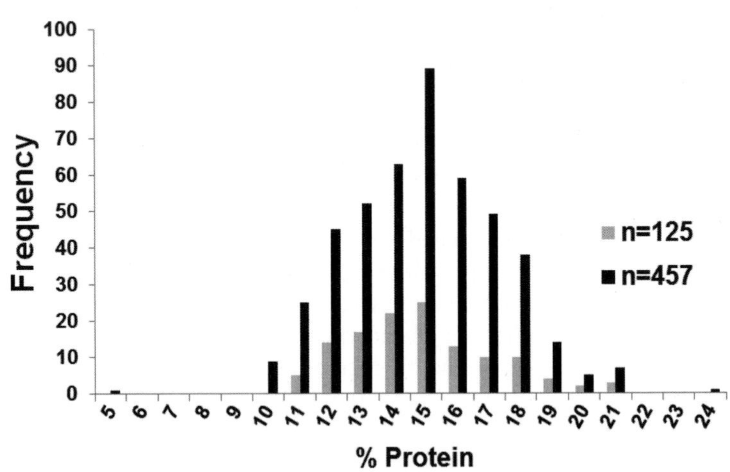

Figure 12.2 Histograms of grain protein concentrations for the larger population of 457 *(black bars)* and reduced set of 138 *(gray bars)* Illinois Protein Strain Recombinant Inbred (IPSRI) lines. Protein concentrations were measured by near-infrared reflectance in 2007.

in the IPSRIs. For example, one use of the 50k SNP array on a population derived from parental lines unrelated to B73 found only 14,000 informative markers. This number could yet be reduced in the IPSRIs. Genotyping by sequencing (GBS), which involves genome complexity reduction using restriction enzymes (Elshire *et al.*, 2011), has been developed more recently and applied to diverse maize populations. GBS of the IPSRIs will likely be useful in assessing patterns of genetic diversity and possibly identify regions with significantly increased LD that could indicate targets of selection, but genetic map construction (Huang *et al.*, 2009) could be problematic depending on how accurately genotypes can be inferred (on the basis of genetic linkage) when not directly sequenced based on genetic linkage.

High density genotyping not only permits mapping experiments but also allows for comparisons of allele frequencies between the strains and generations, such as was done more recently with the Virginia chicken populations divergently selected for body weight (Johansson *et al.*, 2010). Because seeds from previous generations of the selection experiment can be saved and regenerated, genomic comparisons can be made between cycles spanning an even longer time period. Although seeds from the original population are unavailable, seeds are available from cycle 65 and every subsequent cycle thereafter. In this way, comparisons made between cycle 65 and 105 span 40 cycles of selection (Figure 12.1). Comparisons can also be made between the populations at a given time point. The efficacy of GBS for measuring allele frequencies in populations is unknown because read depth needs to increase to call genotypes confidently when more than two alleles are present. The Illinois Selection Strain populations may require the use of sequence capture approaches to sample genetic variation effectively. For example, a recently described method enriches for targeted subsections of the genome and can generate high coverage sequence data at a relatively low cost (Rohland and Reich, 2012). Collectively, the IPSRIs, inbred protein strains, and cycles are invaluable genetic material for various mapping and functional genomics studies.

Characterization of Zein Genes and Their Expression in Illinois Protein Strains

Because zeins can account for 60% of total seed protein in mature grain (Nelson, 1979), it is expected that differences in zein gene organization and expression are associated with the large changes in grain protein concentration observed in the Illinois Protein Strains. Significant differences in the alcohol-soluble fraction of endosperm protein was observed between IHP and ILP in 1922 (Showalter, 1922). Subsequent studies with methods of increasing resolution for individual zein classes and polypeptides confirmed this initial observation (Hansen *et al.*, 1946; Schneider *et al.*, 1952; Lorenzoni *et al.*, 1978; Wilson, 1992; Below *et al.*, 2004) and demonstrated by SDS-PAGE analysis the most dramatic difference in kernel protein accumulation in the Illinois Protein Strains is in the amount of α-zeins produced (Bhattramakki *et al.*, 1996; Below *et al.*, 2004).

Both the 22-kDa and 19-kDa α-zein clusters have been fully sequenced in the B73 inbred, and the 22-kDa cluster has also been sequenced in the BSSS53 inbred (Song *et al.*, 2001), facilitating more detailed studies of zein gene expression. The results of these studies illustrate the dynamic nature of the α-zein genes, which are rapidly evolving in terms of relative chromosomal position, copy number, and expression (see Chapter 8). The zeins are the most highly expressed genes in the endosperm. The 19-kDa and 22-kDa α-zein genes are expressed in a coordinated fashion, beginning around 12 days after pollination (DAP), peaking around 16 DAP, and continuing throughout development, with RNA expression patterns closely matching protein accumulation (Langridge *et al.*, 1982; Woo *et al.*, 2001). Despite a large copy number, only a fraction of α-zein gene coding sequences produce transcripts, indicating the presence of pseudogenes interspersed with functional genes (Woo *et al.*, 2001; Song and Messing, 2003; Feng *et al.*, 2009). The α-zein genes exhibit variation for the number of expressed zein gene family members among genotypes as well as the relative expression levels of genes shared by genotypes (Feng *et al.*, 2009; Miclaus *et al.*, 2011). Crosses among divergent 22-kDa zein haplotypes produce a range of nonadditive gene expression responses, suggesting ample regulatory variation on which selection for changes in zein accumulation may act.

Global RNA profiling approaches are useful for elucidating genome-wide expression variation that may be contributing to divergent protein phenotypes. A microarray study conducted by our laboratory compared gene expression of developing seeds between the inbred genotypes IHP1 and ILP1. These data are deposited at the NCBI Gene Expression Omnibus (GEO) (accession GSE 18006). Nearly 43,000 probes representing ~30,000 genes showed robust fluorescent signals after data processing and normalization using LIMMA with the R package (Wettenhall and Smyth, 2004). Figure 12.3 plots the expression ratio of IHP1 relative to ILP1 versus average probe signal intensity for this experiment. Following statistical tests for differential expression and Bonferroni correction for false discovery rate, 2.2% of features exhibited robust higher expression in IHP1 compared with ILP1, and 1.6% were more strongly expressed in ILP1. Among the 96 probes annotated as α-zeins on the array (red dots), many exhibit the highest expression levels in the experiment and are strongly upregulated in IHP1. Although also showing relatively strong expression levels, most probes annotated as β-zeins, δ-zeins, and γ-zeins (blue dots) are not differentially expressed between IHP1 and ILP1. All non-zein genes are plotted in gray.

Validation and further analysis are currently ongoing for zein gene expression and other differentially expressed genes that represent candidate targets of long-term selection for grain protein concentration. Owing to its inherent advantages over microarrays for assaying expression of closely related genes and alleles and providing a broader and unbiased survey of gene expression, we have conducted RNASeq experiments to profile gene expression changes in the Illinois Selection strains. Through comparisons of the selected populations at different cycles of the experiment, where multiple alleles and substantial heterozygosity remains, we will more thoroughly document both sequence and regulatory variation in response to long-term selection.

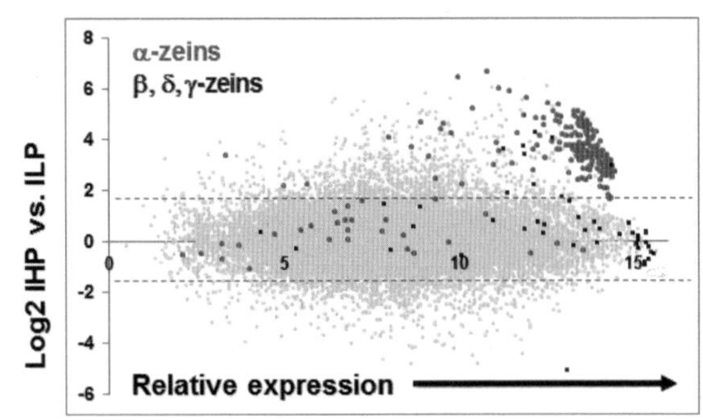

Figure 12.3 Relative zein gene expression levels in 16 DAP seeds as determined by microarray. Plotted is the expression ratio of IHP1 relative to ILP1 versus average probe signal intensity. *Dashed lines* represent cutoffs for features exhibiting statistically significant differential intensity (*t*-test, P value < .05; FDR adjusted q value < 0.05). The features annotated as α-zeins are plotted in red, and β-zeins, δ-zeins, and γ-zeins are plotted in blue. All other features are plotted in gray. (For color detail, see color plate section.)

Contribution of Zein Regulatory Factor *Opaque2* to Observed Responses to Selection in Illinois Protein Strains

Both prior QTL mapping studies and RNA profiling experiments suggest that many genes contribute to the dynamic and prolonged responses to selection for grain protein concentrations. However, an alternative hypothesis that cannot yet be excluded is that quantitative variation in the activities of a few key regulatory genes may explain phenotypic changes. A prime candidate for such a regulatory gene is *Opaque2*. *Opaque2* encodes a bZIP transcription factor that activates 22-kDa α-zein gene expression by binding to a highly conserved promoter element (Schmidt *et al.*, 1992). *o2* also interacts with the *Prolamin-box Binding Factor* (*PBF*), a Dof zinc finger DNA binding protein that binds to a highly conserved motif (the prolamin box, 5′-TGTAAAG-3′) found in the promoters of all zein genes and genes encoding prolamins from other cereals (see Chapters 8 and 9) (VicenteCarbajosa *et al.*, 1997; Hwang *et al.*, 2004).

Figure 12.4 shows the phenotypic impacts of introgressing the *o2-R* null mutant allele (Bernard *et al.*, 1994) on seed protein concentrations in the IHP1 and B73 inbred backgrounds. The mutant ear exhibits the characteristic opaque kernel phenotype in the IHP1 background (Figure 12.4*A*). Using estimates from near-infrared reflectance measurements of grain, the *o2-R* mutation reduced total protein from 27.0% to 22.4% in IHP1 and from 8% to 7% in B73. As expected, 22-kDa α-zein protein was significantly reduced in both IHP1 and B73 (Figure 12.4*B*). The 4.6% reduction conditioned by the *o2-R* allele in IHP1 likely represents the maximum possible genetic effect that could be associated with *o2* in the Illinois Protein Strains.

The *o2* locus is highly polymorphic (Henry and Damerval, 1997), and previously developed marker assays for *o2* were used to assess genetic variation at the *o2* locus among the Illinois Protein Strains. Approximately 48 individuals from ILP_{65}, IHP_{65}, ILP_{100}, IHP_{100}, and $IRHP_{100}$ were genotyped using SSR marker umc1066 (maizeGDB), which amplifies the highly polymorphic region of the *o2* gene described by Hartings *et al.* (1994). Polymorphic variants were observed at cycle 65 and 100 in IHP, but all other populations were fixed for a single common variant (data not shown). The two classes of *o2* genotypes in IHP exhibited a significant difference in mean grain protein

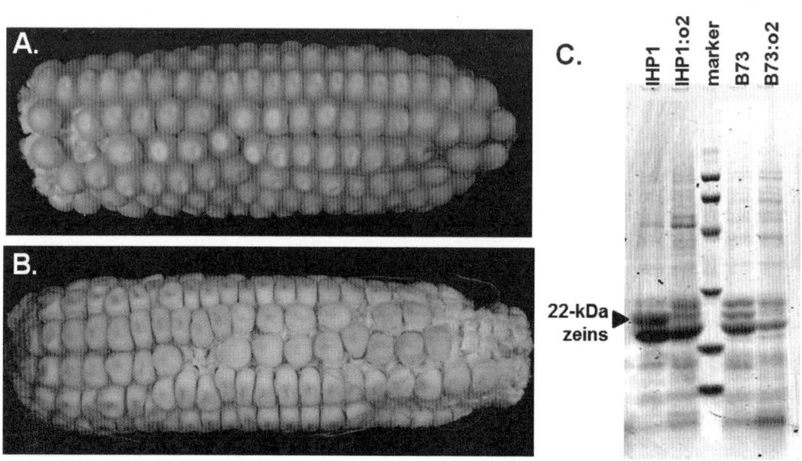

Figure 12.4 Phenotypes of *opaque2* introgressions into IHP1 and B73 inbred lines. *A*, Wild-type IHP1 ear. *B*, IHP1 ear homozygous for the *o2* mutation. *C*, SDS-PAGE of total seed proteins from IHP1, B73, and o2 introgressions into each of these genetic backgrounds. The relative migration of the 22-kDa α-zeins is indicated. (For color detail, see color plate section.)

concentration (32% versus 28%, *t* test $P < .01$). The results of this preliminary experiment suggest that variation at the *Opaque2* locus may contribute to the extreme high protein concentrations in IHP but not the other selected protein strains. Extending such surveys of genetic variation at a genome scale is essential to identify candidate targets of selection versus background genetic drift.

Major Effect QTL May Explain IRHP Phenotype

The dramatic decline of protein concentration in IRHP differs from the steady responses observed in the other protein strains. It took only 15 generations of reverse selection in IRHP to revert nearly 50 generations of gain in IHP, the fastest rate of change in phenotype for any of the selected strains (Figure 12.1*A*). It is likely that this rapid evolutionary change in IRHP may be mediated by variation at relatively few loci with major phenotypic effects. This hypothesis may seem unlikely at first, given the fact that when IRHP was initiated, none of the individuals in cycle 48 of IHP exhibited protein concentrations <15%. However, when considering that the experiment has been conducted using a breeding scheme that deliberately maintains genetic variation and reduces inbreeding, it is unlikely that all alleles that increase grain protein have reached fixation. The continued positive response to selection within IHP (Dudley *et al.* 2007) verifies that alleles favorable to the low protein phenotype remained. We know from the previously described genotyping with SSR markers that IHP1 and IRHP1 are most closely related to each other, sharing alleles at 81% of the marker loci tested compared with the 50% shared between IHP1 and ILP1. It appears that selection in IRHP acted on fewer loci, yet was still rapid and effective.

The low protein phenotype of IRHP differs from that of ILP. Most known mutations that dramatically reduce α-zein accumulation are associated with opaque kernel phenotypes or soft endosperm texture. While ILP kernels also appear opaque, IRHP kernel composition and quality are similar to what is observed following the breeding efforts for Quality Protein Maize where the *opaque2* mutation reduces α-zeins, but modifications in other proteins ameliorate the problem of soft endosperm texture. If relatively few loci regulate the IRHP1 phenotype, identifying these genes may facilitate their use in current elite maize germplasm.

The genetic architecture of the response to selection in IRHP was investigated directly by the generation and evaluation of a population derived from crossing the IRHP1 and IHP1 inbred lines, resulting in a population segregating for protein concentration. The inheritance of protein concentration is well documented to exhibit a strong maternal effect (Tsai *et al.*, 1990). To control for maternal effects, F_1 hybrids between IHP1 and IRHP1 were backcrossed to IRHP1 followed by self-pollination of BC_1 plants, and a population of 115 BC_1S_1 progeny was created. Although the BC_1S_1 ears are either homozygous IRHP1 or heterozygous and segregating in a 1:2:1 ratio, the protein concentration of these ears follows that of the maternal plant. The BC_1S_1 ears segregate similar to the BC_1 plants from which they are derived and are expected at any given locus to segregate in a 1:1 ratio for homozygous IRHP1 versus heterozygous IRHP1/IHP1 genotypes. Grain protein concentrations measured in this population are plotted in a frequency histogram in Figure 12.5. The population is characterized by a bimodal distribution: 51 individuals exhibited <13% protein, with a mean of 10.6% protein, whereas the remaining 64 individuals exhibited protein >13% with a mean of 14.5%. These observations are consistent with equal segregation into two phenotypic classes of greater than or less than 13% protein (χ^2 test 1.47, $P = .22$, df $= 1$). The bimodal distribution strongly supports the hypothesis of one or a few genes with major effects, with respect to IRHP1. If the trait was controlled by a larger number of genetic factors, a more continuous distribution would be expected with fewer numbers of individuals belonging to more phenotypic classes. The BC_1S_1

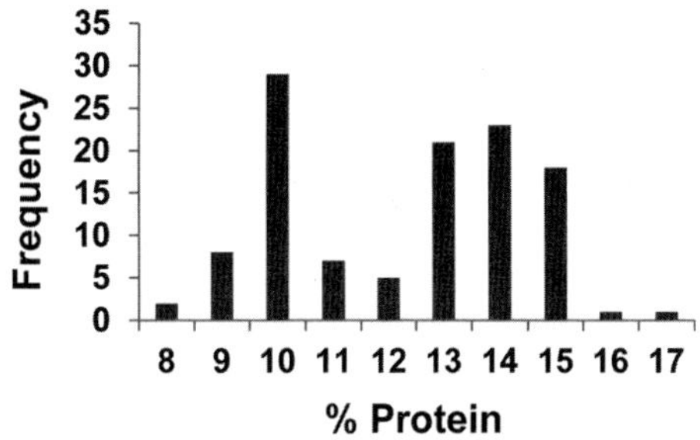

Figure 12.5 Protein concentration frequency histogram of 115 BC_1S_1 individuals from the cross of IHP1 and IRHP1.

population analyzed here represents a good starting point for coarse genetic mapping studies. A larger BC_1S_1 population was generated in summer 2010 for additional fine-mapping. This information will be used to identify gene candidates associated with the more desirable reduced zein phenotype of IRHP1.

Zein Promoter-Reporter Lines to Investigate Regulation of 22-kDa α-Zein Gene Expression in Illinois Protein Strains

The studies summarized in this chapter clearly indicate that the zein synthesis pathway has been one target of selection for altered grain protein concentration in the Illinois Protein Strains. Additional investigations of the molecular mechanisms responsible for these differences may offer insights into various questions related to the control of grain protein concentration. Approaches for studying individual zein gene expression are complicated by the high number and sequence similarity of the α-zein genes and the presence of pseudogenes. Also confounding analysis is the quantitative inheritance of grain protein concentration, where tens to hundreds of genetic factors have been estimated to contribute to the protein concentration differences between IHP and ILP. One approach for tracking expression of individual zein genes is through the use of reporter genes. When fused to a zein gene promoter, the expression of the reporter gene can be measured in a rapid, quantitative, and potentially nondestructive manner to estimate expression of the zein gene of interest.

The most widely used reporter gene for plants is the *GUS* gene. The GUS enzyme has many desirable properties for use as a reporter gene; it can be detected and quantified by histochemical and spectrophotometric assays, is highly sensitive, and does not interfere with normal cellular metabolism (Jefferson *et al.*, 1987). One of the earliest examples of using transgenic promoter-reporter lines to characterize gene expression in maize was the fusion of the *E. coli* β-glucuronidase (*GUS*) reporter gene to a 27-kDa γ-zein promoter (Russell and Fromm, 1997). *GUS* activity was shown to be proportional to transgene RNA expression, inherited in a stable manner, and did not affect endogenous zein gene expression. However, disadvantages of *GUS* are that diffusion of the *GUS* reaction products limits cellular resolution, and assaying *GUS* activity requires destruction of the tissue sample. Fluorescent proteins such as green fluorescent protein (GFP) from jellyfish, modified derivatives of GFP with different spectral properties, and the monomeric red fluorescent protein from reef coral (DsRed) have gained popularity as reporter genes in plants because they overcome some of the limitations associated with the *GUS* reporter gene (Stewart, 2006). Of particular interest for the study of seed gene expression is the ability to visualize the DsRed protein in maize under standard white light, where the kernels appear pink to red in color (Wenck *et al.*, 2003). This property enables rapid monitoring of gene expression during seed development. Additionally, the relative fluorescent intensity associated with reporter gene expression may be quantified by direct imaging without destruction of the sample.

We have evaluated a collection of transgenic lines produced by other laboratories where zein promoters have been used to drive expression of either the *GUS* or *DsRed* reporter genes. These transgenic lines have been crossed to the Illinois Protein Strain inbreds as well as the reference genotype B73 to assess the impacts of genetic background on the regulation of zein promoter activities. *GUS* reporter lines for the 27-kDa γ-zein and both the 22-kDa and 19-kDa α-zeins were initially obtained from Pioneer Hi-Bred International. The results for the *GUS* lines are described by Salas (2008) and can be briefly summarized as being qualitatively consistent with what is known about the regulation of zein gene expression. GUS staining of the Z27 construct was prevalent throughout the entire endosperm of IHP and ILP, but no variation in the level of expression was observed between them. GUS staining of the Z19 and Z22 constructs was found in the peripheral endosperm, consistent with the findings of Woo *et al.* (2001). As expected, IHP also stained darker than ILP for both the Z19 and the Z22 GUS transgenes.

The fluorescent protein promoter-reporter lines employed here are a part of a series of transgenic reporter maize lines that were created as part of a joint project conducted by Jackson's laboratory at Cold Spring Harbor, Sylvester's laboratory at the University of Wyoming, and the Plant Transformation Center at Iowa State University (Mohanty *et al.*, 2009) (http://maize.tigr.org/cellgenomics/index.shtml). Derived from the coral reef species, *Discosoma,* a monomeric version of the tetrameric DsRed protein termed mRFP1 (Campbell *et al.*, 2002) was fused to the C-terminus of the 22-kDa α-zein gene, *Floury2.* To preserve proper tissue and temporal regulation of the *Floury2*-mRFP1 fusion protein, the construct includes an intact α-zein gene driven by native flanking regulatory elements, including ~2000 bp of the *Floury2* gene promoter and 1000 bp of 3′ sequence. Inclusion of the native regulatory sequences may allow these lines to provide more information about regulatory elements acting at the level of transcription and elements involved in processing, localization, and turnover of either zein mRNA or protein.

When transformed into a HI-II background, initial analysis of the mRFP kernels by Wenck *et al.* (2003) revealed the ability of the mRFP protein to be visualized under white light, the kernels appearing pink in color. This property can most likely be attributed to the high level of expression of the *Floury2* gene, the relatively high stability of zein proteins, and the use of the monomeric RFP that does not require multimerization to emit fluorescence (Campbell *et al.*, 2002). *Floury2* has been shown to account for ~20% of all 22-kDa α-zein expression in inbred variety B73 with only one gene expressed more (Song and Messing, 2003; Feng *et al.*, 2009). The ability to visualize *FL2-mRFP* expression under white light allows for the quick nondestructive monitoring of *Floury2* gene expression during seed development.

Regulatory Changes in *FL2-mRFP* Expression When Crossed to Illinois Protein Strains

Three independent *Agrobacterium*-mediated transformation events (47, 52, and 172) for *FL2-mRFP* were used as the donor lines for backcrossing into the four Illinois Protein Strain inbreds and the B73 inbred. Analysis of *FL2-mRFP* expression in the kernels produced from these crosses provides information about the genetic inheritance of the *FL2-mRFP* transgene, its response to genetic background, and its expression throughout development.

Regardless of genetic background, stage of backcrossing, and transgenic event, approximately half of the kernels on an ear are white, and the other half exhibit the red-to-pink coloration associated

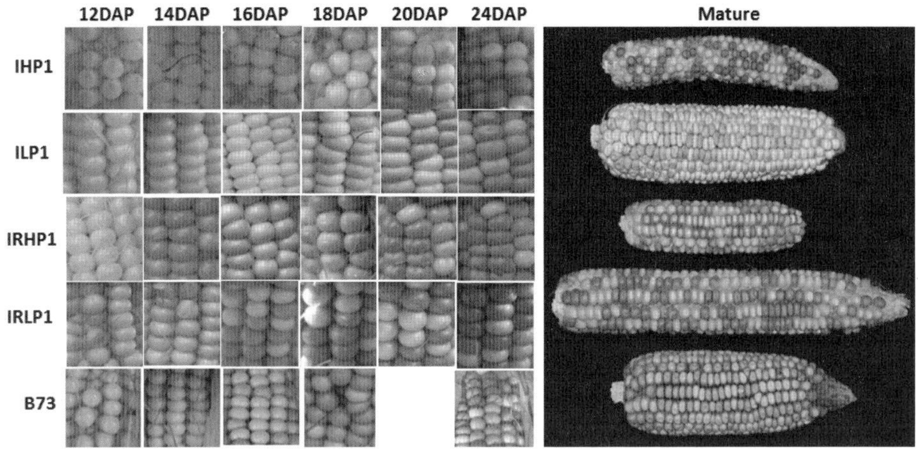

Figure 12.6 Photographs of kernels produced following backcrosses of *FL2-mRFP* transgenic events to the Illinois Protein Strains inbreds (TG event 52) and B73 (TG event 47). Photographs were taken at regular intervals during the period of zein accumulation from 12–24 days after pollination (DAP). (For color detail, see color plate section.)

with the *FL2-mRFP* expression. The 1:1 ratio of white to pink kernels indicates that each transgene is inherited as a single dominant genetic locus. Additionally, *FL2-mRFP* expression is consistent among the three transgenic events, despite having integrated in different locations in the maize genome. This observation demonstrates that location within the tandemly duplicated 22-kDa zein gene cluster is not important for proper regulation of the *FL2-mRFP* reporter gene.

Because the *FL2-mRFP* reporter lines provide a nondestructive visual assay for zein gene expression, transgenic seeds from the crosses to the IPS and B73 were monitored for the onset and progression of red coloration during seed development. Zein accumulation is known to begin at ~10 DAP, peak at ~16 DAP, and continue throughout grain fill. To document *FL2-mRFP* expression throughout development, BC3 kernels from reciprocal crosses of the Illinois Protein Strains and all three transgenic reporter lines were photographed at various time points from 8–24 DAP, where the same ear was used for the entire series of photographs (Figure 12.6). No differences were apparent due to transgenic event or direction of cross. For this reason, only ears from event 52 are shown using the Illinois Protein Strains as the female parents. From these photographs, *FL2-mRFP* expression can be visualized beginning 12 DAP in IHP and IRLP and 14 DAP in ILP, IRHP, and B73 and increases throughout development in all genotypes. Mature ears produced from a similar set of crosses are shown to the right. These results indicate that *FL2-mRFP* expression exhibits spatial and temporal patterns expected for endogenous α-zein genes.

FL2-mRFP expression of BC3 ears also corresponds with known levels of zein protein accumulation. It is the strongest in IHP1, followed by IRHP1, B73, and ILP1. IRLP1 kernels demonstrate much darker pink coloration than is expected given the low protein concentration of this strain. However, because the IRLP1 conversion is not yet complete, it is possible that alleles from the donor parent remain that increase *FL2-mRFP* expression.

To corroborate further mRFP pink coloration intensities with gene expression, zein accumulation and relative expression of zein and *FL2-mRFP* transgene expression using quantitative reverse transcriptase polymerase chain reaction (qRT-PCR) were examined at 24 DAP in the pink seeds from BC3 ears in IHP1 and ILP1 backgrounds. As expected, zein accumulation was higher in kernels from the IHP1 background compared with ILP1 (Figure 12.7). To determine if gene expression

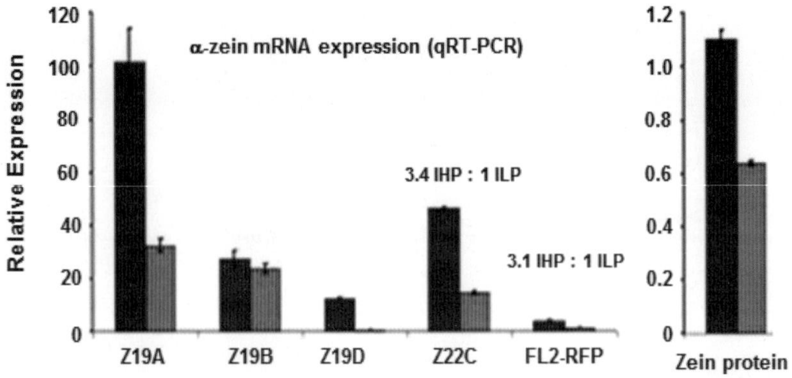

Figure 12.7 Relative expression of the three classes of 19 kDa (z1A, z1B, z1D) and single class of 22 kDa (z1C) α-zein genes in inbred protein strains IHP1 *(black bars)* and ILP1 *(gray bars)* and relative expression of the *FL2-mRFP* transgene when backcrossed (BC3) into IHP1 and ILP1. Total zein protein is also plotted on the right.

patterns correlate with zein protein, primers were designed to anneal to common sequences of each subfamily of the 19-kDa (Z19A, Z19B, and Z19D) and 22-kDa (Z22C) α-zeins and the *FL2-mRFP* transgene, and qRT-PCR was performed. Relative Z19B RNA expression was similar between IHP1 and ILP1, but Z19A, Z19D, and Z22C expression was much higher in IHP1 compared with ILP1. The relative expression of *FL2-mRFP* was also higher in the IHP1 background than ILP1, exhibiting a ratio of (3.1:1). Because *Floury2* belongs to the Z1C subfamily, it is not surprising that Z22C genes were expressed at a similar ratio (3.4:1). These results demonstrate that *FL2-mRFP* protein accumulation follows transgene expression and that *FL2-mRFP* transgene expression is predictive of the entire subfamily of 22-kDa α-zeins. The transgene will not only be useful as a reporter for tracking individual zein gene expression but may represent the whole class of 22-kDa α-zeins.

Regulation of *FL2-mRFP*

As described earlier, the *o2-R* recessive mutant was introgressed to IHP1 to create a near-isogenic line. Here, we use this IHP1: *o2-R* genotype to test whether the *FL2-mRFP* transgene is transcriptionally activated by o2, as are endogenous 22-kDa α-zein genes. Following four backcrosses of the *FL2-mRFP* transgene to IHP1, plants carrying the *FL2-mRFP* transgene were then backcrossed twice to IHP1:*o2-R* to generate ears segregating for both the *FL2-mRFP* transgene and *o2-R*. The resulting ear is shown in Figure 12.8. Kernels homozygous for the *o2-R* mutant allele exhibit the characteristic chalky, opaque kernel phenotype compared with their wild-type counterparts. Approximately 50% of the kernels contain the *FL2-mRFP* transgene, as visualized by the pink coloration of the kernels. Within the transgenic kernels, approximately 25% are also homozygous for *o2-R* and appear both opaque and light pink in color. The reduction in pink coloration indicates decreased *FL2-mRFP* expression and zein protein accumulation. These results demonstrate that the *FL2-mRFP* transgene is regulated by *o2* in a similar fashion as the endogenous *floury2* gene. Collectively, the aforementioned studies indicate that the *FL2-mRFP* transgene behaves like an endogenous zein gene, and the *FL2-mRFP* reporter lines will be a useful tool for further study of 22-kDa α-zein gene regulation.

The *FL2-mRFP* promoter-reporter lines described in this chapter are being used in numerous ongoing genetic experiments. Once fully introgressed into their respective inbred backgrounds,

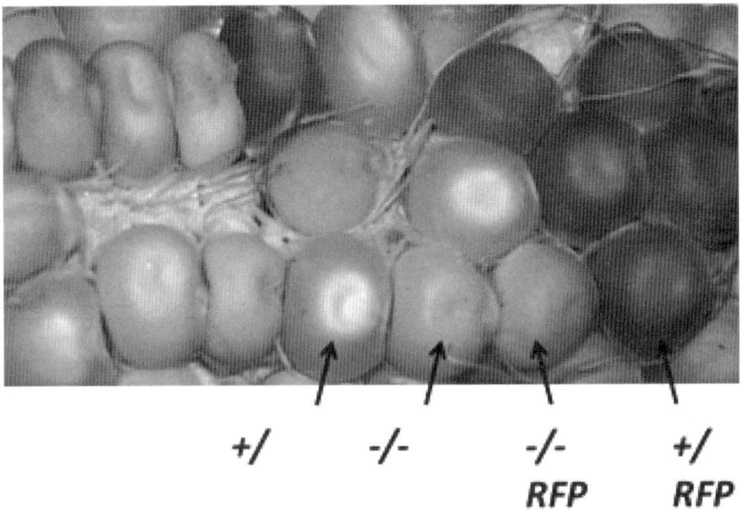

Figure 12.8 Photograph of an IHP1 ear segregating for the *o2* mutation and the *FL2-mRFP* transgene. The *o2* mutant phenotype is chalky and opaque in appearance (–/–) compared with wild-type kernels (+ /). Kernels containing both the *o2* mutation and the *FL2-mRFP* transgene (–/–; RFP) illustrate reduced expression of *FL2-mRFP*, as detected by a significant decrease in pink coloration of the kernels compared with kernels containing only the *FL2-mRFP* transgene, which are much darker (+ /; RFP). (For color detail, see color plate section.)

the potential impacts of gene dosage and imprinting on the maternal inheritance of grain protein concentration will be investigated through reciprocal crosses. A genetic mapping study is also underway using the *FL2-mRFP* transgene activity as a phenotype. A near-isogenic line of *FL2-mRFP* in the B73 inbred background, which produces grain with ~7% protein, was crossed to the previously described IPSRI mapping population. The resulting ears demonstrate a wide range of pink coloration. A method for quantification of *FL2-mRFP* expression from high-resolution digital images is in development. The advantages of using *FL2-mRFP* expression as a phenotype is that it specifically tracks regulatory variation for *floury2* 22-kDa α-zeins, rather than seed protein concentration as a whole, which depends on other genetic and environmental factors. Finally, the IHP1:*FL2-mRFP* reporter line has been crossed to a mutagenized (ethyl methanesulfonate) population of IHP1 plants to screen ears visually for mutations with altered *FL2-mRFP* expression and protein concentration. These and other studies combining genomics advances with the Illinois Long Term Selection experiment will continue not only to provide insight into how the maize genome has responded to artificial selection but also may identify genes and regulatory changes that could further improve maize productivity and grain composition.

Acknowledgments

We gratefully acknowledge the efforts and perseverance of the researchers who have served as principal investigators of the Illinois Long-Term Selection Experiment since 1896, including Cyril G. Hopkins, Louie Smith, Clyde Woodworth, Earl Leng, Denton E. Alexander, Robert J. Lambert, and especially John Dudley. In addition, we thank George Singletary from Pioneer Hybrid for providing GUS reporter lines and Dave Jackson's laboratory at Cold Spring Harbor for providing the FL2-mRFP reporter lines. We acknowledge the sources of funding that have made this research possible, including the Illinois Council for Food and Agricultural Research, the National Science Foundation (Award DBI 0501700), and the United States Department of Agriculture (NIFA Award 2011-67013-30041). Finally, we are grateful for the support of Christine Lucas's graduate research through the Robert J. Lambert graduate fellowship.

References

Alleman M, Doctor J (2000). Genomic imprinting in plants: observations and evolutionary implications. Plant Mol Biol 43: 147–161.

Alrefai R, Berke TG, Rocheford TR (1995). Quantitative trait locus analysis of fatty-acid concentrations in maize. Genome 38: 894–901

Below FE, Seebauer JR, Uribelarrea M, Schneerman MC, Moose SP (2004). Physiological changes accompanying long-term selection for grain protein in maize. Plant Breed Rev 24: 133–151.

Bernard L, Ciceri P, Viotti A (1994). Molecular analysis of wild-type and mutant alleles at the opaque-2 regulatory locus of maize reveals different mutations and types of o2 products. Plant Mol Biol 24: 949–959.

Bhattramakki D, Sachs MM, Kriz AL (1996). Expression of genes encoding globulin and prolamin storage proteins in kernels of Illinois long term chemical selection strains. Crop Sci 36: 1029–1036.

Bird AP, Wolffe AP (1999). Methylation-induced repression—belts, braces, and chromatin. Cell 99: 451–454.

Campbell RE, Tour O, Palmer AE, Steinbach PA, Baird GS, Zacharias DA, Tsien RY (2002). A monomeric red fluorescent protein. Proc Natl Acad Sci U S A 99: 7877–7882.

Chaudhuri S, Messing J (1994). Allele-specific parental imprinting of dzr1, a posttranscriptional regulator of zein accumulation. Proc Natl Acad Sci U S A 91: 4867–4871.

Chia JM, Song C, Bradbury PJ, Costich D, de Leon N, Doebley J, Elshire RJ, Gaut B, Geller L, Glaubitz JC, Gore M, Guill KE, Holland J, Hufford MB, Lai J, Li M, Liu X, Lu Y, McCombie R, Nelson R, Poland J, Prasanna BM, Pyhajarvi T, Rong T, Sekhon RS, Sun Q, Tenaillon MI, Tian F, Wang J, Xu X, Zhang Z, Kaeppler SM, Ross-Ibarra J, McMullen MD, Buckler ES, Zhang G, Xu Y, Ware D (2012). Maize HapMap2 identifies extant variation from a genome in flux. Nat Genet 44: 803–807.

Clark RM, Wagler TN, Quijada P, Doebley J (2006). A distant upstream enhancer at the maize domestication gene tb1 has pleiotropic effects on plant and inflorescent architecture. Nat Genet 38: 594–597.

Dembinski E, Rafalski A, Wisniewska I (1991). Effect of long-term selection for high and low protein-content on the metabolism of amino-acids and carbohydrates in maize kernel. Plant Physiol Biochem 29: 549–557.

Dijkhuizen A, Dudley JW, Rocheford TR, Haken AE, Eckhoff SR (1998). Near-infrared reflectance correlated to 100-g wet-milling analysis in maize. Cereal Chem 75: 266–270.

Dudley JW (ed) (1977). *Seventy-six Generations of Selection for Oil and Protein Percentage in Maize*, Iowa State University Press, Ames, Iowa.

Dudley JW (1994). Linkage disequilibrium in crosses between Illinois maize strains divergently selected for protein percentage. Theor Appl Genet 87: 1016–1020.

Dudley JW, Clark D, Rocheford TR, LeDeaux JR (2007). Genetic analysis of corn kernel chemical composition in the random mated 7 generation of the cross of generations 70 of IHP x ILP. Crop Sci 47: 45–57.

Dudley JW, Dijkhuizen A, Paul C, Coates ST, Rocheford TR (2004). Effects of random mating on marker-QTL associations in the cross of the Illinois High Protein X Illinois Low Protein maize strains. Crop Sci 44: 1419–1428.

Dudley JW, Lambert RJ (2004). 100 generations of selection for oil and protein in corn. Plant Breed Rev 24: 79–110.

Egger G, Liang G, Aparicio A, Jones PA (2004). Epigenetics in human disease and prospects for epigenetic therapy. Nature 429: 457–463.

Elshire RJ, Glaubitz JC, Sun Q, Poland JA, Kawamoto K, Buckler ES, Mitchell SE (2011). A robust, simple genotyping-by-sequencing (GBS) approach for high diversity species. PLoS One 6: e19379.

Feng LN, Zhu J, Wang G, Tang YP, Chen HJ, Jin WB, Wang F, Mei B, Xu ZK, Song RT (2009). Expressional profiling study revealed unique expressional patterns and dramatic expressional divergence of maize alpha-zein super gene family. Plant Mol Biol 69: 649–659.

Ganal MW, Durstewitz G, Polley A, Berard A, Buckler ES, Charcosset A, Clarke JD, Graner EM, Hansen M, Joets J, Le Paslier MC, McMullen MD, Montalent P, Rose M, Schon CC, Sun Q, Walter H, Martin OC, Falque M (2011). A large maize (Zea mays L.) SNP genotyping array: development and germplasm genotyping, and genetic mapping to compare with the B73 reference genome. PLoS One 6: e28334.

Gehring M, Choi Y, Fischer RL (2004). Imprinting and seed development. Plant Cell 16 Suppl: S203–S213.

Goldman IL, Rocheford TR, Dudley JW (1993). Quantitative trait loci influencing protein and starch concentration in the Illinois long-term selection maize strains. Theor Appl Genet 87: 217–224.

Goldman IL, Rocheford TR, Dudley JW (1994). Molecular markers associated with maize kernel oil concentration in an Illinois high-protein x illinois low-protein cross. Crop Sci 34: 908–915.

Hansen DW, Brimhall B, Sprague GF (1946). Relationship of zein to the total protein in corn. Cereal Chem 23: 329–335.

Hartings H, Lazzaroni N, Rossi V, Riboldi GR, Thompson RD, Salamini F, Motto M (1995). Molecular analysis of *opaque-2* alleles from *Zea mays* L. reveals the nature of mutational events and the presence of a hypervariable region in the 5′ part of the gene. Genet Res Camb 65: 11–19.

Henry AM, Damerval C (1997). High rates of polymorphism and recombination at the Opaque-2 locus in cultivated maize. Mol Gen Genet 256: 147–157.

Hoener IR, DeTurk EE (1938). The absorption and utilization of nitrate nitrogen during vegetative growth by Illinois High Protein and Illinois Low Protein corn. J Amer Soc Agron 30: 232–243.

Hopkins CG (1899). Improvement in the chemical composition of the corn kernel. Ill Agr Exp Stn Bull 55: 205–240.

Hopkins CG, Smith LH, East EM (1903). The structure of the corn kernel and the composition of its different parts. Univ Illinois Agricultural Experiment Station Bull 87.

Huang X, Feng Q, Qian Q, Zhao Q, Wang L, Wang A, Guan J, Fan D, Weng Q, Huang T, Dong G, Sang T, Han B (2009). High-throughput genotyping by whole-genome resequencing. Genome Res 19: 1068–1076.

Hwang YS, Ciceri P, Parsons RL, Moose SP, Schmidt RJ, Huang N (2004). The maize O2 and PBF proteins act additively to promote transcription from storage protein gene promoters in rice endosperm cells. Plant Cell Physiol 45: 1509–1518.

Jefferson RA, Kavanagh TA, Bevan MW (1987). GUS fusions: beta-glucuronidase as a sensitive and versatile gene fusion marker in higher plants. EMBO J 6: 3901–3907.

Johansson AM, Pettersson ME, Siegel PB, Carlborg O (2010). Genome-wide effects of long-term divergent selection. PLoS Genet 6: e1001188.

Lambert RJ, Alexander DE, Mejaya IJ (2004). Single kernel selection for increased grain oil in maize synthetics and high-oil hybrid development. Plant Breed Rev 24: 153–175.

Langridge P, Pintortoro JA, Feix G (1982). Zein precursor messenger-RNAs from maize endosperm. Mol Gen Genet 187: 432–438.

Lauria M, Rupe M, Guo M, Kranz E, Pirona R, Viotti A, Lund G (2004). Extensive maternal DNA hypomethylation in the endosperm of Zea mays. Plant Cell 16: 510–522.

Laurie CC, Chasalow SD, LeDeaux JR, McCarroll R, Bush D, Hauge B, Lai C, Clark D, Rocheford TR, Dudley JW (2004). The genetic architecture of response to long-term artificial selection for oil concentration in the maize kernel. Genetics 168: 2141–2155.

Lorenzoni C, Alberi P, Viotti A, Soave C, Di Fonzo N, Gentinetta E, Maggiore T, Salamini F (1978). Biochemical characterization of high and low protein strains of maize. In: Miflin BU, Zoschke M (eds). *Carbohydrate and Protein Synthesis*, Commission of the European Communities, Brussels, Belgium, pp 173–197.

Lund G, Ciceri P, Viotti A (1995). Maternal-specific demethylation and expression of specific alleles of zein genes in the endosperm of Zea mays L. Plant J 8: 571–581.

Miclaus M, Xu JH, Messing J (2011). Differential gene expression and epiregulation of alpha zein gene copies in maize haplotypes. PLoS Genet 7: e1002131.

Mohanty A, Luo A, DeBlasio S, Ling XY, Yang Y, Tuthill DE, Williams KE, Hill D, Zadrozny T, Chan A, Sylvester AW, Jackson D (2009). Advancing cell biology and functional genomics in maize using fluorescent protein-tagged lines. Plant Physiol 149: 601–605.

Moose SP, Dudley JW, Rocheford TR (2004). Maize selection passes the century mark: a unique resource for 21st century genomics. Trends Plant Sci 9: 358–364.

Nelson OE (1979). Genetic control of polysaccharide and storage protein synthesis in the endosperms of barley, maize, and sorghum. Adv Cereal Sci Technol 3: 41–71.

Reggiani R, Soave C, Di Fonzo N, Gentinetta E, Salamini F (1985). Factors affecting starch and protein content in developing endosperms of high and low protein strains of maize. Genet Agric 39: 221–232.

Rohland N, Reich D (2012). Cost-effective, high-throughput DNA sequencing libraries for multiplexed target capture. Genome Res 5:939–946.

Russell DA, Fromm ME (1997). Tissue-specific expression in transgenic maize of four endosperm promoters from maize and rice. Transgenic Res 6: 157–168.

Salas A (2008). Regulation of zein transgene expression in response to long term selection for grain protein concentration in maize. M.S. thesis. University of Illinois at Urbana-Champaign.

Schmidt RJ, Ketudat M, Aukerman MJ, Hoschek G (1992). Opaque-2 is a transcriptional activator that recognizes a specific target site in 22-kd-zein genes. Plant Cell 4: 689–700.

Schneider E, Earley E, DeTurk E (1952). Nitrogen fractions of the component parts of the corn kernel as affected by selection and soil nitrogen. Agron J 44: 161–169.

Sene M, Thevenot C, Hoffmann D, Benetrix F, Causse M, Prioul JL (2001). QTLs for grain dry milling properties, composition and vitreousness in maize recombinant inbred lines. Theor Appl Genet 102: 591–599.

Showalter MF, Carr RH (1922). Characteristic proteins in high- and low-protein corn. Am Chem Soc J 44: 2019–2023.

Smith LH (1908). Ten generations of corn breeding. Ill Agricultural Experiment Station Bull 128: 457–575.

Song RT, Llaca V, Linton E, Messing J (2001). Sequence, regulation, and evolution of the maize 22-kD alpha zein in gene family. Genome Res 11: 1817–1825.

Song RT, Messing J (2003). Gene expression of a gene family in maize based on noncollinear haplotypes. Proc Natl Acad Sci U S A 100: 9055–9060.

Springer NM, Ying K, Fu Y, Ji T, Yeh CT, Jia Y, Wu W, Richmond T, Kitzman J, Rosenbaum H, Iniguez AL, Barbazuk WB, Jeddeloh JA, Nettleton D, Schnable PS (2009). Maize inbreds exhibit high levels of copy number variation (CNV) and presence/absence variation (PAV) in genome content. PLoS Genet 5: e1000734.

Stewart CN (2006). Go with the glow: fluorescent proteins to light transgenic organisms. Trends Biotechnol 24:155–162.

Tsai CL, Dweikat I, Tsai CY (1990). Effects of source supply and sink demand on the carbon and nitrogen ratio in maize kernels. Maydica 35: 391–397.

Uribelarrea M, Below FE, Moose SP (2004). Grain composition and productivity of maize hybrids derived from the Illinois protein strains in response to variable nitrogen supply. Crop Sci 44: 1593–1600.

Uribelarrea M, Moose SP, Below FE (2007). Divergent selection for grain protein affects nitrogen use efficiency in maize hybrids. Field Crops Res 100: 82–90.

VicenteCarbajosa J, Moose SP, Parsons RL, Schmidt RJ (1997). A maize zinc-finger protein binds the prolamin box in zein gene promoters and interacts with the basic leucine zipper transcriptional activator Opaque2. Proc Natl Acad Sci U S A 94: 7685–7690.

Wenck A, Pugieux C, Turner M, Dunn M, Stacy C, Tiozzo A, Dunder E, van Grinsven E, Khan R, Sigareva M, Wang WC, Reed J, Drayton P, Oliver D, Trafford H, Legris G, Rushton H, Tayab S, Launis K, Chang YF, Chen DF, Melchers L (2003). Reef-coral proteins as visual, non-destructive reporters for plant transformation. Plant Cell Rep 22: 244–251.

Wettenhall JM, Smyth GK (2004). limmaGUI: a graphical user interface for linear modeling of microarray data. Bioinformatics 20: 3705–3706.

Willmott DB, Dudley JW, Rocheford TR, Bari AL (2006). Effect of random mating on marker-QTL associations for grain quality traits in the cross of Illinois High Oil x Illinois Low Oil. Maydica 51: 187–199.

Wilson CM (1992). Zein diversity in Reid, Lancaster and Illinois Chemical corn strains revealed by isoelectric focusing. Crop Sci 32: 869–873.

Wittkopp PJ (2005). Genomic sources of regulatory variation in cis and in trans. Cell Mol Life Sci 62: 1779–1783.

Woo YM, Hu DWN, Larkins BA, Jung R (2001). Genomics analysis of genes expressed in maize endosperm identifies novel seed proteins and clarifies patterns of zein gene expression. Plant Cell 13: 2297–2317.

Woodworth CM, Leng ER, Jugenheimer RW (1952). Fifty generation of selection for oil and protein in corn. Agron J 44: 60–65.

Zheng P, Allen WB, Roesler K, Williams ME, Zhang S, Li J, Glassman K, Ranch J, Nubel D, Solawetz W, Bhattramakki D, Llaca V, Deschamps S, Zhong GY, Tarczynski MC, Shen B (2008). A phenylalanine in DGAT is a key determinant of oil content and composition in maize. Nat Genet 40: 367–372.

13 Machine Vision for Seed Phenomics

Jeffery L. Gustin and A. Mark Settles

Introduction

Seeds have a broad array of morphological, biochemical, and biological phenotypes. For many crops, the seed is the desired end-product for breeding and cultivation with the seed phenome directly related to product quality. In seed production, it is critical to have viable seeds that germinate and establish well in the field (Aquila, 2009). For seed crops, composition, morphology, and resistance to pathogens are important seed traits that greatly influence the value of a farmer's harvest (Shewry, 2007; Nesi *et al.*, 2008; Fox and Manley, 2009; Pinzi *et al.*, 2009). Most seed traits are complex to score, and destructive analytical approaches are often needed to obtain quantitative data. For example, determining germination rates typically requires a sample from a seed lot to be germinated under controlled conditions and germinated seedlings are scored visually. For biochemical assays such as protein content, a sample of a seed lot is typically ground to a fine powder, dried, and combusted in a C/N analyzer to determine nitrogen content. In these examples, labor-intensive manual counting or seed grinding and chemical analysis is required for quantifying a single seed phenotype.

With the extensive development of genomics technologies, our ability to quantify genetic information and RNA transcripts has become automated with sophisticated computational analysis tools. Even though the seed phenome is related to genetic information, quantifying and studying seed phenotypes is still highly dependent on manual counting and analytics. Systematic and automated phenotyping technologies, also known as phenomics, provide excellent opportunities to align more closely the level of detail in genome and phenotype information to understand better the relationships between genotype and phenotype.

Machine vision approaches are attractive alternatives to traditional phenotyping. Machine vision combines electromagnetic imaging and spectroscopy technologies with computation to replace manual or destructive analytical approaches. Machine vision technologies range from analysis of optical digital images to various types of spectroscopy and nuclear magnetic resonance (NMR) imaging. The types of data that can be extracted from machine vision approaches vary on the wavelengths of light used to image seeds and the preparation of the seed samples for the imaging technology (Figure 13.1). The throughput of these technologies depends on the level of resolution of the images and image acquisition designs. Higher throughput technologies can be automated and provide massive savings in time and cost to obtain quantitative data (Armstrong, 2006). Some machine vision platforms are considered to be complete substitutes for conventional analytical chemistry (Conway and Earle, 1963; Blanco and Villarroya, 2002; Rodriguez-Saona and Allendorf,

Seed Genomics, First Edition. Edited by Philip W. Becraft.

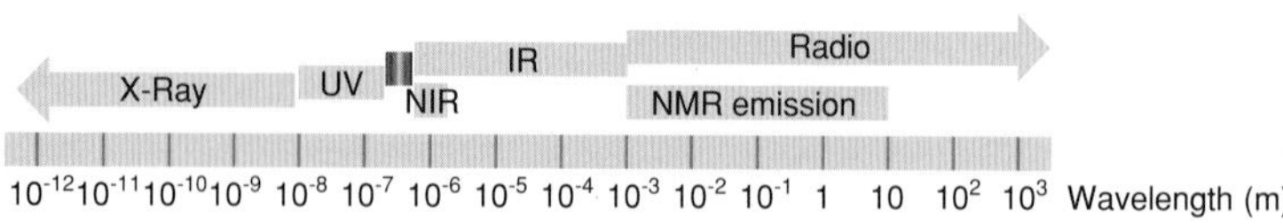

Figure 13.1 Electromagnetic spectrum showing wavelength ranges for machine vision technologies. (For color detail, see color plate section.)

2011). Other imaging technologies provide spatial resolution of quantitative traits that would be impossible to gather without a machine vision platform (Kim *et al.*, 2006).

In this chapter, we highlight machine vision technologies as they have been applied to seed phenotype analyses. The chapter is organized based on spectral type and discusses phenotype features that have been extracted by imaging seeds within each spectral range (Figure 13.1). We start with short-wavelength, high-energy x-rays and end with a discussion of magnetic resonance imaging (MRI), which uses longer wavelength radio waves.

High-Energy Imaging: X-ray Tomography and Fluorescence

The high energy of x-rays allows two types of imaging of biological samples: x-ray attenuation and x-ray fluorescence (Stuppy *et al.*, 2003; Dhondt *et al.*, 2010; Moore *et al.*, 2010; Burghardt *et al.*, 2011). At lower irradiation levels, the differential absorption or attenuation of x-rays by biological tissues can be used to produce radiographs. Similar to laser confocal microscopy, a series of radiographs at serial planes of section can be reconstructed into a three-dimensional image of the sample via computed tomography (CT) scan. The resolution of CT is related to the level of irradiation and focus of the x-ray source. High-powered, monochromatic x-ray sources from synchrotron radiation can be used to obtain three-dimensional images of plant tissues at <1 μm resolution (Stuppy *et al.*, 2003; Dhondt *et al.*, 2010). There are only a few synchrotrons available worldwide for high-resolution x-ray tomography, but laboratory high-resolution CT scanners, commonly known as micro-CT scanners, can also be used for analysis of plant materials. Micro-CT scanners use a polychromatic x-ray source and have lower power translating to slightly lower resolution at similar scanning times between the two methods (Burghardt *et al.*, 2011). In both cases, the level of x-ray exposure needed for high-resolution imaging of seed or plant tissue causes DNA damage, and the technique cannot be considered a live-cell approach. However, CT provides the advantage of being nondestructive; samples can be imaged and then processed for other analytical techniques.

Micro-CT provides many advantages over conventional microscopy. The approach does not require sectioning and can be applied to dry or fossilized samples. Surface images similar to scanning electron micrographs are produced. In addition, samples can be digitally sectioned in any plane once the three-dimensional reconstruction is complete allowing comparison of tomograms with traditional micrographs (Smith *et al.*, 2009). X-ray attenuation is directly related to the density of the sample, and various attenuation cutoffs or thresholds can be used to isolate tissues with similar density and visualize the outer surface, air spaces, or specific structures within the sample. The tomogram can also be used to calculate accurately tissue volume, surface area, porosity, and densities of specific regions within a sample. For density measurements, the polychromatic x-ray sources used in laboratory micro-CT scanners are subject to beam hardening, in which tissues absorb a disproportionate amount of long-wavelength "soft" x-rays leading to errors in the calculated density of thin versus thick samples (Burghardt *et al.*, 2011). The geometry and size of samples need to be similar to compare density measurements directly.

In plants, x-ray tomography has primarily been used to image root systems and vascular systems (Brodersen *et al.*, 2011; Flavel *et al.*, 2012; Hunter *et al.*, 2012). The main application of CT in seeds has been to phenotype fossils for morphological comparison with extant species. In the last 5 years, synchrotron tomography has become an important technique to image the anatomy of small fossilized seed to place plants within anatomically based phylogenies (Friis *et al.*, 2007; Mendes *et al.*, 2008a, 2008b; Chen *et al.*, 2009a, 2009b; Friis *et al.*, 2009). Micro-CT has also been used to study the changes in pore size that occur during coffee bean roasting and to model moisture loss during maize grain drying (Pittia *et al.*, 2011; Takhar *et al.*, 2011). However, the approach has yet to be extensively used for phenotyping live plant seeds even though there is potential for highly quantitative studies. Micro-CT images of maize kernels can distinguish the major anatomical features of a mature kernel: embryo, vitreous endosperm, starchy endosperm, and pericarp (Takhar *et al.*, 2011). Figure 13.2 shows cross sections of micro-CT images of *opaque2* (*o2*) and normal kernels. The *o2* mutant disrupts expression of zein seed storage proteins and starch granule packing with a loss of the hard, vitreous endosperm (see Chapters 8, 9, and 12) (Gibbon and Larkins, 2005).

The hard x-rays from some synchrotron light sources are of sufficient power to complete synchrotron x-ray fluorescence microscopy (SXFM). In SXFM, a monochromatic x-ray beam is focused onto the tissue to eject core shell electrons from the atoms within the sample (Lombi *et al.*, 2011a). When higher shell electrons replace the vacancies, each element emits characteristic wavelengths of light allowing the quantification of biologically relevant ions. Tomography can also be completed with SXFM to produce three-dimensional images of elemental distributions within a tissue. SXFM of barley grains has shown that micronutrients including iron (Fe), copper (Cu), manganese (Mn), potassium (K), and zinc (Zn) all accumulate in localized areas of the seed with the bulk of the accumulation in the embryo and vascular bundle (Lombi *et al.*, 2011b). In *Arabidopsis* seeds, SXFM was used to visualize directly Fe distribution in a mutant of the *VACUOLAR IRON TRANSPORTER 1* (*VIT1*) gene (Kim *et al.*, 2006). Even though *vit1* mutant and wild-type seeds accumulate equivalent

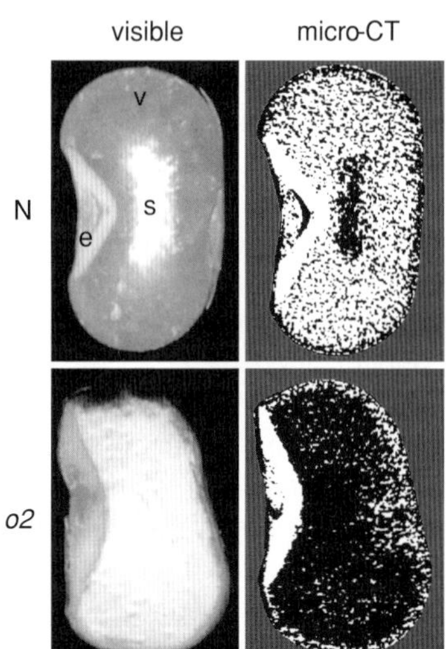

Figure 13.2 Comparison of seed cross sections from normal (N) and *opaque2* (*o2*) mutant. Visible panels are hand sections made after completing CT scans. Micro-CT scans have an identical threshold for attenuation with white pixels indicating regions above the density threshold and black pixels indicating regions below. *v*, vitreous endosperm; *s*, starchy endosperm; *e*, embryo. (For color detail, see color plate section.)

levels of Fe, SXFM tomograms showed that Fe was not as highly localized to provascular regions in *vit1* compared with wild-type.

These examples illustrate how attenuation and fluorescence x-ray CT can yield very high-resolution seed phenotype data that would be difficult to obtain with other analytical techniques. Attenuation provides an alternative to traditional microscopy, whereas x-ray fluorescence provides high resolution of ion concentrations. Equivalent ion quantification technologies require sectioning or dissection of the seed and would not give the same level of three-dimensional resolution (Moore *et al.*, 2010). Data acquisition and processing times are faster for x-ray imaging than traditional microscopy, with scans taking minutes to hours (Dhondt *et al.*, 2010). However, rendering the tomogram is computationally intense and requires significant researcher input. X-ray CT approaches cannot be considered exceptionally high-throughput, and published examples of micro-CT on seeds contain very small sample sizes (Kim *et al.*, 2006; Lombi *et al.*, 2011b; Pittia *et al.*, 2011; Takhar *et al.*, 2011).

Optical Imaging: Visible Spectrum

Optical imaging relies on the transmission and reflectance properties of the visible spectrum. Visible light perceived by the human eye extends from 380–770 nm. The natural resonance frequencies of atoms in a material lead to differential absorption, transmission, and reflectance of the visible spectrum. Chemical bonds influence atomic resonance frequencies leading to some biological compounds having specific visible reflectance such as chlorophylls and carotenoids. Seed color and morphology are very important traits that humans have used to direct crop production either consciously or unconsciously for at least 10,000 years (Purugganan and Fuller, 2009). Seed size, color, and shape remain relevant targets for current breeding strategies as well as genomic studies.

Digital cameras and scanners provide inexpensive image acquisition for machine vision phenotyping of seeds (Dell'Aquila, 2007, 2009). Digital imaging devices convert visible light into electrical signals through proportional conversion of photons into electrons based on the photoelectric effect. Although digital image sensors propagate light intensity information, the sensor does not distinguish the light wavelength. Color is produced by filtering or spectral separation and integrating signals from major color wavelengths: red, green, and blue. A digital camera gives an approximation of the true color in the image. The features from an optical image can be related to a seed's genetic and biological origin, biochemical composition, and potential for future propagation. Currently, digital imaging is used at both commercial and research levels to improve seed phenotyping accuracy and efficiency.

Two important commercial applications of digital imaging include seed identification and testing seed germination. Seed lots are graded based on grain quality and manually inspected for the presence of potentially noxious weed species (USDA, 2009). Automated image-based platforms are being tested as alternatives for manual assessments. Seed identification projects have found that multivariate mathematical models can be trained to identify weed seeds accurately by using data extracted from images of the seed (Chtioui *et al.*, 1996; Granitto *et al.*, 2005). In one example, a seed identification system identified ~98% of weed seeds from 236 different species by using an optimized set of just four image features (Granitto *et al.*, 2005). Seed identification is also important for determining the final grade and price of grain lots. For example, wheat grain lots are graded by class and subclass. Machine vision approaches have been tested for their ability to discriminate between different wheat grain classes and similar species such as rye, barley, and oats (Majumdar and Jayas, 2000a, 2000b, 2000c, 2000d; Paliwal *et al.*, 2003; Arefi *et al.*, 2011; Mebatsion *et al.*,

2012). These CCD camera-based platforms can be 95%–100% accurate at identifying similar grains. Accuracy improves as mathematically derived features such as Hue moment invariants and Fourier transformations are included in the feature set. The images can also be used to identify damaged grain and foreign material, each of which is important in assigning the final grade of the seed lot (Paliwal *et al.*, 2003).

Seedling establishment and vigor in the field is an important trait for yield. Uniform establishment reduces weed pressure, allows for reduced application of inputs, and promotes more uniform flowering and harvest (Ashraf and Foolad, 2005). Establishment is correlated with the speed, proportion, and uniformity of seed germination under stressful environments, and seed germination assays involving multiple stressful environments are a standard approach to characterizing seed lot quality (Bingham *et al.*, 1994). Germination assays either count seedlings during a predetermined timeframe for a mean germination time or count root radical emergence, which is an earlier phase of germination (Matthews *et al.*, 2011). Machine vision platforms using digital cameras or flatbed scanners can capture both types of germination scores, but root radical emergence is more amenable to machine vision approaches.

The French National Seed Testing Facility (Algers, France) runs a semiautomated seed phenotyping platform using Jacobsen tables for environmental control with four static cameras per table (Wagner *et al.*, 2011). Computer control systems dictate periodic image acquisition, and because samples are imaged within the environment in which they are grown, individual seed germination profiles can be traced from imbibition to early seedling stage. Basic measurements such as area and XY position are automatically extracted from each seed by the image processing program ImageJ. This facility has imaged >170,000 seeds from plants such as tomato, alfalfa, sunflower, garlic, and maize since 2005 (Ducournau *et al.*, 2005; Wagner *et al.*, 2011). Similar machine vision approaches have been developed for melon and lettuce seed vigor tests (Sako *et al.*, 2001; Marcos *et al.*, 2006).

Germination assays for very small seeds such as *Arabidopsis* are also possible (Joosen *et al.*, 2010, 2012). In this case, seeds are germinated in clear plastic trays on blue filter paper, and the trays are moved from the growth chamber to the camera for each image. Automated image processing extracts multiple germination parameters such as maximum germination and area under the germination curve. The platform was used to phenotype *Arabidopsis* recombinant inbred lines (RILs) for quantitative trait loci (QTL) germination under salt stress. This study found two QTL for maximum germination under salt stress that overlapped with prior studies as well as novel QTL (Quesada *et al.*, 2002; Clerkx *et al.*, 2004; Ren *et al.*, 2010). By completing the phenotyping with a machine vision platform, multiple traits such as time to 50% germination and area under the germination curve could be scored for QTL mapping.

Machine vision platforms can also be used to characterize seed morphology. In maize, kernel hardness and flour quality are associated with the proportion of the kernel that is vitreous. Vitreousness is a qualitative trait that is difficult to score and is traditionally scored by eye using a light box (Fox and Manley, 2009). In an effort to make this trait more quantitative, Erasmus and Taylor (2004) developed an optical imaging system to measure visible light attenuation through maize kernels. The calibration to physically dissected kernels suggested a correlation that could roughly sort kernels by vitreousness. A similar approach has been used to phenotype Quality Protein Maize RILs to map modifiers of the *o2* mutant (Holding *et al.*, 2008, 2011).

In wheat, both a larger grain size and a more spherical shape are considered desirable characteristics for the kernel (Evers *et al.*, 1990). A commercialized digital camera system (MARVIN seed analyzer) was recently used to characterize the underlying genetic framework of wheat grain morphology (Gegas *et al.*, 2010). Morphological features for size and shape were extracted from six diverse, recombinant doubled-haploid populations as well as primitive and modern elite varieties.

The dataset was used to describe the morphological variability and relationships within a broad wheat population and identify QTL associated with agronomic traits. Grain size and shape were found to be largely independent features, and the genetic structure for determining grain size was quite variable between the different germplasm sources. These observations suggest great potential for selecting grain size and shape as independent traits in wheat and illustrate how machine vision platforms can complete thousands of meticulous measurements through image analysis algorithms.

Flatbed scanners provide high-resolution images at low cost making them an effective tool for imaging small seeds. *Arabidopsis* seed areas can be accurately extracted from scanner images allowing large-scale screens for seed size (Herridge *et al.*, 2011). By examining two RIL populations, Herridge *et al.* (2011) found four and five QTL that explained 43% and 44% of the total seed weight variation. Only one common QTL was identified showing that *Arabidopsis* accessions control seed size with multiple genetic pathways. Subtle seed size mutants were also detected with one mutant having only a 10% reduced seed area. Using a similar flatbed scanner imaging system, *Arabidopsis* seed size variation has been related to seedling root growth to identify a direct physiological relationship between seed size and seedling establishment that is independent of genetic variation (Brooks *et al.*, 2010).

So far, optical imaging of seeds has primarily been used to score seeds for traits of interest. Key advantages of machine vision scoring are higher throughput and the ability to score multiple traits from single images using image processing algorithms. In several examples, optical imaging of seeds has enabled discovery of complex relationships and condition-specific QTL that were practically possible only through implementation of a machine vision approach (Brooks *et al.*, 2010; Gegas *et al.*, 2010; Herridge *et al.*, 2011; Joosen *et al.*, 2012). Machine vision based on optical imaging is not yet used extensively for selecting seeds with desired morphology. However, visible wavelengths are frequently incorporated with near-infrared (NIR) spectroscopy to remove undesirable seeds or to select for seeds of interest (Osborne, 2006).

Resonance Absorption: Infrared Spectrum

The infrared (IR) spectrum has lower energy than visible light. Chemical bonds absorb IR light to alter the vibrations and rotations between the atoms within a molecule. The bonds within organic molecules have characteristic resonance frequencies that absorb specific light wavelengths (Rodriguez-Saona and Allendorf, 2011). Resonance frequencies are typically within the mid-infrared (mid-IR) from 2500–25,000 nm. The same bonds can absorb at overtone frequencies, which are shorter wavelengths that are multiples of the resonance frequencies. The overtone wavelengths are close to the visible spectrum (750–2500 nm) and termed NIR. Although molecules absorb mid-IR and NIR based on the same chemical properties, the energy inherent to the light spectrum significantly changes the spectroscopy apparatus and potential applications in seed phenotyping.

Mid-IR is more readily absorbed by biological macromolecules. At the same intensity of mid-IR and NIR light, the mid-IR does not penetrate as thick of a sample (Rodriguez-Saona and Allendorf, 2011). Mid-IR is typically used as an analytical chemistry technique for thin samples to identify spectral fingerprints and chemical composition. Many mid-IR applications focus on food processing rather than seed phenotyping. Mid-IR spectral fingerprints can be used to differentiate milling fractions of wheat and to track protein structural changes during soybean heat processing (Barron, 2011; Samadi and Yu, 2011). With high-energy synchrotron light sources, mid-IR can be used to characterize thin sections of seeds to identify compositional differences between tissue layers in both soybean and maize seeds (Yu *et al.*, 2004; Pietrzak and Miller, 2005).

NIR is not as readily absorbed, and thicker samples are needed to obtain spectra. For seeds, NIR data typically are collected on ground meal or bulk samples of whole seeds. However, single-seed hyperspectral imaging is also possible (Manley *et al.*, 2009; Zhu *et al.*, 2011). NIR spectra have broad absorption peaks and do not provide a specific fingerprint of chemical composition (Osborne, 2006; Cozzolino, 2009). To correlate the spectra to specific chemical composition, multivariate statistical approaches known as chemometrics are used (Blanco and Villarroya, 2002). Chemometrics is an empirical approach in which spectra are collected on a set of test samples that are quantified with other analytical techniques. The spectral and analytical data are related to each other using multiple linear regression, principal component regression, or partial least-squares regression. Artificial neural networks are also sometimes used for relating IR spectral data to chemical composition. Chemometric calibrations are made by collecting spectra from samples followed by determining their composition with reference analytical methods. NIR predictions contain error from both a reference method and spectra collection and cannot be as accurate as the reference method.

Even with this caveat, NIR has been widely adopted from research laboratories to grain elevators to predict seed traits (Osborne, 2006). In maize, calibrations exist for multiple seed constituents, including protein, oil, starch, and moisture (Orman and Schuman, 1991; Berardo *et al.*, 2009); fatty acids (Yang *et al.*, 2009); amino acids (Fontaine *et al.*, 2002); carotenoids (Brenna and Berardo, 2004); and amylose-to-amylopectin ratios (Campbell *et al.*, 1997, 1999, 2002). More complex grain quality traits are correlated with chemical composition, and NIR has been used to predict wet milling starch yields (Paulsen *et al.*, 2003a, 2003b; Paulsen and Singh, 2004), kernel hardness (Robutti, 1995; Correa *et al.*, 2002; Ngonyamo-Majee *et al.*, 2008), and fungal infection (Pearson *et al.*, 2001; Dowell *et al.*, 2002). Although many of these calibrations have been developed using different NIR spectrometers and wavelength regions, these studies demonstrate a wide range of traits that could be estimated from a single NIR reading. The ability to predict multiple seed phenotypes with a single, nondestructive spectroscopic measurement provides massive cost savings over reference analytical methods that are usually limited to a small number of traits. More recently, bulk NIR was used to phenotype the maize Nested Association Mapping panel for kernel composition (Cook *et al.*, 2012). This collection of 25 RIL populations identified candidate genes that control seed composition and included genes known to affect protein, starch, or oil accumulation along with hundreds of new candidates. The chemical phenotypes of the >60,000 samples would not have been practical to score without NIR spectroscopy.

NIR spectroscopy can also be applied to single seeds. Conventional ground and whole-grain NIR use multiple seeds to estimate average composition for the seed sample. In segregating populations, seeds with different composition are averaged, reducing the ability to discriminate for quality traits. Single-seed selections allow more rapid breeding gain or can be used to sort seeds for phenotypes of interest such as protein quantity in soybean or spinach seed viability (Lee *et al.*, 2010; Olesen *et al.*, 2011). Single-seed NIR has been most successful for predicting oil and protein content in seeds with relatively uniform distribution of constituents, such as rapeseed (Velasco *et al.*, 1999a; Velasco and Mollers, 2002; Hom *et al.*, 2007; Niewitetzki *et al.*, 2010), wheat (Delwiche *et al.*, 1998; Delwiche and Hruschka, 2000) and sunflower achenes (Velasco *et al.*, 1999b, 2004). More recently, single-seed calibrations have been developed for oil quality in oilseed rape (Niewitetzki *et al.*, 2010).

Single-seed NIR has been more challenging for large seeded crops such as maize. Maize kernels have large embryos resulting in an asymmetrical distribution of protein, starch, and oil. The asymmetry gives distinct NIR spectra when the germinal or abgerminal side of the kernel is presented to the spectrometer (Orman and Schumann, 1992; Weinstock *et al.*, 2006; Janni *et al.*, 2008). Maize kernels are also variable in size and shape and present differing surface areas and path lengths.

Transmittance spectroscopy can account for asymmetries by collecting spectra from light passing through whole kernels. However, attempts to predict moisture and oil using single-kernel transmittance have had mixed success (Finney and Norris, 1978; Orman and Schumann, 1992; Cogdill *et al.*, 2004).

An alternative approach is to obtain consistent reflectance NIR spectra such as spectra collected specifically from the abgerminal side of the kernel or hyperspectral imaging in which NIR data are selected from relevant sections of the kernel (Baye *et al.*, 2006; Weinstock *et al.*, 2006). Although these approaches yield acceptable predictions, they are low throughput. The kernels need to be hand-placed to ensure a specific kernel orientation or spacing. Spectral acquisition systems that obtain average reflectance over the whole kernel surface have been shown to be robust and have much greater throughput (Armstrong, 2006; Janni *et al.*, 2008). The Armstrong (2006) single-seed NIR system has been used to survey natural variation of maize and common bean (Spielbauer *et al.*, 2009; Hacisalihoglu *et al.*, 2010). The primary innovations that enable high-throughput NIR data collection are the use of an indium-gallium-arsenide (InGaAs) spectrometer with a fast integration time and a tube that illuminates a seed and captures a spectrum while it falls through the tube (Figure 13.3). Calibrations for this single-seed apparatus have been developed for protein, oil, starch, and seed weight for soybean, maize, and common bean indicating that single-seed NIR is practical for large-seeded crops (Armstrong, 2006; Spielbauer *et al.*, 2009; Tallada *et al.*, 2009; Hacisalihoglu *et al.*, 2010).

In contrast to micro-CT and optical imaging, mid-IR and NIR imaging technologies provide information on chemical composition of seeds. NIR is very amenable to developing high-throughput assays for seed sorting or composition but requires chemometric approaches to predict seed composition. By contrast, mid-IR spectroscopy and microscopy provide more direct information regarding chemical composition but can be used only on thin sections or ground seed material. NIR is generally limited to predicting the major macromolecules within seeds, but minor constituents or nonorganic molecules can be predicted if they have strong correlation with major components.

Figure 13.3 Armstrong (2006) single-kernel grain analyzer collects individual seed weight on a microbalance and then NIR reflectance spectrum as a seed falls through the illuminated light tube.

Resonance Emission: Nuclear Magnetic Resonance

NMR spectroscopy and MRI use strong magnetic fields to polarize the nuclear spin on specific isotopes (Kockenberger *et al.*, 2004; Van As, 2007). Magnetized isotopes can absorb and re-emit radiofrequency pulses at their specific resonance frequencies. In biological samples, NMR typically measures ^{1}H or ^{13}C isotopes. Chemical bonds between H and C atoms influence nuclear spin and broaden the NMR resonance frequency, reducing the signal and increasing noise. When molecules are in a fluid state, the anisotropic interactions in chemical bonds are averaged out, and NMR signals can be detected for liquids, such as water or lipid, within biological samples. Crystalline and globular macromolecules such as cellulose, starch, or protein are not easily detected. ^{1}H NMR generally reports the water within developing seeds and lipid storage bodies in a mature, dry seed. The natural abundance of ^{13}C is low, and ^{13}C NMR requires labeling with specific molecules enriched for high ^{13}C content.

In seeds, ^{13}C NMR is primarily used for metabolic flux analysis as a diagnostic and analytic tool (Eisnereich and Bacher, 2007). NMR spectroscopy identifies the specific location of ^{13}C atoms within purified metabolites after a labeling period. NMR spectroscopy data are also quantitative, yielding estimates of the relative amounts of the isotope variants of a metabolite. Combined, ^{13}C-labeling data can be used to infer the enzymatic reactions that most likely gave rise to the isotope combinations observed for a metabolite. Similar to mid-IR spectroscopy, NMR flux analysis requires destructive sampling and intense researcher input for interpretation, placing this approach outside of machine vision strategies.

In contrast, direct ^{1}H NMR spectroscopy measurements can be made to quantify total oil or moisture content in seeds (Conway and Earle, 1963). Oil measurements are possible on seeds from various plants including maize and soybean (Alexander *et al.*, 1967; Collins *et al.*, 1967). The accuracy and repeatability of NMR oil measurements are so high that NMR is a standard method for measuring oil content of mature seeds. NMR measurements are routine for assessing seed oil contents to screen diverse germplasm (Wang *et al.*, 2010, 2011; Horn *et al.*, 2011), for mapping seed oil QTL (Song *et al.*, 2004), and to characterize transgenic germplasm (Shen *et al.*, 2010). Maize seed NMR has been used successfully to clone a major QTL for oil accumulation, which encodes a type I acyl-CoA:diacylglycerol acyltransferase or DGAT1 (Zheng *et al.*, 2008). NMR oil measurements are also very accurate at the single seed level, and single-seed NMR has been successfully used to increase genetic gain in selections for high oil maize lines (Alexander, 1982; Pamin *et al.*, 1986; Song *et al.*, 1999).

MRI of seeds can yield two-dimensional or three-dimensional images for quantitative information about the water and oil distribution within the seed. Two-dimensional MRI has been used to follow moisture loss during wheat kernel drying (Ghosh *et al.*, 2006), and in recent years, MRI has been applied to living, developing seeds to follow oil deposition in barley and endosperm development in pea (Neuberger *et al.*, 2008, 2009). When the developing endosperm is in a liquid state, it is possible to resolve peaks for major metabolites in three dimensions (Neuberger *et al.*, 2009). By engineering custom double coils to detect both ^{1}H and ^{13}C isotopes and pulse labeling of developing seeds, the uptake and metabolism of ^{13}C-sucrose was imaged in developing barley seeds (Melkus *et al.*, 2011; Rolletschek *et al.*, 2011). These live seed imaging approaches open the possibility for developing much more accurate metabolic flux maps that incorporate the movement of metabolites within seed tissues.

In many ways, NMR and x-ray imaging are comparable. Both technologies allow high-resolution tomography and quantification of the chemical composition of a seed. However, the technologies are most appropriate for very different types of samples. NMR detects chemical compounds, whereas x-ray fluorescence detects elements. NMR images and quantifies chemicals in a liquid state, whereas micro-CT needs significant density contrast within the sample to visualize internal structures effectively. NMR spectroscopy also shares some common features to NIR predictions. Similar to NIR, NMR oil measurements are fast and can be completed with minimal sample preparations at various scales down to single seeds. However, NMR is more limited in the chemical phenotypes that can be determined from mature, dry seeds.

Conclusion

Characterizing the seed phenome is one of the major challenges to developing a comprehensive understanding of gene function in plants. Machine vision technologies are a rich resource for characterizing seed phenotypes and defining the phenotype space for a species. By using spectroscopy and electromagnetic radiation–based imaging, machine vision technologies provide quantitative data about the seed phenome. A key advantage of machine vision is the ability to reanalyze or calculate new phenotypes from pre-existing raw images or spectroscopy data. For example, new predictions can be made from pre-existing NIR data as new calibrations become available. In contrast, manual measurements or chemical assays are usually limited to the specific measurements made at the time by the investigator. There is no single machine vision technology that can provide both structural and chemical information about plant seeds at all stages of development. The decision to use a particular technology and portion of the electromagnetic radiation spectrum is made based on the traits that are viewed as important for the process that is under study. Many of these technologies are nondestructive or minimally invasive and can be used in series to gain a more complete quantification of a seed phenotype. Combining these quantitative phenotype measures with genomics information should help elucidate gene function within the seed.

Acknowledgments

We thank Gary Peter and Alejandro Walker (University of Florida) for providing training and access to a micro-CT scanner and to Paul Armstrong (USDA-ARS) and John Baier (University of Florida) for building and maintaining the single-kernel NIR analyzer at the University of Florida. We apologize to the authors of relevant research articles that were not highlighted in this chapter because of space constraints. The authors' research on machine vision technologies is supported by grants from the NSF Plant Genome Research Program (IOS-1031416) and the Vasil-Monsanto Endowment.

References

Alexander DE (1982). The use of wide-line NMR in breeding high-oil corn. J Am Oil Chem Soc 59: A284.

Alexander DE, Silvela L, Collins FI, Rodgers RC (1967). Analysis of oil content of maize by wide-line NMR. J Am Oil Chem Soc 44: 555–558.

Aquila AD (2009). Digital imaging information technology applied to seed germination testing: a review. Agron Sustain Dev 29: 213–221.

Arefi A, Motlagh AM, Teimourlou RF (2011). Wheat class identification using computer vision system and artificial neural networks. Int Agrophys 25: 319–325.

Armstrong PR (2006). Rapid single-kernel NIR measurement of grain and oil-seed attributes. Appl Eng Agric 22: 767–772.

Ashraf M, Foolad MR (2005). Pre-sowing seed treatment—a shotgun approach to improve germination, plant growth, and crop yield under saline and non-saline conditions. Adv Agron 88: 223–271.

Barron C (2011). Prediction of relative tissue proportions in wheat mill streams by Fourier transform mid-infrared spectroscopy. J Agr Food Chem 59: 10442–10447.

Baye TM, Pearson TC, Settles AM (2006). Development of a calibration to predict maize seed composition using single kernel near infrared spectroscopy. J Cereal Sci 43: 236–243.

Berardo N, Mazzinelli G, Valoti P, Lagana P, Redaelli R (2009). Characterization of maize germplasm for the chemical composition of the grain. J Agric Food Chem 57: 2378–2384.

Bingham IJ, Harris A, Macdonald L (1994). A comparative-study of radicle and coleoptile extension in maize seedlings from aged and unaged seed. Seed Sci Techn 22: 127–139.

Blanco M, Villarroya I (2002). NIR spectroscopy: a rapid-response analytical tool. TRAC Trend Anal Chem 21: 240–250.

Brenna OV, Berardo N (2004). Application of near-infrared reflectance spectroscopy (NIRS) to the evaluation of carotenoids content in maize. J Agric Food Chem 52: 5577–5582.

Brodersen CR, Lee EF, Choat B, Jansen S, Phillips RJ, Shackel KA, McElrone AJ, Matthews MA (2011). Automated analysis of three-dimensional xylem networks using high-resolution computed tomography. New Phytol 191: 1168–1179.

Brooks TLD, Miller ND, Spalding EP (2010). Plasticity of Arabidopsis root gravitropism throughout a multidimensional condition space quantified by automated image analysis. Plant Physiol 152: 206–216.

Burghardt AJ, Link TM, Majumdar S (2011). High-resolution computed tomography for clinical imaging of bone microarchitecture. Clin Orthop Relat Res 469: 2179–2193.

Campbell KG, Bergman CJ, Gualberto DG, Anderson JA, Giroux MJ, Hareland G, Fulcher RG, Sorrells ME, Finney PL (1999). Quantitative trait loci associated with kernel traits in a soft x hard wheat cross. Crop Sci 39: 1184–1195.

Campbell MR, Brumm TJ, Glover DV (1997). Whole grain amylose analysis in maize using near-infrared transmittance spectroscopy. Cereal Chem 74: 300–303.

Campbell MR, Yeager H, Abdubek N, Pollak LM, Glover DV (2002). Comparison of methods for amylose screening among amylose-extender (ae) maize starches from exotic backgrounds. Cereal Chem 79: 317–321.

Chen JY, Bottjer DJ, Davidson EH, Li G, Gao F, Cameron RA, Hadfield MG, Xian DC, Tafforeau P, Jia QJ, Sugiyama H, Tang R (2009a). Phase contrast synchrotron x-ray microtomography of Ediacaran (Doushantuo) metazoan microfossils: phylogenetic diversity and evolutionary implications. Precambrian Res 173: 191–200.

Chen JY, Bottjer DJ, Li G, Hadfield MG, Gao F, Cameron AR, Zhang CY, Xian DC, Tafforeau P, Liao X, Yin ZJ (2009b). Complex embryos displaying bilaterian characters from Precambrian Doushantuo phosphate deposits, Weng'an, Guizhou, China. Proc Natl Acad Sci U S A 106: 19056–19060.

Chtioui Y, Bertrand D, Dattee Y, Devaux MF (1996). Identification of seeds by colour imaging: comparison of discriminant analysis and artificial neural network. J Sci Food Agric 71: 433–441.

Clerkx EJM, El-Lithy ME, Vierling E, Ruys GJ, Blankestijin-De Vries H, Groot SPC, Vreugdenhil D, Koornneef M (2004). Analysis of natural allelic variation of Arabidopsis seed germination and seed longevity traits between the accessions Landsberg erecta and Shakdara, using a new recombinant inbred line population. Plant Physiol 135: 432–443.

Cogdill RP, Hurburgh CR, Rippke GR (2004). Single-kernel maize analysis by near-infrared hyperspectral imaging. Trans ASAE 47: 311–320.

Collins FI, Alexander DE, Rodgers RC, Silvela L (1967). Analysis of oil content of soybeans by wide-line NMR. J Am Oil Chem Soc 44: 708–710.

Conway TF, Earle FR (1963). Nuclear magnetic resonance for determining oil content of seeds. J Am Oil Chem Soc 40: 265–268.

Cook JP, McMullen MD, Holland JB, Tian F, Bradbury P, Ross-Ibarra J, Buckler ES, Flint-Garcia SA (2012). Genetic architecture of maize kernel composition in the nested association mapping and inbred association panels. Plant Physiol 158: 824–834.

Correa CES, Shaver RD, Pereira MN, Lauer JG, Kohn K (2002). Relationship between corn vitreousness and ruminal in situ starch degradability. J Dairy Sci 85: 3008–3012.

Cozzolino D (2009). Near infrared spectroscopy in natural products analysis. Planta Med 75: 746–756.

Dell'Aquila A (2007). Towards new computer imaging techniques applied to seed quality testing and sorting. Seed Sci Technol 35: 519–538.

Dell'Aquila A (2009). Development of novel techniques in conditioning, testing and sorting seed physiological quality. Seed Sci Technol 37: 608–624.

Delwiche SR, Graybosch RA, Peterson CJ (1998). Predicting protein composition, biochemical properties, and dough-handling properties of hard red winter wheat flour by near-infrared reflectance. Cereal Chem 75: 412–416.

Delwiche SR, Hruschka WR (2000). Protein content of bulk wheat from near-infrared reflectance of individual kernels. Cereal Chem 77:86–88.

Dhondt S, Vanhaeren H, Van Loo D, Cnudde V, Inze D (2010). Plant structure visualization by high-resolution x-ray computed tomography. Trends Plant Sci 15: 419–422.

Dowell FE, Pearson TC, Maghirang EB, Xie F, Wicklow DT (2002). Reflectance and transmittance spectroscopy applied to detecting fumonisin in single corn kernels infected with Fusarium verticillioides. Cereal Chem 79: 222–226.

Ducournau S, Feutry A, Plainchault P, Revollon P, Vigouroux B, Wagner MH (2005). Using computer vision to monitor germination time course of sunflower (Helianthus annuus L.) seeds. Seed Sci Technol 33: 329–340.
Eisenreich W, Bacher A (2007). Advances of high-resolution NMR techniques in the structural and metabolic analysis of plant biochemistry. Phytochemistry 68: 2799–2815.
Erasmus C, Taylor JRN (2004). Optimising the determination of maize endosperm vitreousness by a rapid non-destructive image analysis technique. J Sci Food Agric 84: 920–930.
Evers AD, Cox RI, Shaheedullah MZ, Withey RP (1990). Predicting milling extraction rate by image analysis of wheat grains. Asp Appl Biol 25: 417–426.
Finney EE, Norris KH (1978). Determination of moisture in corn kernels by near-infrared transmittance measurements. Trans ASAE 21: 581–584.
Flavel RJ, Guppy CN, Tighe M, Watt M, McNeill A, Young IM (2012). Non-destructive quantification of cereal roots in soil using high-resolution X-ray tomography. J Exp Bot 63: 2503–2511.
Fontaine J, Schirmer B, Horr J (2002). Near-infrared reflectance spectroscopy (NIRS) enables the fast and accurate prediction of essential amino acid contents. 2. Results for wheat, barley, corn, triticale, wheat bran/middlings, rice bran, and sorghum. J Agric Food Chem 50: 3902–3911.
Fox G, Manley M (2009). Hardness methods for testing maize kernels. J Agric Food Chem 57: 5647–5657.
Friis EM, Crane PR, Pedersen KR, Bengtson S, Donoghue PCJ, Grimm GW, Stampanoni M (2007). Phase-contrast x-ray microtomography links cretaceous seeds with Gnetales and Bennettitales. Nature 450: 549–552.
Friis EM, Pedersen KR, Crane PR (2009). Early cretaceous mesofossils from Portugal and Eastern North America related to the Bennettitales-Erdtmanithecales-Gnetales group. Am J Bot 96: 252–283.
Gegas VC, Nazari A, Griffiths S, Simmonds J, Fish L, Orford S, Sayers L, Doonan JH, Snape JW (2010). A genetic framework for grain size and shape variation in wheat. Plant Cell 22: 1046–1056.
Ghosh PK, Jayas DS, Gruwel MLH, White NDG (2006). Magnetic resonance image analysis to explain moisture movement during wheat drying. Trans ASABE 49: 1181–1191.
Gibbon BC, Larkins BA (2005). Molecular genetic approaches to developing quality protein maize. Trends Genet 21: 227–233.
Granitto PM, Verdes PF, Ceccatto HA (2005). Large-scale investigation of weed seed identification by machine vision. Comput Electron Agric 47: 15–24.
Hacisalihoglu G, Larbi B, Settles AM (2010). Near-infrared reflectance spectroscopy predicts protein, starch, and seed weight in intact seeds of common bean (Phaseolus vulgaris L.). J Agr Food Chem 58: 702–706.
Herridge RP, Day RC, Baldwin S, Macknight RC (2011). Rapid analysis of seed size in Arabidopsis for mutant and QTL discovery. Plant Methods 7: 3.
Holding DR, Hunter BG, Chung T, Gibbon BC, Ford CF, Bharti AK, Messing J, Hamaker BR, Larkins BA (2008). Genetic analysis of opaque2 modifier loci in quality protein maize. Theor Appl Genet 117: 157–170.
Holding DR, Hunter BG, Klingler JP, Wu S, Guo X, Gibbon BC, Wu R, Schulze JM, Jung R, Larkins BA (2011). Characterization of opaque2 modifier QTLs and candidate genes in recombinant inbred lines derived from the K0326Y quality protein maize inbred. Theor Appl Genet 122: 783–794.
Hom NH, Becker HC, Mollers C (2007). Non-destructive analysis of rapeseed quality by NIRS of small seed samples and single seeds. Euphytica 153: 27–34.
Horn PJ, Neogi P, Tombokan X, Ghosh S, Campbell BT, Chapman KD (2011). Simultaneous quantification of oil and protein in cottonseed by low-field time-domain nuclear magnetic resonance. J Am Oil Chem Soc 88: 1521–1529.
Hunter CT, Kirienko DH, Sylvester AW, Peter GF, McCarty DR, Koch KE (2012). Cellulose synthase-like d1 is integral to normal cell division, expansion, and leaf development in maize. Plant Physiol 158: 708–724.
Janni J, Weinstock BA, Hagen L, Wright S (2008). Novel near-infrared sampling apparatus for single kernel analysis of oil content in maize. Appl Spectrosc 62: 423–426.
Joosen RVL, Arends D, Willems LAJ, Ligterink W, Jansen RC, Hilhorst HWM (2012). Visualizing the genetic landscape of Arabidopsis seed performance. Plant Physiol 158: 570–589.
Joosen RVL, Kodde J, Willems LAJ, Ligterink W, van der Plas LHW, Hilhorst HWM (2010). GERMINATOR: a software package for high-throughput scoring and curve fitting of Arabidopsis seed germination. Plant J 62: 148–159.
Kim SA, Punshon T, Lanzirotti A, Li LT, Alonso JM, Ecker JR, Kaplan J, Guerinot ML (2006). Localization of iron in Arabidopsis seed requires the vacuolar membrane transporter VIT1. Science 314: 1295–1298.
Kockenberger W, De Panfilis C, Santoro D, Dahiya P, Rawsthorne S (2004). High resolution NMR microscopy of plants and fungi. J Microsc 214: 182–189.
Lee JD, Shannon JG, Choung MG (2010). Selection for protein content in soybean from single F(2) seed by near infrared reflectance spectroscopy. Euphytica 172: 117–123.
Lombi E, Hettiarachchi GM, Scheckel KG (2011a). Advanced in situ spectroscopic techniques and their applications in environmental biogeochemistry: introduction to the special section. J Environment Qual 40: 659–666.

Lombi E, Smith E, Hansen TH, Paterson D, de Jonge MD, Howard DL, Persson DP, Husted S, Ryan C, Schjoerring JK (2011b). Megapixel imaging of (micro)nutrients in mature barley grains. J Exp Bot 62: 273–282.

Majumdar S, Jayas DS (2000a). Classification of cereal grains using machine vision: I. morphology models. Trans ASAE 43: 1669–1675.

Majumdar S, Jayas DS (2000b). Classification of cereal grains using machine vision: II. color models. Trans ASAE 43: 1677–1680.

Majumdar S, Jayas DS (2000c). Classification of cereal grains using machine vision: III. texture models. Trans ASAE 43: 1681–1687.

Majumdar S, Jayas DS (2000d). Classification of cereal grains using machine vision: IV. combined morphology, color, and texture models. Trans ASAE 43: 1689–1694.

Manley M, Williams P, Nilsson D, Geladi P (2009). Near infrared hyperspectral imaging for the evaluation of endosperm texture in whole yellow maize (Zea mays L.) kernels. J Agric Food Chem 57: 8761–8769.

Marcos J, Bennett MA, Mcdonald MB, Evans AF, Grassbaugh EM (2006). Assessment of melon seed vigour by an automated computer imaging system compared to traditional procedures. Seed Sci Technol 34: 485–497.

Matthews S, Wagner MH, Ratzenboeck A, Khajeh-Hosseini M, Casarini E, El-Khadem R, El-Yakhlifi M, Powell AA (2011). Early counts of radicle emergence during germination as a repeatable and reproducible vigour test for maize. Seed Test Int 141: 39–45.

Mebatsion HK, Paliwal J, Jayas DS (2012). Evaluation of variations in the shape of grain types using principal components analysis of the elliptic Fourier descriptors. Comput Electron Agric 80: 63–70.

Melkus G, Rolletschek H, Fuchs J, Radchuk V, Grafahrend-Belau E, Sreenivasulu N, Rutten T, Weier D, Heinzel N, Schreiber F, Altmann T, Jakob PM, Borisjuk L (2011). Dynamic C-13/H-1 NMR imaging uncovers sugar allocation in the living seed. Plant Biotechnol J 9: 1022–1037.

Mendes MM, Friis EM, Pais J (2008a). Erdtmanispermum juncalense sp nov., a new species of the extinct order Erdtmanithecales from the Early Cretaceous (probably Berriasian) of Portugal. Rev Palaeobot Palyno 149: 50–56.

Mendes MM, Pais J, Friis EM (2008b). Raunsgaardispermum lusitanicum gen. et sp nov., a new seed with in situ pollen from the Early Cretaceous (probably Berriasian) of Portugal: further support for the Bennettitales-Erdtmanithecales-Gnetales link. Grana 47: 211–219.

Moore KL, Schroder M, Lombi E, Zhao FJ, McGrath SP, Hawkesford MJ, Shewry PR, Grovenor CRM (2010). NanoSIMS analysis of arsenic and selenium in cereal grain. New Phytol 185: 434–445.

Nesi N, Delourme R, Bregeon M, Falentin C, Renard M (2008). Genetic and molecular approaches to improve nutritional value of Brassica napus L. seed. CR Biol 331: 763–771.

Neuberger T, Rolletschek H, Webb A, Borisjuk L (2009). Non-invasive mapping of lipids in plant tissue using magnetic resonance imaging. Methods Mol Biol 579: 485–496.

Neuberger T, Sreenivasulu N, Rokitta M, Rolletschek H, Gobel C, Rutten T, Radchuk V, Feussner I, Wobus U, Jakob P, Webb A, Borisjuk L (2008). Quantitative imaging of oil storage in developing crop seeds. Plant Biotech J 6: 31–45.

Ngonyamo-Majee D, Shaver RD, Coors JG, Sapienza D, Correa CES, Lauer JG, Berzaghi P (2008). Relationships between kernel vitreousness and dry matter degradability for diverse corn germplasm. I. Development of near-infrared reflectance spectroscopy calibrations. Animal Feed Sci Tech 142: 247–258.

Niewitetzki O, Tillmann P, Becker HC, Mollers C (2010). A new near-infrared reflectance spectroscopy method for high-throughput analysis of oleic acid and linolenic acid content of single seeds in oilseed rape (Brassica napus L.). J Agric Food Chem 58: 94–100.

Olesen MH, Shetty N, Gislum R, Boelt B (2011). Classification of viable and non-viable spinach (Spinacia oleracea L.) seeds by single seed near infrared spectroscopy and extended canonical variates analysis. J Near Infrared Spect 19: 171–180.

Orman BA, Schumann RA (1991). Comparison of near-infrared spectroscopy calibration methods for the prediction of protein, oil, and starch in maize grain. J Agric Food Chem 39: 883–886

Orman BA, Schumann RA (1992). Nondestructive single-kernel oil determination of maize by near-infrared transmission spectroscopy. J Am Oil Chem Soc 69: 1036–1038.

Osborne BG (2006). Applications of near infrared spectroscopy in quality screening of early-generation material in cereal breeding programmes. J Near Infrared Spect 14: 93–101.

Paliwal J, Visen NS, Jayas DS, White NDG (2003). Cereal grain and dockage identification using machine vision. Biosyst Eng 85: 51–57.

Pamin K, Compton WA, Walker CE, Alexander DE (1986). Genetic-variation and selection response for oil composition in corn. Crop Sci 26: 279–282.

Paulsen MR, Mbuvi SW, Haken AE, Ye B, Stewart RK (2003a). Extractable starch as a quality measurement of dried corn. Appl Engin Agric 19: 211–217.

Paulsen MR, Pordesimo LO, Singh M, Mbuvi SW, Ye BY (2003b). Maize starch yield calibrations with near infrared reflectance. Biosyst Engin 85: 455–460.

Paulsen MR, Singh M (2004). Calibration of a near-infrared transmission grain analyser for extractable starch in maize. Biosyst Engin 89: 79–83.

Pearson TC, Wicklow DT, Maghirang EB, Xie F, Dowell FE (2001). Detecting aflatoxin in single corn kernels by transmittance and reflectance spectroscopy. Trans ASAE 44: 1247–1254.

Pietrzak LN, Miller SS (2005). Microchemical structure of soybean seeds revealed in situ by ultraspatially resolved synchrotron Fourier transformed infrared microspectroscopy. J Agr Food Chem 53: 9304–9311.

Pinzi S, Garcia IL, Lopez-Gimenez FJ, de Castro MDL, Dorado G, Dorado MP (2009). The ideal vegetable oil-based biodiesel composition: a review of social, economical and technical implications. Energ Fuel 23: 2325–2341.

Pittia P, Sacchetti G, Mancini L, Voltolini M, Sodini N, Tromba G, Zanini F (2011). Evaluation of microstructural properties of coffee beans by synchrotron X-ray microtomography: a methodological approach. J Food Sci 76: E222–E231.

Purugganan MD, Fuller DQ (2009). The nature of selection during plant domestication. Nature 457: 843–848.

Quesada V, Garcia-Martinez S, Piqueras P, Ponce MR, Micol JL (2002). Genetic architecture of NaCl tolerance in Arabidopsis. Plant Physiol 130: 951–963.

Ren ZH, Zheng ZM, Chinnusamy V, Zhu JH, Cui XP, Iida K, Zhu JK (2010). RAS1, a quantitative trait locus for salt tolerance and ABA sensitivity in Arabidopsis. Proc Natl Acad Sci U S A 107: 5669–5674.

Robutti JL (1995). Maize kernel hardness estimation in breeding by near-infrared transmission analysis. Cereal Chem 72: 632–636.

Rodriguez-Saona LE, Allendorf ME (2011). Use of FTIR for rapid authentication and detection of adulteration of food. Annu Rev Food Sci Techn 2: 467–483.

Rolletschek H, Melkus G, Grafahrend-Belau E, Fuchs J, Heinzel N, Schreiber F, Jakob PM, Borisjuk L (2011). Combined noninvasive imaging and modeling approaches reveal metabolic compartmentation in the barley endosperm. Plant Cell 23: 3041–3054.

Sako Y, McDonald MB, Fujimura K, Evans AF, Bennett MA (2001). A system for automated seed vigour assessment. Seed Sci Technol 29: 625–636.

Samadi, Yu P (2011). Dry and moist heating-induced changes in protein molecular structure, protein subfraction, and nutrient profiles in soybeans. J Dairy Sci 94: 6092–6102.

Shen B, Allen WB, Zheng PZ, Li CJ, Glassman K, Ranch J, Nubel D, Tarczynski MC (2010). Expression of ZmLEC1 and ZmWRI1 increases seed oil production in maize. Plant Physiol 153: 980–987.

Shewry PR (2007). Improving the protein content and composition of cereal grain. J Cereal Sci 46: 239–250.

Smith SY, Collinson ME, Rudall PJ, Simpson DA, Marone F, Stampanoni M (2009). Virtual taphonomy using synchrotron tomographic microscopy reveals cryptic features and internal structure of modern and fossil plants. Proc Natl Acad Sci U S A 106: 12013–12018.

Song TM, Kong F, Li CJ, Song GH (1999). Eleven cycles of single kernel phenotypic recurrent selection for percent oil in Zhongzong no. 2 maize synthetic. J Genet Breeding 53: 31–35.

Song XF, Song TM, Dai JR, Rocheford T, Li JS (2004). QTL mapping of kernel oil concentration with high-oil maize by SSR markers. Maydica 49: 41–48.

Spielbauer G, Armstrong P, Baier JW, Allen WB, Richardson K, Shen B, Settles AM (2009). High-throughput near-infrared reflectance spectroscopy for predicting quantitative and qualitative composition phenotypes of individual maize kernels. Cereal Chem 86: 556–564.

Stuppy WH, Maisano JA, Colbert MW, Rudall PJ, Rowe TB (2003). Three-dimensional analysis of plant structure using high-resolution X-ray computed tomography. Trends Plant Sci 8: 2–6.

Takhar PS, Maier DE, Campanella OH, Chen G (2011). Hybrid mixture theory based moisture transport and stress development in corn kernels during drying: validation and simulation results. J Food Eng 106: 275–282.

Tallada JG, Palacios-Rojas N, Armstrong PR (2009). Prediction of maize seed attributes using a rapid single kernel near infrared instrument. J Cereal Sci 50: 381–387.

USDA (2009). *Inspecting Grain: Practical Procedures for Grain Handlers,* Federal Grain Inspection Service, Washington, DC.

Van As H (2007). Intact plant MRI for the study of cell water relations, membrane permeability, cell-to-cell and long distance water transport. J Exp Bot 58: 743–756.

Velasco L, Mollers C (2002). Nondestructive assessment of protein content in single seeds of rapeseed (Brassica napus L.) by near-infrared reflectance spectroscopy. Euphytica 123: 89–93.

Velasco L, Mollers C, Becker HC (1999a). Estimation of seed weight, oil content and fatty acid composition in intact single seeds of rapeseed (Brassica napus L.) by near-infrared reflectance spectroscopy. Euphytica 106: 79–85.

Velasco L, Perez-Vich B, Fernandez-Martinez JM (1999b). Nondestructive screening for oleic and linoleic acid in single sunflower achenes by near-infrared reflectance spectroscopy. Crop Sci 39: 219–222.

Velasco L, Perez-Vich B, Fernandez-Martinez JM (2004). Use of near-infrared reflectance spectroscopy for selecting for high stearic acid concentration in single husked achenes of sunflower. Crop Sci 44: 93–97.

Wagner MH, Demilly D, Ducournau S, Durr C, Lechappe J (2011). Computer vision for monitoring seed germination from dry state to young seedlings. Seed Testing Int 142: 49–51.

Wang ML, Morris JB, Pinnow DL, Davis J, Raymer P, Pederson GA (2010). A survey of the castor oil content, seed weight and seed-coat colour on the United States Department of Agriculture germplasm collection. Plant Genet Res 8: 229–231.

Wang ML, Morris JB, Tonnis B, Pinnow D, Davis J, Raymer P, Pederson GA (2011). Screening of the entire USDA castor germplasm collection for oil content and fatty acid composition for optimum biodiesel production. J Agric Food Chem 59: 9250–9256.

Weinstock BA, Janni J, Hagen L, Wright S (2006). Prediction of oil and oleic acid concentrations in individual corn (Zea mays L.) kernels using near-infrared reflectance hyperspectral imaging and multivariate analysis. Appl Spectrosc 60: 9–16.

Yang XH, Guo YQ, Fu Y, Hu JY, Chai YC, Zhang YR, Li JS (2009). Measuring fatty acid concentration in maize grain by near-infrared reflectance spectroscopy. Spectrosc Spectral Analysis 29:106–109.

Yu PQ, McKinnon JJ, Christensen CR, Christensen DA (2004). Imaging molecular chemistry of pioneer corn. J Agric Food Chem 52: 7345–7352.

Zheng P, Allen WB, Roesler K, Williams ME, Zhang S, Li J, Glassman K, Ranch J, Nubel D, Solawetz W, Bhattramakki D, Llaca V, Deschamps S, Zhong GY, Tarczynski MC, Shen B (2008). A phenylalanine in DGAT is a key determinant of oil content and composition in maize. Nat Genet 40: 367–372.

Zhu DZ, Wang K, Zhang DY, Huang WJ, Yang GJ, Ma ZH, Wang C (2011). Quality assessment of crop seeds by near-infrared hyperspectral imaging. Sensor Lett 9: 1144–1150.

Index

Seed Genomics, First Edition. Edited by Philip W. Becraft.